W0079395

Acta Physica Austriaca

Supplementum XIV

Proceedings of the
XIV. Internationale Universitätswochen für Kernphysik 1975
der Karl-Franzens-Universität Graz
at Schladming (Steiermark, Austria)
24th February—7th March 1975

Sponsored by
Bundesministerium für Wissenschaft und Forschung
Steiermärkische Landesregierung
Sektion Industrie der Kammer der
gewerblichen Wirtschaft für Steiermark
International Centre for Theoretical Physics, Triest

1975

Springer-Verlag
Wien New York

Electromagnetic Interactions and Field Theory

Edited by Paul Urban, Graz

With 107 Figures

1975

Springer-Verlag
Wien New York

Organizing Committee

Chairman

Prof. Dr. Paul Urban
Vorstand des Institutes für Theoretische Physik
der Universität Graz

Committee Members

Doz. Dr. H. Latal
Doz. Dr. N. Pucker
Dr. A. Mas-Parareda
Dr. W. Plessas
Dr. L. Pittner

Secretary

M. Krautilik

Softcover reprint of the hardcover 1st edition 1975

Library of Congress Cataloging in Publication Data

Internationale Universitätswochen für Kernphysik der Karl-Franzens-Universität Graz, 14th, Schladming, Austria, 1975.
Electromagnetic interactions and field theory.

(Acta physica Austriaca. Supplementum ; 14)
1. Electromagnetic interactions--Congresses.
2. Quantum field theory--Congresses. I. Urban, Paul Oskar, 1905- II. Austria. Bundesministerium für Wissenschaft und Forschung. III. Title. IV. Series.
QC794.I55 1975 530.1'4 75-19399

ISBN-13:978-3-7091-8426-4 e-ISBN-13:978-3-7091-8424-0
DOI: 10.1007/978-3-7091-8424-0

CONTENTS

Acta Physica Austriaca, Suppl. XIV, 1 — 3 (1975)

OPENING ADDRESS

by

P. URBAN
Institut für Theoretische Physik
Universität Graz

It is a great pleasure for me to welcome you most cordially at our fourteenth International Winter School in Schladming. About onehundred and sixty scientists from seventeen countries participate this year, which number gives us confidence that our efforts to provide an interesting program of lectures were successful.

In summer last year when we finalized the program of this year's meeting we couldn't anticipate that in November the surprising discovery of new long-lived particles would stir up the physicists and would even produce headlines in the newspapers. Since the original reports about the new particles a wealth of scientific papers to their interpretation has appeared and further experimental data are continuously being published. We are now in the fortunate position that among our lecturers we invited some experts especially to this topic of utmost actuality which will present us with the newest results and attempts at interpretations both from the experimental and theoretical viewpoint.

As is well known these particles have been discovered as narrow resonances in the electron-positron channel by which fact the already highly interesting physics of storage rings gained further impetus. Theoreticians, who comprise the main part of our participants, will get the opportunity to obtain some insight into the experimental problems of this area, but of course also lectures on the current theoretical ideas about the interpretation of the experimental data are scheduled.

Another area of electromagnetic interactions comprises electro- and photoproduction; i.e. the production of a number of particles through inelastic scattering by electrons or photons. There two regions are to be distinguished where different theoretical methods are employed: on the one hand production at high energies where asymptotic approximations can be made, and on the other hand at lower energies where resonances dominate. We will hear lectures also about other aspects of inelastic lepton scattering, especially in connection with new concepts of gauge theories, as the appearance of neutral currents.

It is a great honor for us that one of the founders of quantum electrodynamics, Prof. Schwinger, has accepted our invitation and will talk extensively here in Schladming about his formulations of quantum field theory and recent results in the description of deep inelastic processes. Also purely quantum electrodynamical question will have their say: here an interesting area of research is the interaction with intense electromagnetic fields as they could be achieved e.g. by means of lasers.

The second topic of this year's meeting is concerned with somewhat more abstract problems: both structural questions in quantum field theory and stochastic processes will be treated. In addition some aspects of mathematical physics will be discussed.

As always some seminars will supplement the main lectures and thus complete the scientific program. I hope that also this year we will be successful in providing new insights, this time mainly into the problems of electromagnetic interactions and field theory and wish to all of you a successful and pleasant stay here in Schladming.

Acta Physica Austriaca, Suppl. XIV, 5—87 (1975)

PHOTON HADRON INTERACTION IN THE RESONANCE REGION+

by

H. ROLLNIK
Physikalisches Institut der
Universität Bonn

I. INTRODUCTION

In these lectures I shall describe the progress of our understanding of the photon hadron interaction in the resonance region. I will address myself to the fundamental properties of the electromagnetic hadron current and its behaviour under symmetries as well as to a detailed description of photon couplings of hadronic states. The progress to date is based on an impressive accumulation of experimental information about photon hadron processes and its phenomenological interpretation and has led to a remarkable improvement of our insight in the electromagnetic structures of hadrons.

Most of the new results have been obtained for

+ Lecture given at XIV. Internationale Universitätswochen für Kernphysik, Schladming, Austria, February 24 - March 7, 1975.

baryon states and real photons. Therefore I shall concentrate mainly on processes where real photons interact with nucleons and shall deal with other topics, especially with virtual photons, only if it helps to clarify the general picture. However, the next section II is devoted to a rather general kinematical discussion of the photon transition operator by which we hope to give a sound basis for the application of light cone algebra ideas. Then we turn to concrete questions starting with a short description of the present evidence for time reversal invariance and the absence of exotic pieces in the hadronic current. This information comes mainly from the $\Delta(1232)$-energy region. What we can learn on the $\gamma N\Delta$-coupling and how accurate our knowledge is will be described in section IV. The much more complicated situation for the higher isobars will be dealt with in the following section where I shall try to explain the basic ingredients of the so called "multipole analyses", their results and shortcomings, without burdening the reader with too many technical details. There is only one physical scheme by which the many photon-baryon couplings thus obtained have been interpreted with increasing success: the quark model and related symmetry approaches. The last part of the lectures are devoted to a detailed exposition of their consequences for photon hadron physics.

II. THE LIGHT-LIKE ELECTRIC DIPOLE OPERATOR

Let us start with a general, mainly kinematical discussion of the transition amplitude for the photonic excitation

$$\gamma + A \to B$$

where A should be a single baryon state, mostly a nucleon, characterized by its 4-momentum p, helicity and intrinsic quantum numbers which we shall not specify for the moment. B might be a general, possibly composite, hadronic state with total 4-momentum p'. The S-matrix element for this process is given by

$$\langle B(p')|S|\gamma(k),\ A(p)\rangle=(-i)\langle B(p')|e\int d^4x J^\mu(x)A_\mu(x)|A(p)\rangle \quad (1)$$

where the hadronic part $e\cdot J^\mu(x)$ of the electromagnetic current and the 4-potential A_μ for a (real or virtual) photon of momentum k enters:

$$A_\mu(x) = \varepsilon_\mu\, e^{-ik\cdot x} \;; \qquad k\cdot\varepsilon = 0\ . \quad (1a)$$

We shall use (1) only in the one photon approximation. All particles states are taken to be normalized in a Lorentz covariant way[1,2].

$$\langle p'|p\rangle = (2\pi)^3\ 2p^o\ \delta^3(\vec{p}-\vec{p}\,') \quad (1b)$$

One usually uses translational invariance to obtain from (1)

$$\langle B(p')|S|\gamma(k),A(p)\rangle = -i(2\pi)^4\delta^4(p'-p-k)e\cdot\varepsilon_\mu j^\mu(p,p') \quad (2)$$

with

$$j^\mu(p,p') = \langle B(p')|J^\mu(o)|A(p)\rangle \quad (3)$$

For specific processes the spin structure of $\varepsilon_\mu j^\mu$ can be made explicit either by using a covariant decomposition or by working with helicity amplitudes. As examples for the first method we list in table I the well-known formulae for the form factor.

Table I Invariant decomposition of the $\frac{1}{2}^+ - \frac{1}{2}^+$ and the $\frac{1}{2}^+ - \frac{3}{2}^+$ current matrix elements

a) $\gamma + \frac{1}{2}^+ \rightarrow \frac{1}{2}^+$

$$<N'(p',s')|J^\mu(o)|N(p,s)> = \bar{u}(p',s')[F_1(q^2)\gamma^\mu + + i\frac{F_2(q^2)}{M+M'}\sigma^{\mu\nu}q_\nu]u(p,s)$$

$$q = p'-p; \quad p^2 = M^2; \quad p'^2 = M'^2; \quad \bar{u}(p,s)u(p,s) = 2M$$

b) $\gamma + \frac{1}{2}^+ \rightarrow \frac{3}{2}^+$

$$\varepsilon_\mu <\Delta(p',s')|J^\mu(o)|N(p,s)> = \bar{u}_\nu(p',s')\gamma_5[\frac{G_M^\Delta(q^2)}{M+M_\Delta}\gamma_\mu + + \frac{G_E^\Delta(q^2)}{(M+M_\Delta)^2}(p+p')_\mu + \frac{G_S^\Delta(q^2)}{(M+M_\Delta)^2}q_\mu]u(p,s)F^{\mu\nu} \quad .$$

<u>Notation</u> For Dirac spinors and matrices, see Bjorken and Drell[3].

$u_\nu(p',s')$ is the Rarita - Schwinger spinor obeying $(\not{p}' + M_\Delta)\, u_\nu = 0$

$$p'^{\nu} u_\nu = \gamma^{\nu} u_\nu = 0 ,$$

$F_{\mu\nu} = q_\mu\, \varepsilon_\nu - q_\nu\, \varepsilon_\mu$ is the electromagnetic field tensor.

Table II <u>Helicity amplitude for $\gamma + N \rightarrow N + \pi$ and its partial wave decomposition</u>

Work in the c.m. of the π - N system, energy W, angle θ and use helicity states for the nucleon $|p,\nu\rangle$ and photon polarisation vectors $\varepsilon_\mu^{(\lambda)}$ corresponding to definite helicity ($\lambda = \pm 1$):

$$\Gamma_{\nu';\lambda\nu}(w,\theta) = \langle N(p',\nu'),\pi(q)|\,\varepsilon_\mu^{(\lambda)}\, J^\mu(0)\,|N(p,\nu)\rangle$$

		Total initial helicity:
$H_1 = F_1$ (single flip)	$= \Gamma_{-\frac{1}{2};1,-\frac{1}{2}}$;	$\lambda = \frac{3}{2}$
$H_2 = N$ (non flip)	$= \Gamma_{-\frac{1}{2};1,+\frac{1}{2}}$;	$\lambda = \frac{1}{2}$
$H_3 = D$ (double flip)	$= \Gamma_{1\frac{1}{2};1,-\frac{1}{2}}$;	$\lambda = \frac{3}{2}$
$H_4 = F_2$ (single flip)	$= \Gamma_{+\frac{1}{2};1,+\frac{1}{2}}$;	$\lambda = \frac{1}{2}$

$$H_1 = \frac{1}{\sqrt{2}} \sin\theta \cos\frac{1}{2}\theta \sum_{\ell=1}^{\infty} (B_{\ell+} - B_{(\ell+1)-})(P''_{\ell} - P''_{\ell+1})$$

$$H_2 = \sqrt{2} \cos\frac{1}{2}\theta \sum_{\ell=0}^{\infty} (A_{\ell+} - A_{(\ell+1)-})(P'_{\ell} - P'_{\ell+1})$$

$$H_3 = \frac{1}{\sqrt{2}} \sin\theta \sin\frac{\theta}{2} \sum_{\ell=1}^{\infty} (B_{\ell+} + B_{(\ell+1)-})(P''_{\ell} + P''_{\ell+1})$$

$$H = \sqrt{2} \sin\frac{\theta}{2} \sum_{\ell=0}^{\infty} (A_{\ell+} + A_{(\ell+1)-})(P'_{\ell} + P'_{\ell+1})$$

$$E_{\ell+} = \frac{1}{\ell+1}(A_{\ell+} + \frac{\ell}{2} B_{\ell+}); \; M_{\ell+} = \frac{1}{\ell+1}(A_{\ell+} - \frac{\ell+2}{2} B_{\ell+})$$

for $\ell \geq 1$)

$$E_{\ell-} = \frac{1}{\ell}(-A_{\ell-} + \frac{\ell+1}{2} B_{\ell-}); M_{\ell-} = \frac{1}{\ell}(A_{\ell-} + \frac{\ell-1}{2} B_{\ell-})$$

for $\ell \geq 2$)

---

decomposition of the nucleon current and the nucleon-Δ-isobar transition matrix element. Table II illustrates the second method by giving the helicity amplitudes for pion photoproduction $\gamma + N \rightarrow \pi + N$ and their partial wave development into "multipole" amplitudes.

For later use we now rewrite (1) in another way by using light cone coordinates[4]. For any 4-vector a^μ we introduce

$$a_+ = \frac{1}{2}(a^o + a^3); \; a_- = a^o - a^3; \; \vec{a}_\perp = (a^1, a^2)$$

$$a \cdot b = a_+ b_- + a_- b_+ - \vec{a}_\perp \cdot \vec{b}_\perp \quad . \tag{4}$$

Here the 3-direction has been singled out. We take the photon momentum $\vec{k}$ along the <u>negative</u> 3-direction, so that

$$\vec{k}_\perp = 0; \quad k_+ = \frac{1}{2}(k^o - |\vec{k}|); \qquad k_- = k^+ + |\vec{k}| \quad . \tag{5}$$

For real photons ($k^2 = 0$) the + component of the momentum vanishes:

$$k_+ = 0, \qquad k_- = 2k^o \tag{5a}$$

while for virtual spacelike photons ($k^2 < 0$) one has

$$k_+ < 0, \qquad k_- > 2k^o$$

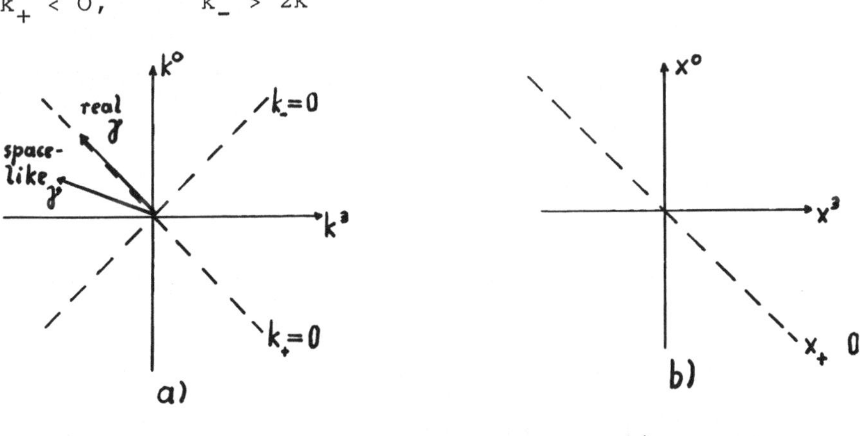

Lightcone coordinates a) k-space b) x-space

By rewriting $d^4x = dx_+ \, dx_- \, d^2x_\perp$ and using translational invariance for the x_+-direction only we find from (1) after integrating over dx_+ with $q = p' - p$

$$\langle B|S|\gamma A\rangle = -2\pi \; i \; \delta(q_- - k_-) \langle B| \int d^4x \, \delta(x_+) e \varepsilon_\mu J^\mu(x) e^{-ik\cdot x} |A\rangle \; . \tag{6}$$

Here the integration is restricted to the hyperplane $x_+=0$. While this is a good feature for applying lightcone algebra the integral contains "bad" operators. By the notion "good" or "bad" we as usual classify operators according to their behaviour under a boost along the positive 3-direction. For a 4-vector one has

$$e^{iuK_3}: \quad \begin{aligned} & a'_+ = e^{u} a_+ ; \; a'_- = e^{-u} a_- \\ & \vec{a}'_\perp = \vec{a}_\perp \; . \end{aligned} \tag{7}$$

Current or tensor densities which transform like a_+ are called "good" densities, if they transform like $\vec{a}_\perp$ or a_- they are "bad" or even "terrible". The significance of this distinction can be most easily seen by considering the behaviour of light-like charges

$$F = \int d^4x \, \delta(x_+) \; V(x) \tag{8}$$

under K_3-boost. Using $U^{-1}(\Lambda) \; V(x) \; U(\Lambda) = V'(\Lambda^{-1}x)$ one finds

$$e^{-iu\,K_3} \, F \, e^{iu\,K_3} = \int d^4x \; \delta(x_+) \; V' \; (\hat{x})$$

$$= \int d^4x\ \delta(\hat{x}_+)\ e^{-u}\ V'(\hat{x}) \tag{8a}$$

where for short $\hat{x} = (e^{-u}\ x_+, e^{+u}\ x_-,\ \vec{x}_\perp)$.

Therefore charges for good densities are invariant under the considered boost while bad densities can be made to vanish by going to the limit $u \to \infty$.

The integrand of (6)

$$\varepsilon_\mu\ J^\mu = \varepsilon_+\ J_- + \varepsilon_-\ J_+ - \vec{\varepsilon}_\perp \cdot \vec{J}_\perp \tag{9}$$

contains one good term and three bad ones. For real photons, see equ. (5a), one can choose

$$\varepsilon_+ = \varepsilon_- = 0 \tag{9a}$$

and is left with the transverse components $\vec{J}_\perp$ only. As has been stressed frequently in the literature use can be made of charge conservation to replace $\vec{J}_\perp$ by J_+[5]. For that purpose start from (3) which can be considered as a function of $P = \frac{1}{2}(p + p')$ and $q = p' - p$. The continuity equation reads

$$q_\mu j^\mu(P,q) = q_+ j_-(P,q) + q_- j_+(P,q) - \vec{q}_\perp\ \vec{j}_\perp(P,q) = 0. \tag{10}$$

By energy momentum conservation $k = q$ and by our choice of the photon direction ($\vec{k}_\perp = 0$) we need to know $\vec{j}_\perp$ only for vanishing transverse momentum transfer. From (10) we have

$$\vec{j}_\perp (P,q) \Big|_{\vec{q}_\perp=0} = \frac{\partial}{\partial \vec{q}_\perp} (q_+ j_- + q_- j_+) \ . \tag{11}$$

This relation is valid also for virtual photons but the r.h.s. still contains bad operators. For real photons one has however $q_+ = k_+ = 0$ and arrives at the desired result

$$\vec{j}_\perp (P,q) \Big|_{\vec{q}_\perp=0} = q_- \frac{\partial}{\partial \vec{q}_\perp} j_+(P,q) \Big|_{\substack{q_+=0 \\ \vec{q}_\perp=0}} \tag{12}$$

where q_- is given by $q_- = k_- = 2\,k^0$. Therefore the transition matrix can be written in the form

$$\langle B|S|\gamma A\rangle = (-i)(2\pi)^4 \delta^4(k+p-p')\, 2k^0\, e\, \vec{\varepsilon}_\perp \cdot \frac{\partial j_+}{\partial \vec{q}_\perp} \Big|_{\substack{q_+ = 0 \\ q_- = 2ko \\ \vec{q}_\perp = 0}} \ . \tag{13}$$

The well-known quantum mechanical correspondence

$$i \frac{\partial}{\partial \vec{q}_\perp} \leftrightarrow \vec{x}_\perp$$

suggests that we look for a connection of the derivative of j_+ with the matrix elements of a light-like electric dipole operator

$$\vec{D}_\perp = \int d^4x\ \delta(x_+)\ \vec{x}_\perp\ J_+(x) \ . \tag{14}$$

In fact by sandwiching this operator one can formally arrive at

$$<B(p')|\vec{D}_\perp|A(p)> = \int d^4x \ \delta(x_+)\vec{x}_\perp \ e^{iq\cdot x} \ j_+(P,q) =$$

$$= (2\pi)^3 i \ \delta(q_+)(\frac{\partial}{\partial\vec{q}_\perp} \ \delta^2(\vec{q}_+)) \ j_+(P,q) \tag{15}$$

$$= (2\pi)^3 \delta(q_+) \ \delta^2(\vec{q}_\perp)(-i) \ \frac{\partial j_+}{\partial \vec{q}_\perp} \ (P,q)$$

so that

$$<B(p')|S|\gamma(k)A(p)> = 2\pi\delta(q_- - k_-)2k^o e<B(p')|\vec{D}_\perp\cdot\vec{\varepsilon}_\perp|A(p)> \ . \tag{16}$$

This result however contains a contradiction: While the l.h.s. is translational invariant the dipole operator $\vec{D}$ changes under a translation according to

$$e^{-iP.a} \ \vec{D}_\perp \ e^{+iP.a} = \vec{D}_\perp + Q \ \vec{a}_\perp \tag{17}$$

where we used the lightcone expression for the electric charge

$$Q = \int d^4x \ \delta(x_+) \ J_+(x) \quad . \tag{18}$$

Equ. (17) is known already from classical electrodynamics: The electric dipole moment changes under translations if the total charge of the system does not vanish.

In quantum mechanics the dipole operator connects therefore states with different transverse momenta. It is useful to exhibit this fact explicitly. For that purpose we consider a normalizable one particle state $|\Psi\rangle$ (a wave packet!) and its wave function

$$\Psi(p,\vec{p}_\perp) = \langle p,\vec{p}_\perp|\Psi\rangle$$

where we restrict ourselves for simplicity to a spinless particle. The states $|p\ \vec{p}_\perp\rangle$ with $p \equiv p_+$ are constructed in the usual way[6]. By applying $\vec{D}_\perp$ one finds by standard techniques

$$\vec{D}_\perp \Psi(p,\vec{p}_\perp) = \int \frac{dq\,d^2q_\perp}{2q}\delta(p-q)(-i)\ \frac{\partial}{\partial\vec{q}_\perp}\delta^2(\vec{p}_\perp-\vec{q}_\perp)\ \cdot$$

$$\cdot\langle p,\vec{p}_\perp|\,J_+\ |q,\vec{q}_\perp\rangle\Psi(q,\vec{q}_\perp) \tag{19}$$

$$= \frac{i}{2p}\ \frac{\partial}{\partial\vec{q}_\perp}\ (\langle p\vec{p}_\perp\,|J_+|q,\vec{q}_\perp\rangle\ \ \Psi(q,\vec{q}_\perp)\Big|_{\substack{q\,=\,p\\ \vec{q}_\perp\,=\,\vec{p}_\perp}}\ .$$

In the last step a partial integration has been performed which is fully justified since $|\Psi\rangle$ is a wave packet. For the corresponding manipulation in equ. (15) however one has no good arguments. Indeed by working out (19)

$$\vec{D}_\perp\Psi(p,\vec{p}_\perp) = \frac{i}{2p}(\frac{\partial}{\partial\vec{q}_\perp}j_+)\Psi(p,\vec{p}_\perp) + \underbrace{\frac{i}{2p}\langle p\vec{p}_\perp\,|J_+|\,p,\vec{p}_\perp\rangle}_{Q_\Psi}\ \frac{\partial\Psi}{\partial\vec{p}_\perp} \tag{20}$$

a difference between the dipole operator and

$$\frac{\partial}{\partial \vec{q}_\perp} j_+$$

becomes explicit. Only if the charge Q_ψ of the state ψ vanishes does the second term in (20) not occur.

Thus the contradiction in equ. (16) is due to an unjustified partial integration in (15). The use of wave packets when calculating magnetic moments has already been advocated frequently in the literature[7]. The general problem which is posed by the operator $\vec{D}_\perp$ has been fully recognized in ref. 4. where its solution has also been indicated. We formulate this solution in an explicit way by defining a translational invariant electric dipole operator $\vec{\mathcal{D}}_\perp$. In principle we have only to substract the second term in (2o). This can be done most elegantly by using the boost operators $\vec{E}_\perp$ for the $x_+ = 0$ hyperplane.

Let us recall: In the lightcone formalism the 6 generators $M_{\mu\nu} = - M_{\nu\mu}$ of the homogeneous Lorentz transformations are used in following combinations

$$\begin{aligned}
\vec{E}_\perp &= \vec{M}_{+\perp} = \frac{1}{2} (\vec{M}_\perp^o + \vec{M}_\perp^3) \\
\vec{F}_\perp &= \vec{M}_{-\perp} = \vec{M}_\perp^o - \vec{M}_\perp^3 \\
K_3 &= M_{-+} = M_{o3} \\
J_3 &= M_{12}
\end{aligned} \tag{21}$$

We need the commutation properties of $\vec{E}_\perp$:

$$[\vec{P}_\perp , \vec{E}_\perp] = i\, P\, I \tag{22a}$$

$$[P, \vec{E}_\perp] = 0; \quad [\vec{E}_\perp , \vec{E}_\perp] = 0 \tag{22b}$$

which follow easily from those of $M_{\mu\nu}$, cp. e.g. ref. 8. Here we used the notation (4) for the energy-momentum operator, putting for convenience

$$P \equiv P_+ = \frac{1}{2}(P_o + P_3) \ .$$

Moreover a product of vectors is always understood as a tensor product.

On the other hand the commutation properties of $\vec{D}_\perp$ can be read off from (17)

$$[\vec{P}_\perp , \vec{D}_\perp] = (-i)\, Q\, I \qquad . \tag{23}$$

Equations (22a) and (23) now suggest how the wanted operator $\vec{\mathcal{D}}_\perp$ can be constructed. In fact by defining

$$\vec{\mathcal{D}}_\perp \overset{\text{def.}}{=} \vec{D}_\perp + \frac{Q}{P}\vec{E}_\perp \tag{24}$$

one gets an operator which commutes with $\vec{P}_\perp$. More explicitly $\vec{\mathcal{D}}_\perp$ is given by

$$\underline{\vec{\mathcal{D}}_\perp = \int d^4x\, \delta(x_+)\, (\vec{x}_\perp + \frac{1}{P}\vec{E}_\perp)\, J_+(x)} \qquad . \tag{24a}$$

The second term in (24) in fact compensates the unwanted term in (20). For a proof use the explicit expression

$$\vec{E}_\perp \, \Psi(p,\vec{p}_\perp) = (-i)p \frac{\partial}{\partial \vec{p}_\perp} \Psi(p,\vec{p}_\perp) \tag{25}$$

which follows from (22a). By (25) the operator $\frac{1}{p}\vec{E}_\perp$ is well defined also for $p = 0$.

Combining our results we find the following formula for the transition matrix for real photon decays in terms of good operators[9].

$$\langle B(p')|S|\gamma(k)A(p)\rangle = 2\pi\delta(q_- - k_-)2k^o e\langle B(p')|\vec{D}_\perp \cdot \vec{\varepsilon}_\perp|A(p)\rangle \,. \tag{26}$$

This result is useful for general considerations; for practical calculations one had better work with (13) directly.

As an illustration we give the results for $\frac{\partial j_+}{\partial \vec{q}_\perp}$ in case of the nucleon current, table Ia:

$$\frac{\partial}{\partial \vec{q}_\perp} \langle N(p',s')|J_+(o)|N(p,s)\rangle \Big|_{\vec{q}_\perp = 0} = -F_2(o)\,\eta^\dagger_{s'}\vec{\sigma}_\perp\sigma_3\eta_s \,. \tag{27}$$

Here η_s are Pauli spinors. This formula can be easily understood: Only the anomalous magnetic terms of $\langle N'|J_+|N\rangle$ which are proportional to $\vec{q}_\perp$ survive after differentiating at $\vec{q}_\perp = \vec{0}$.

For angular momentum considerations introduce the operator

$$D^{(+)} = D_{\perp 1} + i\ D_{\perp 2}$$

which corresponds to

$$\frac{\partial j_+}{\partial q_-} = \frac{\partial j_+}{\partial q_1} + i\ \frac{\partial j_+}{\partial q_2} \quad .$$

By angular momentum conservation only the matrix element

$$<p,\vec{p}_\perp;\tfrac{1}{2}|D^{(+)}|p,\vec{p}_\perp;-\tfrac{1}{2}>$$

is different from zero and is given by

$$<\tfrac{1}{2}|D^{(+)}|-\tfrac{1}{2}> \sim 2F_2(0) = 2\kappa;\ \mu_A = \frac{e\kappa}{2M} \qquad (28)$$

apart from δ-functions. Thus the matrix elements of $D^{(+)}$ give the anomalous magnetic moments of the nucleon. Equation (27) is certainly known and used for many years[10]. What we wanted to stress is that we must work with the translational invariant operator $\vec{D}_\perp$ and not with the simple dipole operator (14). It has been tried to give $\vec{D}_\perp$ itself a physical meaning by considering wave packets which are symmetrical in transverse momenta[7], but it seems appropriate to use only operators which have a well defined meaning for all state vectors[4].

III. TEST OF GENERAL SYMMETRY PROPERTIES OF THE HADRONIC CURRENT

The discussion about possible unusual symmetry

properties of the hadronic electromagnetic current has gone on for years. The raised questions can be answered finally only by experiments, either by refining existing measurements or by searching for new independent empirical informations. In the course of this work it may happen that preliminary or incomplete data lead to exciting news and it takes several years of elaborate research to disprove them. In this section we deal with two examples of this kind: violation of time reversal and the presence of exotic isotensor pieces in the hadronic current.

In the theoretical discussion both are frequently considered together. This is motivated mainly by the fact that the $\gamma N\Delta$-vertex is the simplest coupling of a photon to a pair of hadrons for which T-violation and an isotensor term are not forbidden by general arguments[11].

Consequently the investigation of photon induced processes in the Δ (1232) energy region was a main tool to study T- and isospin abnormities. In this section we briefly describe the latest data and their analyses which indicate compatibility with a normal electromagnetic current.

a) Time reversal

The conceptually simplest way to disprove time reversal invariance is by looking for violation of detailed balance, in suitable two body reactions. Candidates - where the $\gamma N\Delta$ vertex is involved - are the photodisintegrations of deuterons and ^{3}He and their reverse

$$\gamma + d \rightleftarrows n + p \tag{29a}$$

$$\gamma + H_e^3 \rightleftarrows p + d \tag{29b}$$

and the photoproduction of negative pions

$$\gamma + n \rightleftarrows \pi^- + p \tag{30}$$

both in the Δ (1232) region. The experiments are admittedly difficult, especially for the last reactions (30). This is reflected in the experimental data reproduced in Fig. 1. The photoproduction cross sections are quite consistent and scatter around the results of a multipole analysis[13] indicated by the full line. On the other hand the pion capture data differ between different laboratories but the latest CERN results (where the experimental errors have been carefully corrected) agree well within their accuracy with the photoproduction points. Also for (29a, b) detailed balance holds. We give in Fig. 2 the cross sections for the latter reactions where the agreement is even more impressive.

Therefore we can conclude: The latest data confirm the general trend which confines T-violation to the K^o - $\bar{K}^o$ complex[15,16].

b) The isotensor current

All fashionable theories start with the assumption that the electromagnetic current behaves like the electric charge under SU(3), i.e. J^μ is the member of an octett. This is also true for generalized symmetry schemes which allow for new degrees of freedom like color or charm. Under this hypothesis J^μ can be written as a sum of an isoscalar and an isovector term

$$J^\mu = J^\mu_S + J^\mu_V \quad . \tag{31}$$

But during recent years the possible presence of higher isospin parts has been vividly discussed[17]. During this discussion model dependent properties of photoproductions amplitudes have been frequently used. But in view of its fundamental significance a check of (31) which does not depend on the special dynamics of the investigated process is highly wanted. It is not easy to find a process which fulfills this condition and is accessible to experimental work[18]. Consider the photoproduction of pions on nucleons:

$$\gamma + N \rightarrow \pi + N \ .$$

If (31) is valid 3 isospin amplitudes are necessary to describe its charge properties:

$$\langle \pi N | J_S | N \rangle \quad : \quad T^{(o)} \quad \text{isoscalar photon,} \quad I_{\pi N} = \tfrac{1}{2}$$

$$\langle \pi N | J_V | N \rangle \quad : \quad T^{(1)} \quad \text{isovector photon,} \quad I_{\pi N} = \tfrac{1}{2}$$

$$T^{(3)} \quad \text{"} \quad \text{"} \quad , \quad I_{\pi N} = \tfrac{3}{2} \qquad (32)$$

where $I_{\pi N}$ denote the isospin of the πN system. Therefore the amplitudes for the four possible charge modes

$$\gamma + p \rightarrow \pi^o + p \quad : \quad T\,(\pi^o\ p) \qquad (33a)$$

$$\gamma + p \rightarrow \pi^+ + n \quad : \quad T\,(\pi^+\ n) \qquad (33b)$$

$$\gamma + n \rightarrow \pi^- + p \quad : \quad T\,(\pi^-\ p) \qquad (33c)$$

$$\gamma + n \rightarrow \pi^o + n \quad : \quad T\,(\pi^o\ n) \qquad (33d)$$

are linearly dependent

$$\sqrt{2}\ (T\ (\pi^o p) - T\ (\pi^o n)) = T\ (\pi^+ n) + T\ (\pi^- p)\ . \qquad (34)$$

To test this relation one needs all four reactions (33). Recent experiments have been done on the neutral reaction (33d), but at present only the moduli of the four amplitudes are determined in a model independent way. Triangle inequalities which can be deduced from (34) are easily fulfilled since the cross sections for the four processes differ only by factors of 1.5. Fortunately the production of neutral pions in the Δ (1232) energy region is almost completely dominated by this isobar in contradistinction to $\pi^{\pm}$ - production where the photon-pion coupling contributes as much as the $\gamma N\Delta$-interaction.

Thus a comparison of reaction (32a) and (32d) tests in a nearly model independent way for the existence of an isotensor current. To understand this possibility in detail let us write

$$T\ (\pi^o p) = T^{(3)} + T^{(3)'} + T^{(1)} + T^{(o)} \qquad (35a)$$

where the notation of (32) has been used and an isotensor amplitude $T^{(3)'}$ (with $I_{\pi N} = \frac{3}{2}$) introduced. By a 180° isospin rotation around the 2-axis one obtains from (35a) for the $\pi^o n$ amplitude

$$T\ (\pi^o n) = T^{(3)} - T^{(3)'} + T^{(1)} - T^{(o)}\ . \qquad (35b)$$

The isovector amplitude $T^{(1)}$ and the isoscalar $T^{(o)}$ are small in comparison with resonating $T^{(3)}$. By using the quantities

$$t = \frac{T^{(3)'}}{T^{(3)}}; \quad x_o = \frac{T^{(o)}}{T^{(3)}}; \quad x_1 = \frac{T^{(1)}}{T^{(3)}} \tag{36}$$

one has for the ratio of differential cross sections

$$R = \frac{d\sigma\ (\pi^o n)}{d\sigma\ (\pi^o p)} = \left|\frac{1 - t + x_1 - x_o}{1 + t + x_1 + x_o}\right|^2 . \tag{37}$$

Since all these ratios can be assumed to be small one obtains for (37)

$$R \simeq 1 - 4\ \mathrm{Re}\ (t + x_o) . \tag{38}$$

Here t is a real quantity by the Watson theorem[19]. For the resonating amplitude $T^{(3)}$ we write

$$T^{(3)} = \frac{\gamma/2}{M_\Delta - W - i\ \frac{\Gamma}{2}} \tag{39}$$

where W = total cm energy, M_Δ and Γ are the parameters of the Δ-isobar and γ determines the photonic decay width of Δ. Now we introduce a modest dynamical assumption

$$T^{(o)} \approx \text{real} . \tag{40}$$

This is true in all models considered so far where the isoscalar term is essentially given by the Born approximation. From (39) and (40) we find

$$\mathrm{Re}\, x_o = \frac{1}{4}\, \alpha_o\, (M_\Delta - W)$$

where α_o is a real parameter with a slow energy dependence. Therefore R is given by

$$R \approx 1 - 4t - \alpha_o\, (M_\Delta - W)\ . \tag{41}$$

It depends linearly on the energy and the difference R $(W = M_\Delta) - 1$ determines directly the isotensor parameter t. Fig. 3 shows the experimental data[20] for this ratio determined by three laboratories. Evidently the data are well described by formula (41). The three lines are obtained by using a more detailed theory for the isoscalar amplitudes[21] and a value

$$t = 0.00 \pm 0.02 \tag{41a}$$

is inferred. Without committing to such dynamical calculation one finds by fitting (41) directly to the data

$$t = -\,0.005 \pm 0.003\ . \tag{41b}$$

(statistical errors only)

A quite different search for an isotensor current has been completed recently in nuclear physics. Following an older proposal[22] the photonic transitions between $I = \frac{3}{2}$ and $I = \frac{1}{2}$ nuclear states with mass number A = 13 have been compared. Combining the results of three independent experiments the following photonic decay widths have been obtained

$$\Gamma(^{13}C^* \rightarrow {}^{13}C + \gamma) = (23.3 \pm 2.7)\ \mathrm{eV}$$
$$\Gamma(^{13}N^* \rightarrow {}^{13}N + \gamma) = (30.8 \pm 5.5)\ \mathrm{eV}\ . \qquad (42)$$

Without an isotensor these widths should be equal. The authors of ref. 25 conclude that (42) could be an indication for a few per cent isotensor contribution. But in view of the experimental difficulties (three experiments are involved!) it is much too early for any reliable conclusion.

On the other hand the nuclear physics test is not burdened with $I = \frac{1}{2}$ background channels as it is in the case in pion photoproduction. Moreover a 1% isotensor effect in a nucleus corresponds to a much higher value for t (equ. (36)) since only a few percent Δ-admixture are expected in a nucleus. Thus nuclear transitions are much more sensitive to an isotensor and the rather small bounds (41a, b) from π-photoproductions calls for high accuracy experiments in nuclear physics.

In summary we still can live without an isotensor electromagnetic current.

IV. THE ELECTROMAGNETIC COUPLINGS OF THE Δ (1232) - ISOBAR

In this and the next section we shall deal with the electromagnetic decay widths of baryon isobars. To define such a width Γ_γ one usually introduces the narrow width approximation, i.e. one describes short lived excited states mathematically by Kets $|N^*\rangle$ like stable particles. Let us adopt to this method. Its principle problems will be discussed in subsection IVe though not solved.

To apply the general formula of section II we describe the nucleon by a Hilbert vector

$$|N;\ p_+,\ \vec{p}_\perp\ ,\nu> \qquad \nu = \pm\,\frac{1}{2} \tag{43a}$$

where $p_+ = \frac{1}{2}\,(p^o + p^3)$, $\vec{p}_\perp = (p^1,\ p^2)$ are used to fix the momentum and ν to fix the spin component of the nucleon. Analogously for the isobar N^*

$$|N^*,\ p_+',\ \vec{p}_\perp{}',\ \nu'> \qquad \nu' = -\ J,\ldots,\ +J \tag{43b}$$

where J denotes the spin of N^*. The electromagnetic coupling between N and N^* is given by the matrix elements of the dipole operator $\vec{D}_\perp$ for real <u>photons</u>

$$<N^*,\ p_+'\vec{p}_\perp{}',\nu'|\vec{D}_\perp\ \ |N;\ p_+\vec{p}_\perp\ ,\nu> \quad . \tag{44}$$

As explained above $\vec{D}_\perp$ commutes with the transverse momentum operator $\vec{P}_\perp$; the same is true for P_+ since $\vec{D}_\perp$ is by construction (equ. (24a)) invariant under the translation $x_- \to x_-' = x_- + a_-$. Because of

$$[\vec{P}_\perp,\ \vec{D}_\perp] = [\vec{P}_+,\vec{D}_\perp] = 0 \tag{45}$$

one can extract δ - functions from the matrix elements (44)

$$<p_+'\vec{p}_\perp{}'\nu'|\vec{D}_\perp\ |p_+\vec{p}_\perp\ \nu>= (2\pi)^3(-i)\,\delta(p_+'-p_+)\,\delta^2(\vec{p}_\perp{}'-\vec{p}_\perp)\cdot$$

$$<p_+\vec{p}_\perp \nu'|\vec{d}_\perp|p_+\vec{p}_\perp \; \nu> \tag{46}$$

which corresponds of course to equ. (15), which now reads

$$<p_+\vec{p}_\perp |\vec{d}_\perp|p_+\vec{p}_\perp> = \left.\frac{\partial j_+}{\partial \vec{q}_\perp}\right|_{\vec{q}_\perp = 0} \quad . \tag{46a}$$

To deal also with spins in a neat way we use

$$D^{(\pm)} = D_{\perp 1} \pm i \, D_{\perp 2} \qquad \text{or} \qquad d^{(\pm)} = d_{\perp 1} \pm i \, d_{\perp 2} \quad . \tag{47}$$

$D^{(\pm)}$ are related by a reflection

$$Y = P \, e^{-\pi} \, J^2$$

with respect to the 1-3 plane:

$$Y^{-1} \, D^{(+)} \, Y = D^{(-)} \quad . \tag{47a}$$

Thus we need only to consider

$$<p_+ \, \vec{p}_\perp \;\; \nu'|d^+|p_+\vec{p}_\perp \nu>$$

which differs from zero only if $\nu' = \nu + 1$.

Thus only $\nu' = \frac{3}{2}$ and $\frac{1}{2}$ occurs for $J \geq \frac{3}{2}$:

$$<\tfrac{3}{2} \; |d^+| \; \tfrac{1}{2}>, \quad <\tfrac{1}{2} \; |d^+| \; -\tfrac{1}{2}> \; {}^{+}$$

+ For $J=\frac{1}{2}$ only the second matrixelement remains.

The corresponding decay widths depend on the momenta. In the N^* rest system $p_+ = \frac{M_*}{2}$, $\vec{p}_\perp = \vec{0}$ one has with our invariant normalisation (1b) for the partial decay width

$$\Gamma_\gamma(\nu + 1 \to \nu) \equiv \Gamma_\gamma^{\nu+1} = \frac{|\vec{k}|}{8M_*^2} |e2k^o \langle\nu+1|d^+|\nu\rangle|^2$$

$$= \frac{(k^o)^3}{2M_*^2} e^2 |\langle\nu+1|d^+|\nu\rangle|^2 \qquad (48)$$

where we used a general formula (ref. 1,2) and equ. (26). Here $|\vec{k}| = k^o$ are the photon momentum and energy. The total γ-width is obtained by

$$\Gamma_\gamma = \frac{(k^o)^3}{2M_*^2} \frac{1}{2J+1} e^2 \left[|\langle\tfrac{3}{2}|d^+|\tfrac{1}{2}\rangle|^2 + |\langle\tfrac{1}{2}|d^+|-\tfrac{1}{2}\rangle|^2 \right] \qquad (49)$$

where the matrix elements have to be taken in the N^* rest-system and

$$k^o = \frac{M_*^2 - M^2}{2M_*} \quad . \qquad (50)$$

In the literature frequently helicity-amplitudes $A_\lambda \equiv A_\lambda^{J^P}$ are used which are connected by the dipole matrix elements through

$$A_\lambda = \sqrt{\frac{\pi}{4} \frac{k^o}{MM_*}} \, e \, \langle \lambda|d^+|\lambda-1 \rangle \qquad (51)$$

so that

$$\Gamma_\gamma = \frac{1}{\pi} k^{o2} \frac{M}{M_*} \frac{2}{2J+1} [|A_{\frac{3}{2}}|^2 + |A_{\frac{1}{2}}|^2] \quad . \tag{52}$$

The amplitudes A_λ or $<\lambda|d^+|\lambda-1>$ are linear combinations of the electric and magnetic transition amplitudes used throughout in atomic and nuclear physics. The coefficients of these relations depend in a somewhat complicated way on spin and parity J^P of the N^*.

On the other hand the dipole matrix elements can be connected with the invariant decompositon of Table I. For the special case of the Δ-isobar one finds in the hadronic Breit system[27] $(\vec{p}_\perp = 0)$

$$<\Delta p_+\vec{0}\tfrac{1}{2}|d^+|Np_+\vec{0}-\tfrac{1}{2}> = \frac{2p_+}{\sqrt{6}(M+M_\Delta)}\left(\frac{M}{M_\Delta}G_M^\Delta(o) + \frac{M_\Delta-M}{2(M_\Delta+M)}G_E^\Delta(o)\right) \tag{53a}$$

$$<\Delta p_+\vec{0}\tfrac{3}{2}|d^+|Np_+\vec{0}\tfrac{1}{2}> = \frac{2p_+}{\sqrt{2}(M+M_\Delta)}\left(G_M^\Delta(o) + \frac{M-M_\Delta}{2(M_\Delta+M)}G_E^\Delta(o)\right) \quad . \tag{53b}$$

Experimentally practically all amplitudes (51) have been determined by analyzing measurements of pion photorproductions. Various degrees of sophistication have been used thereby: a simple Breit-Wigner ansatz[28], a relativistic isobar model[29], a K-matrix formalism[31].

In the spirit of the narrow width approximation we use the unitarity relation (i T = S - 1)

$$2 \text{ Im } T = T^+T \tag{54}$$

to find for photoproduction in a channel with definite values for spin-parity J^P and isospin I:

$$2\langle \pi N, J^P, I, \; |\mathrm{Im}\, T|\gamma N, \lambda\rangle = \langle \pi N, J^P I|T^\dagger|N^* J^P I\rangle\langle N^* J^P I|T|\gamma N, \lambda\rangle . \tag{55}$$

On the left hand side the imaginary part of the partial wave amplitudes of the photoproduction amplitudes occur which according to Table II are given by

$$\mathrm{Im}\, A_{\ell\pm} \quad \text{for} \quad \lambda = \frac{1}{2}; \qquad \mathrm{Im}\, B_{\ell\pm} \quad \text{for} \quad \lambda = \frac{3}{2}$$

where $J = \ell \pm \frac{1}{2}$. The first factors on the r.h.s. of (55) give the π N N^*-vertex while the second factor is just the wanted photo decay amplitude for N^*. By working out the kinematical details one finds in the cm of photoproduction[26]

$$\mathrm{Im}\, A_{\ell\pm}\; (W = M_\Delta) = \mp\, \gamma(\Gamma_\pi)\, C^I_{\pi N}\, A^{J^P,I}_{\frac{1}{2}}$$

$$\mathrm{Im}\, B_{\ell\pm}\; (W = M_\Delta) = \pm\, \gamma(\Gamma_\pi)\, C^I_{\pi N}\, A^{J^P,I}_{\frac{3}{2}} \sqrt{\frac{16}{(2J-1)(2J+3)}} \; . \tag{56}$$

On the r.h.s. the amplitudes (51) enter where we have also explicitly indicated the J^P, I quantum numbers of the N^*. $\gamma(\Gamma_\pi)$ is given by

$$\gamma(\Gamma_\pi) = \left(\frac{1}{\pi(2J+1)} \frac{|\vec{k}|}{|\vec{q}|} \frac{M}{M_*} \frac{\Gamma_\pi}{\Gamma^2}\right)^{\frac{1}{2}} \tag{56a}$$

and the Clebsch Gordan factor $C^I_{\pi N}$ takes account for the branching in the various π N charge states, $|\vec{k}|$ and $|\vec{q}|$ are the cm momenta of the photon and the pion respectively. Γ_π denotes the pion decay width for N^* and Γ its total width.

The quantities γ (Γ_π) are determined by π N phase shift analyses. Thus we need the multipole amplitudes $A_{\ell\pm}$, $B_{\ell\pm}$ at the resonance energy M_* for a determination of $A^{J^P}_\lambda$, I.

In the rest of this chapter we describe the empirical information for the Δ (1232)-isobar.

a) Energy independent multipole analyses in the Δ (1232)-region

Instead of the $A_{\ell\pm}$, $B_{\ell\pm}$ the old fashioned electric and magnetic amplitudes $E_{\ell\pm}$, $M_{\ell\pm}$ (see Table II) are used in the energy region up to E_γ = 450 MeV.

For these low energies only s- and p-wave amplitudes are needed to describe neutral pionproduction. For charged pions higher ℓ-values enter due to the possibility of a "direct production mechanism". But the Born term diagrams of Fig. 5 are completely calculable and take full account of this possibility. Therefore we are left with the following 12 amplitudes

$$E_{o+}^{(o,1,3)}(\tfrac{1}{2}^-),M_{1-}^{(o,1,3)}(\tfrac{1}{2}^+),M_{1+}^{(o,1,3)}(\tfrac{3}{2}^+),E_{1+}^{(o,1,3)}(\tfrac{3}{2}^+) \quad (57)$$

where the superscripts distinguish as in (32) the three isospin channels and J^P of the πN state is given in brackets. The phases of these complex quantities can be assumed to be known from πN scattering from the Watson theorem

$$M_{1+}^{(o,1,3)}(w) = \pm \left|M_{1+}^{(o,1,3)}(w)\right| e^{i\delta_{1+}^{I}(w)} \qquad \begin{array}{l} I = \tfrac{1}{2} \text{ for } (o,1) \\ I = \tfrac{3}{2} \text{ for } (3) \end{array} \quad (58)$$

and analogous relations for the other multipoles. Thus 12 real quantities remain for each energy to be determined by experiments. On the other hand many independent observables for the reaction (33) have been measured: differential cross sections, polarisations of the final nucleon, photon- and target asymmetries. It is therefore possible to determine all 12 multipoles for each energy between 180 and 450 MeV without further commitments to dynamical assumptions. In this sense various "energy independent" multipole analyses have been performed by several groups[32,33,34]. They differ in the number and quality of data points and the input πN phases $\delta_{\ell\pm}^{I}$ which have slightly changed during the years. Moreover in recent work a critical use of data has been made in favor of smooth results[33].

We describe briefly the resulting multipoles for the Δ-partial wave. The photoproduction data are now good enough for a determination of the Δ^+ mass in contrast to

$$M_{\Delta^{++}} = 1230,3 \text{ MeV} \qquad \text{and} \qquad M_{\Delta^o} = 1232.9 \text{ MeV}$$

known from π-N scattering[35]. The recent BD-fit[33] required the mass

$$M_{\Delta^+} = 1231.8 \text{ MeV}$$

of the Δ^+ isobar. For the magnetic dipole amplitude with $I = \frac{3}{2}$ an impressive circle in the Argand diagram was obtained in all fits, see Fig. 6 for the NPS(71) and BD(75) fits. For the determination of the $\gamma N\Delta$-coupling we need the value of Im M_{1+} at the energy where Re $M_{1+}^{(3)}$ vanishes. Table III summarizes these results together with the value predicted by the Chew-Low formula[36]. The electric quadrupole amplitude $E_{1+}^{(3)}$ is small.

Im $M_{1+}^{(3)}$ $[10^{-3} \frac{1}{m_\pi}]$	
37.83 ± 0.17	NPS(71)
38.08	N(71)
37.4	BD(74)
40.7	Chew-Low

Table III. The magnetic dipole amplitude at resonance.

Already in the first analysis NPS[32] a double zero for its imaginary part was indicated, (see Fig. 7a) though at an energy somewhat above M_Δ. By also considering Compton scattering data and performing a simultaneous multipole fit the small $E_{1+}^{(3)}$ was stable only if this double zero occurred exactly at resonance[37]. This is confirmed by the BD analysis[33], see the circles in Fig. 7b. The other symbols give the results for fits aimed at higher resonances which have only qualitative significance in this Δ region. We thus conclude: the E2 amplitude can be represented near resonance by a formula of the type

$$E_{1+}^{(3)} = h\ (w)\ (w - M_\Delta)\ e^{i\delta_\Delta(w)} \tag{59}$$

where h(w) denotes a real function.

b) Conclusions for the $\gamma N\Delta$-coupling

The very good evidence for the vanishing of the electric quadrupole at the Δ-mass confirms the validity of the Becchi-Morpurgo selection rule first derived on the basis of a static quark model[38]. Here one starts from a non-relativistic Hamiltonian for the quark photon interaction

$$H_{elm} = - \sum_{i=p,n,\lambda} e_i \Big[\underbrace{\frac{\vec{p}_i}{m_Q} \cdot \vec{A}(\vec{x}_i)}_{\text{orbital current term}} + \underbrace{\frac{g}{2m_Q} \vec{\sigma}_i \cdot (\nabla x \vec{A}(x_i))}_{\text{spin term}} \Big] \tag{60}$$

where the gyro-magnetic factor g does not depend on the quark index. An electric quadrupole operator is contained in the first-convection current-term. Its matrix elements vanish between states of vanishing orbital angular momentum as the nucleon and the Δ are assumed to be. On the other hand one can predict from (60) the magnetic dipole transition rate in terms of the quark magnetic moment[39,26].

$$\mu = \frac{eg}{2m_Q} \quad .$$

For our dipole matrix elements the result is[40]

$$e\, d_{\frac{3}{2}} \equiv e \, \langle \Delta \tfrac{3}{2} | d^+ | N \tfrac{1}{2} \rangle = - 4\sqrt{\tfrac{2}{3}} \sqrt{MM_\Delta}\, \mu$$

$$e\, d_{\frac{1}{2}} \equiv e \, \langle \Delta \tfrac{1}{2} | d^+ | N - \tfrac{1}{2} \rangle = - \tfrac{4}{3} \sqrt{2} \sqrt{MM_\Delta}\, \mu \; . \qquad (61)$$

The E2 selection rule is reflected by the relation

$$d_{\frac{3}{2}} = \sqrt{3}\, d_{\frac{1}{2}} \qquad (61a)$$

which follows from (56) and the formulae of Table II if one requires $E_{1+} = 0$. In the quark model μ is identified with the magnetic moment of the proton leading to absolute values for d_λ.

$$d_{\frac{3}{2}} = -2\sqrt{\tfrac{2}{3}}\sqrt{\frac{M_\Delta}{M}}\,\kappa_p \,; \quad d_{\frac{1}{2}} = \frac{-2\sqrt{2}}{3}\sqrt{\frac{M_\Delta}{M}}\,\kappa_p \qquad (61b)$$

where $\mu_p = \kappa_p \frac{e}{2M}$.

The corresponding helicity amplitude follows from

$$A_\lambda = e \sqrt{\frac{\pi k^o}{2MM_\Delta}} d_\lambda \quad . \tag{62}$$

On the other hand the experimental value for Im M_{1+} leads by (56) and $A_{1+} = \frac{1}{2} M_{1+}$, $B_{1+} = - M_{1+}$ to

$$A_{\frac{1}{2}} = - \frac{1}{2\gamma} \sqrt{\frac{2}{3}} \text{ Im } M_{1+}^{(3)} \quad . \tag{63}$$

Using table III we obtain the values given in table IV which are about 30% larger than

	$A_{\frac{1}{2}}$ $[10^{-3}\ \text{GeV}^{-\frac{1}{2}}]$	$A_{\frac{1}{2}}^{\text{Quark}}$	Ratio
NPS	- 141 ± 0.6		1.31
N	- 142	- 108	1.31
BD	- 139		1.29

Table IV. The helicity $\lambda = \frac{1}{2}$ $\gamma N\Delta$-coupling.

the quark model predictions. At the moment we shall not be concerned with refinements of the quark model which could decrease the discrepancy. Rather we ask whether the experimental determinations based on (63) are reliable enough for a critical comparison.

c) Resonance vs. pole parameters

The empirical data for the $\gamma N\Delta$-couplings depend on the pion decay width Γ contained in γ by (56a). In preparing table IV we used the half width of a Breit Wigner parametrisation

$$\Gamma = 112 \text{ MeV}$$

taken from ref. 35. For some time, however, it has been advocated to use the position of the complex pole in the partial wave scattering amplitude as the correct value of mass and width[41]. The imaginary part of the pole position Γ_{pole} is lower than Γ thus decreasing the discrepancy with the quark model, cp. (63) and (56). For a more satisfactory treatment the scattering amplitude

$$f_{1+}(w) = e^{i\delta_{1+}(w)} \sin \delta_{1+}(w)$$

as well as the magnetic dipole amplitude

$$M_{1+}(w)$$

should be continued to the pole position

$$w_{pole} = M_{pole} - i \frac{\Gamma_{pole}}{2}$$

and their residue at w_{pole} be determined. Using a standard Breit-Wigner formula for f_{1+}, for instance

$$\tan\delta_{1+}(w) = \frac{w_R\ \Gamma(q)}{w_R^2 - w^2};\quad \Gamma(q) = \Gamma_R\left(\frac{q}{q_R}\right)^3 \frac{1+(rq_R)^2}{1+(rq)^2} \tag{64a}$$

one finds

$$w_{pole} = 1211 - i\ 49.5\ \text{MeV} \tag{64b}$$

and a residue which corresponds to a complex width[42]

$$\Gamma_{pole} = 98.9 \cdot e^{-i\ 45.4^{\circ}}\ \text{MeV} \tag{64c}$$

For the γ-width determined analogously from the residue of M_{1+} at w_{pole} one obtains

$$\Gamma_{\gamma,pole} = 0.41 \cdot e^{-i\ 42.5^{\circ}}\ \text{MeV}\ . \tag{65}$$

It goes beyond the scope of these lectures to discuss the exact meaning of these complex quantities; let us merely mention that using $\Gamma = 110$ MeV; $\Gamma_\gamma = 0.77$ MeV

$$\left|\frac{\Gamma_{\gamma,pole}}{\Gamma_{pole}}\right| \approx \frac{1}{1.7}\left|\frac{\Gamma_\gamma}{\Gamma}\right| \tag{65a}$$

reducing the γNΔ-coupling by a factor $\sqrt{1.7} = 1.3$ which - presumably by chance - is just the discrepancy encountered in table IV. But we hasten to add that it is very well possible to fit the observed f_{1+} amplitude by a function which has no poles at all[43]. Thus the whole game may be academic leaving us with the unpleasant feeling that we do not know what an unstable particle and its coupling constants really are.

d) Peak- vs. Area-Method

We used in (63) the maximum of Im M_{1+} to determine A_λ. But the resonance structure of Im M_{1+} is rather broad. Therefore it has been proposed to use the area under the resonance curve[44]. Indeed one does such a calculation automatically when evaluating e.g. the Cabibbo-Radicati sum rule where integrations of the following type

$$\int_{\Delta\text{-region}} \langle N|d^-|w\rangle \langle w|d^+|N\rangle \, dw \tag{66}$$

automatically occur. It was reported[45] that by this method one obtains a magnetic $\gamma N\Delta$-coupling which coincides with the quark model value within a factor 0.9 ± 0.1. Therefore we cannot exclude that the static quark model in its simplest version describes the Δ isobar pretty well.

V. MULTIPOLE ANALYSES FOR HIGHER EXCITED NUCLEON STATES

1. The Method

The method used for the Δ (1232) region cannot be applied above 600 MeV say. Here the number of partial waves is too large and due to the opening of production channels other than πN unitarity no longer determines the ratio of real and imaginary parts. In order to get progress more theoretical inputs or/and specific parametrisations of the partial wave amplitudes must be introduced.

Walker[46] has been most successful by using a simple

parametrisation employing good insight in the data and much physical intuition. His ansatz for a multipole amplitude A(w) can be written as

$$A(w) = A^{\text{Electric Born}}(w) + A^{\text{Res}}(w) + A^{\text{Back}}(w) \qquad (67).$$

(To avoid clumsy expressions we disregard spins and drop quantum numbers in our description of the method.) Here the electric Born amplitude contains only those parts of the Born diagram, fig. 5, where the photon couples to the convection current of pions and nucleons and not to their magnetic moments. For the resonating part $A^{\text{Res}}(w)$ a Breit Wigner formula is used. The background $A^{\text{Background}}$ is treated simply by hand: one chooses numerical values for the amplitudes at energy points separated by 150 to 200 MeV and interpolates by a polynomial. Here of course the experience of a livelong work with photoproduction enters. The resulting multipole amplitudes give an impressive fit up to c.m. energies w = 2 GeV for all measured quantities. Certainly such a procedure most not lead to a unique solution. Therefore other analyses have been carried through where the arbitrariness of the ansatz (67) has been reduced[47,48,49]. First one can use analyticity - instead of the no longer useful unitarity - to connect real and imaginary parts. The generally accepted tools are fixed momentum transfer dispersion relations

$$A(s,t) = A^{\text{Born}}(s,t) + \frac{1}{\pi}\int_{s_0}^{\infty} ds'\left[\frac{1}{s'-s} \pm \frac{1}{s'-u}\right] \text{Im}\, A(s',t). \quad (68)$$

Because of (68) one needs only to parametrize the imaginary

part where the resonance structure should show up most prominently. The disadvantages of (68) are two fold

1) one must integrate up to infinite energies,
2) the integration goes partly over an unphysical region.

For problem 1) one considers the high energy part $\Lambda \leq s' < \infty$ separately. Here Im A is either put equal to zero[48] if Λ is large enough or approximated by some effective resonance[47] (which must not necessarily correspond to an existing isobar) or by using some Regge pole parametrization which fits smoothly to the $s' \leq \Lambda$ amplitudes[48,49]. For the resonance region $s' < \Lambda$ one first develops into partial waves

$$\mathrm{Im}\, A(s,t) = \sum_{\ell} a_{\ell}(s)\, P_{\ell}(\cos\theta) \tag{69}$$

and introduces a fit ansatz for $a_{\ell}(s)$ which contains resonant and background terms. In detail it may look quite differently.

a) additive background[49]

$$a_{\ell}(s) = r_{\ell}(s) + b_{\ell}(s)$$

where r_{ℓ} is taken to be a Breit-Wigner function and $b_{\ell}(s)$ a smooth function.

b) multiplication background[50]

$$a_\ell(s) = e^{2i\delta_\ell^b(s)} \cdot r_\ell(s)$$

where δ_ℓ^b describes a background phase which in practice is determined by an analysis of the structure of the πN amplitude.

c) K-matrix formalism[47]

Here A is considered as a part of a 3 channel matrix T, where the channels are πN, γN and an artificial third channel which describes effectively all other open channels. Resonance and background are added for the K-Matrix.

$$\frac{1}{K} = \frac{1}{T} - iq \qquad (q = \text{momentum matrix})$$

but keeping after matrix inversion only the terms of T which are linear in the photon couplings.

Before starting computer work with one of this schemes one must worry about problem 2). Fig. 8 illustrates its origin. For any t-value (except one!) the s'-integration leaves the physical region for energies smaller than certain s_1. For $s_o \leq s' < s_1$ one could define A(s,t) by the expansion (69). There are, however, two limitations: For high ℓ and large t the Legendre polynomials can lead to rapid unphysical increase and the series expansion (69) may be basically wrong because of strong singularities in A(s,t). Their possible locations can be inferred from a Mandelstam representation and are depicted in fig. 9 as the cross-hatched regions. If the (negative) t value is smaller than t_s one must integrate directly through the singularities. The value of t_s is about -1.1 GeV for neutral pion production corresponding to a photon energy of about E_γ = 900 MeV. For charged pions t_s is about -1.7 GeV which occurs at $E_\gamma \approx 1.2$ GeV. On the other hand reflections of singularities for positive t values may cause convergence troubles in (69). They are expected[48] for neutral pions at $E_\gamma \approx 1.2$ GeV and for charged ones at $E_\gamma \approx$ 900 MeV.

These singularities are expected to be especially dangerous for s' varying over the Δ (1232) region. By a sophisticated conformal mapping method one can try to estimate the influence of these singularities on the partial wave amplitudes. Fig. 10 illustrates how the real part of an electric dipole amplitude E_{0+} is changed by these singularities[51]. If we disregard the dip at about E_γ = 1 GeV strong changes occur only for $E_\gamma > 1.3$ GeV. In the concrete analyses these effects have not yet been taken into account systematically. In the MOR analysis[47] a term of the form

$$\Delta \, \mathrm{Im} \, A(s',t) = C \, \theta(t_s - t) \, \delta(s - M_\Delta^2) \, (t_s - t)$$

was added to the imaginary parts with a constant C to be determined in the fitting program.

In view of these complexities one might simply accept the results if they describe the data well without bothering about analyticity and convergence problems. In view of the lack of uniqueness one can, however, feel better if the multipoles fulfill general principles by construction.

2) The Results

a) The Δ (1232)-region

Table V shows the results of the different fits together with that of Table IV for γNΔ coupling

P_{33} (1232)	$A^{V_3}_{\frac{1}{2}}$	$A^{V_3}_{\frac{3}{2}}$
NPS[32]	- 141 ± 0.6	- 244 ± 1
PDG[52]	- 141 ± 3	- 259 ± 7
MW[46]	- 140 ± 6	- 254 ± 7
KMORR[47]	- 143	- 257
Quark model	- 108	- 187

Table V. γNΔ coupling of various high energy fits

The agreement of the energy independent analyses with the analyses mainly aimed at the higher resonances is impressive. The differences in the helicity $\frac{3}{2}$ amplitudes are due to the fact that we put in the first row the electric quadrupole exactly equal to zero.

b) The second resonance region: $P_{11}(1470)$, $D_{13}(1520)$, $S_{11}(1535)$

Here a grouping of three $I = \frac{1}{2}$ isobars can be seen in πN scattering. Table VI contains the photonic couplings

$P_{11}(1470)$	$A^p_{\frac{1}{2}}$	$A^n_{\frac{1}{2}}$
PDG[52]	-76 ± 20	40 ± 45
MW[46]	-70 ± 23	-43 ± 35
KMORR[47]	-80	0
NW[51a]	-136	35
Quark model	27^*	-18^*

Units: 10^{-3} $\text{GeV}^{-\frac{1}{2}}$

$D_{13}(1520)$	$A^p_{\frac{1}{2}}$	$A^p_{\frac{3}{2}}$	$A^n_{\frac{1}{2}}$	$A^n_{\frac{3}{2}}$
PDG[52]	-8 ± 18	177 ± 15	-76 ± 13	-124 ± 10
MW[46]	-6 ± 6	165 ± 11	-66 ± 10	-188 ± 13
KMORR[47]	-19	170	-70	-128
NW[51a]	2	184	-87	-126
Quark model	-34	109^*	-31	-109^*

$S_{11}(1535)$	$A^{p}_{\frac{1}{2}}$	$A^{p}_{\frac{3}{2}}$	$A^{n}_{\frac{1}{2}}$	$A^{n}_{\frac{3}{2}}$
PDG[52]	98 ± 25		-42 ± 13	
MW[46]	63 ± 13		-51 ± 21	
KMORR[47]	89		-52	
NW[51a]	78		- 9	
Quark model	156		108	

Table VI $\gamma N\ N^*$ in 2nd resonance region.

All fits agree with the dominance of D_{13} (1520) which was concluded from the angular distribution already years ago. It is instructive to look for the total width, Table VII.

	$P_{11}(1470)$	$D_{13}(1520)$	$S_{11}(1535)$
KMORR[47]	6.4	29.3	7.9
Quark model	0.73*	13.0*	24.3

Table VII. Total intensity (10^{-3} GeV^{-1}).

By comparing the different fits one observes rather stable couplings for the $D_{13}(1520)$ with small errors for the dominant $\lambda = \frac{3}{2}$. For neutron targets this dominance is not clear. Moreover the coupling is no longer purely isovector, which would require opposite value for protons and neutrons. The P and S coupling seem to increase in the later and more elaborate fits.

In view of the large number of data available and of the fact that the methods should work fine in this energy region one might not be satisfied especially with the determination of the S_{11} and P_{11} coupling. (even sign discrepancies for $A_{\frac{1}{2}}$ (P^{o}_{11})!).

In the tables VI and VII Quark model predictions are listed where D_{13} and S_{11} have a simple interpretation as the lowest orbital momentum L = 1 excitation coupling with the quark spin $S = \frac{1}{2}$ to the two spin values of these isobars D_{13} (1525) = $^2P_3$; S_{11}(1535) = $^2P_1$. For the photon amplitudes in general both terms of the quark Hamiltonian[60] contribute.

If by some selection rule only one term occurs or if both terms enter with the same sign the predictions are more reliable. This has been denoted by * following MOR[47]. Starred or unstarred the signs of the quark amplitudes agree for D_{13} and S_{11} with the experimental ones. The magnitudes differ appreciably. The model requires a dominance of S_{11} which is qualitatively wrong. For the P_{11} even the signs seem to be wrong, if one interpretes P_{11} as a first radial excitation state as is mostly done.

c) Higher resonances

After this experience we shall expect only qualitative agreement between the different analyses themselves and if compared with the quark model. The table VIII containing a compilation from the London conference confirms this expectation.

We concentrate on two special cases. The $D_{13}(1690)$ and $D_{15}(1670)$ are interpreted as the ${}^4P_3$ and ${}^4P_5$ quark states. For these ${}^4P_J$ levels the Moorhouse selection rule holds: Photonic couplings do vanish for the charged isobars.

Present evidence is found in table IX. The rule

$D_{13}(1690)$	$A^P_{\frac{1}{2}}$	$A^P_{\frac{3}{2}}$	$D_{15}(1670)$	$A^P_{\frac{1}{2}}$	$A^P_{\frac{3}{2}}$
PDG	-60 ± 60	40 ± 40		22 ± 10	25 ± 12
MW	0 ± 34	0 ± 29		10 ± 13	42 ± 24
KMORR	22	61	7	7	17
DLR	-1 ± 55	-21 ± 35		7 ± 19	15 ± 17
Quark model	0	0		0	0

Table IX. Test of the Moorhouse selction rule

seems to hold up to a possible violation for D^+_{15} (1670). As the second example we consider the Copley-Karl-Obryk (COK) Selection rule. This rule deals with [56] states with $L > 0$. Here $I = \frac{1}{2}$ states with helicity $\lambda = \frac{3}{2}$ can be constructed by coupling the quark spin $S = \frac{1}{2}$ (which is the only quark spin for $I = \frac{1}{2}$ possible) with orbital angular momentum. For such a state only the orbital current part of (60) can cause transitions to the ground state. But this term contains the total charge as a factor, hence vanishes for the neutral partners.

This rule is relevant for the L = 2 states $P_{13}(1770) = {}^2D_3$ (quark) and $F_{15}(1680) = {}^2D_5$ (quark). Table X shows that the data scatter around its prediction.

	$F_{15}(1680)$	$P_{13}(1770)$
PDG	- 15 ± 15	40 ± 60
MW	0 ± 30	0 ± 44
KMORR	- 15	- 23
DLR	- 20 ± 23	- 56 ± 110
Quark model	0	0

Table X $A^n_{\frac{3}{2}}$ for octet L = 2 states

It is interesting to note that the COK-rule is based on the absence of a spin orbit term in the Hamiltonian[26]. Any inclusion of such terms would weaken the rule.

Resonance (mass)	Charge	Heli- city	Quark model	MW	KMORR	DLR
P33(1232)	+	1/2	-108*	-140±6	-143	-144±4
		3/2	-187*	-254±7	-257	-249±7
P11(1434)	+	1/2	27*	-70±23	-80	-78±14
	o	1/2	-18*	-43±35	0	62±26
S11(1505)	+	1/2	156	63±13	89	52±24
	o	1/2	-108	-51±21	-52	-35±23
D13(1514)	+	1/2	-34	-6±6	-19	3±14
		3/2	109*	165±11	170	183±14
	o	1/2	-31	-66±10	-70	-81±21
		3/2	-109*	-118±13	-128	-155±22
D15(1665)	+	1/2	0*	10±13	7	7±19
		3/2	0*	42±24	17	15±17
	o	1/2	-38*	4±15	-43	-62±30
		3/2	-53*	-9±30	-90	-64±17
F15(1682)	+	1/2	-10	-8±11	-25	21±22
		3/2	60*	129±16	96	158±14
	o	1/2	30*	8±18	33	16±42
		3/2	0*	0±30	-15	-20±23
P11(1646)	+	1/2	-40*	-68±24	18	-19±62
	o	1/2	10*	48±45	18	-71±49
D33(1649)	+	1/2	88	0±48	79	77±39
		3/2	84*	0±41	61	55±18
S31(1660)	+	1/2	47	105±38	27	-21±23
D13(1680)	+	1/2	0*	0±34	22	-1±55
		3/2	0*	0±29	61	-21±35
	o	1/2	-10*	0±34	73	58±100
		3/2	-40*	0±44	51	-52±72

Resonance (mass)	Charge	Helicity	Quark model	MW	KMORR	DLR
SS11(1688)	+	1/2	0	12±15	52	18±20
	0	1/2	30	-19±22	-55	-2±48
P33(1750)	+	1/2	20	0±38	17	-12±35
		3/2	34	0±33	-53	18±47
F35(1842)	+	1/2	-20	47±67	25	-15±27
		3/2	-90	-28±66	-44	62±39
P13(1850)	+	1/2	100	0±25	26	68±30
		3/2	-30	0±22	-12	-68±43
	0	1/2	-30	0±50	14	42±54
		3/2	0	0±44	-23	-56±110
F37(1940)	+	1/2	-50*	-59±29	-67	-100±17
		3/2	-70*	-93±24	-80	-70±13
P31(1943)	+	1/2	30	-32±65	18	-15±18
D13(1971)	+	1/2				-1±33
		3/2				82±48
	0	1/2				-3±56
		3/2				170±91
P33(2000)	+	1/2	30		-34	
		3/2	-50		-9	

Table VIII. Resonance couplings.

The γN couplings in the usual convention are shown in units of 10^{-3} $GeV^{-1/2}$. No errors are available at the time of writing for KMORR.

VI. SYMMETRY APPROACH TO THE PHOTON-BARYON COUPLINGS

As demonstrated in the last section the quark model is successful in predicting qualitative features of the γ-nucleon couplings: the three quark model selections rules are rather well fulfilled in the data. On the other hand for non vanishing couplings the quantitative agreement was not impressive. In the second resonance region according to this model the $S_{11}(1550)$ should be more strongly coupled than $D_{13}(1520)$ in definite contradiction to the empirical evidence. Inclusion of recoil corrections - by harmonic oscillator wave functions say - would in most cases make things worse, since the absolute values of quark model transition probabilities are generally smaller than the experimental ones. Therefore it is advisable to go back to the group theoretical foundation of the quark model and ask how one could possibly enlarge this frame and thus avoid quantitative clashes. At the Schladming Winterschool 1972 M. Gell-Mann introduced the notions which have turned out to be very fruitful to clarify the possibilities of the quark model. His distinction between "constituent" and "current" quark loosens the connection between the quark wave functions and the transition operator thus making theoretical predictions more flexible. Concretely more terms are added to the Hamiltonian (60) for the γ-hadron interaction[7,54] the relative strengths of which are not specified. It is left to further dynamical calculations with relativistic quark models to tighten the frame again by predicting the coefficients[55].

1) The electric current in the $SU(6)_W$ frame

To provide the basis for the further development let us recall:

a) Hadrons are considered as irreducible representations of a "constituent quark" group $SU(6)_{const.} \times O(3)$. No explicit expressions for the generators of this group are given. We know about it only via its representations. For example the baryon states span the representations:

$$\{[56], L = 0\}, \{[70], L = 1\}, \{[56], L = 2\}$$

b) The "current quark group" can be constructed from the (directly or indirectly) measurable densities

vector	$V^{\mu}_{\alpha}(x)$	
axialvector	$A^{\mu}_{\alpha}(x)$	$\alpha = 0,1,\ldots, 8$
tensor	$V^{\mu\nu}_{\alpha}(x) = V^{\nu\mu}_{\alpha}(x)$.	

They give the following good light-like charges

$$F^{i}_{\alpha} = \int d^4x \; \delta(x_+) \; F^{i}_{\alpha}(x) \qquad (i = 0,1,2,3) \tag{70}$$

with

$$F^{0}_{\alpha} = V^{+}_{\alpha}(x) = \frac{1}{2}(V^{0}_{\alpha}(x) + V^{3}_{\alpha}(x)) \qquad \hat{=} \; q^{\dagger}_{+}(x) \; \frac{\lambda\alpha}{2} \; q_{+}(x) \tag{70a}$$

$$F^{1}_{\alpha} = V^{+,2}_{\alpha}(x) = \frac{1}{2}(V^{02}_{\alpha} + V^{32}_{\alpha}) \qquad \hat{=} \; q^{\dagger}_{+}(x) \; \frac{\lambda\alpha}{2}\gamma_1\gamma_5 q_{+}(x) \tag{70b}$$

$$F^{2}_{\alpha} = V^{+,1}_{\alpha}(x) = \frac{1}{2}(V^{01}_{\alpha} + V^{31}_{\alpha}) \qquad \hat{=} \; q^{\dagger}_{+}(x) \; \frac{\lambda\alpha}{2}\gamma_2\gamma_5 q_{+}(x) \tag{70c}$$

$$F^3_\alpha = A^+_\alpha(x) = \frac{1}{2}(A^o_\alpha + A^3_\alpha) \qquad \hat{=}\ q^\dagger_+(x)\ \frac{\lambda\alpha}{2}\sigma_3 q_+(x) \ . \qquad (70d)$$

$$\alpha = 0,1,..,8$$

On the r.h.s. of these equations the bilinear expressions of quark spinors with the same transformation properties are listed. q_+ is defined in terms of the quark Dirac spinor by

$$q_+(x) \ := \frac{1+\alpha_3}{2}\, q(x) \ ; \qquad \alpha_3 = \begin{bmatrix} 0 & \sigma_3 \\ \sigma_3 & 0 \end{bmatrix} .$$

For i = 0 the internal index α runs over 1 through 8, for i = 1,2,3 the index α takes nine values. The resulting 35 charges (70) fulfill the commutation relations of a $SU(6)_W$ algebra which is just $SU(6)_{W,\text{current}}$. Being light-like charges they have simple Lorentz-group properties: All commute with $P = \frac{1}{2}(P^o + P^3)$, the transverse momenta $\vec{P}_\perp$, with the 3-direction boost K_3 and the transverse lightcone boosts $\vec{E}_\perp$

$$[F^i_\alpha, P] = [F^i_\alpha, \vec{P}_\perp] = [F^i_\alpha, K_3] = [F^i_\alpha, \vec{E}_\perp] = 0 \ . \qquad (71a)$$

With respect to rotations J_3 F^o_α and F^3_α are scalars

$$[J_3, F^o_\alpha] = [J_3, F^3_\alpha] = 0 \qquad (71b)$$

while

$$F^{(\pm)}_\alpha \ := F^1_\alpha \pm i\, F^2_\alpha \qquad (71c)$$

carry angular momentum ±1

$$[J_3, F_\alpha^{(\pm)}] = \pm F_\alpha^{(\pm)} \quad . \tag{71d}$$

It is useful to consider two subgroups of $SU(6)_{w,current}$.

The operators

$$F_\alpha^o, \; F_\alpha^3 \qquad (\alpha = 1,\dots,8)$$

generate a SU(3) x SU(3) subalgebra. The remaining longitudinal operator F_o^3 can be considered as quark spin S_3

$$S_3 = F_o^3 \quad . \tag{72a}$$

Defining a quark orbital angular momentum by

$$L_3 := J_3 - S_3 \tag{72b}$$

one is led to a group

$$SU(3) \times SU(3) \times O(2) \tag{73}$$

generated by

$$F_\alpha^o + F_\alpha^3 \qquad F_\alpha^o - F_\alpha^3 \qquad L_3 \quad . \tag{73a}$$

Its representations are denoted by

$$[(A,B)S_3, \quad L_3] \tag{74}$$

where A and B are the dimensions of the first and second SU(3) groups respectively.

A second important subalgebra is defined by

$$\vec{W} = \begin{bmatrix} F_o^1 \\ F_o^2 \\ F_o^3 \end{bmatrix} \quad . \tag{75}$$

These three operators obey the commutation relations of a SU(2) algebra. Its irreducible representations can therefore be characterized by an integer or half integer number $W = 0, \frac{1}{2}, 1, \ldots,$ and their basis vectors specified by (W, W_3). The generators (70) have the following W-spin properties

$$\begin{aligned} F_\alpha^o \quad & \hat{=} \ (W = 0,\ W_3 = 0) \\ F_\alpha^1 \pm i\, F_\alpha^2 \ & \hat{=} \ (W = 1,\ W_3 = \pm 1) \\ F_\alpha^3 \quad & \hat{=} \ (W = 1,\ W_3 = 0) \quad . \end{aligned} \tag{76}$$

Note that the W-spin vectors combine the vector quantities $F_\alpha^{1,2}$ and the pseudoscalar F^3, a fact known as W-S spin flip[59]. The electric charge $Q^\alpha = F_3^o + \frac{1}{\sqrt{3}} F_8^o$ can be considered as a member of the adjoint representation of SU(3) x SU(3). With the notation (74) it can be characterized by

$$Q \mathrel{\hat{=}} \{[35];[(8,1) + (1,8)]_{S_3=0},\ L_3 = 0 \,|\, W = 0\} \quad . \tag{77}$$

On the other hand Q behaves as a W-spin scalar. These preparations have been necessary to describe the $SU(6)_{w,current}$ properties of the electric dipole operator

$$\vec{D}_\perp = \int d^4x\,\delta(x_+)[\vec{x}_\perp + \frac{1}{p}\vec{E}_\perp](V_3^+(x) + \frac{1}{\sqrt{3}} V_8^+(x)) \quad . \tag{78}$$

Using current algebra and the commutators (71a,b) one sees that $\vec{D}_\perp$ belongs also to a [35] representation. Its behaviour under the subgroup SU(3) x SU(3) is given by

$$D^{(+)} = D_{\perp 1} + iD_{\perp 2} \mathrel{\hat{=}} \{[35],[(8,1) + (1,8)]_o, L_3 = 1 \big| W = 0\} \ . \tag{79}$$

As indicated the quark spin S_3 vanishes due to the simple U(3) property

$$[F_o^3, D^{(+)}] = 0 \ .$$

Important physical consequences have been concluded from this fact. With respect to W-spin also the dipole is a scalar.

Because of its Lorentz properties (71a) the light-like charges F_α^i do not change the momentum components $p = \frac{1}{2}(p^o + p^3)$ and $\vec{p}_\perp$. Consequently the vacuum state is annihilated $F_\alpha^i|0\rangle = 0$ and moving particle states with the same momentum $(p, \vec{p}_\perp)$ can form a representation of $SU(6)_{w,current}$. Since the mass operators $P^\mu P_\mu$ do not commute with the F_α^i such a representation can contain particle states with different masses and also states with different numbers of particles.

The simplest possibility would be that moving hadrons belong to irreducible representations of $SU(6)_{current}$. This can be ruled out by phenomenological arguments[10,53] but also by a simple "a priori"-reasoning[56]. By phenomenology one has argued as follows: The lowest baryon states should be classified according to

$$N, \Delta \varepsilon \{ [56], \ldots, L_3 = 0 \}$$

in view of the successes of the static quark model. But the dipole operator $\vec{D}_\perp$ carries $L_3 = \pm 1$, thus

$$<N \,|\vec{D}_\perp|\, N> = <N \,|\vec{D}^\perp|\Delta> = 0 \;.$$

Therefore the anomalous magnetic moments of the nucleons (cp. (27)) and $\gamma N\Delta$-coupling should vanish. By Hey and Bell an impressive "theoretical" argument was given[56]: Consider the $\{[35], L_3 = 0\}$ meson representation! It contains the pions and ρ-mesons. Because $L_3 = 0$ the angular momentum J_3 coincides with $S_3 = W_3 = F_3^0$. The good component J^+ of the electric current is a W-spin scalar, therefore

$$<\rho^+;p',\lambda'|J^+|\rho^+;p,\lambda> = 0 \qquad \text{for} \quad \lambda \neq \lambda' \tag{80}$$

where λ is the eigenvalue of J_3 or W_3 $(\lambda, \lambda' = \pm 1, 0)$. On the other hand the covariant decomposition of the current matrix elements reads

$$<\rho^+;p',\lambda'|J^\mu|\rho^+;p,\lambda> = I_1^\mu F_1(t) + I_2^\mu F_2(t) + I_3^\mu F_3(t) \tag{80a}$$

where $t = (p' - p)^2$; $q = p'-p$ and

$$I_1^\mu = -(p+p')^\mu \varepsilon'^* . \varepsilon ; \quad I_2^\mu = (\varepsilon'^* . q)\varepsilon^\mu - (\varepsilon \cdot q)\varepsilon'^{*\mu} \tag{80b}$$

$$I_3^\mu = (p+p')^\mu \ (\varepsilon'^* . q) \ (\varepsilon \cdot q) \ .$$

ε and ε' denote the polarization vectors of the ρ^+-meson. By specializing (69) to $\lambda' = -1$, $\lambda = +1$ and $\lambda' = 0$, $\lambda = +1$ one finds

$$F_1(t) = F_2(t) = F_3(t) = 0$$

for all $t \neq 0$. Usual analyticity implies

$$F_1(0) = 0 \ .$$

But $F_1(0)$ gives the electric charge, thus $F_1(0) = 1$. This contradiction could only be avoided if the $\{[35], L_3=0\}$ is not realized physically.

2) General single quark dipole operator

Let us try the next simple possibility: The hadronic $SU(6)_{constituent}$ multiplets can be connected with irreducible $SU(6)_{current}$ representations by a unitary transformation[57]. Such transformations V have been studied intensively for free quarks[7,58]. In this case the "Melosh transformation" V must be a single quark operator. Therefore one could also expect that the transformed dipole operator

$$\vec{D}_\perp' = V^{-1} \vec{D}_\perp V \tag{81}$$

shares this property. We shall deal with an argument against this supposition later. Let us analyze in this subsection the hypothesis:

$\vec{D}'$ = single quark operator with SU(3) properties of the charge Q. (82)

This assumption allows us to write

$$D'^{(+)} = \int d^4x \; \delta(x_+) \; q_+^\dagger(x) \; M^{(+)} \; q_+(x) \tag{82a}$$

where the Dirac matrix $M^{(+)}$ can depend on x_-, $\vec{x}_\perp$, $\frac{\partial}{\partial x_-}$, $\frac{\partial}{\partial \vec{x}_\perp}$ and should guarantee that the integral carries angular momentum $J_3 = +1$. First we describe the behaviour of q(x) under $SU(6)_W$. Decompose

$$q_+(x) = q_\uparrow(x) + q_\downarrow(x) \tag{83}$$

with

$$q_{\uparrow\downarrow}(x) = \frac{1\pm\sigma_3}{2} q_+(x) = \frac{1\pm\gamma_5}{2} q_+(x) \tag{83a}$$

where $q_+ = \frac{1}{2}(1 + \alpha_3)\, q$ and $\sigma_3 = \gamma_5 \, \alpha_3$ has been used. The spin up and spin down spinors have a simple $SU(6)_W$ behaviour which we describe by the SU(3) x SU(3) classification

$$q_\uparrow \hat{=} (3,1)_{\frac{1}{2}} \;; \qquad q_\downarrow \hat{=} (1,3)_{-\frac{1}{2}}$$
$$\bar{q}_\uparrow \hat{=} (\bar{3},1)_{-\frac{1}{2}} ; \qquad \bar{q}_\downarrow \hat{=} (1,\bar{3})_{+\frac{1}{2}} \quad . \tag{83b}$$

Therefore the transformation properties of the direct product $\bar{q}_+ \otimes q_+$ are given by

$$\bar{q}_+ \otimes q_+ \hat{=} (8,1)_o \oplus (1,8)_o \oplus (1,1)_o \oplus (1,1)_o \oplus$$
$$\oplus (3,\bar{3})_1 \oplus (\bar{3},3)_{-1} \quad .$$

The singlet pieces must not occur because of the assumed nature of $\vec{D}_\perp'$.

Using this result for (82) and assuming adequate L_3 properties of M^+ one obtains

$$\begin{aligned} D'^{(+)} &= A \; \{[(8,1) + (1,8)]_o , L_3 = +1\} \\ &+ B \; \{(3,\bar{3})_1 , L_3 = 0\} \\ &+ C \; \{[(8,1) - (1,8)]_o , L_3 = +1\} \\ &+ D \; \{(\bar{3},3)_{-1} , L_3 = +2\} \quad . \end{aligned} \tag{84}$$

In each term we have indicated the transformation properties of the matrix M^+ by the quantum number L_3. The four pieces of $D'^{(+)}$ are written in such a way that they show simple W-spin behaviour

$$[(8,1) + (1,8)]_o \leftrightarrow S = 1,\ S_3 = 0 \text{ or } W = 0,\ W_3 = 0$$
$$(85a)$$
$$[(8,1) - (1,8)]_o \leftrightarrow S = 0,\ S_3 = 0 \text{ or } W = 1,\ W_3 = 0\ .$$

Here we have used the S-W-spin flip[59].

$$(3,\bar{3})_{+1} \leftrightarrow S = 1,\ S_3 = +1 \quad \text{or} \quad W = 1,\ W_3 = +1$$
$$(85b)$$
$$(\bar{3},3)_{-1} \leftrightarrow S = 1,\ S_3 = -1 \quad \text{or} \quad W = 1,\ W_3 = -1\ .$$

By using (84) and (85) the $SU(6)_W$ transformation properties of a single quark dipole operator are given by[54]

$$\begin{aligned} D'^{(+)} &= A\ \{35,\ W = 0,\ \ W_3 = 0;\ \ L_3 = +1\} + \\ &+ B\ \{35,\ W = 1,\ \ W_3 = 1;\ \ L_3 = 0\} + \\ &+ C\ \{35,\ W = 1,\ \ W_3 = 0;\ \ L_3 = +1\} + \\ &+ D\ \{35,\ W = 1,\ \ W_3 = -1;\ \ L_3 = +2\}\ . \end{aligned} \qquad (86)$$

Let us describe the physical meaning of these four terms;

A is a W-spin scalar and transforms like the original dipole operator $D^{(+)}$, it is a purely <u>orbital term,</u>

B carries no orbital momentum but spin and can be regarded as a generalized <u>spin term</u>

C and D depend on the orbital and spin angular momentum. These can be considered as generalized <u>spin orbit terms</u>. A simple example for C result from spin orbit coupling $\vec{\sigma}\cdot\vec{L}$ via minimal electromagnetic coupling[26]

$$e\ \vec{\sigma} \times \vec{r}$$

C occurs explicitly in the socalled ^3P_o model.

The four pieces in (86) can take account for all observed photonic matrix elements. B is responsible for nucleon magnetic moments and the $\gamma N\Delta$-coupling

$$<[56],\ L_3 = 0|D'^{(+)}|[56],\ L_3 = 0> \ :\ B\ . \tag{87a}$$

The photonic decays of the odd parity states ([70], 1^-) depend on three terms

$$<[70],\ L_3 = +1|D'^{(+)}|[56],\ L_3 = 0> \ :\ A,B,C\ . \tag{87b}$$

For {[56], 2^+} all of them can cooperate. Since B is a spin term the Becchi-Morpurgo selection rule (no E2 for $\Delta \to N + \gamma$) remains valid. One can also show[54,60] that Moorhouse's rule holds for (87b), only the COK rule is violated by the spin orbit terms C and D.

In the calculation of transition matrix elements one reduced matrix element occurs for each term in (86) and remains undetermined. This leads to the mentioned flexibility. As an illustration we mention the results of such a fit[62] to the experimental γNN^* amplitudes of Metcalf and Walker[46]. Here one was forced to keep all three pieces A,B,C with a large spin-orbit term C

$$A = 8.25; \qquad B = 2.15; \qquad C = 4.24 \tag{88}$$

In this way the trouble of the simple quark model with the $S_{11}(1535)$ and $D_{13}(1520)$ has been overcome, see table XI. Moreover all photonic couplings of the negative parity {[70], 1^-} baryon multiplet have been very well

$D_{13}(1520)$	$A^p_{1/2}$	$A^p_{3/2}$	$A^n_{1/2}$	$A^n_{3/2}$
MW[46]	-6 ± 6	165 ± 11	-66 ± 10	-118 ± 13
A,B,C Fit[62]	-2.4	158	-41	-122
$S_{11}(1535)$	$A^p_{1/2}$		$A^n_{1/2}$	
MW[46]	63 ± 13		-51 9 21	
A,B,C Fit[62]	50		-58	

Table XI. $SU(6)_W$-Fit for the 2nd resonance region

fitted[62]. Independent work[63] does not obtain such good results but also indicates that the C-term exists.

3) Exotic terms in the generalized dipole operator

Osborn[8,64] has raised considerable doubts about the consistency of hypothesis (82) in view of the two pieces contained in the original $\vec{D}_\perp$:

$$\vec{D}_\perp = \vec{D}_\perp + \frac{Q}{P} \vec{E}_\perp \quad . \tag{89}$$

Formally one obtains by applying the Melosh transformation (81)

$$\vec{D}_\perp' = V^{-1} \vec{D}_\perp \dot{V} + \frac{Q}{P} V^{-1} \vec{E}_\perp V \tag{90}$$

where SU(3) invariance of V has been used. But there is no reason for V to commute with $\vec{E}_\perp$. In fact by using an explicit expression for V of the type

$$V = e^{iY}; \; Y = \int d^4x \; \delta(x_+) \; q_+^\dagger(x) \vec{\gamma}_\perp \cdot \vec{\partial}_\perp F(\partial_\perp^2) \; q_+(x) \quad . \tag{91}$$

One sees immediately using[65]

$$\vec{E}_\perp = \int d^4x \; \delta(x_+) \vec{x}_\perp \; q_+^\dagger \; (i \frac{\partial}{\partial x_-})^{-1} \; q_+ \tag{92}$$

that $\vec{E}_\perp$ cannot be invariant under V. But then the properties of the second term in (89) are rather ugly. By evaluating the transformated $\vec{E}_\perp$ Osborn obtained an expression with the following $SU(6)_W$ properties:

$$V^{-1}\,\vec{E}_\perp\,V \hat{=} \{[35],\ W = 1,\ W_3 = \pm 1,\ L_3 = 0\} \quad .$$

Thus one encounters in (90) a product representation

$$QV^{-1}\,\vec{E}_\perp V \simeq \{[35],\ W = 0\} \quad x \quad \{[35],\ W = 1\} \tag{93}$$

where after expanding the r.h.s. with respect to irreducible representations also "exotic", non [35] pieces will occur. These may be interpreted physically as "many quark" operators containing quadrilinear and higher order expressions in the quark fields. Already the "Q-part" of the original dipole operator (89) has formally a $q_+^\dagger \ldots q_+ \; q_+^\dagger \ldots q_+$ structure by equ. (18) and (92). Thus the result (92) is not surprising but other than $Q\vec{E}_\perp$ the operator (93) has also nonvanishing matrix elements between different particle states. It can especially contribute to the $\Delta \to N + \gamma$ transitions. Fortunately the new operator vanishes for neutral particles ($Q = 0$); thus decays like

$$\begin{aligned} &\Delta^0(1232) \to N^0(939) + \gamma \\ &N^0(1520) \to N^0(939) + \gamma \end{aligned} \tag{94}$$

are not affected. Moreover for the Δ-decay only the B term, (see 87a , contributes. Therefore the relevant dipole matrix elements can be related by $SU(6)_W$ Clebsch-Gordan coefficients[66]. In order to make this explicit we first remark that the matrix elements (47b) of the dipole operator are proportional to the momentum component p_+ due to its behaviour under the K_3-boost which follows from the K_3-invariance of $\vec{D}_\perp$ and equ. (46):

$$<p_+\vec{p}_\perp \lambda|d^+|p_+\vec{p}_\perp \lambda-1> = p_+ F_\lambda; \qquad \lambda = \frac{1}{2}, \frac{3}{2} . \tag{95}$$

This general property is illustrated by equs. (53a,b). The K_3 invariant quantities F_λ are related by the $SU(6)_W$ Clebsch-Gordan coefficients. The same holds for the dipole matrix elements only in the Lorentz frame where the p_+ are equal. Experimentally the decays (94) are compared in the rest frame of the initial particle where p_+ is given by $\frac{1}{2} M_{initial}$. Therefore one finds[66]

$$<\Delta^o, \frac{3}{2}|d^+|n, \frac{1}{2}> = \sqrt{\frac{3}{2}} \frac{M_\Delta}{M_n} <n, \frac{1}{2}|d^+|n, -\frac{1}{2}> . \tag{96}$$

The mass ratio is due to the p_+ dependence of (95). With (28) this relation leads to the following value of the matrix element defined in (61)

$$d_{\frac{3}{2}} = 2 \sqrt{\frac{3}{2}} \frac{M_\Delta}{M_n} \kappa_n \quad . \tag{96a}$$

Using $\kappa_n/\kappa_p = -\frac{2}{3}$ which we consider in our context as an experimental rule one arrives at

$$d_{\frac{3}{2}} = -2 \sqrt{\frac{2}{3}} \frac{M_\Delta}{M_n} \kappa_p \quad . \tag{96b}$$

This result improves the quark model value of equ. (61b) and table IV by a factor $\sqrt{\frac{M_\Delta}{M_n}} = 1.15$.

Analogously all conclusions from (87b) for neutral isobars remain valid. Troubles must be expected only if the charge differs from zero and the simple electric dipole moment $\vec{D}_{\perp}$ is not translational invariant.

But more detailed dynamical considerations would be necessary to estimate the order of magnitude of the new terms[64].

ACKNOWLEDGEMENTS

In preparing this lecture the author has benifited from many discussions with Dr. W. Pfeil, P. Nölle and H. Wessel.

REFERENCES

1. P. D. B. Collins and E. J. Squires, Springer Tracts in Modern Physics, Vol. 45 (1968).
2. V. de Alfaro et al., Currents in Hadron Physics, North Holland 1973.
3. J. D. Bjorken and S. D. Drell, Relativistic Quantum Mechanics.
4. In preparing the following considerations we made ample use of the clarifying papers by H. Osborn, especially Nuclear Physics B80 (1974), 90, where also reference to earlier work on light cone formulation can be found.
5. See e.g. J. Weyers, Constituent Quarks and Current Quarks, Louvain Summer School 1973.
6. One can use the boost

$$|p,\vec{p}_\perp, s> = e^{-\frac{i}{p}\vec{p}_\perp\cdot\vec{E}_\perp} e^{-i\ln\frac{p}{p_o}K_3} |p_o, \vec{o}, s>$$

where $\vec{E}_\perp$ denotes the boost in the transverse direction and the ket on the r.h.s. describes a reference state. Compare e.g. ref. 4.

7. H.J. Melosh, Cal. Tec. Ph. D Thesis (1973) unpublished; H. J. Melosh, Phys. Rev. D9 (1974) 1095. See also ref. 5.

8. H. Osborn, Nuclear Physics B38 (1972) 429.

9. Though $\vec{E}_\perp$ and P do not commute with K_3 separately the combination $\frac{1}{P}\vec{E}_\perp$ does!

10. It plays an important role in applications of current algebra especially in connection with the Cabibbo-Radicati sum rule (N. Cabibbo and L.A. Radicati, Phys. Letters 19 (1966), 697) and the analysis of magnetic moments in $SU(6)_W$ (R. Dashen and M. Gell-Mann in Proc. of the 3rd Coral Gables Conf. on Symmetry Principles at High Energies 1968).

11. For a systematic analysis in the frame of dispersion theory cp. A Donnachie and G. Shaw Phys. Rev. D5, (1972) 1117.

12. W. Bayer et al., Physics Letters (to be published).

13. W. Pfeil and D. Schwela, Nucl. Phys. B 45 (1972) 379.

14. Berkeley-Caltec Collaboration, C.A. Heusch et al.

15. For a more detailed review about an exotic electromagnetic current see A. Donnachie, Springer Tracts in Modern Physics 63 (72) 121.

16. The recent experimental bounds on the electric dipole

moment of the neutron

$$|d_n| \lesssim e \cdot 5 \cdot 10^{-24} \mathrm{cm}$$

(see review talk by K. Kleinknecht at the XVIIth International Conf. on High Energy Physics, London 1974) also start to restrict the T-violating phase ϕ (see ref. 15) of the photon hadron interaction:

$$|\sin \phi| \lesssim \frac{1}{10} .$$

17. It was initiated by the observation of A. I. Sanda and G. Shaw (Phys. Rev. Lett. 24, (1970) 1310) that the photoproduction of π^--mesons off neutrons did not fit easily in the accepted dispersion theoretic scheme.
18. For a review see A. Pais, Phys. Rev. D5, 1170 (72).
19. K. H. Watson, Phys. Rev. 95, 228 (1954). This theorem is based on unitarity and time reversal invariance which we now assume to hold.
20. Fig. 3, summarizes data from Frascati (Di Capua et al., LNC 8 Nr. 11 (1973) 692) Daresbury (R.W. Cliff et al., Phys. Rev. Letters 33, 1500 (74)) Bonn (A. Christ, Dissertation Bonn 1974) obtained by measuring the ratio

$$\frac{d\sigma\ (\gamma + n \rightarrow \pi^0 + n)}{d\sigma\ (\gamma + p \rightarrow \pi^0 + p)}$$

on deuteron targets. In this ratio deuteron binding effects are assumed to cancel.

21. G. Shaw, Daresbury Study Weekend, July 1973, DL/R32.

22. R.J. Blin-Stoyle, Phys. Rev. Lett. 23, (1969) 535.

23. G. Wittwer et al., Physics Letters 30B (69) 634.

24. S.S. Hanna et al., Phys. Rev. 168 (1968) 1169.

25. P. Hinterberger et al., Bonn ISKP-Preprint (1975). The author thanks Dr. Hinterberger for informing him about this iso-tensor test.

26. This notation has been introduced by L.A. Copley, G. Karl, and E. Obryk, Nucl. Phys. B13, 303 (69). See also the mini review in Particle Data Group Phys. Letters 50B, (1974) 121.

27. S. P. de Alwis and J. Stern, Nucl. Phys. B77, 509 (1974).

28. R.H. Dalitz and D.G. Sutherland, Phys. Rev. 146, (1966) 1180.

29. W. Pfeil, Dissertation Bonn (1968) (unpublished) Y.T. Kim, Diploma Thesis Bonn (1969) (unpublished) Cp also Ref. 30.

30. H. Rollnik, Photoproduction of Pseudoscalar Mesons in Methods of Subnuclear Physics, Vol.

31. R. G. Moorhouse, H. Oberlack and A.H. Rosenfeld, Phys. Rev. D9, (1974) 1.

32. P. Noelle, W. Pfeil and D. Schwela, Nucl. Phys. B26 (1971) 461; P. Noelle, Diploma Thesis Bonn PI-2-92 (1971) (unpublished) (N).

33. F.A. Berends and D.L. Weaver, Nucl. Phys. B30, (1971) 575 (BW); F.A. Berends and A. Donnachie, Nucl. Phys. B84 (1975) 342 (BD).

34. Yu. M. Alekhsandrow et al., Sov. J. Nucl. Phys. 12

(1971) 416; Nucl. Phys. B45 (72), 589.

35. J. R. Carter et al., Nucl. Phys. B58, (73) 378.

36. G.F. Chew and F. E. Low, Phys. Rev. 101, 1571 (1956); more detailed discussions of the Chew-Low Theory can be found in ref. 30.

37. W. Pfeil, H. Rollnik, and S. Stankowski, Nucl. Phys. B73 (1974) 166.

38. C. Becchi and G. Morpurgo, Phys. Letters 17, (1965) 352.

39. R. H. Dalitz, Lectures at the Grenoble Summer School (1965).

40. In our normalisation (2) the d_λ are dimensionless quantities thus the mass factors in (61) are appropriate.

41. For a compilation, see Particle Data Group, Phys. Lett., 39B (72).

42. Complex residues of this kind have been considered e.g. by J.S. Ball et al., Phys. Rev. D7, 2789 (1973). The numbers quoted in the text have been determined independently. I am grateful to Dr. W. Pfeil and H. Schneider for their assistance.

43. G. Calucci, L. Fonda, and G. C. Ghirardi, Phys. Rev. 166 (68) 1719; L. Fonda, G. C. Ghirardi, and G.L. Shaw, Phys. Rev. D8 (73) 353.

44. H.F. Jones and M.D. Scadron, ICTP/72/5.

45. F.J.Gilman and I. Karliner, SLAC-PUB-1382 (1974).

46. R.L. Walker, Phys. Rev. 182, (1969) 1729; W.J. Metcalf and R.J. Walker, Nucl. Phys. B76, (1974) 253.

47. R.G.Moorhouse and H. Oberlack, Phys. Lett. 43B, 44 (73);

R.G. Moorhouse, H. Oberlack, and A.H. Rosenfeld, Phys. Rev. D9 (1974)1.G. Knies, R.G. Moorhouse, and H. Oberlack, Phys. Rev. D9 (74) 2680; G. Knies, H. Oberlack, A. Rittenberg, A.H. Rosenfeld, M.Boydanski, and G. Smadje, Phys. Rev. D10 (74) 2778.

48. R.C.E. Devenish, D.H. Lyth, and W.A. Rankin, DNPL/ P109 (1971); R.C.E. Devenish, D.H. Lyth, and W. A. Rankin, Physics Lett. 36B, (1971) 394; R.C.E. Devenish, D.H. Lyth, and W. A. Rankin, Physics Lett.47B (1971) 53.

49. R. L. Crawford, Glasgow Preprint (1975).

50. P. Noelle, Dissertation Bonn (1975).

51. H. Wessel, Dissertation Bonn (1975).

51a.These numbers are preliminary; they are calculated by the authors from ref. 50,51.

52. PDG: average of various analyses of ref. 47, 48, 49 in their earlier stages done by the particle data group, Physics Lett. 50B, 124 (1974).

53. M. Gell-Mann, Schladming Lectures 1972.

54. A. J. Hey and J. Weyer, Phys. Lett. 48B, 69 (1974).

55. There are already many developments towards such models a) R.P. Feynman, M. Kislinger, and F. Ravndal, Phys. Rev. D3, 2706 (1971); R. G. Lipes, Phys. Rev. D5, (1972) 2849; b) A. Böhm, H. Joos, and M. Krammer, Nucl. Phys. B51 (1973) 397 Schladming lectures 1973:

c) Le Yaouanc et al., Phys. Rev. D9, 2636 (1974);
d) P.J. Walters, A.M. Thomson and F.D. Gault, J.Phys. A7 (74) 1681.

56. J.S. Bell and A.J.G. Hey, CERN Preprint TH 1882.

57. This must not be true generally. As an example remem-

ber that orbital angular momentum and spin generate the same algebra but their eigenstates are not connected by a unitary transformation.

58. Instead of quoting the vastliterature we refer to the Schladming lectures 1973 by J.S. Bell.

59. For a recent representation see H. Lipkin, Phys. Reports 8C, (73) 173.

60. A. Love and D. V. Nanopoulos, Sussex preprint (1973).

61. J.L. Rosner and W. P. Petersen, Phys. Rev. D7, (73) 747.

62. R. J. Cashmore, A.J.G. Hey and P.J. Litchfield, Nucl. Phys. B (to be published).

63. F.J. Gilman and I. Karliner, SLAC-PUB-1382 (74).

64. H. Osborn, Nucl. Phys. B80, 113 (74).

65. Formula (80) is a solution, of the commutation relations (22 a,b).

66. The $SU(6)_W$ coefficients can be conveniently read off from Table IV of ref. 63.

FIGURE CAPTIONS

Fig. 1: Detailed balance test for $\gamma + n \rightleftarrows \pi^- + p$ at a photon energy of $E_\gamma = 350\text{-}360$ MeV. Data are represented as cross sections for $\gamma + n \rightarrow \pi^- + p$ for convenient comparison[12].

Fig. 2: Detailed balance check for $\gamma + He^3 \rightleftarrows p + d$ according to ref. 14. The full line gives a simultaneous polynomial fit to both data.

Fig. 3: The cross section ratio

$$\frac{d\sigma\ (\gamma + n \rightarrow \pi^{o} + n)}{d\sigma\ (\gamma + p \rightarrow \pi^{o} + p)}$$

for various energies and angles.

Fig. 4: Search for an isotensor current by nuclear transitions.

Fig. 5: Born diagrams for pion photoproduction.

Fig. 6: Argand plot for $M_{1+}^{(3)}$
a) NPS-fit (1971) ref.32; b) BD-fit (1975) ref.33.

Fig. 7: The $E_{1+}^{(3)}$ multipole
a) Real and imaginary part of N(71) ref.32;
b) Real part of BD (75) ref. 33.

Fig. 8: Fixed t-integration in the s-t plane.

Fig. 9: Spectral functions boundaries.

Fig.10: The electric dipole amplitude E_{o+} generated by the dispersion integral over the resonating magneting dipole M_{1+}. Do: without taking the s-u channel into account; D: with their influence, B: Born approximation.

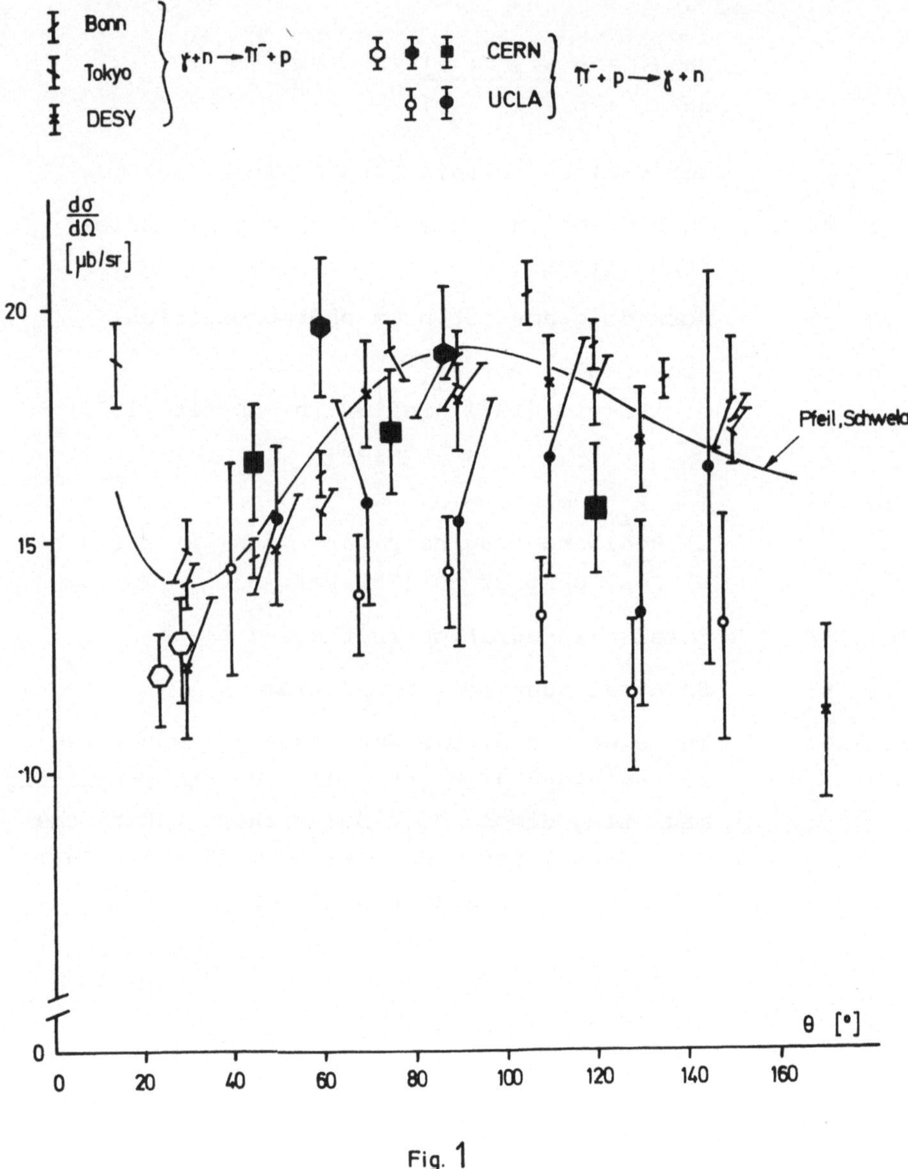
Bonn
Tokyo
DESY
γ+n → π⁻+p
CERN
UCLA
π⁻+p → γ+n
dσ/dΩ
[μb/sr]
20
15
10
0
Pfeil, Schwela
θ [°]
0
20
40
60
80
100
120
140
160

Fig. 1

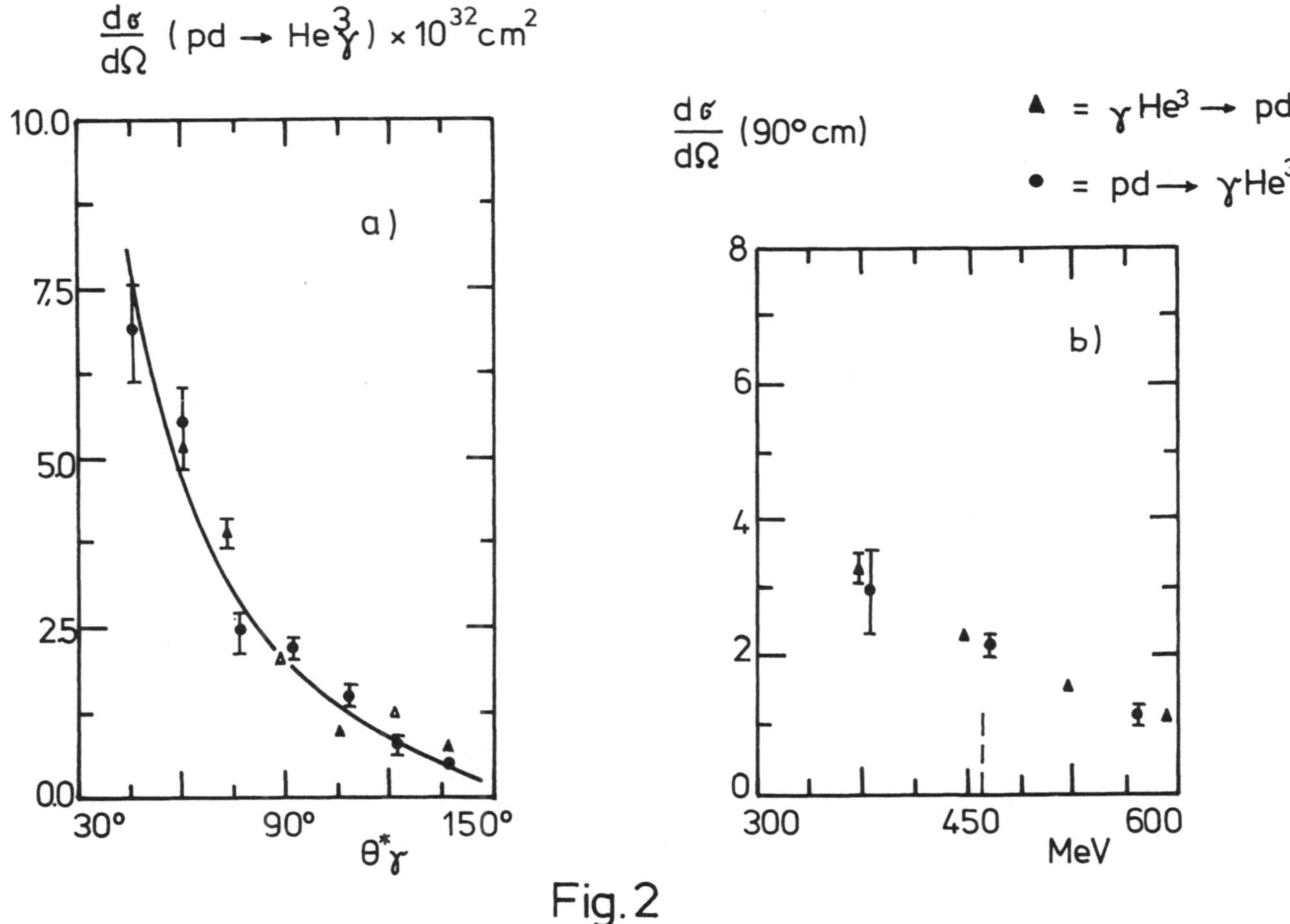
dσ/dΩ (pd → He³γ) × 10³² cm²
10.0
7.5
5.0
2.5
0.0
30°
90°
150°
θ*γ
a)
dσ/dΩ (90° cm)
▲ = γHe³ → pd
● = pd → γHe³
8
6
4
2
0
300
450
600
MeV
b)

Fig. 2

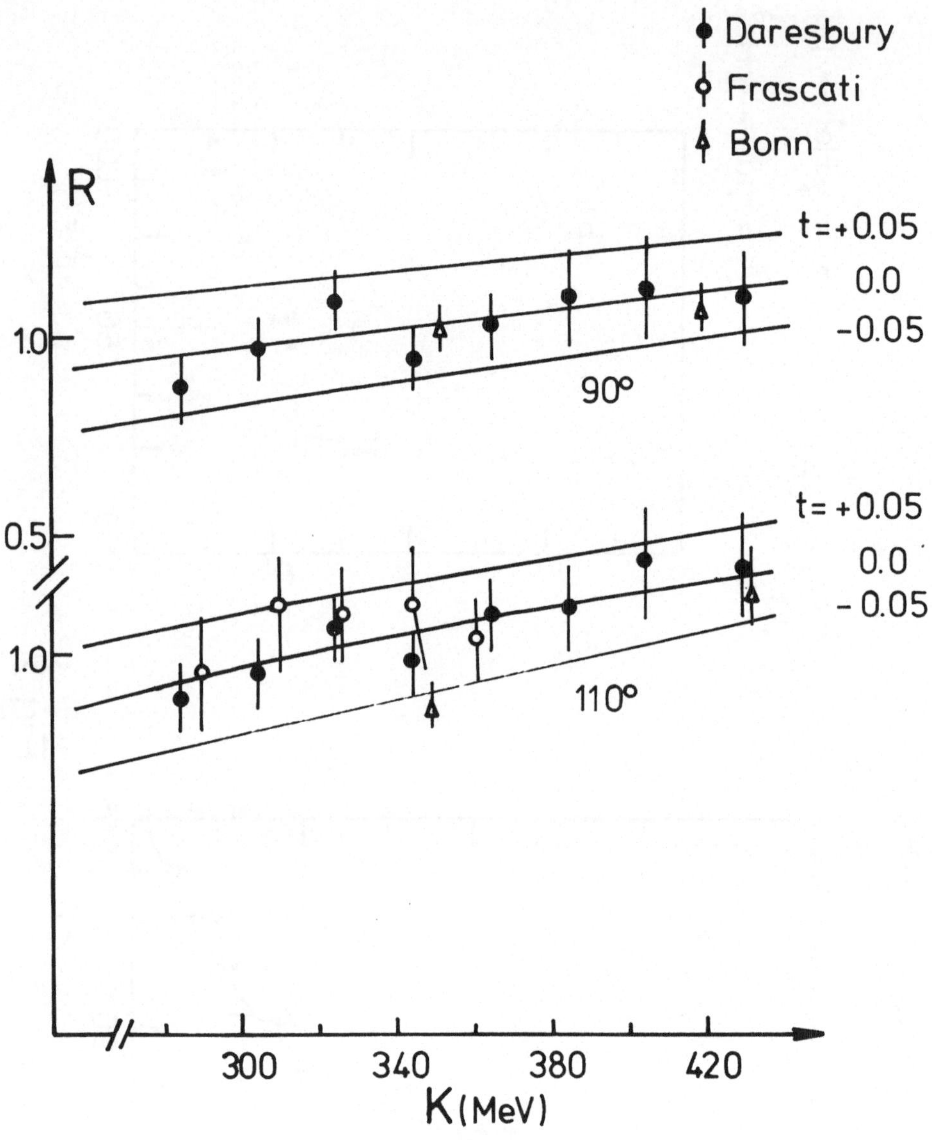

Daresbury
Frascati
Bonn
R
t=+0.05
0.0
-0.05
1.0
90°
t= +0.05
0.0
- 0.05
0.5
1.0
110°
300
340
380
420
K(MeV)

Fig. 3

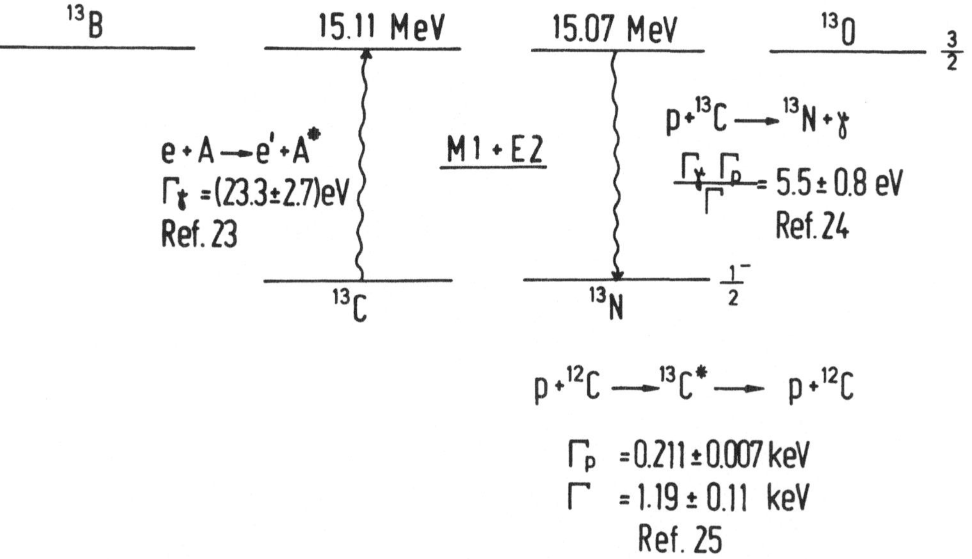

Fig. 4 SEARCH FOR AN ISOTENSOR CURRENT BY NUCLEAR TRANSITIONS

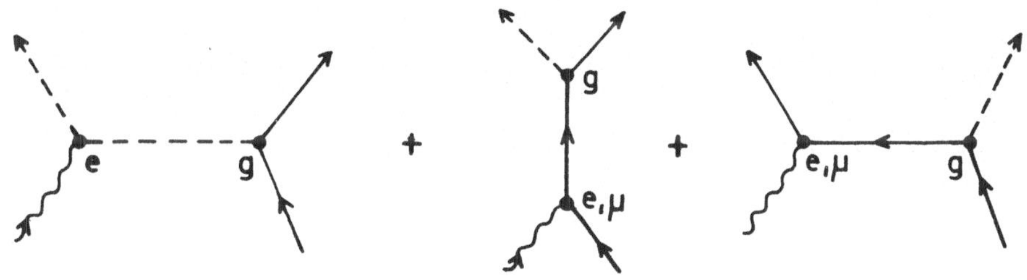

Fig. 5

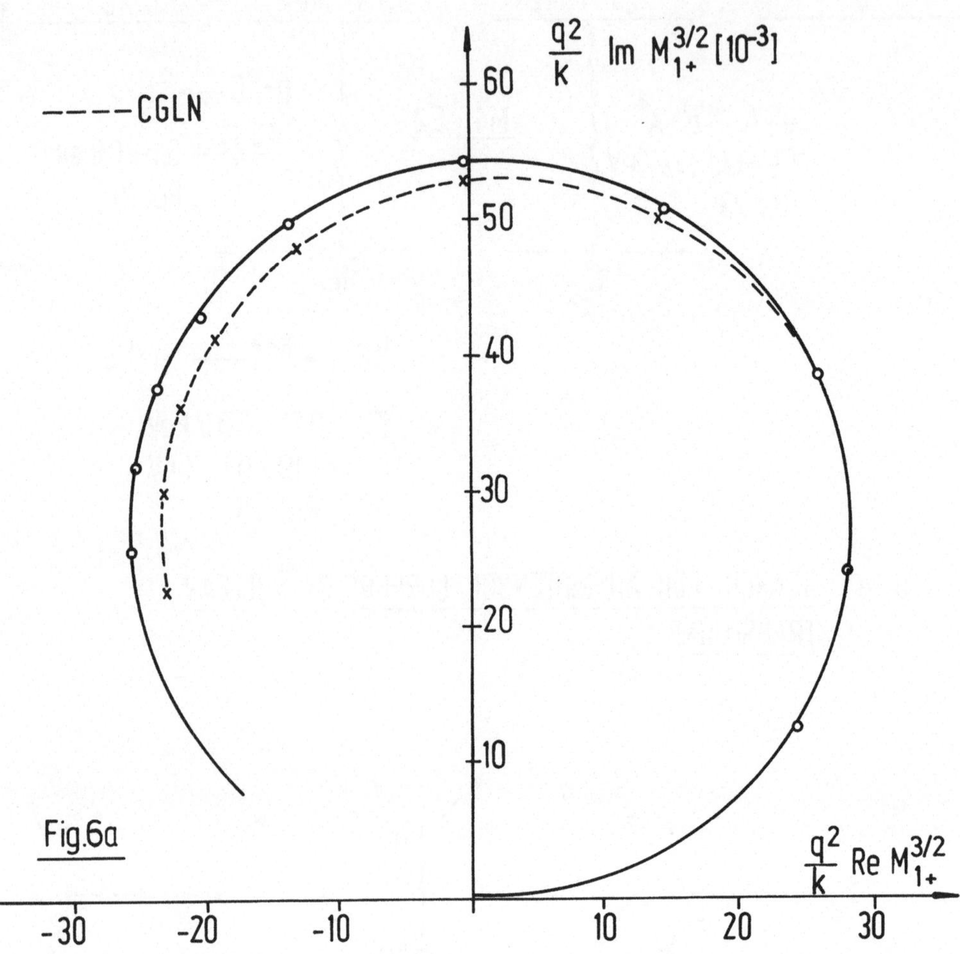
$\frac{q^2}{k}$ Im $M_{1+}^{3/2}$ [10^{-3}]
$\frac{q^2}{k}$ Re $M_{1+}^{3/2}$
----CGLN
60
50
40
30
20
10
-30
-20
-10
10
20
30

Fig.6a

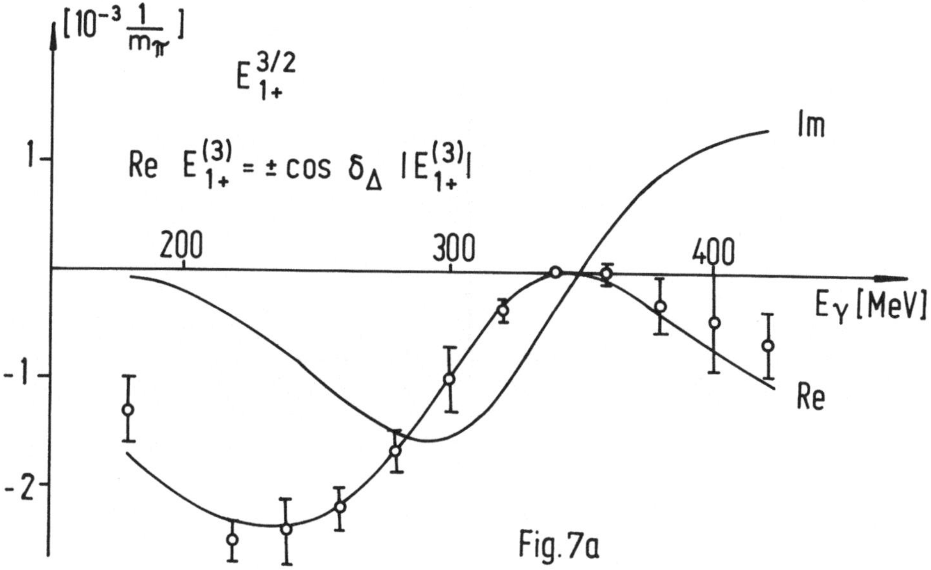

$[10^{-3}\frac{1}{m_\pi}]$
$E^{3/2}_{1+}$
$\mathrm{Re}\ E^{(3)}_{1+} = \pm \cos\ \delta_\Delta\ |E^{(3)}_{1+}|$
1
-1
-2
200
300
400
E_γ [MeV]
Im
Re

Fig.7a

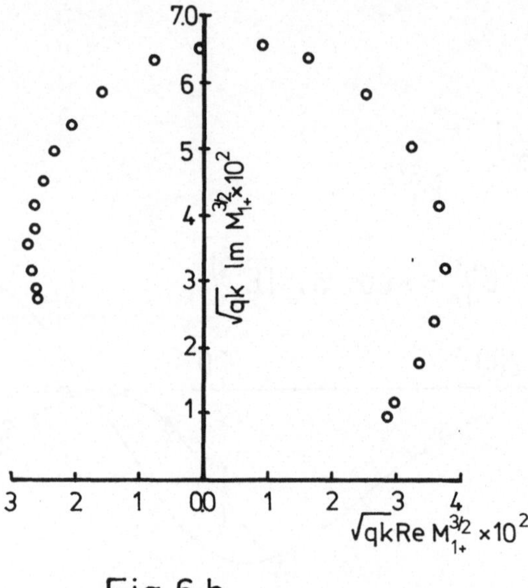
7.0
6
5
4
3
2
1
3
2
1
0.0
1
2
3
4
$\sqrt{qk}$ Im $M^{3/2}_{1+} \times 10^2$
$\sqrt{qk}$ Re $M^{3/2}_{1+} \times 10^2$

Fig. 6 b

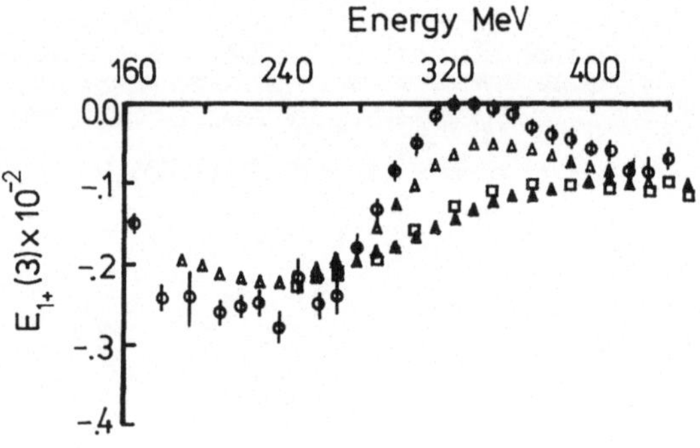
Energy MeV
160
240
320
400
0.0
-.1
-.2
-.3
-.4
$E_{1+}(3) \times 10^{-2}$

Fig. 7 b

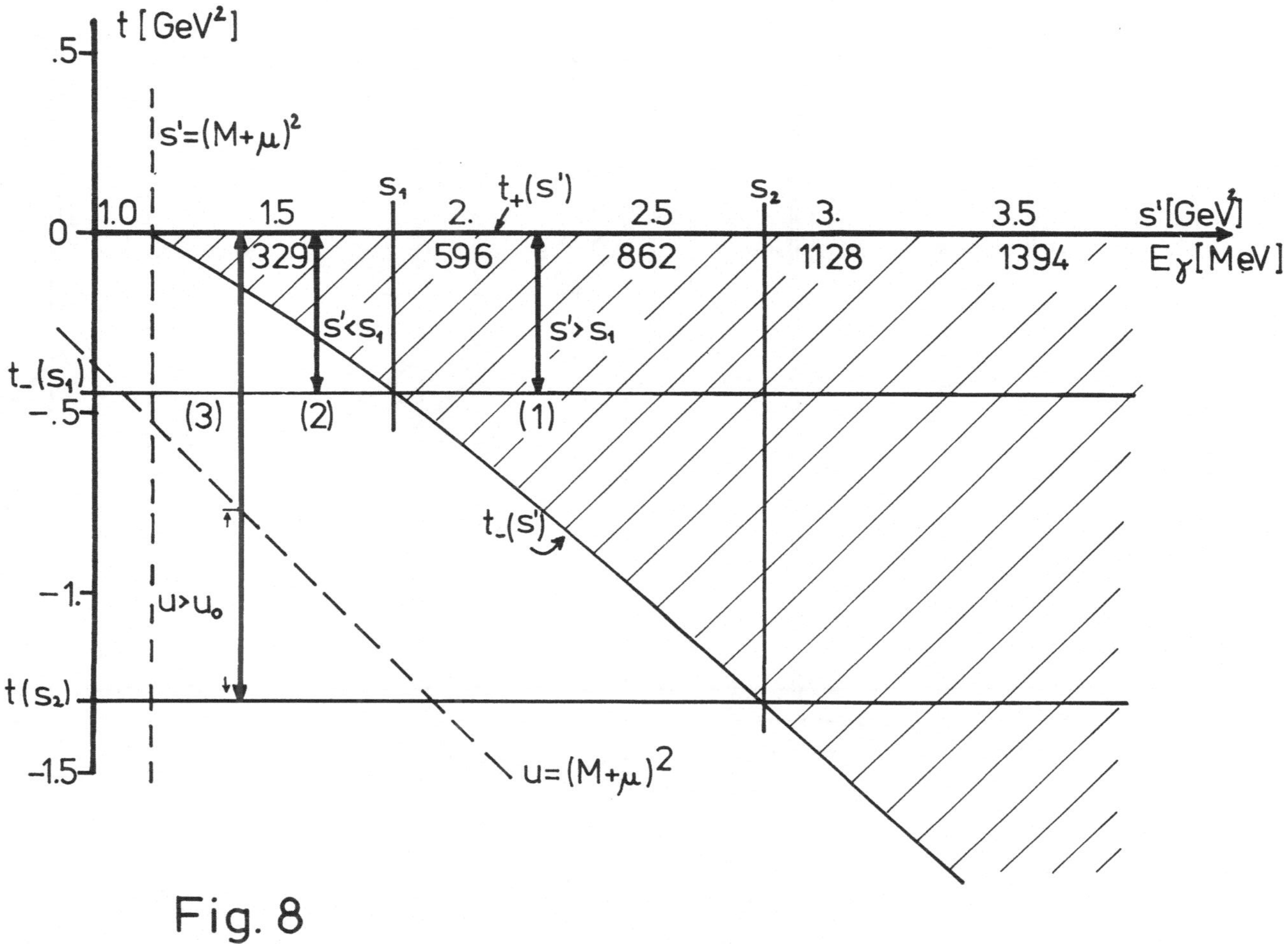

t [GeV²]
.5
0
-.5
-1.
-1.5
s'=(M+μ)²
1.0
1.5
2.
2.5
3.
3.5
s'[GeV²]
329
596
862
1128
1394
E_γ [MeV]
s₁
s₂
t₊(s')
t₋(s')
t₋(s₁)
t(s₂)
s'<s₁
s'>s₁
(3)
(2)
(1)
u>u₀
u=(M+μ)²

Fig. 8

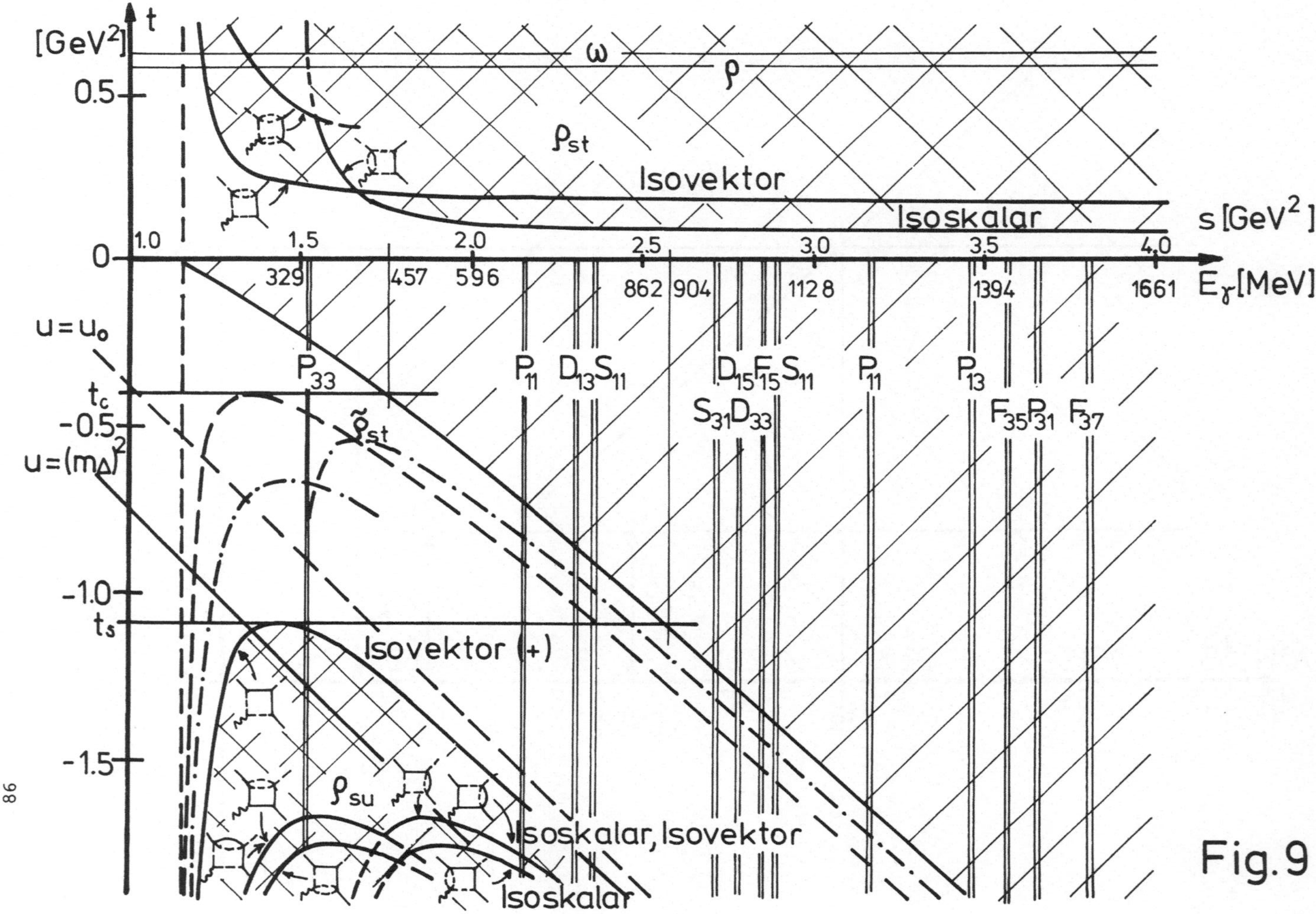

t
[GeV^2]
0.5
0
−0.5
−1.0
−1.5
ω
ρ
ρ_{st}
Isovektor
Isoskalar
s [GeV^2]
1.0
1.5
2.0
2.5
3.0
3.5
4.0
329
457
596
862
904
1128
1394
1661
E_γ [MeV]
$u=u_0$
t_c
$u=(m_\Delta)^2$
t_s
P_{33}
P_{11} D_{13} S_{11}
D_{15} F_{15} S_{11}
S_{31} D_{33}
P_{11}
P_{13}
F_{35} P_{31} F_{37}
$\tilde{\rho}_{st}$
Isovektor (+)
ρ_{su}
Isoskalar, Isovektor
Isoskalar

Fig.9

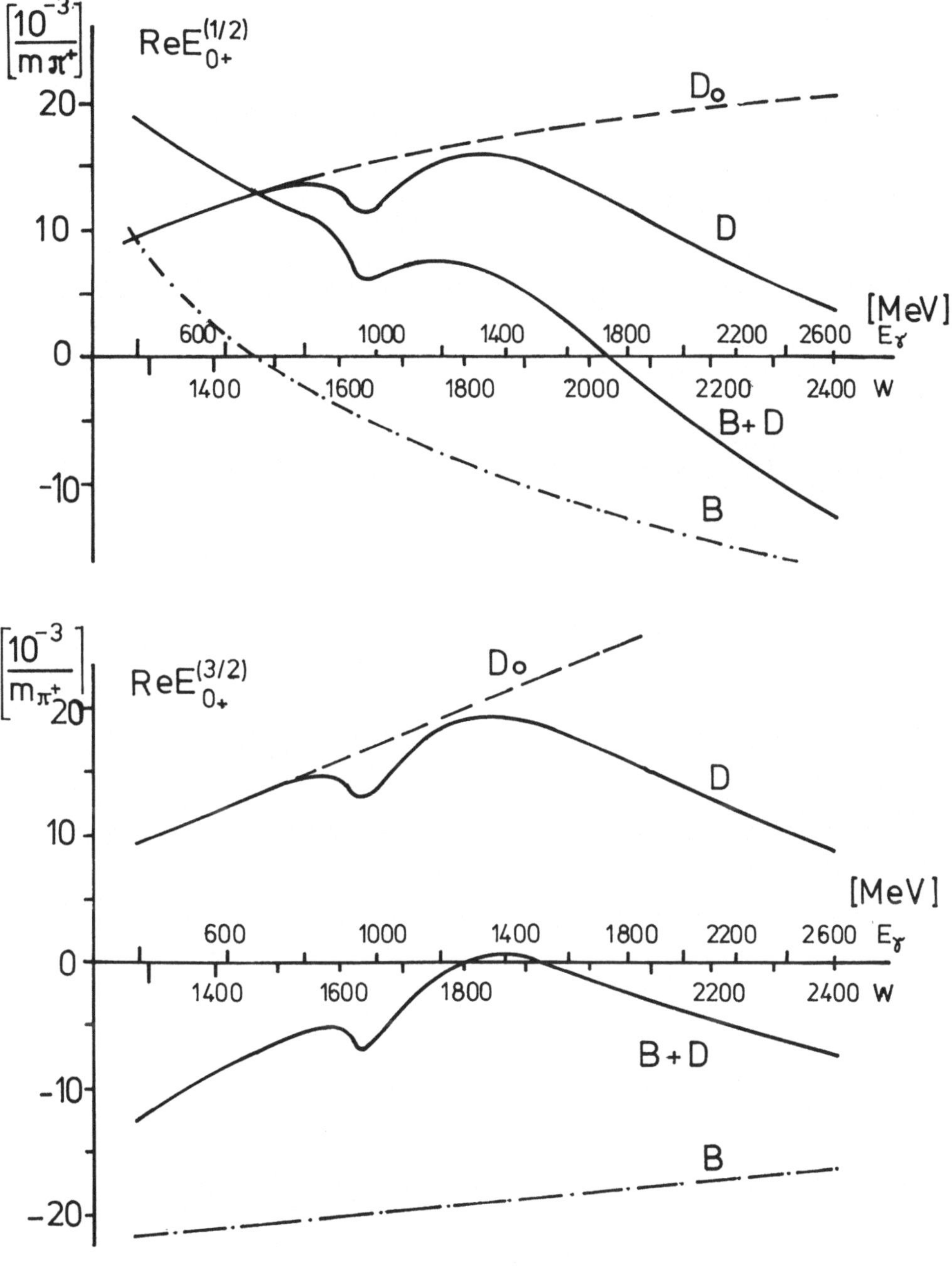

$\left[\frac{10^{-3}}{m_{\pi^+}}\right]$
$\mathrm{Re}E_{0+}^{(1/2)}$
20
10
0
-10
D_0
D
B+D
B
[MeV]
600
1000
1400
1800
2200
2600
E_γ
1400
1600
1800
2000
2200
2400
W
$\left[\frac{10^{-3}}{m_{\pi^+}}\right]$
$\mathrm{Re}E_{0+}^{(3/2)}$
20
10
0
-10
-20
D_0
D
B+D
B
[MeV]
600
1000
1400
1800
2200
2600
E_γ
1400
1600
1800
2200
2400
W

Fig. 10

Acta Physica Austriaca, Suppl. XIV, 89—142 (1975)

TWO-BODY AND QUASI-TWO-BODY ELECTRO- AND PHOTOPRODUCTION AT HIGH ENERGIES+)

by

A. BARTL
Institut für Theoretische Physik
Universität Wien

I. INTRODUCTION

The field of two-body photoproduction is a rather old one. In fact, single pion photoproduction off hydrogen is experimentally studied since more than two decades and also some of the theoretical ideas go back to that time. But this field is still a very progressive one. Within the last ten years our experimental knowledge about photoproduction of pions and other pseudoscalar and vector mesons at high energies has been largely improved and by now also some of the corresponding electroproduction experiments have been performed at various laboratories. The experimental study of these exclusive electron reactions has been stimulated by the behaviour of the total hadronic electro-

---

+) Lecture given at XIV. Internationale Universitätswochen für Kernphysik, Schladming, Austria, February 24 - March 7, 1975.

production cross section in the deep inelastic region. There it was observed that the cross section falls off very slowly with k^2 and one wanted to see wether the exclusive channels at high energies show such a behaviour too. By high energy we mean an energy above the resonance region, $s > 5(\text{GeV}/c)^2$, and during the course of my lectures I shall confine myself to such high energy reactions. The resonance range will be covered by the lectures of Professor Rollnik.[1]

Quite general, the theory of two body reactions is not in a good shape. One has at best a classification concept for a large variety of phenomena. Also a number of theoretical models have been constructed. But these models are only partially able to give a satisfactory description of the phenomena. A true theory of two body elementary particle reactions at high energies is missing. This is mainly due to the fact that we do not have a fundamental theory of strong interactions at all. But even if we would have such a complete theory of strong interactions at our disposal, we probably still would have to query about the mechanisms which control a certain reaction. So to study two body reactions is interesting in itself and with a suitable physical picture about the reaction mechanisms we may hope to learn about the structure of elementary particles themselves.

Reactions which are induced by photons provide the most stringent tests for models of high energy amplitudes. In most cases the experimental data are more accurate than in pure strong interaction physics and the amplitudes constructed from the models must obey a number of theoretical constraints due to the zero mass of the photon. Furthermore, real (and virtual) photons may be polarized and asymmetries may be measured whose comparison with model predictions is then much more decisive than comparison of cross sections only. In electron induced experiments the mass of the virtual photon can be varied and this can give new information about the structure of the model amplitudes.

Two-body reactions may be either diffractive or peripheral depending on the number of angular momentum partial waves which are involved. To the diffractive type of reactions there belong the elastic or quasielastic processes like photoproduction of neutral vectormesons (ρ^o,ω, ϕ...), whereas e.g. photoproduction of pseudoscalar mesons proceeds peripherally. Peripheral reactions are usually analysed in terms of t-channel exchanges. We shall mainly follow this approach here although during the last years via duality a number of s-channel pictures has been developed. The analysis of diffractive processes along the lines of t-channel exchanges has led to the concept of the pomeron. Due to the spin of the photon the study of photoproduction of vector mesons turns out to be an important tool to investigate the properties of the pomeron.

The main characteristic features of diffractive reactions are

a) Differential cross section $\frac{d\sigma}{dt}$ is independent of s (apart some possible factors of ln s) at fixed t in the small t region ($|t| < 0.8\ (GeV/c)^2$)

b) Diffraction peak in forward direction, i. e.

$$\frac{d\sigma}{dt} = A\ e^{Bt}$$

for small t with e.g. $\frac{d\sigma}{dt}(t = 0) \approx 100\ \mu b/GeV^2$ and the slope parameter $B \approx 6\text{-}8\ GeV^{-2}$ for ρ^o production.[2]

c) s-channel helicity conservation (This is experimentally confirmed up to now for vector meson photoproduction and not for diffractive hadron reactions in general)

On the other hand, pseudoscalar meson photoproduction processes, which proceed peripherally, show a very rich structure. This is true even for the t dependence of the differential cross sections as can be seen from the so called Diebold plot[3] (see fig.1). The quantity $(s-m^2)^2\frac{d\sigma}{dt}$ (m is the nucleon mass) turns out to be almost energy independent. More precisely, if one parametrizes the differential cross section as

$$\frac{d\sigma}{dt} = F(t)\ s^{2\alpha_{eff}(t)-2}$$

then the value of $\alpha_{eff}(t)$ which one obtaines lies between -0.2 and +0.2 in the t range shown. The various structures of $\frac{d\sigma}{dt}$ are attributed to the t channel exchanges which are involved. So e.g. the sharp forward spike in $\pi^+ n$ and $\pi^- p$ production and the narrow forward dip in $\pi^- \Delta^{++}$ production is due to π exchange. The minimum in $\pi^o p$ production at $t \approx -0.5 (GeV/c)^2$ may be attributed to ω exchange etc (the very narrow forward spike in $\pi^o p$ production is due to the Primakoff effect which we shall not discuss here). In the range $0.2\ (GeV/c)^2 < |t| < 1.0\ (GeV/c)^2$ all these differential cross sections show a t dependence roughly like e^{3t}. There seems to be a further similarity among the different pseudoscalar meson photoproduction processes concerning the polarized photon asymmetry Σ. This quantity can be measured with linearely polarized photons and is defined as

$$\Sigma = \frac{\frac{d\sigma_{\perp}}{dt} - \frac{d\sigma_{\parallel}}{dt}}{\frac{d\sigma_{\perp}}{dt} + \frac{d\sigma_{\parallel}}{dt}} \tag{1}$$

where $\frac{d\sigma_{\perp}}{dt}$ $(\frac{d\sigma_{\parallel}}{dt})$ is the differential cross section for production with photons linearely polarised perpendicular (parallel) to the production plane. Whereas Σ may show various structures in the near forward direction, for $|t| > 0.5$ $(GeV/c)^2$ and at high energies Σ is positive and close to unity for all pseudoscalar meson photoproduction reactions measured up to now (see f.i. ref. 4).

In general photon induced reactions follow the same patterns as pure hadronic reactions. For electron induced reactions this is true in the region of high s and moderate t and k^2 (k^μ the four-momentum of the virtual photon). This analogy stems from the fact that the photon developes a hadronic component (mainly built up of neutral vector mesons) which dominates the photon hadron reactions in this specified kinematical region.

As an example of two-body electroproduction we consider the process $ep \to e\pi^+n$. In this case the cross sections

$$\frac{d\sigma_U}{dt} + \varepsilon\,\frac{d\sigma_L}{dt}\,,\quad \frac{d\sigma_T}{dt}$$

and $\frac{d\sigma_I}{dt}$ have been measured in the region $|k^2| < 0.7$ $(GeV/c)^2$, $|t| < 0.2$ $(GeV/c)^2$ and $s \approx 5$ GeV^2, where $\frac{d\sigma_U}{dt}$ and $\frac{d\sigma_L}{dt}$ are the transverse unpolarized and longitudinal cross section, respectively

$$\frac{d\sigma_T}{dt} = \frac{d\sigma_\parallel}{dt} - \frac{d\sigma_\perp}{dt}$$

and $\frac{d\sigma_I}{dt}$ is a longitudinal-transverse interference term, ε is the degree of longitudinal polarisation of the virtual photon. In the kinematic region considered large values of $\frac{d\sigma_U}{dt} + \varepsilon\frac{d\sigma_L}{dt}$ have been observed which are attributed to large values of $\frac{d\sigma_L}{dt}$. This is just the opposite to what one knows from the deep inelastic region where $\frac{\sigma_L}{\sigma_U} \approx 0.15$.

II. TWO SIMPLE MODELS FOR CHARGED PION PRODUCTION

For the processes $\gamma p \to \pi^+ n$ and $\gamma p \to \pi^- \Delta^{++}$ there exist two old and simple models, the so-called electric Born term models, which we shall discuss very briefly now. These models work reasonably well for large s and $|t| \lesssim 4\mu^2$ (μ is the π mass), and any other more complicated model has to be related in this kinematic region to the electric Born model. For the process $\gamma p \to \pi^+ n$ it is given by the graphs of fig. 2. One pion exchange alone would not be gauge invariant, but inclusion of the orbital current part of the nucleon exchange diagram (Fig. 2b) gives a gauge invariant matrix element.[5] From this one can calculate the differential cross sections for real photons with polarisation parallel and perpendicular to the scattering plane, $\frac{d\sigma_\parallel}{dt}$ and $\frac{d\sigma_\perp}{dt}$, respectively, and in the kinematic region of $s \gg m^2$ and $|t| \lesssim 4\mu^2$ one obtains

$$\frac{d\sigma_{\parallel}}{dt} = \frac{e^2 g^2}{16\pi(s-m^2)^2} \left(1 - \frac{2t}{t-\mu^2}\right)^2 \qquad (2)$$

$$\frac{d\sigma_{\perp}}{dt} = \frac{e^2 g^2}{16\pi(s-m^2)^2} \quad . \qquad (3)$$

Here e is the electric charge, g is the π-N coupling constant, $s = (k + p_1)^2$ and $t = (k-q)^2$. In the forward direction ($t = 0$) we have $\frac{d\sigma_{\parallel}}{dt} = \frac{d\sigma_{\perp}}{dt}$, and the numerical values are such that $(s-m^2)^2 \frac{d\sigma_{\parallel}}{dt} = 260$ μb. GeV^2 at t = 0. This is only about 20 % below the experimental value. Whereas $\frac{d\sigma_{\perp}}{dt}$ is more or less constant in the t intervall considered, $\frac{d\sigma_{\parallel}}{dt}$ shows a very strong t-dependence, decreasing from its maximum value at t = 0 to $\frac{d\sigma_{\parallel}}{dt} = 0$ at $t = -\mu^2$ owing to a destructive interference between the pion exchange and the nucleon exchange contribution. Consequently the polarised, photon asymmetry defined in eq. (1) rises from $\Sigma = 0$ at $t = 0$ to $\Sigma = 1$ at $t = -\mu^2$. Such a behaviour is also exhibited by the experimental results,[6,7] as shown in fig.3.

An analogous model can be constructed for $\pi^- \Delta^{++}$ photoproduction.[8] The Feynman diagrams are given in fig. 4. According to the spin 3/2 of the Δ now also a contact term appears because of gauge invariance. From the nucleon exchange and the Δ exchange only the orbital current parts are taken into account as this constitutes a minimal gauge invariant extension of one pion exchange. Again $\frac{d\sigma_{\parallel}}{dt}$ shows a strong variation with t, but now a constructive interference between the one pion exchange and the other contributions appears. A comparison with the experimental data[7,9a] is shown in fig. 5.

In the region $|t|>4\ \mu^2$ these electric Born term models fail completely. So they have to be modified or to be substituted by different models. There are attempts to obtain a better t dependence by adding more resonances in the s channel (and u channel); see e.g. ref. 10. On the other hand one can start with Reggeized pion exchange and can add further Regge poles. The role of the nucleon exchange as a background term is then assumed by Regge cut contributions (see e.g. ref. 11).

In the case of charged pion electroproduction up to now mainly these electric Born models, supplemented with some modifications, have been applied to explain the experimental data. For π^+n electroproduction the electric Born model gives rise to a large σ_L. But without further modifications it gives $\sigma_U + \varepsilon\ \sigma_L$ too small by more than 50 % (see ref. 12 and references quoted therein). To $\pi^+\Delta^o$ electroproduction the electric Born model was applied by Berends and Gastmans.[13] These authors obtained good agreement with experiments although this model fails for the real photoproduction case. A detailed study of the electric Born model for $\pi^-\Delta^{++}$ electroproduction at threshold is given in ref. 9b.

From eq. (2) and (3) one can see that pion exchange contributes only to $\frac{d\sigma_{\parallel}}{dt}$ and not to $\frac{d\sigma_{\perp}}{dt}$. This is a special case of Stichel's theorem[14] which was originally formulated for pion photoproduction and which has been generalized by various authors to other pseudoscalar meson production processes. According to this theorem, in the limit of large s, natural parity exchange in the t-channel contributes only to $\frac{d\sigma_{\perp}}{dt}$, whereas $\frac{d\sigma_L}{dt}$ and $\frac{d\sigma_{\parallel}}{dt}$ receive contributions from unnatural parity t-channel exchange only. This theorem is very important in the study of reaction mechanisms. So f.i. from the experimental fact that $\Sigma \approx 1$ for $|t| > 0.5$ $(GeV/c)^2$ we can conclude that in this t range natural parity t-channel exchanges play the dominant role. Unnatural parity exchanges are important in the small-t range only.

III. KINEMATICS OF SINGLE PION ELECTROPRODUCTION

In the near future polarized lepton beams and polarized proton targets shall become available experimentally. Therefore, we shall derive in this section a general formula for the differential cross section for the scattering of a polarized electron on a polarized proton, when the polarization of the final electron and the recoiling nucleon remain undetected. We shall express the cross section in terms of bilinear combinations of helicity amplitudes. In so doing we shall follow the method outlined in refs. 12 and 15. For the case that also the polarization of the recoil nucleon is detected we refer to ref. 16 (such a measurement might become feasible in e.g. $K^+\Lambda$ electroproduction where the polarization state of the Λ can be observed through its decay).

In the one photon exchange approximation the electron process

$$ep \rightarrow e\pi N \qquad (4)$$

reduces to a virtual photon process

$$\gamma p \rightarrow \pi N \qquad (5)$$

where the "mass" of the virtual photon is

$$k^2 = - 4 E E' \sin^2 \frac{\psi_L}{2} . \tag{6}$$

Here E and E' are the energies of the incoming and outgoing electron, respectively, and ψ_L is the electron scattering angle, all quantities measured in the laboratory system. The amplitude for process (4) is then given by

$$T = e \bar{u} (\ell_2) \gamma^\mu u(\ell_1) \frac{1}{k^2} \langle p_2, q |J_\mu| p_1 \rangle \tag{7}$$

where q, p_1 and p_2, ℓ_1 and ℓ_2 are the four-momenta of the pion, the ingoing and outgoing nucleon, the ingoing and outgoing electron, respectively, $k = \ell_1 - \ell_2$ is the four-momentum of the virtual photon. The cross section of process (4) is then proportional to the quantity

$$\sum_{\text{spins}} |T|^2 = \frac{e^2}{k^4} L_{\mu\nu} T^{\mu\nu} \tag{8}$$

with

$$L_{\mu\nu} = \sum_{\text{spins}} (\bar{u}(\ell_2) \gamma_\mu u(\ell_1))^* \bar{u}(\ell_2) \gamma_\nu u(\ell_1) \tag{9}$$

$$T_{\mu\nu} = \sum_{\text{spins}} \langle p_2, q |J_\mu| p_1 \rangle^* \langle p_2, q |J_\nu| p_1 \rangle . \tag{10}$$

The polarization state of the incoming electron is characterized by a four-vector s^μ which obeys $s^2 = -1$ and $s \cdot \ell_1 = 0$. In the rest system of the electron s^μ reduces to the four-vector $(0,\vec{n})$, where $\vec{n}$ is the polarization vector. After a boost along $\vec{\ell}_1$, s^μ becomes

$$s^\mu = \left(\frac{\vec{\ell}_1\vec{n}}{m_e}, \ \vec{n} + \frac{\vec{\ell}_1(\vec{\ell}_1\vec{n})}{m_e(E+m_e)}\right) \tag{11}$$

(m_e denotes the electron mass), which for $|\vec{\ell}_1| \to \infty$ reduces to

$$s^\mu = \frac{\xi}{m_e}\,\ell_1^\mu \ , \qquad \xi = \frac{(\vec{\ell}_1\vec{n})}{|\vec{\ell}_1|} \quad . \tag{12}$$

Here ξ measures the degree of longitudinal polarization. Neglecting terms of the order m_e/E one obtains for $L_{\mu\nu}$ from eq. (9)

$$L_{\mu\nu} = \frac{1}{2m_e^2}\,(\ell_{1\mu}\,\ell_{2\nu} + \ell_{1\nu}\,\ell_{2\mu} + \tfrac{1}{2}k^2\,g_{\mu\nu} + $$

$$+ \, i\,\xi\,\varepsilon_{\mu\nu\rho\sigma}\,k^\rho \ell_1^\sigma) = L^s_{\mu\nu} + L^{as}_{\mu\nu} \quad . \tag{13}$$

The asymmetric part of the lepton current tensor, $L^{as}_{\mu\nu}$, occurs only if the initial lepton is polarized. In the limit $m_e/E \to 0$ only the longitudinal polarization survives which means that the small electron mass precludes experiments with transversally polarized electron beams at higher energies.

As we are working in the one photon approximation it is convenient to carry over the information content of $L_{\mu\nu}$ into the polarisation state of the virtual photon. This can be done by introduction of the spin density matrix of the virtual photon

$$\rho_{ik} = \varepsilon_\mu^*(i)\,\varepsilon_\nu\,(k) L^{\mu\nu}\,\frac{2m_e^2(1-\varepsilon)}{-k^2} = \rho_{ik}^s + \rho_{ik}^{as} \tag{14}$$

$\varepsilon_\mu(i)$ is the polarization vector of the virtual photon. In the helicity basis (i,k = + 1, -1, 0) and in the Breit frame ($k^0 = 0$, $\vec{k} \parallel \vec{p}_1$) the spin density matrix of the virtual photon has the form

$$\rho_{ik}^s = \frac{1}{2}\begin{pmatrix} 1 & -\varepsilon e^{-2i\phi} & \sqrt{\varepsilon(1+\varepsilon)}e^{-i\phi} \\ -\varepsilon e^{2i\phi} & 1 & -\sqrt{\varepsilon(1+\varepsilon)}e^{i\phi} \\ \sqrt{\varepsilon(1+\varepsilon)}e^{i\phi} & -\sqrt{\varepsilon(1+\varepsilon)}e^{-i\phi} & 2\,\varepsilon \end{pmatrix} \tag{15}$$

$$\rho_{ik}^{as} = \frac{\xi}{2} \begin{bmatrix} \sqrt{1-\varepsilon^2} & 0 & \sqrt{\varepsilon(1-\varepsilon)}\ e^{-i\phi} \\ 0 & -\sqrt{1-\varepsilon^2} & \sqrt{\varepsilon(1-\varepsilon)}e^{i\phi} \\ \sqrt{\varepsilon(1-\varepsilon)}e^{i\phi} & \sqrt{\varepsilon(1-\varepsilon)}e^{-i\phi} & 0 \end{bmatrix} \tag{16}$$

We are referring to a right handed coordinate system whose z-axis is in the direction of $\vec{k}$ and whose y-axis is in the direction of $\vec{k} \times \vec{q}$. ϕ is the angle between the electron plane and the hadron plane according to the convention used e.g. by Berkelman[17]. The quantity ε can be expressed through quantities in the laboratory system as follows[18]

$$\varepsilon = \left[1 + \frac{2|\vec{k}_L|^2}{-k^2} \tan^2 \frac{\psi_L}{2}\right]^{-1} . \tag{17}$$

A closer inspection of eq. (15) shows that already for unpolarized electrons (i.e. $\xi = 0$) the virtual photon has a transverse polarization to degree ε with the polarization vector in the electron plane.[19] In addition there is a longitudinal polarization with the longitudinal to transverse intensity ratio equal to ε. If f.i. the incident electron is right polarized (i.e. $\xi = 1$) then the virtual photon is in a pure polarization state which is a superposition of a transverse elliptic component and a longitudinal component.

For the evaluation of $T_{\mu\nu}$, eq. (10) we shall now express the hadronic matrix element of the electromagnetic current, $\langle p_2, q|J_\mu|p_1\rangle$, through the CMS helicity amplitudes of the virtual photon process (5), namely

$$\varepsilon_\mu(\lambda)\langle p_2, q|J^\mu|p_1\rangle = \langle\lambda_2|T|\lambda_1,\lambda\rangle \tag{18}$$

where λ, λ_1, and λ_2 denote the helicities of the virtual photon, the ingoing nucleon and outgoing nucleon, respectively. For the transverse helicity amplitudes we adopt the conventional notation defined by Walker[20]

$$\begin{aligned} H_1 &= \langle -\tfrac{1}{2}|T|-\tfrac{1}{2},1\rangle \quad , \quad & H_2 &= \langle -\tfrac{1}{2}|T|\tfrac{1}{2},1\rangle \\ H_4 &= \langle \tfrac{1}{2}|T|\tfrac{1}{2},1\rangle \quad , \quad & H_3 &= \langle \tfrac{1}{2}|T|-\tfrac{1}{2},1\rangle \;. \end{aligned} \tag{19}$$

The following linear combinations are even more convenient,

$$h^N_\pm = \frac{1}{\sqrt{2}}\,(H_4 \pm H_1) \;,\; h^F_\pm = \frac{1}{\sqrt{2}}\,(H_3 \mp H_2) \tag{20}$$

as $h^{N,F}_+$ correspond to incident photons polarized perpendicular to the scattering plane and $h^{N,F}_-$ correspond to photon polarization parallel to the scattering plane. F and N refer to baryon flip and non-flip respectively. The longitudinal helicity amplitudes are

$$h^N_o = \langle\tfrac{1}{2}|T|\tfrac{1}{2},0\rangle \;,\; h^F_o = \langle\tfrac{1}{2}|T|-\tfrac{1}{2},0\rangle \quad . \tag{21}$$

These longitudinal amplitudes vanish for $k^2 \to 0$.

The polarization state of the target nucleon is described by the spin density matrix

$$\rho_p = \frac{1}{2}\,(1 + \vec{P}\vec{\sigma}) \tag{22}$$

with $\vec{\sigma}$ the Pauli spin matrices and $\vec{P}$ the polarization of the target nucleon.

Finally the electroproduction cross section for process (4) is written as

$$\frac{d^3\sigma}{d\Omega_L dE' d\Omega} = \Gamma_t \frac{d\sigma}{d\Omega}\,(s, k^2, \varepsilon, \Theta, \phi) \tag{23}$$

where Γ_t is the photon flux factor

$$\Gamma_t = \frac{\alpha}{2\pi^2} \frac{E'}{E} \frac{K}{-k^2} \frac{1}{1-\varepsilon} \tag{24}$$

$K = (s-m^2)/2m$ is the equivalent real photon lab energy, $d\Omega_L$ is the solid angle of the final electron in the lab system and $d\Omega = \sin\Theta\, d\Theta\, d\phi$ is the solid angle of the pion in the hadronic CMS. The virtual photoproduction cross section is

$$\frac{d\sigma}{d\Omega} = \frac{q\sqrt{s}}{K\,m} \sum_{\substack{\lambda_1, \lambda_1', \lambda_2 \\ \lambda, \lambda'}} \langle\lambda_2|T|\lambda_1', \lambda'\rangle^* \langle\lambda_2|T|\lambda_1, \lambda\rangle \langle\lambda|\rho|\lambda'\rangle \langle\lambda_1|\rho_p|\lambda_1'\rangle \tag{25}$$

where q is the pion CMS momentum. The cross section eq. (25) is composed of four parts,

$$\frac{d\sigma}{d\Omega} = \frac{d\sigma_o}{d\Omega} + \frac{d\sigma_e}{d\Omega} + \frac{d\sigma_t}{d\Omega} + \frac{d\sigma_{et}}{d\Omega} \quad . \tag{26}$$

If neither the electron nor the target is polarized then one has only $\frac{d\sigma_o}{d\Omega}$,

$$\frac{d\sigma_o}{d\Omega} = \frac{d\sigma_U}{d\Omega} + \varepsilon\,\frac{d\sigma_L}{d\Omega} + \varepsilon\,\frac{d\sigma_T}{d\Omega}\cos 2\phi + \sqrt{\tfrac{1}{2}\varepsilon(1+\varepsilon)}\,\sigma_I\,\cos\phi \tag{27}$$

and in a well known manner it can be expressed through the amplitudes eq. (2o) and (21):

$$\frac{d\sigma_U}{d\Omega} = \frac{q\sqrt{s}}{Km}\,\frac{1}{2}(|h_+^N|^2 + |h_+^F|^2 + |h_-^N|^2 + |h_-^F|^2)$$

$$\frac{d\sigma_T}{d\Omega} = \frac{q\sqrt{s}}{Km}\,\frac{1}{2}(|h_-^N|^2 + |h_-^F|^2 - |h_+^N|^2 - |h_+^F|^2) \tag{28}$$

$$\frac{d\sigma_L}{d\Omega} = \frac{q\sqrt{s}}{Km}\,(|h_o^N|^2 + |h_o^F|^2)\,,\quad \frac{d\sigma_I}{d\Omega} = \frac{q\sqrt{s}}{Km}\,2\mathrm{Re}(h_o^N\,h_-^{N*} + h_o^F\,h_-^{F*})\ .$$

If the initial electron is polarized, also the term $\frac{d\sigma_e}{d\Omega}$ in eq. (26) is present:

$$\frac{d\sigma_e}{d\Omega} = -\xi\,\frac{q\sqrt{s}}{Km}\,\sqrt{2\varepsilon(1-\varepsilon)}\,\sin\phi\quad \mathrm{Im}\,(h_o^N h_-^{N*} + h_o^F h_-^{F*}) \tag{29}$$

$\frac{d\sigma_e}{d\Omega}$ and $\frac{d\sigma_I}{d\Omega}$ measure imaginary part and real part of the same quantity. Notice that these two cross sections are interference terms between longitudinal and transverse <u>parallel</u> polarization amplitudes.

If the target is polarized then also $\frac{d\sigma_t}{d\Omega}$ occurs and this part of the cross section reads

$$\frac{d\sigma_t}{d\Omega} = \frac{q\sqrt{s}}{Km} \{P_x[-\sqrt{2\varepsilon(1+\varepsilon)}\sin\phi \; ImX_1 - \varepsilon\sin 2\phi ImX_2]$$

$$-P_y[ImY_1 + \varepsilon\cos 2\phi \; ImY_2 + 2\varepsilon ImY_3 + \sqrt{2\varepsilon(1+\varepsilon)}\cos\phi \; ImY_4]$$

$$+P_z[\varepsilon\sin 2\phi \; ImZ_2 + \sqrt{2\varepsilon(1+\varepsilon)} \; \sin\phi \; ImZ_1]\} \; . \tag{30}$$

Here we have introduced the following abbreviations:

$$X_1 = h_o^F h_+^{N*} + h_o^N h_+^{F*} \qquad X_2 = h_-^F h_+^{N*} + h_-^N h_+^{F*}$$

$$Y_1 = h_+^N h_+^{F*} + h_-^N h_-^{F*} \qquad Y_2 = h_-^N h_-^{F*} - h_+^N h_+^{F*} \tag{31}$$

$$Y_3 = h_o^N h_o^{F*} \qquad Y_4 = h_o^N h_-^{F*} - h_o^F h_-^{N*}$$

$$Z_1 = h_o^N h_+^{N*} - h_o^F h_+^{F*} \qquad Z_2 = h_-^N h_+^{N*} - h_-^N h_+^{F*}$$

If both electron and target are polarized one has also σ_{et}:

$$\sigma_{et} = -\xi \frac{q\sqrt{s}}{Km}\{-P_x[\sqrt{2\varepsilon(1-\varepsilon)}\cos\phi \mathrm{Re}X_1 + \sqrt{1-\varepsilon^2}\ \mathrm{Re}X_2] +$$

$$+ P_y \sqrt{2\varepsilon(1-\varepsilon)}\ \sin\phi\ \mathrm{Re}Y_4 + \qquad (32)$$

$$+ P_z[\sqrt{1-\varepsilon^2}\ \mathrm{Re}Z_2 + \sqrt{2\varepsilon(1-\varepsilon)}\ \cos\phi\ \mathrm{Re}Z_1]\} .$$

These formulae present the most general form (within the one-photon approximation) for the cross section of single pion electroproduction with polarized beam and polarized target. Hence they are also valid e.g. in the resonance region but they are especially suitable for high energy scattering, as in this limit $h_+^{N,F}$ get contributions only from natural parity exchange in the t-channel and $h_-^{N,F}$ and $h_o^{N,F}$ only from unnatural parity exchange.

Note in particular that target polarization can give much more information than electron beam polarization. Information on the various pieces of the cross section can be extracted through measurements of suitable asymmetries. So f.i. for target polarization along the y-axis one can define the asymmetry

$$T = \left.\frac{\sigma(P_y = +1) - \sigma(P_y = -1)}{\sigma(P_y = +1) + \sigma(P_y = -1)}\right|_{\text{integr.over}\phi} = -2\ \frac{\mathrm{Im}\ Y_1 + 2\varepsilon \mathrm{Im} Y_3}{N} \qquad (33)$$

where the subscript "integr. over ϕ" indicates that the ϕ integrated cross sections have to be inserted and

$$N = |h_+^N|^2 + |h_+^F|^2 + |h_-^N|^2 + |h_-^F|^2 + 2\varepsilon(|h_o^N|^2 + |h_o^F|^2) . \quad (34)$$

For real photoproduction ($k^2 = 0$) the quantity T corresponds to the target asymmetry. In an analogous way other asymmetries may be defined for extracting the other quantities. Note also that in real photoproduction Im Y_2 is proportional to the recoil polarization.

IV. REGGE MODELS FOR PSEUDOSCALAR MESON PRODUCTION

a) Single Pion Production

Here the following reactions are possible:

$$\gamma p \to \pi^+ n \quad (35)$$

$$\gamma n \to \pi^- p \quad (36)$$

$$\gamma p \to \pi^o p \quad (37)$$

$$\gamma n \to \pi^o n \quad (38)$$

where the photon γ may be real or virtual. Most experimental information is available for the real photonprocesses (35) and (37). In isospin space we may decompose the amplitude into three parts (remember that the photon contains an isoscalar and an isovector part):

$$T = T^{(o)}\ \tau^{i} + T^{(+)}\delta_{3i} + T^{(-)}\ \tfrac{1}{2}[\tau^{i},\tau^{3}]\ . \qquad (39)$$

The index i refers to the isospin of the pion. $T^{(o)}$ contains the isoscalar photon contributions and so it can have only (I = 1, G = +1) exchanges in t channel. $T^{(+)}$ and $T^{(-)}$ stem from the isovector photon and, according to their symmetry properties, $T^{(+)}$ receives only contributions from (I = 0, G = -1) t-channel exchanges, whereas to $T^{(-)}$ only (I = 1, G = -1) t-channel exchanges contribute. The amplitudes for the reactions (35)-(38) then have the following isospin structure:

$$\begin{aligned} T\ (\pi^{+}n) &= -\sqrt{2}\ (T^{(o)} + T^{(-)}) \\ T\ (\pi^{-}p) &= \sqrt{2}\ (T^{(o)} - T^{(-)}) \\ T\ (\pi^{o}p) &= T^{(o)} + T^{(+)} \\ T\ (\pi^{o}n) &= -T^{(o)} + T^{(+)}\ . \end{aligned} \qquad (40)$$

In the following table 1 we list the Regge poles[21] which contribute to the various amplitudes to leading order in s. Here H is the isoscalar member of the B nonet. As its state is unclear, we have denoted it with a question mark. Similarly, the resonance interpretation of the A_1 is not clear. It turns out, however, that at least in real photoproduction the A_1 contribution can be neglected. For the Reggeization of the π some care has to be taken according to the special role of one pion exchange due to gauge invariance. Table 1 shows that our choice of amplitudes is already well adapted to the problem, as at most one Regge pole contributes to each single amplitude.

TABLE 1

Ampl.	Overall hel.flip	Reggepoles (O)	(+)	(-)	Cuts (O)	(+)	(-)
h_+^F	O , 2	ρ	ω	A_2	ρ,B	ω,H?	π,A_2
h_+^N	1	ρ	ω	A_2	ρ	ω	A_2(small)
h_-^F	O , 2	B	H?	π	ρ,B	ω,H?	π,A_2
h_-^N	1	-	-	A_1	-	-	A_1(small)
h_o^F	1	B	H?	π	B	H?	π (small)
h_o^N	O	-	-	A_1	-	-	A_1

We shall first discuss real photoproduction of $\pi^+ n$. In this case the most important contributions come from the π and the A_2 pole, i.e.

$$\frac{8\pi}{\sqrt{s}} h_-^{F(-)} = - \frac{\sqrt{2}\, eg\pi\alpha_\pi'}{\Gamma(\alpha_\pi(t)+1)} \frac{t}{s_o} \frac{1 + e^{-i\pi\alpha_\pi(t)}}{2\sin\pi\alpha_\pi(t)} \left(\frac{s}{s_o}\right)^{\alpha_\pi(t)-1} \tag{41}$$

$$\frac{8\pi}{\sqrt{s}} h_+^{N(-)} = \frac{e\kappa g^E \pi\alpha'}{\sqrt{2}\Gamma(\alpha(t))} \sqrt{-t} \frac{1 + e^{-i\pi\alpha(t)}}{2 \sin\pi\alpha(t)} \left(\frac{s}{s_o}\right)^{\alpha(t)-1} \tag{42}$$

$$h_+^{F(-)} = \sqrt{-t} \frac{g^M}{g^E} h_+^{N(-)} \quad . \tag{43}$$

Furthermore, $h_-^{N(-)} = 0$ since, as usual, A_1 exchange is neglected. In equ. (41) to (43), $\alpha_\pi(t) = \alpha'_\pi(t-\mu^2)$ is the π trajectory, $\alpha(t) = 2 + \alpha'.(t - m^2_{A2})$ is the A_2 trajectory, $s_o \approx 1$ GeV^2 is a scale parameter, $e.\kappa$ is the $\gamma\pi A_2$ coupling constant and g^E and g^H are the A_2-N coupling constants. Due to the Γ-functions in the denominators of equ.(41)-(43) the amplitudes exhibit the so-called non sense wrong-signature (NWS) zeros. We assume that no further t dependence is contained in the Regge residue functions.

From pure hadronic reactions one knows that the ρ-A_2 and the π-B Regge poles are exchange degenerate. Together with vector meson dominance this implies[11) 22)]

$$\text{Im } f(A_2) = 3 \text{ Im } f(\rho), \quad \text{Im } f(\pi) = 3 \text{ Im } f(B) \tag{44}$$

and with equ. (40) this yields the Regge pole amplitudes for $\pi^+ n$ photoproduction

$$\frac{8\pi}{\sqrt{s}} h_-^F = \frac{2eg\pi\alpha'_\pi}{\Gamma(\alpha_\pi(t)+1)} \frac{t}{s_o} \frac{1 + 2e^{-i\pi\alpha_\pi(t)}}{3 \sin\pi\alpha_\pi(t)} \left(\frac{s}{s_o}\right)^{\alpha_\pi(t)-1} \tag{45}$$

$$\frac{8\pi}{\sqrt{s}} h_+^N = - \frac{e\kappa g^E \pi\alpha'}{\Gamma(\alpha(t))} \sqrt{-t} \frac{1 + 2e^{-i\pi\alpha(t)}}{3 \sin\pi \alpha(t)} \left(\frac{s}{s_o}\right)^{\alpha(t)-1} \tag{46}$$

$$h_+^F = \sqrt{-t} \frac{g^M}{g^E} h_+^N \tag{47}$$

$\alpha_\pi(t)$ and $\alpha(t)$ are now the π-B and the ρ-A_2 exchange degenerate Regge trajectories, respectively, and g^E and g^M now denote the electric and the magnetic ρ-N coupling constant, respectively, i.e. $2mg^M/g^E \approx 3.7$.

Eqs. (41) to (43) or (45) to (47) show that all Regge pole amplitudes vanish for $t \to 0$. This is due to the helicity flip which for real photons is always present at the γ-π vertex. Consequently all Regge pole contributions in pseudoscalar meson photoproduction vanish in the forward direction. But on the contrary experiment shows a sharp maximum for the differential cross section in the case of charged pion photoproduction. This means that Regge poles alone are not sufficient for the description of these processes and additional contributions are necessary. Within the usual Regge picture these additional contributions are due to Regge cuts. Actually, charged pion photoproduction was one of the first places were the existence of Regge cuts was strongly demanded for phenomenological reasons.

Within the conventional absorbed Regge model these Regge cuts are generated by absorption i.e. by additional elastic scattering in the initial and in the final state. This can be seen in the following simple way. At high energies the partial wave decomposition of a helicity amplitude H_i goes over into a Fourier-Bessel transform in impact parameter space,

$$H_i(s,b) = \frac{1}{2q^2} \int_0^\infty \Delta d\Delta \quad J_n(b\Delta)\; H_i(s,t), \tag{48}$$

where $\Delta = \sqrt{-t}$ and $n = |\lambda - \lambda_1 + \lambda_2|$ is the overall helicity flip. For simplicity the elastic scattering occuring in the initial and in the final state is described by pure imaginary pomeron exchange. Then the absorption correction to an amplitude H_i is obtained by multiplying the Regge pole amplitude H_i^{pole} (s,b) by an absorbing function f(b), for which one conventionally uses

$$f(b) = 1 - C \exp\{-\frac{b^2}{2A}\} \quad . \tag{49}$$

Assuming for simplicity initial state and final state absorption to be the same, one obtains for the Regge cut contribution[11)]

$$H_i^{cut}(s,b) = - C \exp\{- b^2/2A\}\; H_i^{pole}\;(s,b) \tag{50}$$

and

$$H_i^{cut}(s,t) = \tfrac{1}{2}CA \int_0^\infty dt'\; H_i^{pole}(s,t')\exp\{\tfrac{1}{2}A(t+t')\}I_n(A\sqrt{tt'}) \tag{51}$$

where I_n is the modified Bessel function. In pion photoproduction one needs large cut contributions, i.e. $C \approx 1$.

From π-N scattering one has A $\approx$ 8 GeV^{-2}. Due to the function I_n, large absorption corrections occur only for n = 0. In table 1 we have, therefore, also listed the helicity flip for each amplitude and the corresponding Regge pole-pomeron (RP) cuts which one has to expect according to this procedure. At small t these cut contributions play the analogous role as the nucleon exchange in the electric Born term model.

The model thus presented can explain the experimental data only for $|t| < 0.2$ $(GeV/c)^2$. For larger values of $|t|$ a number of deficiencies arise. First of all, the Regge pole amplitudes (eq. (45) to (47)) alone would predict equal cross sections for π^+n and π^-p production at t $\approx$ -0.6 $(GeV/c)^2$ owing to a NWS zero of the ρ Regge pole. Experimentally, however, the difference between the π^+n and π^-p cross sections shows just a maximum at approximately this t value. As a way out of this difficulty, large RP cuts from a more complicated absorption[11] or additional RR cut contributions[22] have to be invoked, or a ρ Regge residue without a NWS zero may be used.[23] In the region of t $\approx$ -1.0 $(GeV/c)^2$, however, all types of absorbed Regge models have difficulties, because large interferences between pole and cut terms occur leading to a wrong t dependence. Therefore, these models are unable to reproduce all data in a completely satisfactory way. Some examples of Regge model fits are given in fig. 6 to 8 for the differential cross section and polarized target asymmetry of π^+n photoproduction and for the π^-/π^+ ratio.

The transition to pion electroproduction could be accomplished in the most simple way by extrapolating an already existing photoproduction model to $k^2 < 0$. For the case of $\pi^+ n$ electroproduction this was tried in ref. 12 on the basis of the photoproduction model of Worden.[11] The k^2 dependence of the transverse helicity amplitudes was assumed to be given by $m_\rho^2/(m_\rho^2-k^2)$ as suggested by simple ρ-pole dominance. No additional longitudinal couplings have been introduced. The longitudinal amplitude h_o^F is then completely determined by π and B exchange, whereas $h_o^N \approx 0$ neglecting A_1 exchange. The results are shown in fig. 9 where it is clearly seen that $\frac{d\sigma_U}{dt} + \varepsilon \frac{d\sigma_L}{dt}$ is too small by about 50 %. Furthermore, the zero of $\frac{d\sigma_I}{dt}$, which is due to the well known zero of h_-^F, occurs at the wrong value of t. So one must conclude that either the k^2 dependence as given by simple ρ-pole dominance is wrong, or additional couplings or further cut contributions are necessary which vanish for $k^2 = 0$. Some evidence for a variation of the cut contribution with k^2 has been presented in a recent analysis of π- and ρ^o-production in ref. 25.

In the case of π^o photoproduction ω exchange gives the most important contribution. π exchange is not possible. Since the ω-N coupling is mainly non-flip, the amplitude h_+^N dominates and there is a forward dip. The dip at $t \approx -0.5$ $(\mathrm{GeV/c})^2$ is probably explained by a NWS zero of the ω Regge pole contribution.[11,23]

From table 1 one can see that in π^o electroproduction $\frac{d\sigma_L}{dt}$ is expected to be small, since to the longitudinal amplitudes only B exchange can contribute which, however, couples only to isoscalar photons. Therefore, measurements of π^o electroproduction could give further information on additional longitudinal couplings or additional cut contributions to the amplitudes at $k^2 < 0$. Furthermore, the mechanism which generates the dip at $t \approx -0.6\ (GeV/c)^2$ might be better understood when information about π^o electroproduction becomes available.

b) $K^+\Lambda$ and $K^+\Sigma^o$ Production

These processes have been less extensively studied, since one expects that a model of pion production may be applied with only small modifications to K^+ production too. There is, however, a remarkable difference between K^+ and π^+ production: As the mass of the K^+ is much larger than that of the π^+, the pole at $t = \mu^2$ is much more distant from the physical region in the kaon than in the pion case. Consequently, there is no spike in the forward direction, but at higher energies a dip is present. Also the electric Born term model does not work for K^+ production.

Absorbed Regge models for $K^+\Lambda$ and $K^+\Sigma^o$ photoproduction, based on K exchange and exchange degenerate $K^*(890)$ - $K^{**}(1420)$ exchange with the absorption corrections calculated in an analogous way as discussed in the previous section have been given by Levy et al.[26] and Goldstein et al.[27] Good agreement was obtained for the differential cross sections of ref. 28 (see fig. 10), and for the polarized photon asymmetries of ref. 29, whereas the data on the Λ recoil polarization[30] have been qualitatively reproduced. Addition of a K_B Regge pole (the strange analogy of the B) exchange degenerate with the K causes only small numerical changes.[31]

Levy et al. also tried to extrapolate their model to K^+ electroproduction[32] in a straightforward manner quite analogous to that described in the previous section. Their results disagree with experiment as the cross section $\frac{d\sigma_U}{dt} + \varepsilon \frac{d\sigma_L}{dt}$ turns out to be too small by a factor of two. One can show that a larger K-exchange contribution alone can not account for this discrepancy in a consistent way. Since $\frac{d\sigma_T}{dt}$ is small, additional unnatural parity exchange is needed. In ref. 32 two possibilities for such an additional contribution have been studied. One possibility is K_B exchange with a new coupling breaking exchange degeneracy with the K, the other one is to introduce a new K_A (1^{++}) - K_A (2^{--}) trajectory (the strange analogy to the A_1). Both contributions vanish for $k^2 = 0$. Both possibilities are able to reproduce the experimental data of ref. 33 in a satisfactory way, but the present situation does not allow to discriminate between them. If the proposed picture turns out to be true, then the B or the A_1 trajectory should play an analogous role in pion electroproduction. Furthermore, extraction of information about the K electromagnetic form factor from K electrproduction would become much more difficult or at least strongly model dependent.

V. MODIFIED BORN TERM MODELS

In section II it was mentioned that the electric Born term model, supplemented with a pion form factor as given by vector meson dominance and with a nucleon electric form factor as given by the dipole fit for the Sachs form factors, and applied to π^+n electroproduction cannot account for the experimental data. Gutbrod and Kramer[34] proposed the following modification of the model: At high energies the intermediate nucleon in the Feynman diagram fig. 2b is far off mass shell. Its electromagnetic form factors then need not to have the same k^2 behaviour as for an on-shell nucleon. The authors of ref. 34 speculate that the form factor $F_1(k^2)$ for an off-shell nucleon might fall off with k^2 very slowly similar to the total cross section in the deep inelastic region. Furthermore, they also assume a damping in the t behaviour which might have its origin in further s-channel exchanges. Leaving $F_1(k^2)$ as a free parameter to be determined by a fit to experiment, for each value of k^2 respectively, they are able to describe the experimental cross sections in a very satisfactory way. In particular, the position of the zero of $\frac{d\sigma_I}{dt}$ is reproduced at the correct t value. This is due to the structure of h_-^F which now approximately looks like

$$\frac{8\pi}{\sqrt{s}}\, h_-^F \approx \frac{eg}{s-m^2}\,[F_1(k^2) - F_\pi(k^2)\,\frac{2t}{t-\mu^2}] \quad . \tag{52}$$

If now $F_1(k^2) > F_\pi(k^2)$ for $k^2 < 0$, then the zero in h_-^F moves to larger values of t for increasing $|k^2|$. Similar effects occur in the model of ref. 35 based on a Veneziano B_5-function. Whether the cross section $\frac{d\sigma_U}{dt}$ shows indeed such a behaviour of a slow decrease with $|k^2|$ is up to now an unsolved question. An experimental determination of $\frac{d\sigma_U}{dt}$ and $\frac{d\sigma_L}{dt}$ separately would be necessary.

An attempt to understand the success of the electric Born term model for charged pion photoproduction was made by Barbour et al.[10,36] by the use of fixed-t dispersion relations. Similar considerations have been presented in ref. 37 on the basis of the dual absorption model.[38] All these approaches are related to the well known dispersion techniques.[39] Barbour et al assume the imaginary parts of the amplitudes to be given by resonance dominance at low energies and by simple Regge pole exchange at high energies. The real parts are then calculated from fixed-t dispersion relations. The Born terms are already included and survive for $|t| < 0.1\ (GeV/c)^2$ at high energies, because the dominant resonances do not contribute in this region. For $0.1 < |t| < 0.8\ (GeV/c)^2$ strong cancellations occur between the Born terms and low energy resonances yielding $\frac{d\sigma}{dt}$ and Σ with the right magnitude. A similar analysis was carried out for $\pi^{\pm}\Delta$ photoproduction.[40]

VI. PHOTO-AND ELECTROPRODUCTION OF NEUTRAL VECTOR MESONS

a) Kinematics of Vector Meson Photoproduction

As the produced vector meson has spin $J^P = 1^-$, the kinematics of vector meson photoproduction is more complicated than that of pseudoscalar meson photoproduction. But through a measurement of the angular distribution of its decay products, the spin density matrix elements of the vector meson can now be determined too. These spin density matrix elements then provide additional information about the reaction mechanisms of vector meson production. If the initial photon is linearly polarized, then at high energies certain combinations of the vector meson spin density matrix elements receive contributions from natural or unnatural t-channel exchanges only. This is the analogy of Stichel's theorem in pseudoscalar meson photoproduction and it is stated in table 2 below[41,42] for linear photon polarization perpendicular and parallel to the production plane.

TABLE 2

	γ polarization		
$\rho^{11} = \rho_{11} - \rho_{1-1}$	U	N	$\sin^2\theta^* \cos^2\phi^*$
$\rho^{22} = \rho_{11} + \rho_{1-1}$	N	U	$\sin^2\theta^* \sin^2\phi^*$
$\rho^{33} = \rho_{oo}$	U	N	$\cos^2\theta^*$
$-2\ \mathrm{Re}\rho^{13} = \rho_{1o} - \rho_{-1o}$	U	N	$\sin 2\theta^* \cos\phi^*$

Here ρ_{11}. etc denote the spin density matrix elements ρ_{ik} in the helicity basis (i,k = +1, -1, O) and N and U indicate the contributions which these combinations of density matrix elements receive at high energies from natural and unnatural parity t-channel exchange, respectively. Table 2 holds for unpolarized targets as well as for target polarization normal to the production plane. In the last column we have denoted also the characteristic angular distribution of the subsequent decay of the vector meson, where e.g. for a two-body decay ϕ^* is the angle between the production plane and the decay plane and θ^* is the polar angle of one of the decay particles measured relatively to the negative direction of the recoil nucleon in the rest system of the vector meson (this corresponds to the so-called helicity system). To be more specific, the decay angular distribution of the vector meson which was produced by a linearly polarized photon can be worked out to be (details are in ref. 41 and 42)

$$W(\phi,\theta^*,\phi^*)=\frac{3}{4\pi}\{\frac{1}{2}(1-\rho^{o}_{oo})+\frac{1}{2}(3\rho^{o}_{oo}-1)\cos^2\theta^*-\sqrt{2}\mathrm{Re}\rho^{o}_{1o}\sin 2\theta^*\cos\phi^*$$
$$-\rho^{o}_{1-1}\sin^2\theta^*\cos 2\phi^*-P_\gamma\,(\cos 2\phi\,[\rho^{1}_{11}\sin^2\theta^* + \rho^{1}_{oo}\cos^2\theta^*$$
$$-\sqrt{2}\,\mathrm{Re}\rho^{1}_{1o}\,\sin 2\theta^*\,\cos\phi^* - \rho^{1}_{1-1}\,\sin^2\theta^*\,\cos 2\phi^*]$$
$$-P_\gamma\,\sin 2\phi[\sqrt{2}\,\mathrm{Im}\rho^{2}_{1o}\sin 2\theta^*\sin\phi^*+\mathrm{Im}\rho^{2}_{1-1}\sin^2\theta^*\sin 2\phi^*]\}\tag{53}$$

where ϕ is the angle between the photon polarization vector and the scattering plane (compare with equ. (15)) and P_γ is the degree of linear photon polarization. The density matrix elements ρ^{o}_{ik} describe the vector meson in case of an unpolarized photon beam and ρ^{1}_{ik} and ρ^{2}_{ik} result from the linear polarization of the photon. These density matrix elements are thus determined by the experimental (θ^*,ϕ^*) distribution. They can be used to examine the production mechanism. For example, the relative strength of the contributions σ^N, σ^U from natural parity t-channel exchange, respectively, to the cross section can be measured through the quantity P_σ,

$$P_\sigma = \frac{\sigma^N - \sigma^U}{\sigma^N + \sigma^U}\tag{54}$$

From table 2 one can see that to leading order in energy P_σ is given by

$$P_\sigma = 2\,\rho^1_{1-1} - \rho^1_{oo} \quad . \tag{55}$$

In a similar way from the knowledge of the ρ_{ik} information can be obtained about the helicity structure of the production amplitudes. If f.i. the production mechanism is such that s-channel helicity conservation holds then in the helicity system the density matrix elements should have the following structure:

$$\rho^o_{oo} = \rho^o_{1o} = \rho^o_{1-1} = \rho^1_{oo} = \rho^1_{11} = \rho^1_{1o} = \rho^2_{1o} = 0$$

$$\rho^o_{11} = \frac{1}{2}$$

$$\rho^1_{1-1} = \pm\frac{1}{2} \qquad \text{for } J^P = 0^+\ (0^-) \text{ exchange} \tag{56}$$

$$\rho^2_{1-1} = \mp\frac{i}{2}$$

<u>b) ρ^o Production</u>

According to the diffractive nature of this process pomeron exchange is the dominant contribution. f exchange and A_2 exchange are also possible (the latter only for isoscalar photons like also π exchange) but they are less important at high energies as they drop approximately like $1/\sqrt{s}$ relative to the pomeron contributions. Following our considerations about the kinematics of ρ^o photoproduction, this process should be a good candidate for detailed investigations of the properties of the pomeron. The parity asymmetry, P_σ, as given by eq. (55) has been measured in ref. 43 at different energies for $|t| \leq 0.5$ $(GeV/c)^2$ and turns out to be $P_\sigma \approx 1.0$ within experimental errors. From eq. (54) we conclude that natural parity exchange dominates at high energies, as it should for pomeron exchange. The small fraction of unnatural parity exchange

$$\frac{\sigma^U}{\sigma^N + \sigma^U} = \frac{1}{2}(1-P_\sigma)$$

is found to be consistent with the expected one-pion exchange contribution.[44] There exist now quite accurate experimental determinations of the relevant spin density matrix elements of the ρ^o.[45] In fig. 11 it can be seen that the conditions (56) are practically fulfilled in the helicity system. Therefore, ρ^o photoproduction is predominantly s-channel helicity conserving at the γ-ρ^o vertex, although the helicity flip amplitudes may be of the order of 10% of the helicity conserving ones. Similarly, also in πN scattering one has found evidence for s-channel helicity conservation.[46] There are attempts to compare ρ^o photoproduction with other elastic hadronic reactions.Vector meson dominance plus quark model predicts[47]

$$\frac{d\sigma}{dt}(\gamma p \rightarrow \rho^o p) = \frac{\alpha\pi}{\gamma_\rho^2} \frac{1}{2} \left[\frac{d\sigma^{el}}{dt}(\pi^+ p) + \frac{d\sigma^{el}}{dt}(\pi^- p)\right] \qquad (57)$$

where $\frac{em_\rho}{2\gamma_\rho}$ is the γ-ρ^o coupling constant. With $\gamma_\rho^2/4\pi \approx 0,65$ as measured in the e^+e^- storage ring experiments eq.(57) agrees with the data. A more detailed comparison was made by Chadwick et al[48] on the basis of the dual absorption model.[38] They assume that only pomeron and f exchange contribute so that there is only natural parity t channel exchange. Assuming further the helicity conserving amplitude, T_{11}, to be pure imaginary and the helicity-flip amplitude to be small, they write

$$\frac{\text{Im } T_{o1}}{T_{11}} \approx 2 \text{ Re } \rho_{1o}^N \tag{58}$$

(N refers to natural parity exchange), and then compare $2 \text{ Re } \rho_{1o}^N$ with the corresponding ratio $|F_{+-}|/|F_{++}|$ of the flip to non-flip amplitudes in πN scattering.[46] Both processes show s-channel helicity-flip terms of about equal magnitude and almost energy independent. Then they parametrize the pomeron and f-exchange contribution to the helicity-conserving amplitude for ρ^o photoproduction as

$$\begin{aligned} P(t) &= i\ s\ A_p\ e^{B_p t} \\ \text{Im } f(t) &= A_f\ \sqrt{s}\ e^{B_f t}\ J_o\ (R\sqrt{-t}),\ R \approx 1f \end{aligned} \tag{59}$$

where A_p, A_f, and the slope parameters B_p and B_f are to be fitted. Their results for $\frac{d\sigma}{dt}$ are presented in fig.12 where also the values obtained for the slope parameters are shown. For the pomeron slope parameter they obtain $B_p = (0.59 \pm 0.23) + (0.69 \pm 0.07)$ lns. For comparison also the slope parameter of πN scattering is drawn. So they conclude that if the dual absorption model is correct, then the pomeron amplitude shrinks with increasing s. Furthermore, the s-channel helicity non-conserving terms are then most likely associated with the pomeron itself and not with f-exchange. Going over to ρ^o electroproduction two interesting questions may be asked:

(i) Does the photon become "smaller"?
(ii) Is s-channel helicity conserved for $k^2 < 0$ too?

Several theoretical ideas have been put forward about the "size" of the virtual photon and that it should become "smaller" with increasing $(-k^2)$. The simplest way to see this may be the following. Consider first a real photon. In its interaction with hadrons, the photon developes a hadronic component[49] whose main part is the ρ^o. For a photon beam with energy E travelling through hadronic matter, the difference of the wave numbers of the "pure" component and the hadronic components is $E - \sqrt{E^2 - m_\rho^2} \approx \frac{m_\rho^2}{2E}$ for large E. The characteristic length at which the "pure" photon wave and the hadronic component get out of step is therefore $R \approx \frac{2E}{m_\rho^2}$. Inserting $R \approx 1$ f this gives a characteristic energy $E_c \approx 2$ GeV and for $E > E_c$ the hadronic component of the photon should dominate the photon-hadron interactions. Now for virtual photons the characteristic length where the two components get out of step should be $\approx \frac{2E}{m_\rho^2 - k^2}$ and, therefore, the "pure" pointlike component of the photon should become more important as $|k^2|$ increases. This would mean that the virtual photon-nucleon interaction radius becomes smaller for electroproduction and this should show up in flattening of the diffraction peak of ρ^o production when $(-k^2)$ becomes larger. It can be shown that also the parton picture predicts an analogous behavior.[50] But up to now no confirmation of these ideas has been given by experiments (see f.i. ref. 51.).

What concerns question (ii), the experimental data are consistent with the assumption of s-channel helicity conservation for the transverse as well as for the longitudinal component of the virtual photon up to $k^2 \approx -0.8$ $(GeV/c)^2$ (see f.i. ref. 51.) Furthermore, assuming s-channel helicity conservation, in ref. 51 the longitudinal to transverse cross section ratio is found to be $R=\sigma_L/\sigma_U = 0.23 \pm 0.06$. The k^2 dependence of the cross section is consistent with that given by a simple ρ propagator. This is an interesting result as also this exclusive process shows a different k^2 behaviour than the total cross section in the deep inelastic region. However, ρ^o production becomes a relatively less important part of the total cross section for $k^2 < 0$. The ratio $\frac{\sigma_\rho}{\sigma_{tot}}$ drops from 16% at $k^2 = 0$ to 4% at $k^2 = -1.0$ $(GeV/c)^2$.

c) ω Production

The diffractive component of ω photoproduction is smaller than that of ρ^o photoproduction, since the photon coupling to the ω is much weaker than that to the ρ^o,

$$\frac{\gamma_\rho^2}{4\pi} : \frac{\gamma_\omega^2}{4\pi} = 1:9 \ .$$

But there is now an appreciable unnatural parity exchange, mainly due to one-pion exchange which couples to the isovector photon. The data are consistent with the assumption of s-channel helicity conservation for the diffractive component although A_2 exchange might be important too. The energy dependence and magnitude of the unnatural parity exchange contributions agrees with the predictions of one-pion exchange. The partial decay width $\Gamma_{\omega\to\pi\gamma}$ can be extracted from the behaviour of σ^U and is found to be $\Gamma_{\omega\to\pi\gamma} = 0.8 \pm 0.1$ MeV.[44] This agrees with the values obtained from direct measurements of the decay $\omega \to \pi^0\gamma$. So ω electroproduction could offer a possibility to extract the k^2 dependence of the $\omega\pi\gamma$ form factor. This possibility has been analyzed in detail in ref. 52.

d) ϕ Production

This reaction is very interesting as according to wellknown rules at high energies pomeron exchange should be the only possible reaction mechanism, secondary t-channel exchanges should be negligible.[53] Consequently, photoproduction of ϕ mesons should provide us with an ideal tool to investigate the properties of the pomeron.

In the diffraction region the cross section is parametrized as

$$\frac{d\sigma}{dt} = \left.\frac{d\sigma}{dt}\right|_{t=0} e^{Bt} \quad . \tag{60}$$

A recent experiment[54] gives the results

$$\left.\frac{d\sigma}{dt}\right|_{t=0} =$$

$2.15 \pm 0.2\ \mu b\ (GeV/c)^2$ and $B = 5.4 \pm 0.5\ GeV^{-2}$. From the energy behaviour of the diffraction peak, measured in a number of experiments[55] one finds the energy dependence of the slope parameter B to be consistent with

$$B=B_o+2\alpha' \ln s,\ B_o \approx 4.0 GeV^{-2},\ \alpha'=0,14\pm 0,09\ GeV^{-2}\ . \qquad (61)$$

This means that the pomeron exhibits only a small amount of shrinking, as α' is almost consistent with zero. This value of α' is, however, consistent with that one obtained from pp scattering for $s > 30\ GeV^2$ and which is[56] $\alpha' = 0.10 \pm 0.06\ GeV^{-2}$. On the other hand, vector dominance plus quark model gives the relation

$$\frac{d\sigma}{dt}(\gamma p\to\phi p)=\frac{\alpha\pi}{\gamma_\phi^2}\left[\left(\frac{d\sigma}{dt}(K^+p)\right)^{1/2}+\left(\frac{d\sigma}{dt}(K^-p)\right)^{1/2}-\left(\frac{d\sigma}{dt}(\pi^-p)\right)^{1/2}\right]^2 . \qquad (62)$$

The ϕ photoproduction cross section at $t = 0$ as predicted from eq. (62) is increasing with energy like the pp cross section. But whereas pp scattering shows such a behaviour in the ISR energy range, ϕ photoproduction shows it already at much lower energies. This altogether could be interpreted that ϕ photoproduction has pure pomeron exchange already at rather low energies ($s \geq 10\ GeV^2$) and shows already at such low s values the analogous "asymptotic" behaviour as perhaps it has been seen for pp scattering in the ISR. The small value of α' given in eq. (61) is, however, at variance with the result of Chadwick et al[48] (compare with the discussion following eq. (59)), as these authors predict much larger shrinking for the pomeron.

A further point to be discussed is s-channel helicity conservation. In the experiment of ref. 54 the spin density matrix elements have been measured for ϕ production with unpolarized photons. In the helicity system ρ_{oo} and ρ_{1-1} are vanishing within experimental errors whereas Re ρ_{1o} might assume small values different from zero. This again confirmes s-channel helicity conservation although small helicity flip terms might be present. One has to wait whether these facts will be confirmed by future experiments and at higher energies too.

Factorization might be a further property exhibited by the pomeron. Evidence has been found in hadronic reactions and the tests done so far are generally consistent with the assumption of pomeron factorization though the experimental errors are often large. As has been pointed out by Dass and Fraas[57] pomeron factorization can be tested within the single reaction $\gamma p \to \phi p$ if polarized photons and polarized targets are used.

It will be interesting to see whether the new developments of diffractive scattering in pure hadronic reactions can be applied to high energy photo- and electroproduction of neutral vector mesons too.

REFERENCES

1. H. Rollnik, Photon Hadron Interactions in the Resonance Region, lectures at the 14. Internationale Universitätswochen für Kernphysik, Schladming 1975.

2. G. Wolf, Intern. School of Elementary Particle Physics, Basko Polje, Makarska (1972), and DESY-72/61.

3. R. Diebold, Proc. Boulder Conf. on High Energy Physics (1969).

4. R. Talman, Production of pseudoscalar and vector mesons, Proc. of the 6th Int. Conf. on Electron and Photon Interactions at High Energies, Bonn 1973.

5. H. Harari, Proc. of the Intern. Symp. on Electron and Photon Interactions at High Energies, Daresbury 1969.

6. H. Burfeindt et al, Phys. Lett. 33B, 5o9 (197O).
A. M. Boyarski et al, Phys. Rev. Lett. 2o, 3oo (1968).

7. D. J. Quinn et al, Contribution to the 6th Int. Conf. on Electron and Photon Interactions at High Energies, Bonn 1973, quoted in ref. 4.

8. P. Stichel, M. Scholz, Nuov. Cim. 34, 1381 (1964);
M. P. Locher, W. Sandhas, Zs.f. Physik 195, 461 (1966).

9a) A. M. Boyarski et al., Phys. Rev. Lett. 22, 148 (1969)

b) P. Joos et al, Phys. Lett. 52 B, 481 (1974).

1O. O. Barbour, W. Malone, R. G. Moorhouse, Phys. Rev. D4, 1521 (1971).

11. R. Worden, Nucl. Phys. B37, 253 (1972).

12. A. Bartl, W. Majerotto, Nucl. Phys. B 62, 267 (1973).

13. F. A. Berends, R. Gastmans, Phys. Rev. D5, 2o4 (1972);
G. Kramer, unpublished:
J. P. Ader, M. Capdeville, G. Cohen-Tannondji, Ph. Salin, Nuov. Cim. 56A, 952 (1968).

15. N. Dombey, Rev. Mod. Phys. 41, 236 (1969).

16. A. Actor, University of Heidelberg preprint 1973.

17. K. Berkelman, Proc. of the Int. Symp. on the Electron and Photon Interactions of High Energies, Cornell 1971.

18. L. Hand, Phys. Rev. 129, 1834 (1963).

19. M. Gourdin, Nuov. Cim. 21, 1094 (1961).

20. R. L. Walker, Phys. Rev. 182, 1729 (1969).

21. G. Kramer, Acta Phys. Austr. 40, 150 (1974).

22. G. R. Goldstein, J. F. Owens III, Nucl. Phys. B71, 461 (1974).

23. G. Kramer, Fortschr. d. Physik 19, 725 (1971).

24. H. Genzel et al, report DESY 75/01, January 1975.

25. A.C. Irving, Nucl. Phys. B86, 125 (1975).

26. N. Levy, W. Majerotto, B. J. Read, Nucl. Phys. B55, 493 (1973).

27. G. R. Goldstein, J. F. Owens, J. Rutherford, Nucl. Phys. B53, 197 (1973).

28. A. M. Boyarski et al, Phys. Rev. Lett. 22, 1131 (1969).

29. D. J. Sherden et al, Contribution to the 6th Internat. Conf. on Electron and Photon Interactions at High Energies, Bonn 1973, quoted in ref. 4.

30. G. Vogel et al. Phys. Lett. 40 B, 135 (1973).

31. W. Majerotto, unpublished calculations.

32. A. Bartl, W. Majerotto, Electroproduction of $K^+\Lambda$ and $K^+ \Sigma^o$, to be published in Nucl. Phys. B.

33. T. Azemoon et al, report DESY 74/75, 1974.

34. F. Gutbrod, G. Kramer, Nucl. Phys. B49, 461 (1972).

35. A. Actor, J. G. Körner, J. Bender, Nuov. Cim. 24A, 369 (1974).

36. J. M. Barbour, R. G. Moorhouse, Nucl. Phys. B69, 637 (1974).

37. M. Hontebeyrie, J. Procureur, P. Salin, Nucl. Phys. B55, 83 (1973).

38. H. Harari, Ann. Phys. 63, 432 (1971).

39. J. Engels, W. Schmidt, G. Schwiderski, Phys. Rev. 166, 1343 (1968).

40. J. M. Barbour, W. Malone, Nucl. Phys. B82, 477 (1974).

41. H. Fraas, Nucl. Phys. B71, 314 (1974).

42. K. Schilling, P. Seyboth, G. Wolf, Nucl. Phys. B15, 397 (197o), ibid. B18, 332 (197o).

43. J. Ballam et al, Phys. Rev. D5, 545 (1972).

44. G. Wolf, Proc. of the Intern. Conf. on Electron and Photon Interactions at High Energies, Cornell 1971.

45. J. Ballam et al, Phys. Rev. D7, 315o (1973).

46. G. Höhler, R. Strauss, Z. Physik 232, 2O5 (1969); A. de Lesquien et al. Phys. Lett. 4OB, 277 (1972); G. Cozzika et al., Phys. Lett. 4OB, 281 (1972).

47. H. Joos, Proc. of the Int. Conf. on Elementary Particles, Heidelberg 1967.

48. G. Chadwick, Y. Eisenberg, E. Kogan, Phys. Rev. D8, 16O7 (1973).

49. K. Gottfried, Proc. of the Intern. Conf. on Electron and Photon Interactions at High Energies, Cornell 1971.

5O. J.D. Björken, Proc. of the Intern. Conf. on Electron and Photon Interactions at High Energies, Cornell 1971.

51. P. Joos et al, Contribution to the 17th Intern. Conf. on High Energy Physics, London 1974.

52. H. Fraas, Nucl. Phys. B36, 191 (1972); ibid B49, 253 (1972).

53. P.G.O. Freund, Nuov. Cim. 48A, 541 (1967); V. Barger, D. Cline, Phys. Rev. Lett. 24, 1313 (1970).

54. H. J. Behrend et al., Contribution to the 17th Intern. Conf. on High Energy Physics, London 1974.

55. K. C. Moffeit, Proc. of the Intern. Conf. on Electron and Photon Interactions at High Energies, Bonn 1973.

56. V. Bartenev et al., NAL-PUB-73/54 (1973).

57. G. V. Dass, H. Fraas, DESY 74/58, December 1974.

FIGURE CAPTIONS

Fig. 1: Diebold plot $(s-m^2)^2 \frac{d\sigma}{dt}$ plotted versus (-t) for pseudoscalar meson photoproduction (from ref.3).

Fig. 2: Feynman diagrams for the process $\gamma p \to \pi^+ n$ according to the Born term model.

Fig. 3: $\pi^+ n$ photoproduction. a) Unpolarized differential cross section (from A.M. Boyarski et al, ref.6). b) Differential cross sections for polarized photons (from H.Burfeindt et al. ref.6). c) Polarized photon asymmetry (from D.J. Quinn et al. ref. 7).

Fig. 4: Feynman diagrams for $\gamma p \to \pi^- \Delta^{++}$ according to the Born term model.

Fig. 5: $\pi^- \Delta^{++}$ photoproduction. a) Unpolarized differential cross section (from A.M. Boyarski et al., ref. 9a) b) Polarized photon asymmetry (from D.J. Quinn et al., ref. 7).

Fig. 6: Predictions of the Regge model of ref. 11 for the differential cross section of $\gamma p \to \pi^+ n$ and comparison with experimental data at various energies (fig. taken from ref. 11)

Fig. 7: Polarized target asymmetry for $\gamma p \to \pi^+ n$. Comparison of experimental data with various Regge model predictions: - ref.23, --- ref. 11, -...- ref. 22. (Fig. taken from ref. 24).

Fig. 8: π^- / π^+ ratio (fig. taken from ref. 11).

Fig. 9: $\sigma_U + \varepsilon \sigma_L$, σ_T and σ_I for $\pi^+ n$ electroproduction (fig. from ref. 12).

Fig.10: Predictions of the Regge model of ref. 26 for the differential cross section of $K^+ \Lambda$ photoproduction and comparison with experimental data (fig. taken from ref. 26).

Fig.11: $\gamma p \to \rho^o p$ at 9.3 GeV. ρ^o spin density matrix elements and parity asymmetry P_σ in the helicity system (from ref. 45).

Fig.12: $\gamma p \to \rho^o p$: $\frac{d\sigma}{dt}$ and slope parameters B_p and B_f corresponding to eq. (59). (Fig. taken from ref. 48).

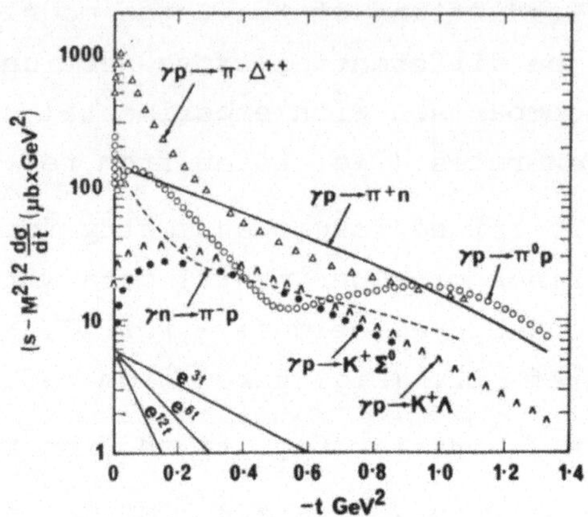
γp→π⁻Δ⁺⁺
γp→π⁺n
γp→π⁰p
γn→π⁻p
γp→K⁺Σ⁰
γp→K⁺Λ
e^{3t}
e^{6t}
e^{12t}
(s − M²)² dσ/dt (μb×GeV²)
−t GeV²

Fig.1

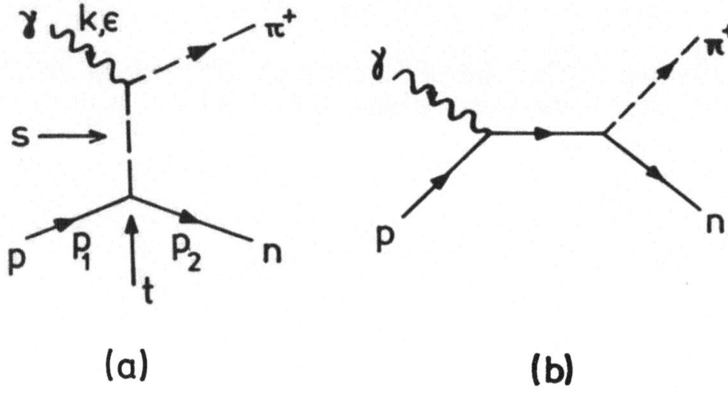
γ
k,ε
π⁺
s
p
p₁
t
p₂
n
(a)
γ
π⁺
p
n
(b)

Fig.2

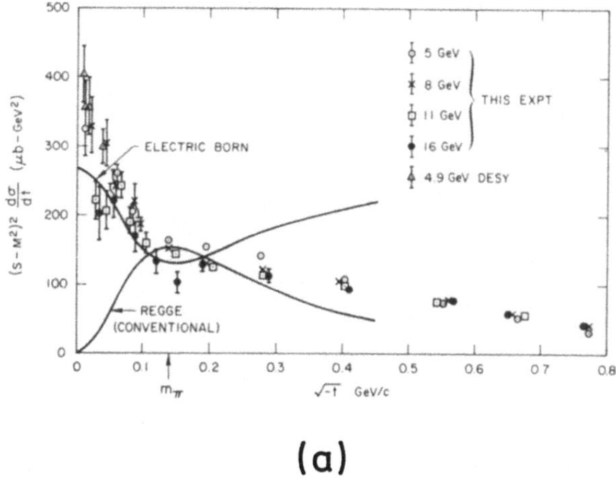
5 GeV
8 GeV
11 GeV
16 GeV
THIS EXPT
4.9 GeV DESY
ELECTRIC BORN
REGGE (CONVENTIONAL)
$(s-M^2)^2 \frac{d\sigma}{dt}$ (μb-GeV^2)
m_π
$\sqrt{-t}$ GeV/c

(a)

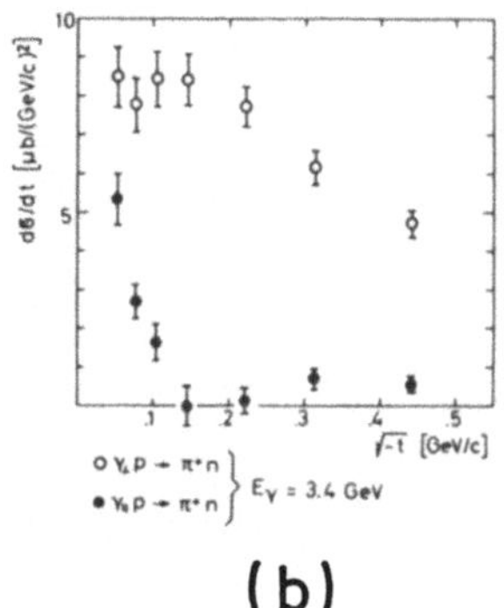
$d\sigma/dt$ [μb/(GeV/c)^2]
$\sqrt{-t}$ [GeV/c]
$\gamma_\perp p \to \pi^+ n$
$\gamma_\parallel p \to \pi^+ n$
E_γ = 3.4 GeV

(b)

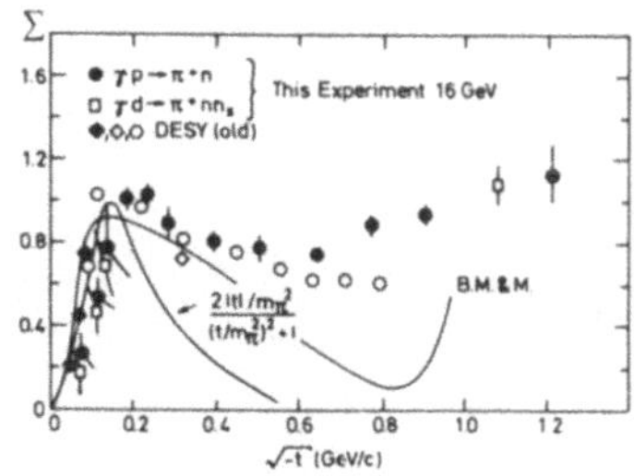
Σ
$\gamma p \to \pi^+ n$
$\gamma d \to \pi^+ n n_s$
DESY (old)
This Experiment 16 GeV
$\frac{2|t|/m_\pi^2}{(t/m_\pi^2)^2+1}$
B.M.&M.
$\sqrt{-t}$ (GeV/c)

(c)

Fig. 3

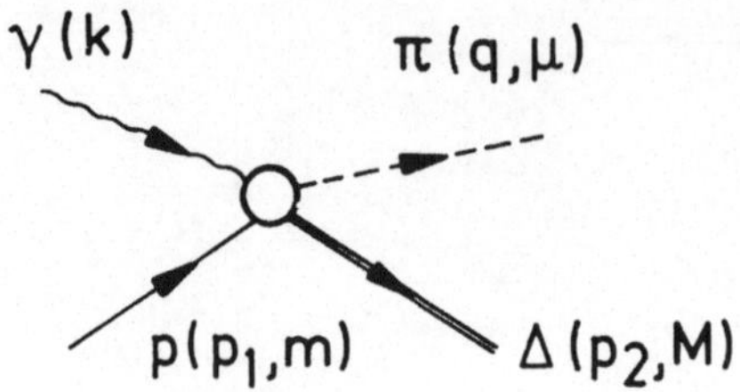

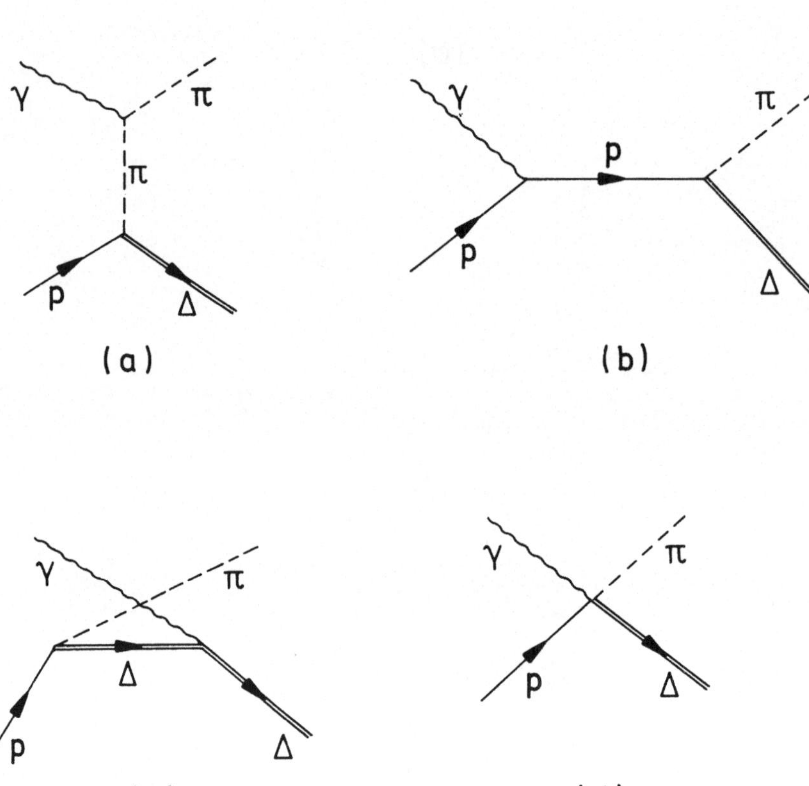

Fig.4

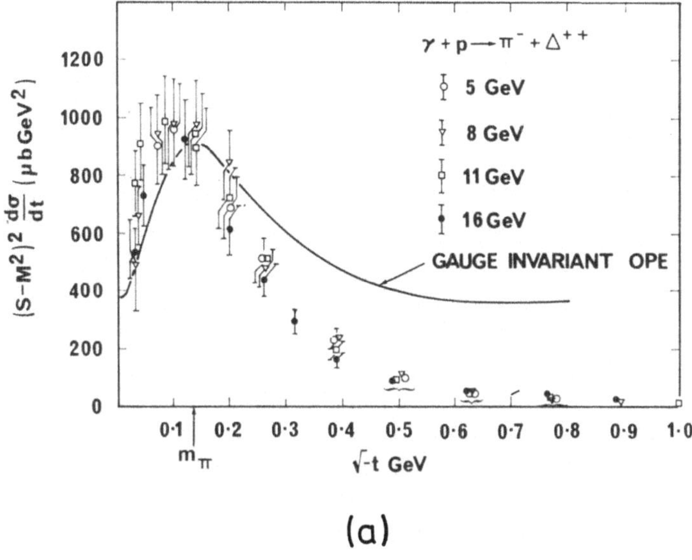
γ + p ⟶ π⁻ + Δ⁺⁺
5 GeV
8 GeV
11 GeV
16 GeV
GAUGE INVARIANT OPE
(S-M²)² dσ/dt (μb GeV²)
√-t GeV
mπ

(a)

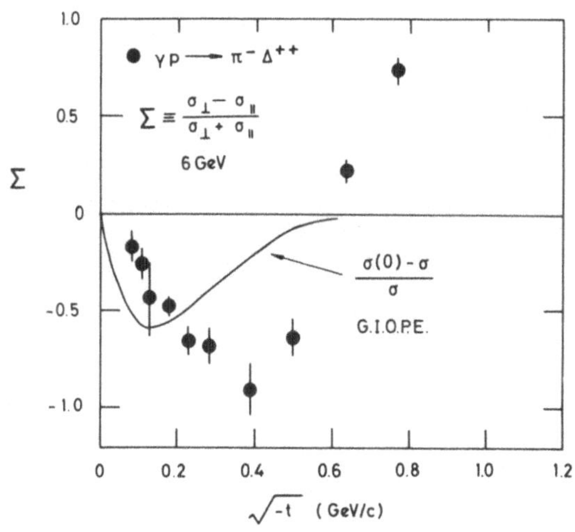
γp ⟶ π⁻ Δ⁺⁺
Σ ≡ (σ⊥ − σ∥)/(σ⊥ + σ∥)
6 GeV
Σ
(σ(0) − σ)/σ
G.I.O.P.E.
√-t (GeV/c)

(b)

Fig. 5

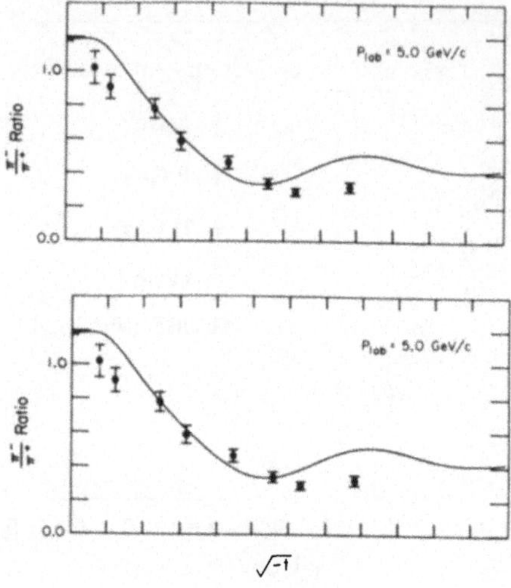

P lab = 5.0 GeV/c
1.0
0.0
π⁻/π⁺ Ratio
P lab = 5.0 GeV/c
1.0
0.0
π⁻/π⁺ Ratio
√-t

Fig. 6

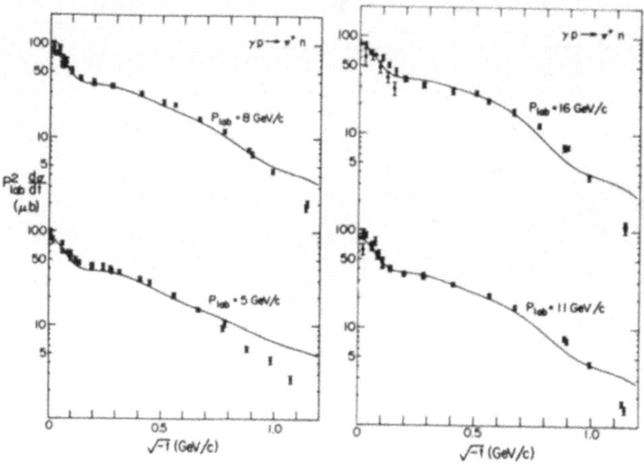

γp → π⁺n
P lab = 8 GeV/c
P lab = 5 GeV/c
γp → π⁺n
P lab = 16 GeV/c
P lab = 11 GeV/c
100
50
10
5
(μb)
0
0.5
1.0
√-t (GeV/c)
√-t (GeV/c)

Fig. 8

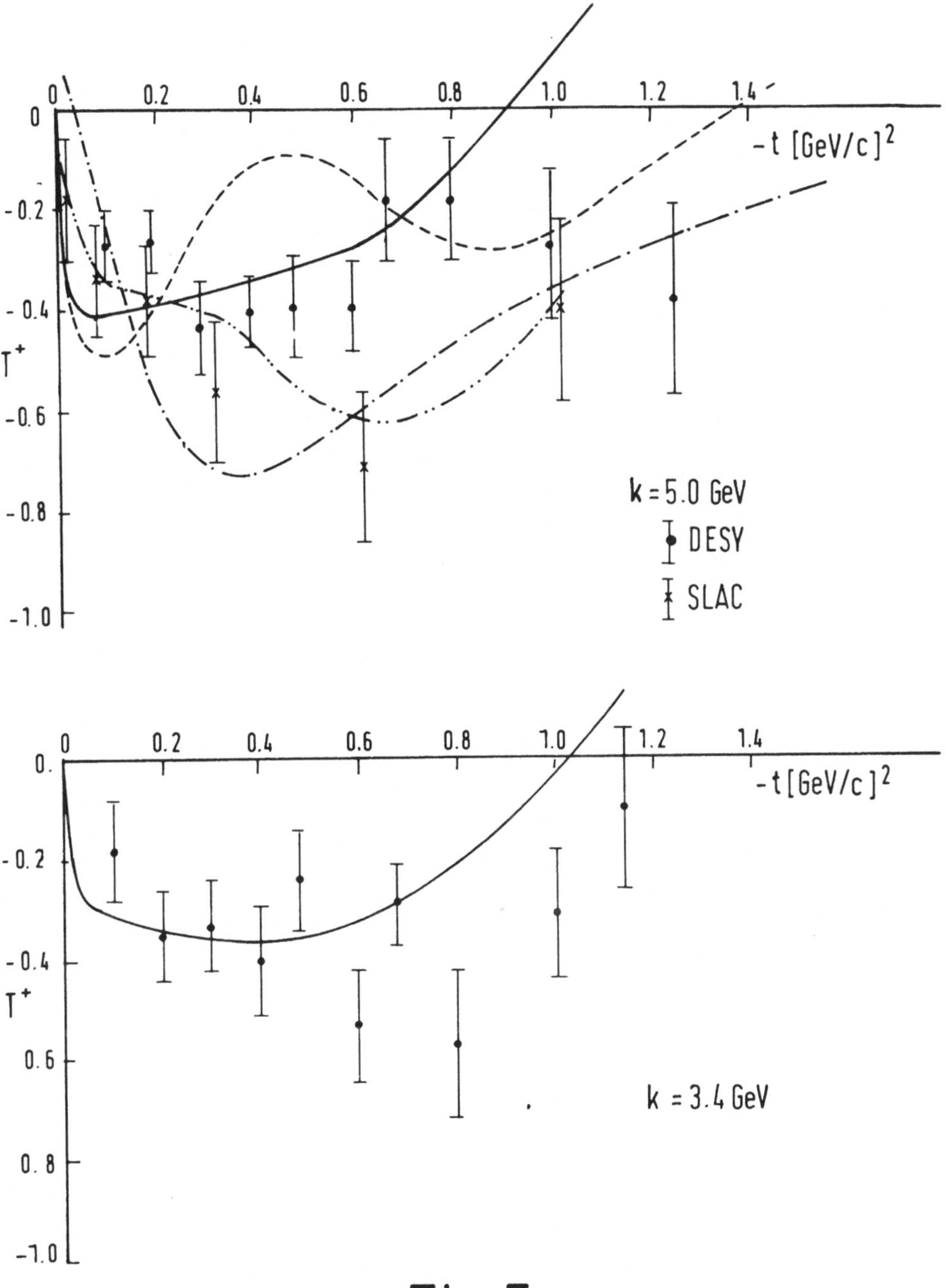

0
0.2
0.4
0.6
0.8
1.0
1.2
1.4
-t [GeV/c]^2
-0.2
-0.4
T^+
-0.6
-0.8
-1.0
k = 5.0 GeV
DESY
SLAC
0.
-0.2
-0.4
0.6
0.8
-1.0
k = 3.4 GeV

Fig. 7

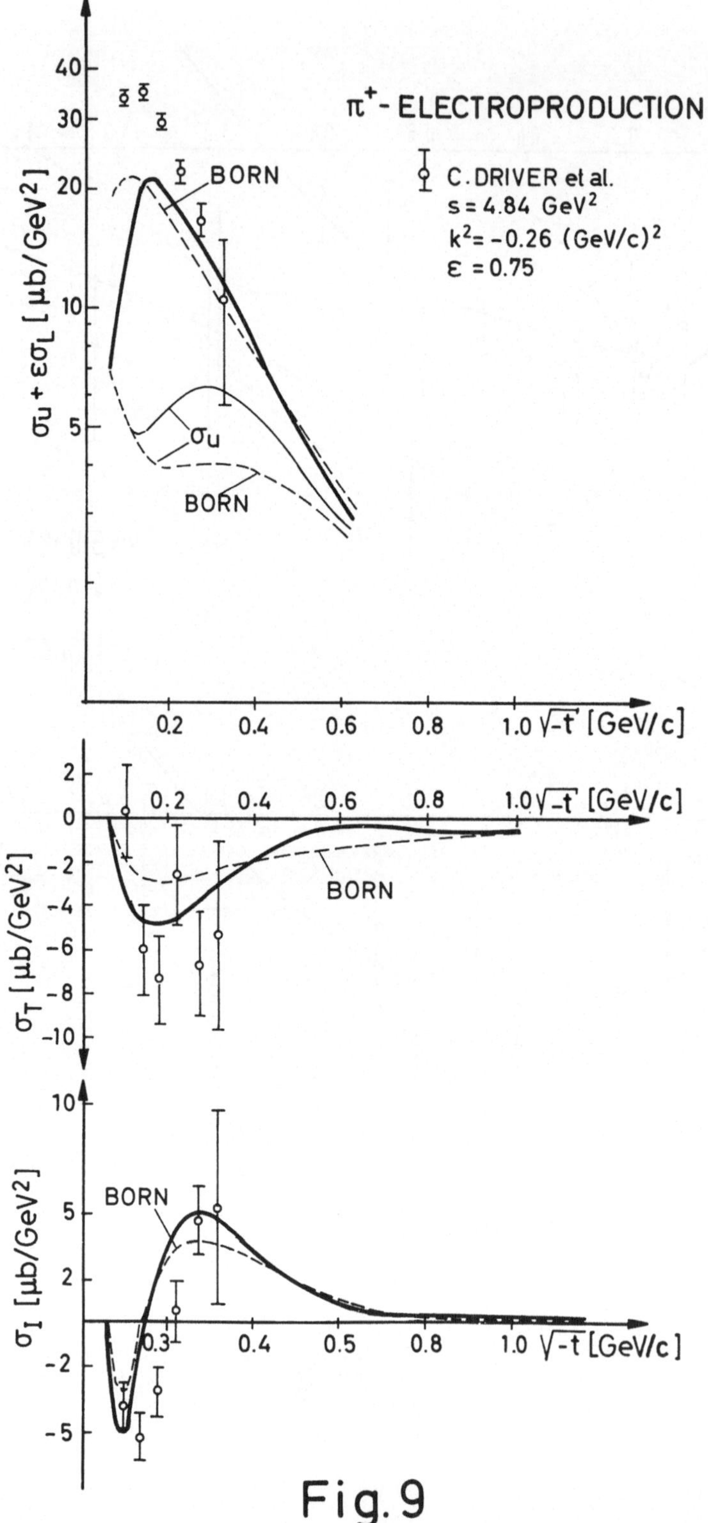

π⁺- ELECTROPRODUCTION
C. DRIVER et al.
s = 4.84 GeV²
k² = -0.26 (GeV/c)²
ε = 0.75
BORN
σu
BORN
σu + εσL [μb/GeV²]
√-t' [GeV/c]
BORN
σT [μb/GeV²]
√-t' [GeV/c]
BORN
σI [μb/GeV²]
√-t [GeV/c]

Fig. 9

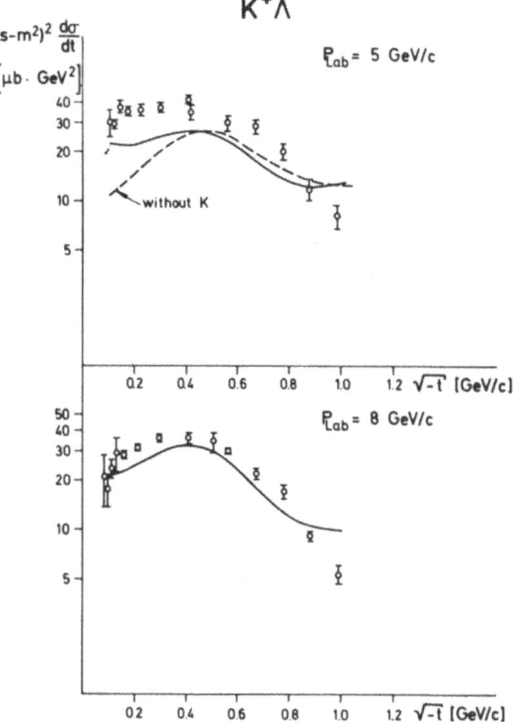
K⁺Λ
(s-m²)² dσ/dt
[μb · GeV²]
P_Lab = 5 GeV/c
without K
√-t [GeV/c]
P_Lab = 8 GeV/c
√-t [GeV/c]

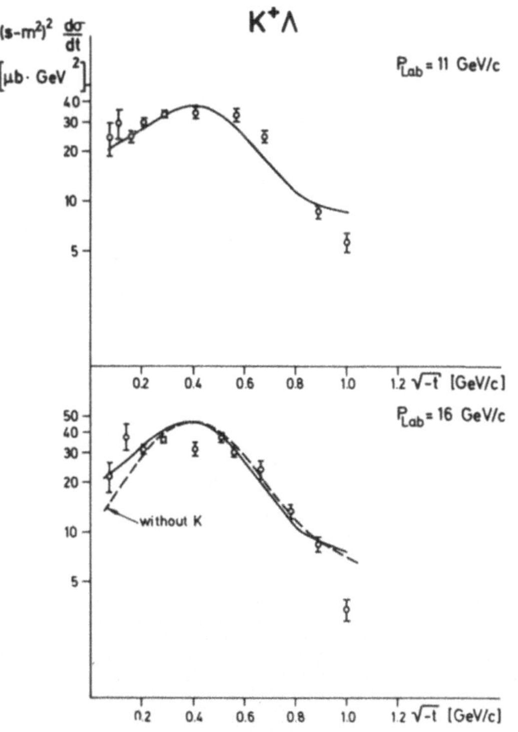
K⁺Λ
(s-m²)² dσ/dt
[μb · GeV²]
P_Lab = 11 GeV/c
√-t [GeV/c]
P_Lab = 16 GeV/c
without K
√-t [GeV/c]

Fig. 10

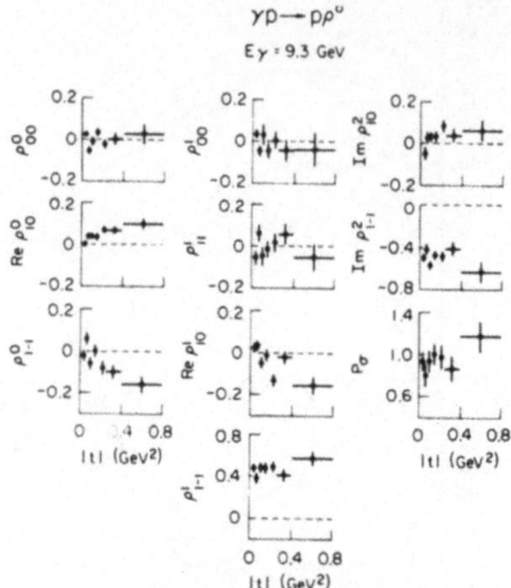

γp → pρ°
Eγ = 9.3 GeV
|t| (GeV²)

Fig.11

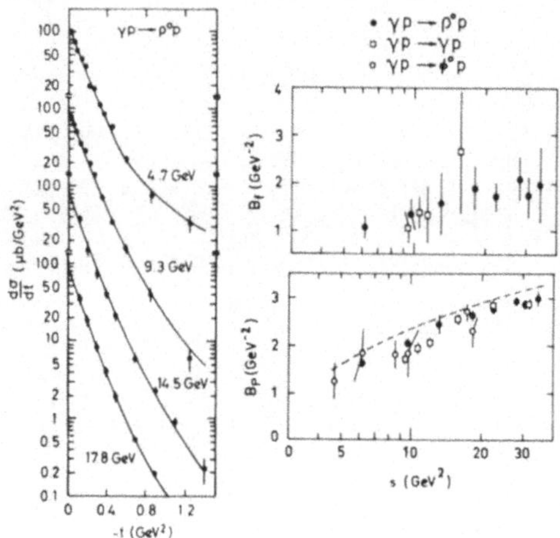

γp → ρ°p
4.7 GeV
9.3 GeV
14.5 GeV
17.8 GeV
dσ/dt (μb/GeV²)
-t (GeV²)
• γp → ρ°p
▫ γp → γp
○ γp → φ°p
s (GeV²)

Fig.12

Acta Physica Austriaca, Suppl. XIV, 143 — 231 (1975)

e^+e^- ANNIHILATION, THE NEW PARTICLES, AND CHARM+

by

J. ELLIS
CERN, Geneva, Switzerland

INTRODUCTION

These notes contain nothing that is not well known to many people. The attempt to give due reference to theoretical papers on the subject was hopeless, and I apologize to people who are not referred to appropriately. The inclusion of something in these notes without acknowledgement does not imply that nobody else thought of it, or even that the author did think of it. It will probably all be irrelevant in a few weeks anyway.

At the time of writing, the prejudice of the author is that some form of "charm" may be the least unlikely explanation for the new phenomena. This prejudice is reflected in the notes, which treat "colour" and other models for the new particles rather slightingly.

+ Lecture given at XIV. Internationale Universitätswochen für Kernphysik, Schladming, Austria, February 24- March 7, 1975.

1. THE NEW PARTICLES[1,2]

In this first lecture we start by listing some of the experimental facts and rumours about the new particles, then make a general phenomenological analysis of widths, branching ratios and so on. We finish by introducing and discussing some of the theoretical explanations proposed for them.

1.1. Experimental facts

Much of the information mentioned here is probably inaccurate and apocryphal, and many of the quoted numbers should not be taken seriously. However, at the moment just a qualitative picture of what is happening is all that is required.

1.1.1. e^+e^- annihilation

Three distinct structures have been seen in this reaction - two very narrow peaks at 3095 ± 4 MeV[2,3] and 3684 ± 4 MeV[4,5] and a broader (200-300 MeV) enhancement at about 4.1 GeV[6]. They seem to have the following properties[7]:

$$\Gamma(3.1 \rightarrow e^+e^-) \approx 5.2 \text{ keV} \approx \Gamma(3.1 \rightarrow \mu^+\mu^-)$$

$$\frac{\Gamma(3.1 \rightarrow e^+e^-)}{\Gamma(3.1 \rightarrow \text{any})} \approx 6.8\%, \quad \Gamma(3.1 \rightarrow \text{any}) \approx 77 \text{ keV} \qquad (1.1)$$

to within 20%.

The angular distribution of the leptons in decays of the 3.1 have no visible forward-backward asymmetry[+] and are consistent with $1 + \cos^2 \theta$, indicating that the 3.1 decays do not violate parity and that it is either vector 1^- or axial 1^+. Also, statistically significant negative interference has been seen at SPEAR in the $e^+e^- \to \mu^+\mu^-$ channel below the peak - as expected from a 1^- state interfering with the 1γ exchange. There is a vast wealth of information[7] on the hadronic decays of the 3.1: an important decay is $2\pi^+ \, 2\pi^-$ + missing neutral, probably a π^0. Specific mesonic decay modes such as $\rho\pi$, $\rho\pi\pi\pi$, $\omega\pi^+\pi^-$, $\pi\pi K\bar{K}$, $K^*K\pi$, $K^{**}K\pi$ have apparently been identified at SPEAR. Very few $2\pi^+ \, 2\pi^-$ decays have been seen. In fact,

$$\frac{\Gamma(3.1 \to 2\pi^+ \, 2\pi^-)}{\Gamma(3.1 \to \mu^+\mu^-)} \approx \frac{\sigma(e^+e^- \to 2\pi^+ 2\pi^-)}{\sigma(e^+e^- \to \mu^+\mu^-)} \qquad (1.2)$$

outside the peak, suggesting that all $2\pi^+ \, 2\pi^-$ decays could come from 1γ intermediate states as in Fig. 1. The five-pion decays are, however, too copious to come from this source, and apparently come from the direct coupling of Fig. 2. Quite generally, since

$$R \equiv \frac{\sigma(e^+e^- \to \text{hadrons})}{\sigma(e^+e^- \to \mu^+\mu^-)} \approx 2.5$$

from data outside the peak, it seems that if the 3.1

[+] Results of a Frascati experiment showing such an asymmetry have not been confirmed by other laboratories.

couples to leptons via the photon and we can neglect interference effects, then

$$\Gamma(3.1 \to \gamma \to \begin{matrix} \nearrow \text{hadrons} \\ \to \mu^+\mu^- \\ \searrow e^+e^- \end{matrix}) \gtrsim 23 \text{ keV} \tag{1.3}$$

leaving O(50) keV for the direct decays of 3.1 → hadrons in the absence of interference. In the decays the final states $p\bar{p}$ and $\Lambda\bar{\Lambda}$ have been seen, though it is not excluded that they may also be accompanied by low momentum γ-rays. These modes have rates both O(100-200) eV. The decays $3.1 \to \pi^+\pi^-$, K^+K^- have apparently not yet been seen, and the DASP experiment at DESY gives upper bounds of 130 eV for each of these modes. From looking at energetic collinear photons produced in the decay, there is no evidence[8] for a $\gamma\gamma$ decay mode (forbidden if the 3.1 has spin = 1). From DESY data on almost collinear energetic photons, there are the following upper limits on neutral decay modes:

$$\frac{\Gamma(3.1 \to \pi^0\gamma)}{\Gamma(3.1 \to e^+e^-)} \lesssim 0.08, \qquad \frac{\Gamma(3.1 \to X\gamma)}{\Gamma(3.1 \to e^+e^-)} \cdot \frac{\Gamma(X \to \gamma\gamma)}{\Gamma(X \to \text{any})} \lesssim 0.2 \tag{1.4}$$

where X is any state with mass $\gtrsim$ 2.6 GeV⁺. The inclusive $K^\pm$ and p spectra in the decay seem to be large for $x = 2E/E_{cm}$ large, with $K^\pm/\pi^\pm \simeq 0.4$ for $x \gtrsim 0.4$ and $\bar{p}/\frac{1}{2}\pi^\pm \simeq 0.3$ for $x \gtrsim 0.7$.

As for the 3.7, it seems[7] to have

⁺ SPEAR data on the decay $3.7 \to 3.1\ \pi^+\pi^-$ indicate that the decays $\Gamma(3.1 \to \text{all neutrals})$ are a small fraction of its total width. This has been assumed in deriving the numbers (1.1) from the experimental data on the 3.1.

$$\Gamma(3.7 \to e^+e^-) \simeq 2.2 \text{ keV} \simeq \Gamma(3.7 \to \mu^+\mu^-) \tag{1.5}$$

$$\frac{\Gamma(3.7 \to 3.1\ \pi^+\pi^-)}{\Gamma(3.7 \to \text{any})} \approx 30\%,\ 200 \text{ keV} < \Gamma(3.7 \to \text{any}) < 800 \text{ keV} \tag{1.6}$$

The total width is uncertain because 3.7 → 3.1 + neutrals decays would produce $\ell^+\ell^-$ pairs from the 3.1 decay which could be confused with those from the direct 3.7 decay. The angular distributions of the products in the 3.7 → 3.1 $\pi^+\pi^-$ decay are exactly as expected from a decay of one 1^- state to another 1^- state by emission of a 0^+ dipion state such as the $\varepsilon(600)$. One would then expect $\Gamma(3.7 \to 3.1\ \pi^0\pi^0) \approx \frac{1}{2}\Gamma(3.7 \to 3.1\ \pi^+\pi^-)$: experimentally, it seems[7] that this relation is not exactly correct. Thus cascade decays such as

$$3.7 \to X + \gamma,\quad X \to 3.1 + \gamma \tag{1.7}$$

or other decays such as $3.7 \to 3.1\eta$ may be present. Decays $3.7 \to X\gamma$ with the emission of a γ having an energy of a few hundred MeV have not yet been seen: systematic searches are now starting.

As for the 4.1 structure, it seems to have a width of ∿(200-300) MeV: integrating over the peak would give $\Gamma(4.1 \to e^+e^-) = O(\text{few})$ keV. Its decay shows no obvious features[7], there being apparently no evidence for 4.1 → 3.7 $\pi\pi$, 3.1 $\pi\pi$ or $\pi^\pm Z^\mp$ with $m_Z \gtrsim 3$ GeV.

1.1.2. Hadronic collisions

So far only the 3.1 has been seen in purely hadronic reactions. It was seen[1] in $p + Be \to 3.1\ (\to \mu^+\mu^-)$ + anything at an energy of 29 GeV with a cross-section difficult to estimate because of the unknown inclusive distribution in p_T and $p_\parallel$, but propably corresponding to $\sigma(p + p \to 3.1 + \text{anything}) \simeq 10^{-34\pm1}\ cm^2$. The process $p + Be \to 3.7\ (\to \mu^+\mu^-)$ + anything is at least 10^{-2} times smaller at this energy[9]: a very much smaller rate is of course to be expected because of the small leptonic branching ratio of the 3.7. The $p + Be \to 3.1$ + anything signal apparently disappears at $\sim$ 20 GeV[7], and the previous experiment which saw a shoulder in the $p + U \to \mu^+\mu^-$ + anything distribution is apparently consistent[10] with a cross-section $O(10^{-34})$ cm^2 at 29 GeV. A more recent[11] experiment, $n + Be \to (\mu^+\mu^-)$ + anything, with an average neutron energy of 250 GeV sees a 3.1 signal possibly of the order of 10^3 larger[+]. The 3.1's produced have apparently larger $\langle p_T \rangle$ than do the ρ^o's produced in the same experiment, and seem to be produced centrally in $x = 2p_\parallel/E_{cm}$.

1.1.3. Photoproduction

The initial results on photoproduction of the 3.1 were negative, with upper limits of $\sim$1 nb at $E_\gamma \simeq 11$ GeV[12++] and $\sim$29 nb at $\langle E_\gamma \rangle \simeq 18$ GeV[13]. But it has now been seen at

---

[+] Analysis of the CERN-Columbia-Rockefeller-Saclay experiment on $pp \to (e^+e^-)$ + anything at the CERN Intersecting Storage Rings (ISR) also gives a cross-section much bigger than that of MIT-BNL.

[++] Rumour has it that this group has now found some events.

Fermilab in $\gamma + Be \rightarrow (\mu^+\mu^-)$ + anything with a cross-section/nucleon = O(20) nb at $E_\gamma \gtrsim 80$ GeV[14]. The cross-section has a forward peak, including a very sharp inner peak consistent with e^{4ot} corresponding to coherent scattering off the entire nucleus. As will be discussed later, this observation suggests very strongly that the 3.1 is coupled to the photon, and scatters diffractively just like a normal hadron at high enough energies as shown in Fig. 3. New experiments at $\langle E_\gamma \rangle \sim 18$ GeV are apparently[15] also seeing $\gamma + N \rightarrow 3.1$ + anything, at a rate smaller by about an order of magnitude than the Fermilab value. Fermilab and SLAC also apparently see $\gamma + N \rightarrow 3.7$ + anything at a considerably smaller rate.

1.2. Phenomenological analysis[16]

We now move on to a phenomenological discussion of the above experimental facts, and a comparison with other particles and processes. It is clear that the 3.1 and 3.7 particles' widths are very small. The leptonic widths are however comparable with those of strong vector mesons: $\Gamma(\phi \rightarrow e^+e^-) \sim 1.3$ keV, $\Gamma(\omega \rightarrow e^+e^-) \sim 0.76$ keV. If we parametrize the 3.1 couplings to leptons by phenomenological axial and vector couplings g_A^μ, g_A^e; g_V^μ, g_V^e, then we find

$$g_A^{\mu\,2} + g_V^{\mu\,2} \approx g_A^{e\,2} + g_V^{e\,2} \approx 0.6 \times 10^{-4} \quad . \tag{1.8}$$

If we suppose that this coupling proceeds via 1γ exchange, then we have $g_A^\mu = g_A^e = 0$, and if the γ-3.1 coupling is

$em^2/2\gamma_{3.1}$ as in Fig. 1, then

$$\frac{1}{\frac{\gamma^2_{3.1}}{4\pi}} \approx 1/2.6 \quad (\text{cf. } \frac{1}{\frac{\gamma^2_{\phi}}{4\pi}} \approx 1/2.5) \ . \qquad (1.9)$$

The corresponding numbers to formulae (1.8) and (1.9) for the 3.7 are a factor of 2-3 lower, and by way of comparison

$$\frac{1}{\gamma^2_{\rho}} \Big/ \frac{1}{\gamma^2_{\rho'}} \approx O(3)$$

if we take ρ' to be the state at $\sim$ 1600 MeV.

How much are the hadronic decays suppressed relative to the decay width of a normal hadron? For comparison we may note

$$\Gamma(\phi \to \text{any}) \approx 4.2 \text{ MeV}, \ \Gamma(\omega \to \text{any}) \approx 10 \text{ MeV} \ .$$

Experience with known states, and with dual and statistical models, might suggest[16] a width of $\sim$ 50 to 500 MeV for a meson with mass 3 GeV. Hence the suppression of $\Gamma(3.1 \to \text{anything})$ is by a factor of 10^3 to 10^4, that of $\Gamma(3.7 \to \text{anything})$ by a factor of 10^2 to 10^3. On the other hand, the 4.1 structure has a characteristic hadronic width.

There has been much discussion and comparison with the suppression of the ϕ coupling to non-strange hadrons.

This latter we can estimate in various ways:

$$\frac{g^2_{\phi\rho\pi}}{g^2_{\omega\rho\pi}} \approx \frac{1}{100}, \frac{\sigma(\pi^- p \to \phi n)}{\sigma(K^- p \to \phi\Lambda)} \approx \frac{1}{60}, \frac{\sigma(\pi^- p \to \phi n)}{\sigma(\pi^- p \to \omega n)} \approx \frac{1}{300} \quad . \tag{1.10}$$

Thus the ϕ coupling to non-strange hadrons seems to be suppressed by a factor of $O(10^2)$. This is usually "explained" by assuming that disconnected quark diagrams such as that of Fig. 4 for $\phi \to 3\pi$ decay are essentially zero. One other similarity to the ω and ϕ should be noted: the predominance of 3, 5, and 7π decay modes and the apparent absence of direct 4π decays suggest that the concept of G parity makes sense for the 3.1, and that it has G = -1.

Simple radiative decays of the 3.1 must also be strongly suppressed. For example, for the radiative decay to a pseudoscalar P,

$$\frac{\Gamma(3.1 \to P\gamma)}{\Gamma(\omega \to \pi^o\gamma)} = \left(\frac{g_{3.1P\gamma}}{g_{\omega\pi^o\gamma}}\right)^2 \left(\frac{m_{3.1}}{m_\omega}\right)^3 \left(\frac{m^2_{3.1} - m^2_P}{m^2_\omega - m^2_\pi}\right) \quad . \tag{1.11}$$

If $m_P << 3.1$, we can use the known rate $\Gamma(\omega \to \pi^o\gamma) \simeq 0.87$ keV to get

$$\Gamma(3.1 \to P\gamma) \approx \left(\frac{g_{3.1P\gamma}}{g_{\omega\pi^o\gamma}}\right)^2 \times 15 \text{ MeV} \quad . \tag{1.12}$$

Since experimentally $\Gamma(3.1 \to \pi^o\gamma) < 0.4$ keV, we must have

$$\left(\frac{g_{3.1\pi^o\gamma}}{g_{\omega\pi^o\gamma}}\right)^2 \lesssim 0.3 \times 10^{-4} \quad ! \tag{1.13}$$

How about the comparison between e^+e^- and hadronic experiments? We can define two ratios R_H and R_L by

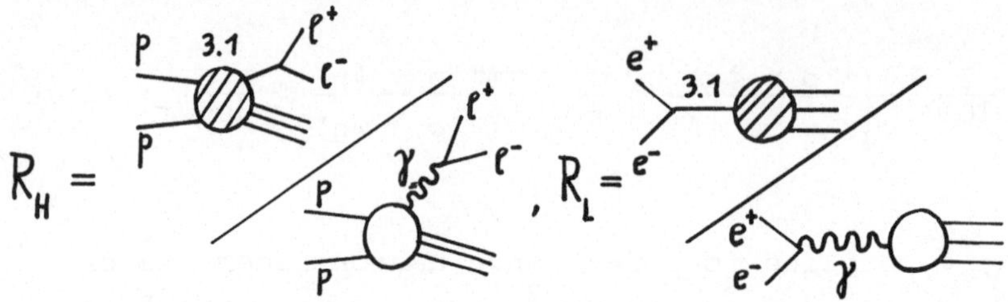

If the production and decay mechanisms of the 3.1 were the same, then we would expect $R_H \simeq R_L$. The comparison has been made[16] for the BNL and e^+e^- experiments. We have

$$R_H = \frac{\text{number of events in peak of Ref. 1}}{\text{number of events/MeV in background}}$$

$$= \frac{242}{5/25} = 1200 \text{ MeV} \ , \tag{1.14}$$

while from the SPEAR data

$$R_L = \frac{\int \sigma_{3.1}(E)\,dE}{\sigma_{\text{background}}} \approx 300 \text{ MeV} \tag{1.14'}$$

yielding $R_H/R_L \simeq 4$. We conclude that production of the 3.1 in the BNL and the decay 3.1 → anything could well go via similar dynamics. On the other hand, the much larger (by $\sim 10^3$) cross-section seen at Fermilab in $n + Be \to 3.1 + \text{anything}$ at $E \gtrsim 250$ GeV suggests that the 3.1 coupling to hadrons is no longer suppressed at such energies.

How about $\Gamma(3.7 \to 3.1\ \pi^+\pi^-)$? According to formula (1.6) this = O(200) keV. If we discuss generally the decay of a vector state V' into a lower mass vector state V emitting two charged pions via an interaction $g_{V'V\pi\pi}\ V'V\pi^+\pi^-$, then

$$\Gamma(3.7 \to 3.1\ \pi^+\pi^-) \approx 14\ \left(\frac{g^2_{3.7,3.1\pi\pi}}{4\pi}\right)\ \text{keV}$$

whereas

$$\Gamma(\rho' \to \rho\pi^+\pi^-) \approx 0.25\ \left(\frac{g^2_{\rho'\rho\pi\pi}}{4\pi}\right)\ \text{MeV} \quad .$$

Using

$$\Gamma(\rho' \to \rho\pi^+\pi^-) \approx 200\ \text{MeV}$$

we find

$$\left(\frac{g_{3.7,3.1\pi\pi}}{g_{\rho'\rho\pi\pi}}\right)^2 \approx \frac{200}{14} \times \frac{0.25}{200} \approx 2 \times 10^{-2} \quad . \qquad (1.15)$$

We conclude that the $3.7 \to 3.1\ \pi^+\pi^-$ decay is suppressed relative to a comparable hadronic width by $\sim 10^{-2}$+.

Finally, we discuss the photoproduction experiments. The cross-section at small t quoted for the Fermilab experiment is so large that it is explicable

+Similar ratios are obtained by assuming that the $(\pi^+\pi^-)$ state is an s-wave resonance ε.

only if the 3.1 couples directly to the photon and is being produced diffractively.

If we assume 3.1 vector meson dominance of the photon, then

$$\sigma_{diff}\ (\gamma N \to 3.1N) \approx \frac{\alpha}{4}\left(\frac{\gamma_{3.1}^2}{4\pi}\right)^{-1} \sigma_{diff}(3.1N \to 3.1N)$$

$$\sigma_{diff}(3.1N \to 3.1N) \approx \frac{1}{16\pi B}\ \sigma_{tot}^2\ (3.1N) \qquad (1.16)$$

where B is the slope of the diffraction peak in 3.1N scattering. From their value of $\sigma_{diff}(\gamma N \to 3.1N)$, the Fermilab group find $\sigma_{tot}(3.1N) \simeq 1$ mb. This cross-section can only be hadronic, but it is suppressed at high energies:

$$\frac{\sigma_{tot}\ (3.1N)}{\sigma_{tot}\ (\phi N)} = O\ (\tfrac{1}{10}) \qquad (1.17)$$

The latest SLAC and Rochester-Cornell data would correspond to a value of σ_{tot} (3.1N) about three times smaller at $E_\gamma \lesssim 20$ GeV.

1.3. Theoretical models

The theoretical response to the new particles has been copious, but far from conclusive. Probably well over 100 papers have appeared in the last two months, and it has only been possible to look at a fraction of them[16]. Accordingly, we will review the range of possibilities only long enough to motivate my own prejudice as to the interpretation of the new phenomena. One of the criteria used is Occam's razor[+]: explanations should ideally give a unified understanding of the 3.1, the 3.7, and hopefully the 4.1 enhancement and the large e^+e^- total cross-section. They should also be consistent with the developed phenomenology of deep inelastic processes, and should ideally have been invented for other reasons before the new particles came along. Suggestions can be classified as follows:

1.3.1. Non-hadronic models

Many people[16] jumped on equation (1.8) and noticed that

$$\frac{g_A^2 + g_V^2}{m_{3.1}^2} \approx 0.6 \times 10^{-5}/m_p^2 \tag{1.18}$$

suggesting a relation to the neutral currents in weak interactions. Notice that for a W boson in a theory of charged weak currents

$$\frac{g_A^2 + g_V^2}{M_W^2} \approx 1.5 \times 10^{-5}/m_p^2 \quad . \tag{1.19}$$

Using the 3.1 as a neutral vector boson gave predictions

+ "Entities should not be unnecessarily multiplied" (Websters Third New International Dictionary). I am deeply indepted to C.H. Llewellyn Smith for this reference.

for neutral current rates in leptonic and semileptonic weak interactions qualitatively in agreement with low-energy experiments. But:

i) What is the 3.7? If it is another weak boson, why is the $3.7 \to 3.1\,\pi\pi$ coupling so large? Even supposing we accept[17] strong pairwise interactions of W bosons with hadrons,

ii) there are apparently[18] no propagator effects with characteristic masses $\sim$ (3-4) GeV seen in the Fermilab neutral current data. Also

iii) it is difficult, if not impossible, to understand the large photoproduction rate.

Discarding weak bosons (including Higgs scalars on account of their spin), some people were led to postulate new medium-strength interactions, possibly violating parity, or time reversal, or charge conjugation, or some combination of these. We assassinate these models with a deft flick of Occam's razor, and move on to the next sub-section.

1.3.2. Hadronic models

It may be possible to understand the new objects as conventional hadrons. We prefer to think not, and go on to consider models involving new hadronic degrees of freedom. There are two obvious ways of extending SU(3), the usual symmetry group of strong interactions. One is SU(3) $\to$ SU(3) x G, where G is some new symmetry, which can be called "colour". The other is SU(3) $\to$ SU(N): $N \geq 4$, which can be called "charm". The best-known example of the first scheme is the Han-Nambu[19] colour scheme (G = SU(3)); another is the Tati[20] scheme (G = SO(3)). The best-known example of the second scheme is traditional charm[21] (a particular model with N = 4), but other possibilities are open (fancy, gentleness, ... N = 5, 6, ...)+. We will again use Occam's razor to slash the possibilities down to Han-Nambu colour - previously proposed in order to understand the particle spectra, quark binding, and the $\pi^o \to 2\gamma$ decay; and SU(4) charm - previously proposed as the simplest way of constructing calculable weak interaction theories. However, many of our remarks about suppression of radiative decays, production in deep inelastic scattering, the dual diagram selection rules, etc., will apply to all "colour" or "charm" schemes.

Han-Nambu colour

In this scheme[22] the "colour" group is SU(3): the ordinary fractionally charged p, n, λ quarks are each replaced by three integrally charged quarks as in Fig.5. The familiar electromagnetic current $J_\mu^{em} = I_\mu^3 + 1/2Y_\mu$ belonging to an $\underline{8}$ representation of SU(3) now becomes

$$J_\mu^{em} = \underset{\text{(uncoloured)}}{\left(I_\mu^3 + \frac{Y_\mu}{2}\right)} + \underset{\text{(coloured)}}{\left(\tilde{I}_\mu^3 + \frac{\tilde{Y}_\mu}{2}\right)} \tag{1.20}$$

where $\tilde{I}$ and $\tilde{Y}$ are the isospin and hypercharge of the new

+ Other schemes are of course possible, e.g. new heavy quarks which are strongly coupled to each other via vector gluons, but not to the familiar quarks. This would again have a symmetry SU(3) x G, but realized very differently from "colour".

SU(3). The uncoloured and coloured pieces of J_μ^{em} belong to $(\underline{8},\underline{1})$ and $(\underline{1},\underline{8})$ representations of SU(3) x SU(3) colour respectively. Ordinary hadrons are supposed to be colour singlets (mesons in $(\underline{8},\underline{1})$ and $(\underline{1},\underline{1})$ representations, etc.).

The proposal is that the 3.1 and 3.7 are colour octet states. There are at least four ways of assigning them.

I) Both 3.1 and 3.7 are in the same $(\underline{1},\underline{8})$ representation, and are $\tilde{\rho}^o$ and $\tilde{\omega}_8$. $\tilde{I}$ = 1 and 0 octet states. In this case, colour symmetry-breaking is very bad: $\Delta m^2/m^2 \sim 40\%$, and it is not clear why colour is conserved so well as to suppress the hadronic decays by $O(10^4)$. For example, you might expect the traditional $(\underline{8},\underline{1})$ vector mesons to acquire large "coloured" components. Such a large breaking of colour symmetry is not impossible theoretically. At the very least, colour symmetry must be broken electromagnetically, e.g. by mass differences. Weak interactions may also violate colour symmetry, and gauge theories suggest this may be the same order of magnitude as that of electromagnetic breaking. A larger breaking from some semi-strong interaction is also possible.

II) The 3.1 and 3.7 are in different $(\underline{1},\underline{8})$ representations - one probably a radial excitation of the other. In this case there is no reason to suppress the hadronic 3.7 → 3.1 $\pi\pi$ decays as required by formula (1.15).

III) The most natural colour scheme is that the 3.1 and 3.7 are in $(\underline{1},\underline{8})$ and $(\underline{8},\underline{8})$ representations, and are ideally mixed so that the 3.1 has only non-strange quarks, the 3.7 only strange quarks. In this case we can postulate that colour symmetry is broken only in the electromagnetic direction. As the photon is both a U-spin and $\tilde{U}$-spin scalar, only U- and $\tilde{U}$-spin scalar states can couple directly to it. In particular only one combination of $\tilde{\phi}^0$ and $\tilde{\omega}$ is coupled to the photon, and 3.1 and 3.7 are the only states in e^+e^- annihilation before the advent of radial excitations[+]. Quark model universality would then give the couplings

$$\gamma_\rho^{-2} : \gamma_\omega^{-2} : \gamma_\phi^{-2} : \gamma_{3.1}^{-2} : \gamma_{3.7}^{-2} = 9 : 1 : 2 : 8 : 4 \qquad (1.21)$$

and the ratio $\Gamma(3.7 \to e^+e^-)/\Gamma(3.1 \to e^+e^-)$ comes out about right. The ratio $\Gamma(3.1 \to e^+e^-)/\Gamma(\phi \to e^+e^-)$ is within the expected range of accuracy, given the large mass differences, and it is historically unclear whether symmetry predictions should apply to γ_V^{-1} or to $m_V^p \gamma_V^{-1}$ for some power $p \neq 0$. Because the 3.7 is strange and the 3.1 non-strange, the usual quark diagram selection rule suppresses $3.7 \to 3.1\ \pi\pi$ decay because the simplest diagram of Fig. 6 is forbidden. The suppression of $3.1 \to$ hadrons is understood from colour symmetry, but the suppression of $3.1 \to$ $\to \gamma$ + hadrons is not. Naively one would expect no suppression relative to $\omega \to \pi^0 \gamma$, a decay rate for $3.1 \to X\gamma$ of O(10) MeV from Eq. (1.9), and a total decay rate $3.1 \to$ $\to \gamma$ + hadrons of O(10-100) MeV. There are arguments for suppressing these decays to some extent: small overlap of $(q\bar{q})$ bound state radial wave functions, coloured vector meson dominance of the photon; but these seem unlikely to

---

[+] If the colour breaking violates $\tilde{U}$-spin as in some weak interaction models, then there should be four coloured vector states in e^+e^-.

suppress the rates by the necessary factor of $\gtrsim 100$. A problem with colour schemes is that they predict a vast spectrum of other coloured states (pseudoscalars, scalars, ...) which should have masses naively expected to start around the mass of the lowest vector state, i.e. ~ 3 GeV, and could be produced together with ordinary hadrons in e^+e^- annihilation above $q^2 = 10$ or in strong decays of the 3.7. Experimentally, there is a threshold around $q^2 \simeq 15$, but there is no news yet of new states in the strong decays of the 3.7 or of the 4.1. Perhaps the coloured vector mesons have especially low masses because of scheme.

IV) The coloured vector mesons mix with coloured vector gluons. This might be a way of making the coloured vector states the lightest. However, there is still a problem: if the 4.1 is identified as a radially excited coloured state, then it should have strong decays into lower coloured states and ordinary hadrons, e.g. $4.1 \to 3.1\ \pi\pi$, and rumour has it that such decays are not yet seen.

The colour scheme also has a problem not associated with e^+e^- annihilation: Why do the deep inelastic electroproduction structure functions not rise far above those of SLAC at Fermilab energies[23] because of the excitation of coloured states? Calculations[24] suggest this rise should be O(100)%. We return to this point in Lecture 3.

While there is nothing conclusive, the general trend of the data seems against the above schemes[+] as possibilities, and in the next lecture we move on to discuss charm.

[+]It may be possible to avoid these problems with strong colour symmetry-braaking chosen in a special way (N. Marinescu and B. Stech, and F.E. Close and J. Weyers, private communications).

2. THE HIDDEN CHARM HYPOTHESIS

Many years ago[21] it was proposed to add a fourth "charmed" quark c to the conventional p, n, λ in order to build attractive and eventually renormalizable theories of weak and electromagnetic interactions. We will discuss this structure in more detail in Lecture 4. Much of what we say now could apply to other "charm" schemes, but we consider for definiteness the usual model in which the charmed quark has Q = 2/3, B = 1/3, I = 0, S = 0, and a new quantum number C = 1, the conventional quarks having C = 0. The resulting quantum number diagram for quarks is given in Fig. 7. The old and new quarks provide a basic SU(4) symmetry for hadrons. As shown in Fig. 8, each nonet of mesons made of p, n, and λ quarks is now replaced by a <u>16</u>-plet, the new mesons being $c\bar{p}$, $c\bar{n}$ (usually called D), $c\bar{\lambda}$ (usually called F), the antiparticles of these, and a "hidden charm" $c\bar{c}$ state (usually called η_c, ϕ_c, f_c, ...). In general the $c\bar{c}$ states can mix with the $\lambda\bar{\lambda}$ and $1/\sqrt{2}\cdot$ $\cdot(p\bar{p} + n\bar{n})$ states.

The hypothesis[16] is that the 3.1 is esentially a pure $c\bar{c}$ state. The 3.7 is then regarded as a radial excitation or daughter state (cf. the ρ-ρ' pair). The 4.1 structure is seen as a second recurrence, broadened because it is sitting above the threshold for producing charmed meson pairs $D\bar{D}$, $F\bar{F}$, etc.[25]. The large $e^+e^- \to$ hadrons cross-section above E_{cm} = 4 is to be due to charmed particle production[+]. The big problem of this

[+] Other options are open. At one extreme the 3.7 state could lie just above the $D\bar{D}$ threshold, with decays suppressed by the phase space. At the other extreme, the 4.1 could be a mixture of $c\bar{c}$ and uncharmed quarks, the $D\bar{D}$ threshold could lie above it, and the 4.1 could decay just into familiar hadrons.

model is of course the very narrow widths of the 3.1 and 3.7. The usual explanation is to use quark diagrams[26]. Normal allowed decays such as $\phi \to K^+K^-$, $4.1 \to D\bar{D}$, correspond to connected diagrams. Disallowed decays, such as $\phi \to 3\pi$, $3.1 \to 3\pi$, $K\bar{K}$, etc., correspond to disconnected diagrams as in Fig. 4. Such diagrams are supposed to be zero as a first approximation, forbidding 3.1 and 3.7 decays to ordinary hadrons and 3.7 → 3.1 + ordinary hadrons in lowest order. However, as we saw in Lecture 1, the suppression of disallowed ϕ decays is $O(10^2)$, whereas the suppression of disallowed 3.1 decays is $O(10^3$ to $10^4)$. We will discuss later whether this improvement in the quark diagram selection rule can be made plausible.

2.1. Charm spectroscopy

The diagram in SU(4) space for a typical meson 16-plet is shown in Fig. 8. To calculate the masses of these states, we try as a first approximation[21] the lowest order SU(4) breaking in quadratic mass formulae. As in gauge theories, the strong interactions are supposed to be SU(4) invariant apart from quark mass terms:

$$\mathcal{L}_{\text{Break}} = m_p\, \bar{p}p + m_n\, \bar{n}n + m_\lambda\, \bar{\lambda}\lambda + m_c\, \bar{c}c \qquad (2.1)$$

$$= m_o\, (u_o + r\, u_8 + r\, R u_{15}) \qquad (2.2)$$

where $u_\alpha = \bar{q}\lambda_\alpha q$ with the λ_α SU(4) generalizations of the usual SU(3) matrices, e.g.

$$\lambda_{15} = \frac{1}{\sqrt{6}} \begin{bmatrix} 1 & 0 & 0 & 0 \\ 0 & 1 & 0 & 0 \\ 0 & 0 & 1 & 0 \\ 0 & 0 & 0 & -3 \end{bmatrix}$$

The parameters r and R describe the relative magnitudes of quark masses: r is the same as the Gell-Mann Oakes Renner[27] parameter c, probably $\simeq -\sqrt{2}$, and $R = m_c - m_p/m_\lambda - m_p$. The breaking of SU(4) comes just from a 15-plet in lowest order; hence in describing mass splittings for a meson 16 = 1 + 15-plet we need, in addition to R, four parameters corresponding to

$$\underline{1} \times \underline{1} = \underline{1}, \quad \underline{1} \times \underline{15} = \underline{15}, \quad \underline{15} \times \underline{1} = \underline{15}, \quad \underline{15} \times \underline{15} \ni \underline{15}.$$

We introduce these via an effective mass term

$$\mathcal{L}_{\text{Break}} = A(\text{Tr}M)^2 + B\ \text{Tr}(M^2) + C\ \text{Tr}(M\Delta M) + D\ \text{Tr}(M)\text{Tr}(\Delta M) \tag{2.3}$$

where M is the meson 16-plet written in matrix form. The parameter B gives the average meson mass^2; A splits singlet and 15-plet masses; C splits the 15-plet masses; and D gives singlet-15-plet mixing. For the vector mesons a fit to the SU(3) nonet implies

$$B \simeq 0, \qquad D \simeq 0,$$

so that the states separate essentially according to their quark content. We then have

$$m^2_{D*} - m^2_\rho \simeq m^2_{F*} - m^2_{K*} \simeq \frac{1}{2}(m^2_{\phi_c} - m^2_\rho) \simeq R(m^2_{K*} - m^2_\rho) \ . \tag{2.4}$$

For the pseudoscalar mesons the situation is more complicated, but irrespective of the singlet-$\underline{15}$-plet mixing parameters we have

$$m_D^2 - m_\pi^2 \simeq m_F^2 - m_K^2 \simeq R(m_K^2 - m_\pi^2) \tag{2.5}$$

implying, if $m_D \simeq 2$ GeV, that $m_F - m_D \simeq 60$ MeV. However, the status of the I = 0 states is unsatisfactory: the apparent octet structure of the lowest pseudoscalars forces a choice of A, B, C, D which gives substantial $c\bar{c}$ content to the X(958). If this were true, then the decay $3.1 \to X\gamma$ would no longer be forbidden by the quark diagram rule, and the width of the 3.1 would be disastrously large. We return to this problem later: in the meantime we assume the validity of (2.4) and (2.5). Identifying the 3.1 with ϕ_c we see from Eq. (2.4) that $m_{D*}, m_{F*} \simeq 1/\sqrt{2} \cdot m_{\phi_c}$, and we expect the pseudoscalars to be somewhat lighter. Therefore m_D, m_F, m_{D*}, and m_{F*} are all around 2.2 GeV. The position of the apparent threshold in $e^+e^- \to$ hadrons is slightly overestimated, suggesting that the mass formula may only work to O(10)%. However, it seems reasonable that the 3.7 may still be below charm threshold and should be narrow.

We can guess other things about the spectrum by analogy with uncharmed spectroscopy. Thus we might expect Regge trajectories approximately linear in $(\text{mass})^2$: if we identify the 3.7 with the ϕ_c', then it seems experimentally that if we identify ρ' with the 1600 state:

$$m_{\phi_c'}^2 - m_{\phi_c}^2 \simeq 2\,(m_{\rho'}^2 - m_\rho^2)$$

Since the $\rho'(1600)$ is approximately degenerate with the g meson with spin 3, we might expect a similar structure for hidden charmed mesons, with an exchange-degenerate vector/tensor meson trajectory with slope $\sim\frac{1}{2}$ GeV^{-2}:

$$\alpha_{\phi_c, f_c}(m^2) \approx -3.8 + \frac{1}{2} m^2 \qquad (2.6)$$

yielding a hidden charmed tensor meson f_c with mass 3.4 to 3.5 GeV. Applying the mass formula (2.4) to charmed meson trajectories, we get $\alpha'_D \approx \alpha'_F \approx \alpha'_{D*} \approx \alpha'_{F*} \approx 1/\sqrt{2}$ and

$$\alpha_D(m^2) \approx -2.8 + \frac{1}{\sqrt{2}} m^2 \qquad (2.7)$$

if $m_D \approx 2$ GeV[+].

In SU(6) and the real world

$$m_K^2 - m_\pi^2 \approx m_{K*}^2 - m_\rho^2 \approx m_\phi^2 - m_{K*}^2 \quad . \qquad (2.8)$$

If we extend this to SU(8) and charmed mesons,

$$m_D^2 - m_\pi^2 \approx m_{D*}^2 - m_\rho^2 \approx m_{\phi_c}^2 - m_{D*}^2 \approx m_{F*}^2 - m_K^2 \quad . \qquad (2.9)$$

[+] Identifying ρ' with the possible $\omega\pi$ state at 1200 MeV would give a different structure: the tensor f_c would be approximately degenerate with the 3.7, the hidden charmed trajectory would be $\alpha_{\phi c, fc}(m^2) \simeq -1.4 + 1/4\ m^2$, and there would be corresponding modifications to equation (2.7) and the subsequent discussion.

Then for $m_D \sim 2$ GeV we get $m_{D*} - m_D \approx 120$ MeV. Bearing in mind the differences (2.6) and (2.7) in trajectory slopes this estimate may easily be wrong by a factor of 2. If we further assume that the $c\bar{c}$ pseudoscalar η_c is essentially pure, and that from ideal mixing and SU(8)

$$m_{\phi_c}^2 - m_{\eta_c}^2 \approx m_\rho^2 - m_\pi^2 \qquad (2.10)$$

which may also easily be wrong by a factor of 2, then $m_{\phi_c} - m_{\eta_c} \approx 90$ MeV and $m_{\eta_c} \approx 3000$ MeV. The observed masses of SU(3) axial mesons might lead us to expect for the axial vectors D_A, F_A, and $A_c$$^+$:

$$m_{D_A}^2 - m_{D*}^2 \approx m_{D*}^2 - m_D^2,\ m_{A_c} \approx 3.3 \text{ GeV} \qquad (2.11)$$

and analogy with the SU(3) scalars suggests

$$\begin{aligned} m_{D_A} &\gtrsim m_{D_S} \gtrsim m_{D*} \\ m_{A_c} &\gtrsim m_{\varepsilon_c} \gtrsim m_{D_c} \end{aligned} \qquad (2.12)$$

for the scalars D_S, F_S, and ε_c. The resulting very rough particle spectra[++] are given in tables[+++]:

[+] There are two of each, with charge conjugation ±.

[++] Charmed baryons are expected to be more massive (see Ref. 25) and we will not discuss them here.

[+++] Similar spectra arise in the dynamical schemes discussed later in this Lecture.

Meson	Quark content	Spin parity	Mass (MeV)	
D	$c\bar{p}$, $c\bar{n}$	0^-	$m_D = O(2000)$	Charmed mesons
F	$c\bar{\lambda}$	0^-	$m_D + 60?$	
D^*	$c\bar{p}$, $c\bar{n}$	1^-	$m_D + 120?$	
F^*	$c\bar{\lambda}$	1^-	$m_D + 180?$	
D_s	$c\bar{p}$, $c\bar{n}$	0^+	$m^*_D = m_D + 150??$	
F_s	$c\bar{\lambda}$	0^+	$m^*_F = m_D + 200??$	
η_c	$c\bar{c}$	0^-	3000?	Hidden charmed mesons
ϕ_c	$c\bar{c}$	1^-	3100	
ε_c	$c\bar{c}$	0^+	3200??	
A_c	$c\bar{c}$	1^+	3300?	
f_c	$c\bar{c}$	2^+	3500?	
g_c	$c\bar{c}$	3^-	3700?	
ϕ_c	$c\bar{c}$	1^-	3700	

2.2. Electromagnetic decays

The most obvious electromagnetic decays of the ϕ_c are $\phi_c \to e^+e^-$, $\phi_c \to \mu^+\mu^-$. Quark model universality and SU(4) symmetry predict γ-vector meson couplings:

$$\gamma_\rho^{-2} : \gamma_\omega^{-2} : \gamma_\phi^{-2} : \gamma_{\phi_c}^{-2} = 9 : 1 : 2 : 8 \quad . \tag{2.13}$$

We saw in Lecture 1 that from the observed decay widths $\gamma_\phi \approx \gamma_{\phi_c}$, so that the γ-ϕ_c coupling is weaker than in the SU(4) symmetry limit, not a disaster in view of the large mass breaking of SU(4) and the uncertainty in what mass factors to include in broken symmetry calculations. We have already noticed that

$$\frac{\gamma_{\phi_c'}^2}{\gamma_{\phi_c}^2} = O\ (2\text{-}3) = \frac{\gamma_{\rho'}^2}{\gamma_\rho^2}$$

if ρ' is the 1600 MeV state.

The abundance of hidden charmed states below the ϕ_c' mass suggests that the 3.7 may have an abundance of radiative decay cascades: $\phi_c' \to f_c\gamma$, $f_c \to \phi_c\gamma$, and so on. The calculation of these is very model-dependent; however, we can give a very rough estimate of

$$\frac{\Gamma(\phi_c \to \eta_c\gamma)}{\Gamma(\omega \to \pi^o\gamma)} = \left(\frac{g_{\phi_c\eta_c\gamma}}{g_{\omega\pi^o\gamma}}\right)^2 \left(\frac{m_{\phi_c}^2 - m_{\eta_c}^2}{m_\omega^2 - m_\pi^2}\right) \left(\frac{m_\omega}{m_{\phi_c}}\right)^3 \tag{2.14}$$

assuming $g_{\phi_c\eta_c\gamma} \approx g_{\omega\pi^o\gamma}$ and from formula (2.10)

$$m_{\phi_c}^2 - m_{\eta_c}^2 \approx m_\omega^2 - m_\pi^2$$

we get

$$\Gamma(\phi_c \to \eta_c\gamma) = O(10)\ \text{keV} \quad . \tag{2.15}$$

A way of calculating the ratio $g_{\phi_c\eta_c\gamma}/g_{\omega\pi\gamma}$ is given by the axial vector anomaly[25] estimates of the pseudo-scalar $\to \gamma\gamma$ decays which work well for $\pi^o \to \gamma\gamma$ and $\eta \to \gamma\gamma$. The anomaly yields

$$\frac{g_{\eta_c\gamma\gamma}}{g_{\pi^o\gamma\gamma}} = \frac{f_\pi}{f_{\eta_c}} \cdot \frac{4\sqrt{2}}{3} \tag{2.16}$$

were f_π and f_{η_c} are defined by

$$<0|A_\mu^\pi|\pi(p)> = -ip_\mu f_\pi , \quad <0|A_\mu^c|\eta_c(p)> = -ip_\mu f_{\eta_c} ,$$

A_μ^c being the charmed axial current. Dominating the photons by hidden charmed vector mesons gives

$$\frac{g_{\phi_c\eta_c\gamma}}{g_{\omega\pi^o\gamma}} = \frac{4\sqrt{2}}{3}\frac{f_\pi}{f_{\eta_c}}\frac{\gamma_{\phi_c}}{\gamma_\omega} \approx \frac{f_\pi}{f_{\eta_c}} \tag{2.17}$$

if we take γ_{ϕ_c} and γ_ω from experiment. Just as $\gamma_{\phi_c}^{-1}$ is smaller than expected from naive SU(4), and $f_\pi \neq f_K$, it may well be that $f_\pi \neq f_{\eta_c}$. But, if we use Eq. (2.16) and $f_\pi \simeq f_{\eta_c}$, then we find $\Gamma(\eta_c \to \gamma\gamma) \sim 300$ keV: this may well be an overestimate. A similar calculation can be made of the $\varepsilon_c\gamma\gamma$ coupling using the canonical trace anomaly[28]. It yields

$$\frac{\Gamma(\varepsilon_c \to \gamma\gamma)}{\Gamma(\varepsilon \to \gamma\gamma)} = \left(\frac{f_\varepsilon}{f_{\varepsilon_c}}\right)^2 \left(\frac{R_c}{R}\right)^2 \left(\frac{m_{\varepsilon_c}}{m_\varepsilon}\right)^3 \tag{2.18}$$

where $R_c/R = 2/3$ is the charmed quark/non-charmed quark ratio in $e^+e^- \to$ hadrons, and

$$\langle 0|\theta^\mu_\mu|\varepsilon\rangle = f_\varepsilon m_\varepsilon^2 \ , \quad \langle 0|\theta^\mu_\mu|\varepsilon_c\rangle = f^2_{\varepsilon_c} \ m^2_{\varepsilon_c}$$

where θ^μ_μ is the trace of the energy momentum tensor. Putting in the very preliminary experimental estimate[29] $\Gamma(\varepsilon \to 2\gamma) \approx 10$ keV and $f_\varepsilon = f_{\varepsilon_c}$, we get $\Gamma(\varepsilon_c \to 2\gamma) =$ $= O(500)$ keV.

2.3. Dynamical schemes

In Sections 2.1 and 2.2 we saw how very naive considerations built up a picture of the spectrum of charmed particles, and gave a limited picture of electromagnetic decays. Many people have gone further than this and tried to construct dynamical schemes based on asymptotically free gauge theories which give particle spectra, radiative decay widths, and try to explain the amazing success of the quark diagram suppression rule. A common ingredient is that because from Eq. (2.5)

$$R = \frac{m_c - m_p}{m_\lambda - m_p} = \frac{m_D^2 - m_\pi^2}{m_K^2 - m_\pi^2} = O(20) \tag{2.19}$$

then if the strange quark mass $m_\lambda = O(100\text{-}300)$ MeV as we expect, then the charmed quark mass must be in the GeV range.

The first scheme, the "charmonium" picture of Appelquist and Politzer[30], proposed that since the ϕ_c has a large mass, the associated gauge theory running coupling constant $\alpha_s(q^2)$ should be much smaller than, for example, at the ϕ mass. They then proposed a picture in which the ϕ_c is an "orthocharmonium" state (cf. orthopositronium) which decays by the emission of three coloured gluons which transmogrify into hadrons, or a photon which gives e^+e^-, $\mu^+\mu^-$, as indicated in Fig. 9.

To calculate the decays

$$\Gamma(3.1 \to \text{hadrons}) = |M_H|^2 |\psi(0)|^2, \quad \Gamma(3.1 \to \ell^+\ell^-) = |M_L|^2 |\psi(0)|^2 \tag{2.20}$$

where $\psi(0)$ is the orthocharmonium wave function at the origin, and M_H and M_L are hadronic and leptonic decay matrix elements. Independently of the wave function they have

$$\frac{\Gamma(3.1 \to e^+e^-)}{\Gamma(3.1 \to \text{hadrons})} = \frac{|M_L|^2}{|M_H|^2} = \frac{\frac{2}{q}\alpha^2}{\frac{2}{q\pi}(\pi^2 - q)\frac{5}{18}\alpha_s^3} \; . \tag{2.21}$$

Taking this ratio as $\sim 1/10$ from experiment gives $\alpha_s \approx 1/5$ to $1/4$. This is a reasonable value in the context of asymptotic freedom, and fits reasonably with the total $e^+e^- \to$ hadrons cross-section for which the asymptotic freedom formula is

$$R = \frac{\sigma(e^+e^- \to \text{hadrons})}{\sigma(e^+e^- \to \mu^+\mu^-)} = 2\,(1 + \frac{\alpha_s}{\pi} + \dots) \tag{2.22}$$

giving R $\approx$ 2 in the background just outside the resonance peak. We can go further and try to use the orthopositronium wave function at the origin in the formulae (2.20). Then we have

$$\Gamma(3.1 \to e^+e^-) = \frac{2}{q}\,\alpha^2\,(\frac{4}{3}\,\alpha_s)^3\,m_c$$

$$\Gamma(3.1 \to \text{hadrons}) = \frac{2}{q\pi}\,(\pi^2 - q)\frac{5}{18}\,\alpha_s^3(\frac{4}{3}\,\alpha_s)^3\,m_c\ . \tag{2.23}$$

Taking m_c = 2 GeV and $\alpha_s \leq 1/4$ gives $\Gamma_L \leq$ 1 keV and $\Gamma_\eta \leq$ 20 keV: certainly in the right ball-park. The scheme can be extended to the η_c ("paracharmonium") and predicts

$$\frac{\Gamma(\phi_c \to \text{hadrons})}{\Gamma(\eta_c \to \text{hadrons})} = \frac{5}{6}\,\frac{2}{q\pi}\,(\pi^2 - q)\;\alpha_s = O(0.01) \tag{2.24}$$

yielding a $\Gamma(\eta_c \to \text{hadrons})$ of the order of 5-10 MeV. The η_c state is wider in this scheme because it can decay by emitting only two gluons as shown in Fig. 9c. The η_c - ϕ_c mass splitting can also be calculated using hyperfine splitting, getting O(10) MeV.

There are some theoretical questions about this model. Quark binding is supposed to be an infrared phenomenon, and the infrared behaviour of gauge theories is supposed to be very different from Coulomb QED. In particular, why should one use the Coulomb formulae (2.23)? And if one does use the formulae, why should the quark-gluon coupling constant used in the decay be the asymptotically free one? The gluons in the binding and decay are almost real and not highly virtual. Another problem[30] is that in a Coulomb potential if the first two vector states have masses 3.1 and 3.7 GeV, then the continuum starts at 3.9 GeV. However, the basic idea that strong interactions become weaker at large $(\text{mass})^2$ is very interesting.

People have tried to go further by generalizing the Coulomb potential. Gauge theories suggest that at large distances the quark-antiquark potential $\propto |\underline{r}|$. Some people[31,32] take a potential of the form

$$V(\underline{r}) = \frac{A}{|\underline{r}|} + B\,|\underline{r}| \tag{2.25}$$

and insert it into a non-relativistic Schrödinger equation to calculate energy levels. Fitting A, B, and m_c to get the 3.1 and 3.7 vector states they find values of m_c justifying a posteriori the use of a non-relativistic wave equation. If the radii $\langle r \rangle$ of these states are calculated they are found to be of the order of a fermi, and our doubts about the use of the short distance form of the gauge theory are strengthened. Many predictions can be made about the spectrum using the potential (2.25): additional vector states with masses $\simeq$ 4.2 GeV, 4.6 GeV, and so on are predicted, together with a congregation of 0^+, 1^+, and 2^+ levels between 3.4 and 3.5 GeV. It is also possible to calculate the cascade radiative decays in such a model.The 3.7 is expected to have a radiative decay width of the order of (200-300) keV, and similarly for the 0^+, 1^+, and 2^+ states.

Another more intuitive approach is taken by the Harvard group[32] who have a branching ratio for radiative decays of the 3.7 of order 1/3, and predict that the 0^+, 1^+, and 2^+ states should decay dominantly into $3.1+\gamma$. It was remarked in Lecture 1 that probably not all the 3.7 decays into the 3.1 go via the emission of two pions. But it remains to be seen what else is going on.

One point of interest is the $3.7 \rightarrow \eta_c \gamma$ decay: this distinctive channel is present in some schemes but strongly suppressed in others. In any case radiative decays should be a rich area of experimental study.

Although the dynamical calculations to date confirm the naive picture of Section 2.1 in general terms, specific details do not yet seem reliable, there being several differences between the schemes of Refs. 31 and 32.

2.4. A simple mixing scheme

In this section we discuss a simple-minded attempt[33] to understand the magnitudes of various couplings of the 3.1 and 3.7. We recall from Lecture 1 the estimates of various suppressions:

$$\frac{\Gamma(3.1 \rightarrow \text{hadrons})}{\Gamma(\text{typical hadron})} = O(10^{-3} \text{ to } 10^{-4}), \quad \frac{\Gamma(3.1 \rightarrow \pi^0\gamma)}{\Gamma(\text{typical radiative decay})} \lesssim O(10^{-4})$$

$$\frac{\sigma(pp \to 3.1 + \text{any})|_{29\text{ GeV}}}{\sigma(pp \to 3.1 + \text{any})|_{250\text{GeV}}} = O(10^{-3}),\ \left(\frac{g_{3.7,3.1\pi\pi}}{g_{\rho'\rho\ \pi\pi}}\right)^2 = O(10^{-2})$$

$$\frac{\sigma_{tot}(3.1N)}{\sigma_{tot}(\phi N)} \simeq O(10^{-1}),\ \frac{\sigma(\gamma N \to 3.1N)|_{18\text{ GeV}}}{\sigma(\gamma N \to 3.1N)|_{80\text{ GeV}}} \simeq O(10^{-1})\ . \tag{2.26}$$

By way of comparison, the suppressions of cross-sections and rates involving ϕ were $O(10^{-2})$. Can we understand simply these various numbers? The lowest order quark diagrams, such as those of Fig. 4, are taken as being disallowed. The first allowed diagrams are single loops like that for $\phi_c \to K^+K^-$ decay shown in Fig. 10a. We can rewrite this suggestively by twisting the loop to get Fig. 10b. We now make a duality transformation on the part of the diagram to the right of the twists, similar to those which identify the resonance and Regge contributions to elastic scattering amplitudes. We then obtain Fig. 11. That the two diagrams of Figs. 10 and 11 are the same is a dynamical assumption. If it is true, then we see that all couplings of $c\bar{c}$ states to states made up of uncharmed quarks take place via an intermediate resonant $\lambda\bar{\lambda}$ or $1/\sqrt{2}\ (p\bar{p} + n\bar{n})$ state, such as ϕ or ω in the case of ϕ_c. Let us now make the assumption that in <u>order of magnitude</u> the transitions $(c\bar{c}) \leftrightarrow (\lambda\bar{\lambda})$, $(\lambda\bar{\lambda}) \leftrightarrow (p\bar{p} + n\bar{n})$, and $(c\bar{c}) \leftrightarrow 1/\sqrt{2}\ (p\bar{p} + n\bar{n})$ are more or less SU(4) symmetric. Possible models for this mixing are the dual model diagram of Fig. 11, the field theoreticians' diagram of Fig. 12a, and the hadron phenomenologists' Fig. 12b. (Notice that we are using the gluon intermediate state differently from Appelquist and Politzer[29]: our gluons do not metamorphose directly into finalstate physical particles.) We further assume that only the lowest vector meson states contribute+. We can now represent the scheme via a $(\text{mass})^2$ matrix

$$\begin{bmatrix} C + \varepsilon & \varepsilon & \varepsilon \\ \varepsilon & \Lambda + \varepsilon & \varepsilon \\ \varepsilon & \varepsilon & N + \varepsilon \end{bmatrix} \tag{2.27}$$

where the matrix is written in a pure $(c\bar{c})$, $(\lambda\bar{\lambda})$, $(N\bar{N}) \equiv 1/\sqrt{2}\,(p\bar{p} + n\bar{n})$ basis, and C, Λ, N are the $(\text{mass})^2$ of the unmixed states, so that in units of GeV^2, $C \simeq 10$, $\Lambda \simeq 1$, $N \simeq 1/2$ for the vector mesons. The quantities ε are supposed to be equal only in order of magnitude. Let us suppose that ε is small in the case of vector mesons. Then diagonalizing the matrix (2.27) we get eigenvectors

$$\begin{array}{ccccc} (c\bar{c}) & + & \frac{\varepsilon}{C-\Lambda}(\lambda\bar{\lambda}) & + & \frac{\varepsilon}{C-N}(N\bar{N}) \\ \frac{\varepsilon}{\Lambda-C}(c\bar{c}) & + & (\lambda\bar{\lambda}) & + & \frac{\varepsilon}{\Lambda-N}(N\bar{N}) \\ \frac{\varepsilon}{N-C}(c\bar{c}) & + & \frac{\varepsilon}{N-\Lambda}(\lambda\bar{\lambda}) & + & \varepsilon(N\bar{N}) \end{array} \tag{2.28}$$

with eigenvalues respectively

$$C + \varepsilon, \quad \Lambda + \varepsilon, \quad N + \varepsilon \quad . \tag{2.29}$$

+This might be due, for example, to the small overlap of radial wave functions in Fig.12a, or to the observed suppression of $\rho' \to 2\pi$ decay, where 2π is expected to be an important contribution to Fig. 12b.

If we look at the SU(3) subsector of (2.28) and (2.29) we see that the ω mass shift is $O(\varepsilon)$. Assuming the ρ and unmixed ω were degenerate we get

$$\varepsilon \approx O\,(m_\omega^2 - m_\rho^2) \approx O\,(\tfrac{1}{30}) \qquad (2.30)$$

in units of GeV^2. We can also estimate ε from the observed suppression of $\phi \to \rho\pi$, which in lowest order takes place only via mixing. We find

$$\frac{g_{\phi\rho\pi}}{g_{\omega\rho\pi}} = O\,(\frac{\varepsilon}{\Lambda - N})$$

giving

$$\varepsilon = O\,(\tfrac{1}{20})\ . \qquad (2.31)$$

The estimates (2.30) and (2.31) agree in order of magnitude, which is all we need: henceforward we take $\varepsilon =$ $= O(1/10)$ for the vector meson system.

Going to the ϕ_C, we see from formulae (2.28) that its admixture of $(\lambda\bar{\lambda})$ and $(N\bar{N})$ is $O[\varepsilon/(C-\Lambda)] = O(\varepsilon/10) =$ $= O(1/100)$[+]. Disallowed amplitudes involving the ϕ_C are suppressed by $O(10^{-2})$, and rates and cross-sections by $O(10^{-4})$. This is consistent with the suppression (2.26) of $3.1 \to$ hadrons and $3.1 \to \pi^0\gamma$. Presumably the production of ϕ_C will also be suppressed by $O(10^{-4})$ if there are no charmed particles in the final state corresponding to the

+ Because C/Λ and C/N are smaller for daugther states, these could in principle be less ideally mixed.

disallowed diagrams (Fig. 13a) and the allowed diagrams (Fig. 13b). Indeed it is persistently rumoured to be well known that in the case of the ϕ:

$$\frac{\sigma(pp \to \phi + \text{non-strange hadrons})}{\sigma(pp \to \phi + \text{any})} = O\ (10^{-1} \text{ to } 10^{-2})\ .$$

From the mass formulae, the MIT-BNL experiment is below the threshold for associated production of ϕ_c and charmed particles, whereas the Fermilab experiment is above. Hence we might expect a rise of 10^2 to 10^4 in the cross-section. Furthermore, at least 99% of the final states at the higher energy should contain charmed particle pairs! Thus a clean way of looking for charm in hadronic collisions may be to trigger on production of the 3.1 state. Later on we will discuss ϕ_c photoproduction using Regge analogues of the twisted loops of Figs. 10 and 11.

We can extend the scheme to other meson multiplets: in general the ε parameter of formulae (2.26) may vary between multiplets, and be of order 1 in some cases. Consider, for example, the pseudoscalars: from the complete absence of ideal mixing at the SU(3) level we conclude $\varepsilon_0 - = O(1)$. But even then $\varepsilon/(C - \Lambda)$, $\varepsilon/(C - N) = O(1/10)$, the X(958) contains only $O(10^{-1})$ $(c\bar{c})$ pairs, and similarly the η_c is ideally mixed to $O(10^{-1})$. In particular its total hadronic decay width is just suppressed by $O(10^{-2})$ and its decay width should be O(few) MeV. The amplitude for $\phi_c \to X\gamma$ should then be suppressed by O(1/10) because of the mixing of the X in Fig. 14a, and $\Gamma(\phi_c \to X\gamma)$ by $O(10^{-2})$. The resulting decay rate is still rather large, but may be reduced by the necessity of ϕ_c dominating the final state γ or some other breaking of SU(4). It is clear, though, that in this scheme we expect $\Gamma(\phi_c \to X\gamma) \gg$ $\gg \Gamma(\phi_c \to \pi^o\gamma)$ ++. Further, because the η_c has $O(1/10)(\lambda\bar{\lambda})$ and $(N\bar{N})$ pairs, we expect its production rate below the associated charmed particle production threshold to be suppressed only by $O(10^{-2})$, and so perhaps

$$\frac{\sigma(pp \to \eta_c + \text{any})}{\sigma(pp \to \phi_c + \text{any})} \gg 1$$

at low energies.

A similar analysis applies to the scalar multiplet, which is also apparently nowhere near ideally mixed at the SU(3) level, implying similar predictions for the ε_c width, production rate, and so on. It also predicts that the amplitude for $3.7 \to 3.1\ \pi^+\pi^-$, assumed to proceed via $3.7 \to 3.1\ \varepsilon$, should only be suppressed by $O(10^{-1})$ because of the ε impurity in Fig. 14b, so that $\Gamma(3.7 \to 3.1\ \pi^+\pi^-)$ is suppressed by $O(10^{-2})$ as seen experimentally (formulae (2.26)).

The same basic idea can also be applied to σ_{total} $(\phi_c N)$. At high energies this proceeds via Pomeron exchange, which in the dual model has the lowest order contribution shown in Fig. 15a. Making a double duality transformation above and below the twists gives Fig. 15b. The Pomeron propagator in the angular momentum plane then contains a factor $1/J - \alpha_P(0)$ with $\alpha_P(0) \approx 1$ fro the loop, and $1/J - \alpha_T(0)$ from the stalks connecting it to the ex-

++As for $\Gamma(\phi_c \to \eta\gamma)$, usual analyses suggest that η is $O(10\%)$ SU(3) singlet, of which some unknown fraction is now expected to be $c\bar{c}$. A suppression of $\Gamma(\phi_c \to \eta\gamma)$ to a rate between that of $\Gamma(\phi_c \to \pi^o\gamma)$ and $\Gamma(\phi_c \to X\gamma)$ might seem not unreasonable.

ternal particles where α_T is the appropriate leading tensor meson trajectory. The Pomeron couplings are supposed otherwise to be SU(4) symmetric. This scheme then gives for the ratios of residues of the Pomeron pole in vector meson-nucleon scattering:

$$\sigma_{total}(\rho N) : \sigma_{total}(\phi N) : \sigma_{total}(\phi_c N)$$
$$= \frac{1}{1-\alpha_f(0)} : \frac{1}{1-\alpha_{f'}(0)} : \frac{1}{1-\alpha_{f_c}(0)} \approx 2 : 1 : \frac{1}{5} \tag{2.32}$$

if we believe the trajectory (2.6). This estimate works very well[34] for the ratios of πN, KN, and ϕN cross-sections giving $\sigma_{total}(\pi N) : \sigma_{total}(KN) : \sigma_{total}(\phi N) = 2 : 3/2 : 1$, to be compared with $\sigma_{total}(\pi N) \approx 23$ mb, $\sigma_{total}(KN) \approx 17$ mb, $\sigma_{total}(\phi N) \approx 13$ mb. It is also not grossly inconsistent with preliminary data on $\sigma_{total}(\phi_c N)$. It is worth noting that in Fig. 15 the Pomeron in $\phi_c N$ scattering is the shadow of charmed particle production+, and that the threshold for production of $D + \bar{D} + N$ in the intermediate state is $E \sim 12$ GeV. Hence it is not surprising that the photoproduction cross-section should increase less rapidly than the cross-section for $pp \to \to \phi_c$ + anything between 18 and 80 GeV.

Thus this simple mixing scheme is consistent with the observed suppressions (2.26). At any rate, the narrow widths of the 3.1 and the 3.7 may not be completely unthinkable.

+ The lowest order dual Pomeron of Fig. 15 is probably hopelessly naive, but a similar thing happens in the multiperipheral model. We return to this point in Lecture 4.

3. e^+e^- ANNIHILATION

We now move on to a discussion of the $e^+e^- \to$ hadrons data and review possible models. Perhaps we will justify the prejudice that the approximate constancy of the total cross-section reflects the crossing of the charmed particle threshold.

3.1. Naive theoretical expectations

Before the advent of the data from CEA[35] and SPEAR[36], the expectation of theoreticians was that the $e^+e^- \to$ hadrons cross-section should scale. The reason was that deep inelastic electroproduction and neutrino production data had suggested that current interactions at large $|q^2|$ had no intrinsic scale. Since a cross-section has the dimensions of $(\text{length})^2$ or $(\text{momentum})^{-2}$, this led to the prediction[+]

$$\sigma(e^+e^- \to \text{hadrons}) \propto \frac{1}{q^2} \quad . \tag{3.1}$$

Applying the same arguments to the inclusive cross-section gave

$$q^2 \frac{d\sigma}{dx}(e^+e^- \to \pi + \text{any}) \to f(x) \;:\; x = \frac{2E}{\sqrt{q^2}} \tag{3.2}$$

so that $(1/\sigma)(d\sigma/dx)$ should be independent of q^2. These predictions were all asymptotic, but experience with deep inelastic scattering, where scaling set in for $|q^2| > 1$-2

[+] The reaction was assumed to proceed predominantly via one-photon exchange.

GeV^2, suggested that data in the CEA-SPEAR I range should exhibit the scaling (3.1) and (3.2). The most popular model for scaling was the parton model, with the photon coupling to point-like objects at large q^2, giving

$$R = \frac{\sigma(e^+e^- \to \text{hadrons})}{\sigma(e^+e^- \to \mu^+\mu^-)} = \sum_{\text{partons}} Q^2 \tag{3.3}$$

corresponding to the diagram in Fig. 16a.

Deep inelastic data suggested the partons were quarks: spectroscopy, quark binding arguments and the $\pi^o \to 2\gamma$ decay rate suggested the Gell-Mann/Zweig colour model, with three identical quarks of each type p, n, λ, and R = 2. The inclusive hadronic cross-section then came from partons decaying into a "cigar" in momentum space oriented along the parton-antiparton separation axis and with limited transverse momenta, as shown in Fig. 17.

The distribution of hadrons in any particular event should then resemble that in a typical high-energy hadronic collision with a central plateau and so on, but with an angular distribution

$$\frac{d\sigma}{d(\cos\theta)} \propto (1 + \cos^2\theta) \tag{3.4}$$

resulting from the spin 1/2 of the partons. The model also expects

$$\frac{d\sigma}{dx}(\pi^+) = \frac{d\sigma}{dx}(\pi^-) = \frac{d\sigma}{dx}(\pi^o) \tag{3.5}$$

and that pions should be the dominant species of particle

produced: hence the charged energy fraction $E_{ch}/E_{tot} \simeq 2/3$. The data of Frascati[37] in the range $2 < q^2 < 10$ suggested that R was indeed O(2) and had the expected charged/neutral energy ratio. But the CEA and SPEAR data completely contradicted the naive expectations (3.1) to (3.4) in the range $10 < q^2 < 25$.

3.2. The CEA[35] and SPEAR[36] data

An almost complete summary of these, with the exception of the latest SPEAR data on R, is given in the Proceedings of last year's London Conference[36]. Let us just summarize the main points:

i) QED works for leptons with cut-off orders of magnitude larger than the q^2 accessible in $e^+e^- \to$ hadrons. Hence weird happenings in $e^+e^- \to$ hadrons cannot be blamed on QED.

ii) The e^+e^- cross-section is large: expressed in terms of R; it rises from ~ 2.5 at $q^2 \simeq 10$ to ~ 5 at $q^2 \simeq 25$, with a (200-300) MeV wide peak at $q^2 \simeq 20$ rising to $R \simeq 6$.

iii) For $x = 2E/\sqrt{q^2} \gtrsim 0.5$, the inclusive charged hadronic distribution is small, and consistent with $q^2 d\sigma/dx$ scaling. For $x \lesssim 0.5$, the distribution is large and rises rapidly with q^2: scaling of $(1/\sigma)(d\sigma/dx)$ is nowhere a good approximation. The average charged particle momentum is roughly constant at ~ 500 MeV.

iv) The angular distribution of charged tracks is consistent with isotropy for $10 \leq q^2 \leq 25$, for $|\cos\theta| \lesssim 0.6$.

v) The mean charged particle multiplicity rises relatively slowly with q^2, with $<n_{ch}> \simeq 4$, and shows no obvious structure.

vi) As a result of (iii) and (v) the fraction E_{ch}/E_{tot} of the total energy carried by charged particles is relatively small, falling to around 0.5 at $q^2 \simeq 25$ - the "energy crisis".

vii) At low momenta $<n_{\pi^\pm}> : <n_{K^\pm}> : <n_p> \simeq 100 : 10 : 1$. As the momentum is increased, $n_{K^\pm}/n_{\pi^\pm}$ increases and is of the order of 1/3 for $x \gtrsim 0.5$ at $\sqrt{q^2} = 4.8$.

3.3. Possible theoretical explanations

There are four possible mechanisms for trying to explain the experimental data:

a) The "hadronic" events may not in fact be due to the production of hadrons alone, the most likely possibibility being the production by the one-photon mechanism of a heavy lepton-antilepton pair which then decay leptonically or semi-leptonically as shown in Fig. 18. This mechanism seems to be ruled out by the scarcity of (μe) coincidences seen at SPEAR, which are all consistent with other sources.

b) Hadrons or leptons may be produced by the two-photon mechanism of Fig. 19. However, by looking at the $e^\pm$ passing through tagging counters near the beam pipes, it was possible to estimate these as small, and the published data should have $\lesssim$ 6% contamination from 2γ processes.

c) The hadrons may indeed be preduced by the one-photon mechanism, but the simple quark parton model has to be modified. We will discuss this possibility in a moment.

d) The hadrons may be produced by some "no-photon" mechanism, e.g. by a direct or crossed channel weak interaction. We shall return later to this possibility.

3.4. One-photon mechanisms

Approaches in the context of one-photon exchange can be grouped into parton and non-parton models. Parton models may either

a) keep the same quark partons, but postulate a very slow approach to scaling, or

b) keep the same quark partons, but give them interactions or form factors violating scaling, or

c) add in new partons, e.g. colour or charm.

First let us discuss[24] the approach to scaling in the parton model. If one took the naive parton model literally, and calculated Fig. 16a in the usual way, then the approach to a constant R would be $O(q^{-4})$ for spin 1/2 partons. In more sophisticated covariant parton models the approach is slower, as $(q^2)^{-p}$ with $p < 1$. The slowest approach is that found in an asymptotically free gauge theory. For three-coloured Gell-Mann/Zweig quarks

$$R = 2\ (1 + \frac{\alpha_s}{\pi} + \dots) \tag{3.6}$$

where $\alpha_s \approx 4\pi/[9 \ln (q^2/\mu^2)]$ with μ an a priori unknown mass parameter. The α_s correction terms in (3.6) correspond to the gluon exchange graphs of Fig. 16b, with an effective gluon coupling α_s. Choosing μ^2 to lie in the typical hadronic range $1/2 < \mu^2 < 2$ yields values of R which are ≤ 3 for $q^2 \geq 5$, as shown in Fig. 20. In fact the data on R for $2.6 \leq \sqrt{q^2} \leq 3.4$ are clearly just in the range expected on the basis of asymptotic freedom.

Many people have suggested other reasons for a slow approach to scaling in e^+e^-. For example:

i) There are strong direct channel resonances with $q^2 > O(\rho, \omega, \phi, \rho', \ldots)$ and scaling can only be expected for q^2 much greater than these values, and hence slower than for $q^2 < 0$;

ii) Only one partial wave contributes to e^+e^- annihilation, this has fewer states than the partial waves in electroproduction, and hence the statistical conspiracy necessary for scaling may be slower in turning up;

iii) There may be threshold effects for the production of massive conventional particles, e.g. $B\bar{B}$, which may delay scaling.

A counter-example to the first two arguments is given[24] by a simple dual model. Following a long tradition[38] we construct a phenomenological current model by replacing external currents by fictitious "leptons" as shown in Fig. 21, taking usual moving trajectories in the hadron-hadron channels, and fixed singularities in the lepton-lepton and lepton-hadron channels. This method has been applied[38] in order to get amplitudes for deep inelastic scattering. Applied to e^+e^- annihilation we have a dual four-point function with a fixed lepton-lepton trajectory $\alpha = K$ and a moving lepton-antilepton trajectory $\alpha \simeq 1/2 + \alpha' q^2$ giving direct channel vector meson poles. This is proportional to

$$\frac{\Gamma(-\frac{1}{2} - \alpha' q^2)}{\Gamma(-K-\frac{1}{2} - \alpha' q^2)} \approx (-\alpha' q^2)^K \left(1 + \frac{K(K+2)}{2\alpha' q^2} + \ldots\right) \tag{3.7}$$

if we use (3.7) as a model for the vacuum polarization $\Pi(q^2)$ and recall that $\sigma(q^2) \propto 1/q^2 \, \mathrm{Im}\Pi(q^2)$ then

$$\frac{R(q^2)}{R(\infty)} = 1 + O\left(\frac{1}{\alpha' q^2}\right) + \ldots \tag{3.8}$$

with α' a normal trajectory slope $\approx 1 \, \mathrm{GeV}^{-2}$. Strict belief in the formula (3.8) would entail $K = 0$, but this may be pushing things too far. Thus we have a rapid approach to scaling independent of whether $q^2 > 0$ or $q^2 < 0$ despite all the direct channel poles and the presence of just one partial wave.

As for the $B\bar{B}$ threshold possibility, SPEAR[36] data indicate that in regions where they have been separated $\langle n_{\bar{p}} \rangle / \langle n_{\pi^+} \rangle \simeq O(1)\%$ essentially independently of energy, so $B\bar{B}$ production seems unlikely to be a large source of scaling violations.

But what of the inclusive distribution $q^2 d\sigma/dx d(\cos\theta)$? The dilemma is that at values of q^2 where R may scale, the inclusive distribution $q^2 d\sigma/dx$ does not scale, and the particles are isotropic rather than having a $(1 + \cos^2\theta)$ angular distribution. This may not be too serious: for scaling at fixed x near 1 it is presumably necessary that the missing $(\mathrm{mass})^2$ m_X^2 be large, bigger than 3 corresponding to $m_X \approx m_g$, for example. Since $m_X^2 \approx (1 - x)\, q^2$, we see that for scaling for $q^2 \approx 9$ we need $x \lesssim 2/3$. Experimentally $q^2 d\sigma/dx$ nevertheless does scale for large x - but even if $q^2 d\sigma/dx$ does scale, then the double differential cross-section $d\sigma/dxd(\cos\theta)$ is probably more sensitive to non-scaling terms, and may well not be $\propto (1+\cos^2\theta)$. Other kinematic effects are present in the parton model at small x: because[24] of the finite width of the hadronic cigar, the inclusive distribution gets substantial corrections at small x:

$$q^2 x \frac{d\sigma}{dx} \approx f_o(x) + \frac{1}{x^2 q^2} f_1(x) + O(\frac{1}{x^4 q^4}) \tag{3.9}$$

where both $f_o(x)$ and $f_1(x)$ are regular as $x \to 0$. These $1/x^2q^2$ terms also violate the expected $(1+\cos^2\theta)$ angular distribution. These effects may be substantial at small x, particularly if $\langle p_T\rangle \approx$ twice the average value of p_T seen in hadronic reactions, as suggested by theoretical models of the "shrinking photon" supposed to cause broadening of p_T and t distributions in deep inelastic electroproduction compared with photoproduction.

We conclude that without some sophistry it is probable that parton models should approximately scale in σ_{tot} for $q^2 \approx 10$, but not necessarily exhibit parton-like features in the inclusive distributions. In this case the rise in R above $q^2 \approx 12$ would come either from giving partons structure, or from postulating new types of partons whose threshold is being passed in the CEA-SPEAR I energy range.

Various authors[39] have proposed that partons may have structure with electromagnetic form factors

$$\bar{p}(k+q)[\gamma_\mu F(q^2) + \sigma_{\mu\nu}q^\nu G(q^2)]p(k) \tag{3.10}$$

either with $F'(0) \neq 0$ giving an electric charge radius, or with $G(0) \neq 0$ giving a magnetic moment. The familiar parton model diagrams are drawn for electroproduction and annihilation as in Fig. 16a. A simple parton charge radius is not able to fit the scaling in deep inelastic and the lack of scaling in e^+e^- simultaneously as it has for the deep inelastic structure function $F_2(x)$, and the annihilation ratio R:

$$F_2(x) \to F_2(x)(1-2a|q^2|), \quad R \to R(1 + 2a\, q^2) \tag{3.11}$$

for the same parameter a. It is possible to reconcile approximate scaling in electroproduction and violations in annihilation by combining charge radii and magnetic moments, but such schemes seem somewhat ad hoc. Also, a simple multiplicative violation of scaling as in (3.11) is not a good model for the apparent small scaling violations in electroproduction. Kogut[40] has suggested, on the other hand, that the renormalization group effective coupling constant α_s may tend to zero for a range in q^2, as in asymptotic freedom (3.6), but that then as yet unknown dynamics drive it away again as q^2 increases, and the correction terms of Fig. 16b in (3.6) start growing. Some of the same remarks about lack of motivation apply to this idea as to the parton form factors.

What of non-parton analysis of $e^+e^- \to$ hadrons in the one-photon approximation? The most interesting of these are probably the various statistical, thermodynamic, and hydrodynamic approaches. Characteristically, these models do not predict the behaviour with energy of the total cross-section, but make predictions for the normalized inclusive distribution $(1/\sigma)(d\sigma/dx)$ and the multiplicity. The main features are:

Statistical[41] models - final-state hadrons populate uniformly all available N-particle phase space (relativistic or non-relativistic).

Thermodynamic[41] models - the spectrum of hadronic states is bootstrapped by requiring the level density of each system to be the same as its constituents, and all exclusive states contribute with equal statistical weight to the inclusive cross-section.

Hydrodynamic[42] models - a blob of very hot and dense pre-matter is formed by the virtual photon in a small volume and then expands and cools according to some hydrodynamical equation of state. When the density falls to typical hadronic values, the matter condenses out into normal hadrons, mainly pions. Opinions differ about the appropriate initial volume and the right equation of state.

Common features of all these models are

$$\frac{1}{\sigma_{tot}} \frac{d\sigma}{dx} \downarrow \text{ at large } x, \qquad \uparrow \text{ at small } x \text{ as } q^2 \uparrow \qquad (3.12)$$

$$\langle n_{ch} \rangle \approx (q^2)^p \;:\; 0 < p < \tfrac{1}{2} \qquad (3.13)$$

and

$$E_{\pi^+} = E_{\pi^0} = E_{\pi^-} \qquad (3.14)$$

The first two of these predictions agree qualitatively with the data: prediction (3.13) falls foul of the "energy crisis". Forgetting this for the moment, we may draw the lesson that at present energies in e^+e^- machines the phase space available to the hadrons may be too small for characteristic parton structure to show up, and qualitative considerations may work well. We should remember[43] that at S ≃ 25 in purely hadronic reactions the inclusive pion distribution is essentially isotropic in the centre of mass and rising rapidly near x = 0, while the total cross-section is already comfortably constant.

3.5. No-photon mechanisms

Motivated by the apparent absence of a swasivious one-photon explanation for the large $e^+e^- \to$ hadrons cross-section, various people discussed new interactions to explain the data. These could be characterized either by extra exchanges in the s-channel such as an intermediate vector boson W[44] or a scalar Higgs meson Φ (Fig. 22a) or[45] by the exchange in the cross-channel of a "hadrolepton" X (Fig. 22b). If the W, Φ, or X mass is sufficiently large, then all these interactions can be Fierz-transformed and written using a phenomenological Lagrangian

$$L_{eff} = \sum_i g_i \, (\bar{L}\gamma_i L) \, (\bar{q}\gamma_i q) \qquad (3.15)$$

where the γ_i are the usual tensors S, P, V, A, and T. An amusing aspect[45] of such schemes is that the non-vector contributions depend in general on the transverse polarization of the colliding beams: for scalars and pseudoscalars

$$\sigma^{S,P}_{pol.} = (1 \pm p^2)\, \sigma^{S,P}_{unpol.} \qquad (3.16)$$

where p is the transverse beam polarization. Theoretically the beams in e^+e^- rings should become transverse polarized while they are circulating, though this has apparently not yet been established by the machine people. They apparently see no time-dependence of the cross-section as would be implied by formula (3.16).

It is also very difficult in the W boson case[45] to reconcile the magnitudes of W couplings necessary to fit the observed violation of scaling in $e^+e^- \rightarrow$ hadrons with the known success of scaling in deep inelastic scattering and the magnitude of purely leptonic weak neutral currents. In general such models must be careful not to contradict other data as well, such as the data on $\mu^\pm$ production and $e^\pm$ electroproduction cross-section differences which are known experimentally to be very small. We refer to Pati and Salam[46] for details.

All the new interactions proposed for the $e^+e^- \rightarrow$ hadrons data, while based on pre-existing theoretical ideas, are much larger than had been expected. Hence we administer them a coup de grâce with Occam's razor, and return to less unconventional models.

3.6. Models with new partons

In Section 3.4 we argued that a simple quark parton model fitted $\sigma_{tot}(e^+e^- \rightarrow \text{hadrons})$ as well as could be expected for $q^2 \leq 12$, and that the violations of parton model predictions for the inclusive distribution should not be worried about too seriously. On the other hand, we argued in Section 3.4 also that simple quark partons could not fit the rise in R from $q^2 = 12$ to $q^2 = 25$, and motivated later in Sections 3.4 and 3.5 the prejudice that there was no good alternative to the simple quark parton scheme for e^+e^- annihilation. Accordingly we now consider adding new degrees of freedom to the quark scheme.

In Lecture 1 we distinguished two possible ways of extending the symmetry group SU(3) : SU(3) $\to$ SU(3) x G or generalized "colour", and SU(3) $\to$ SU(N), $N \geq 4$ or generalized "charm". Which of these is more natural in the light of the e^+e^- and electroproduction data?

Consider first a generalized "colour" scheme, in which each conventional quark q^a is "spectrum analysed" into N coloured quarks q_i^a. They have the advantage that a lot of "coloured" hadronic decays are radiative, which may be useful for the "energy crisis". If we denote the usual threefold quark index by a, and the N-fold colour index by i, then

$$\sum_{i=1}^{N} (Q_i^a)^2 = \sum_{i=1}^{N} (Q_i^a - Q^a)^2 + 2Q^a \sum_{i=1}^{N} (Q_i^a - Q^a) + N(Q^a)^2 \qquad (3.17)$$

and in these models $\sum_{i=1}^{N} Q_i^a = NQ^a$, so that the second term in equation (3.17) vanishes.

Below "colour" threshold $R = N\sum_a (Q^a)^2$ and formula (3.17) shows that it rises to $\sum_{a,i} (Q_i^a)^2 > N\sum_a Q^{a^2}$ above the "colour" threshold. But by the same token, the deep inelastic electroproduction cross-section

$$\propto N \sum_{a} (Q^{a})^{2} F^{a}(x) \tag{3.18}$$

(where $F^{a}(x)$ is the distribution of quarks of type a with fraction $x = q^2/2\nu$ of the target's longitudinal momentum) will also rise, becoming

$$\sum_{a,i} (Q_i^a)^2 F_i^a(x) = \sum_{a,i} (Q_i^a)^2 F^a(x) > N\sum_{a} (Q^a)^2 F^a(x) . \tag{3.19}$$

(We assumed the target was colour symmetric so that the $F_i^a(x)$ were equal for fixed a but varying i.) In general this increase is O(100)% for all values of x. For example, in the Han-Nambu model[19] G = SU(3) is the colour group, and conventional distributions of p and n quarks in the proton and neutron give the following[24] rises in F_2^{ep} and F_2^{en}:

	Rise in F_2^{ep} (%)	Rise in F_2^{en} (%)
$x \approx 1$	50	200
$x \approx 1/3$	67	100
$x \approx 0$	100	100

These increases should presumably appear in electroproduction at values of ν and q^2 comparable to the 20 GeV^2

where e^+e^- scaling is violated. These effects are not apparently seen in the Fermilab[47] muon electroproduction data. What about a generalized "charm" scheme in which we add in new quarks $q^{a'}$? There is no natural resolution of the energy crisis. However, these quarks are expected to show up in the proton and neutron only at small $x \lesssim 0.1$, the so-called "sea" where valence (p and n) and non-valence ($\bar{p}$, $\bar{n}$, λ, $\bar{\lambda}$) quarks are supposed to be roughly equally abundant. In this case electroproduction data should only rise at small x if at all, which is consistent with experiment (we return to this point in Lecture 4). The main features of the above remarks are of course independent of the particular "colour" or "charm" scheme.

In Fig. 20 we have compared the SPEAR I data with simple quark model predictions ($R = 2$, $3\frac{1}{3}$, 4 for Gell-Mann/Zweig, charmed, and Han-Nambu, respectively), the asymptotic freedom version of the Gell-Mann/Zweig model

$$R = 2\left(1 + \frac{4}{9\ \ln\left(\frac{q^2}{\mu_1^2}\right)} + \dots\right) \tag{3.20}$$

where $\frac{1}{2} < \mu_1^2 < 2$, and the asymptotic freedom version of the conventional four-quark charm scheme

$$R = \frac{10}{3}\left(1 + \frac{12}{25\ \ln\left(\frac{q^2}{\mu_2^2}\right)} + \dots\right) \tag{3.21}$$

with $1 < \mu_2^2 < 4$. Both the schemes are of the order of magnitude of the present data - detailed comparison is meaningless in the apparently resonance-dominated region of q^2 seen at present.

In conclusion, the $e^+e^- \to$ hadrons data at present energies may prefer "colour" over "charm", principally because of the "energy crisis", but not conclusively. The apparent absence of a "colour" threshold in electroproduction at comparable q^2 strongly favours "charm" schemes though. However, if it should turn out that there is a further increase in R at $q^2 > 25$, a simple SU(4) charm scheme would probably be untenable. Whether it should be further "charmed" or "coloured" would probably be reflected in the electroproduction cross-sections for $|q^2| > 25$. It is worth noting that Han-Nambu with charm gives R = 6. Anyway, we hope that the detailed study of the conventional charm scheme and of how to see charmed particles experimentally, has now been adequately justified.

4. SEEING CHARM

We saw in the previous lectures that "hidden charm" was not excluded as an explanation of the new narrow resonances. Many of the considerations made were qualitative and did not depend on the details of the "charm" scheme, such as the quantum numbers and weak interactions of the new charmed quark. In this section we specialize to the conventional SU(4) charm scheme, as it appears, for example, in the Weinberg-Salam[48] model of weak interactions, and speculate how it may show up in various experimental situations.

4.1. The decays of charmed particles

The absence so far of heavy leptons and the presence of neutral currents suggest that if you want to make a renormalizable unified gauge theory of weak interactions, then the simplest possibility is the Weinberg-Salam model[48]. Indeed this model is not yet inconsistent with published data on neutral and charged currents in purely leptonic and semileptonic processes[+]. If the absence of strangeness-changing neutral currents is to be accommodated in this scheme, then the simplest possibility is to introduce a fourth charmed quark with quantum numbers Q = 2/3, B = 1/3, I = 0, S = 0, and C = 1, where ordinary quarks have C = 0. In this scheme the hadronic charged current is

$$J^{ch}_{\mu} = \bar{p}\gamma_{\mu}(1-\gamma_5)(n \cos \theta_c + \lambda \sin \theta_c) + \bar{c}\gamma_{\mu}(1-\gamma_5)(-n \sin \theta_c + \lambda \cos \theta_c) \qquad (4.1)$$

where θ_c is the usual Cabibbo angle. The Weinberg-Salam model with charm is the simplest one that has no known defects except the presence of undetermined angles and other parameters. The optimist may hope they will be specified in the next generation of unified field theories[49]. For the moment we assume the validity of (4.1) and pursue[21] its implications for the decay modes of charmed particles. We will assume, as suggested in Lecture 2, that the lightest charmed particles are pseudoscalars. It is not obvious that this is true, and papers exist[50] discussing what happens if the

[+] This remark remains true of the new data presented at the Paris neutrino conference of March 18-2o, 1975.

lightest particles are vectors. The expected mass splittings are such that the decay $D^* \to D\pi$, $F^* \to F\pi\pi$ may well be kinematically forbidden, so that the vectors would decay into pseudoscalars predominantly radiatively.

Let us now look at the decays of the pseudoscalars D and F, much of the analysis being taken from Gaillard, Lee and Rosner[21].

4.1.1. Leptonic decays

The hadronic charm-charging current (4.1) gives a selection rule for leptonic and semileptonic decay amplitudes:

$$\begin{aligned} &\Delta C = \Delta Q = \Delta S = 1, \quad \Delta I = 0 && \propto \cos\theta_C \\ &\Delta C = \Delta Q = 1, \ \Delta S = 0, \Delta I = \tfrac{1}{2} && \propto \sin\theta_C \ . \end{aligned} \tag{4.2}$$

Hence the dominant leptonic decay rates are

$$\left.\begin{aligned} &D^o \to \text{"}K^-\text{"} + \text{leptons} \\ &D^+ \to \text{"}\bar{K}^o\text{"} + \text{leptons} \\ &F^+ \to \text{"}\eta\text{"} + \text{leptons} \end{aligned}\right\} \propto \cos^2\theta_C \tag{4.3}$$

where "$\bar{K}^o$", etc., indicate hadronic final states with the SU(3) quantum numbers of $\bar{K}^o$, etc., but quite possibly multiparticle states with different spin-parity.

The simplest decays are two-body ones akin to $\pi_{\ell 2}$ and $K_{\ell 2}$ decays. Defining[21] f_D and f_F analogously to the f_π of Lecture 2:

$$<0|J^c_\mu|D^+(k)> = if_D \sin\theta_c k_\mu$$
$$<0|J^c_\mu|F^+(k)> = -if_F \cos\theta_c k_\mu \qquad (4.4)$$

then SU(4) symmetry would predict

$$f_D \approx f_F \approx f_K \approx f_\pi$$

which is probably not a good approximation. Experimentally $f_K/f_\pi \approx 1.2$: also, we saw in Lecture 2 that the vector-photon coupling $1/\gamma_{\phi_c}$ was smaller than the SU(4) prediction by a factor of ~ 2. Pressing on regardless, we get

$$\frac{\Gamma(D^+ \to \mu^+\nu)}{\Gamma(K^+ \to \mu^+\nu)} \simeq \frac{m_D}{m_K}, \quad \frac{\Gamma(F^+ \to \mu^+\nu)}{\Gamma(K^+ \to \mu^+\nu)} \approx \frac{m_F}{m_K} \cot^2\theta_c \quad . \qquad (4.5)$$

Taking $\Gamma(K^+ \to \mu^+\nu) \approx 0.5 \times 10^8 \text{ sec}^{-1}$ and $m_F \approx m_D \approx 4m_K$, we get

$$\Gamma(D^+ \to \mu^+\nu) \approx 2 \times 10^8 \text{ sec}^{-1}$$
$$\Gamma(F^+ \to \mu^+\nu) \approx 4 \times 10^9 \text{ sec}^{-1} \quad . \qquad (4.6)$$

Notice that the $(e\nu)$ decays are suppressed by the usual helicity arguments.

The dominant leptonic decays are expected to be multibody; Gaillard, Lee and Rosner[21] give a cute estimate of the inclusive semileptonic decay rate. They regard the decay as resulting from a fundamental process

$$C \to \lambda + L^+ + \nu \qquad \text{or} \qquad C \to n + \lambda^+ + \nu$$

followed by a dressing of the final-state quarks into hadronic matter as shown in Fig. 23a. The sum over final states should then have a rate given by the elementary process whose rate has the same formula as that for the decay:

$$\Gamma_{total} \text{ (charm} \rightarrow L + \nu + \text{hadrons)} = \frac{G_F^2 m_c^5}{196\pi^3} . \tag{4.7}$$

Taking the charmed quark mass $m_c \approx 1.5$ GeV gives a total rate

$$\Gamma_{total} \text{ (charm} \rightarrow L + \nu + \text{hadrons)} \approx 10^{12} \text{ sec}^{-1} . \tag{4.8}$$

Estimates of the specific $D_{\ell 3}$ and $F_{\ell 3}$ decays are consistent with this estimate, giving a few x 10^{11} sec^{-1}, and these may be an appreciable fraction of the total, even if not dominant. Notice that there are no helicity arguments to suppress $(e + \nu)$ + hadrons decays relative to $(\mu + \nu)$ + hadrons.

The dominant decays of charmed particles are expected to be non-leptonic. Discarding γ matrices the effective current-current weak Lagrangian can be written

$$L_{eff} \approx G_F \begin{cases} (\bar{c}\lambda p\bar{n})\cos^2\theta_c + (\bar{c}\lambda p\bar{\lambda} - \bar{c}np\bar{n})\sin\theta_c \cos\theta_c \\ - (\bar{c}np\bar{\lambda})\sin^2\theta_c + \text{herm.conj.} \end{cases} . \tag{4.9}$$

The dominant decays have <u>amplitudes</u> with

$$\Delta C = \Delta S \qquad \propto \cos^2 \theta_c \tag{4.10}$$

$$|\Delta C| = 1, \ \Delta S = 0 \qquad \propto \cos \theta_c \sin \theta_c \ .$$

Hence the dominant non-leptonic decay <u>rates</u> are

$$D^0 \to "\bar{K}^0", \quad D^+ \to "(\bar{K}n\pi)^+", \quad F^+ \to "\pi^+" \quad \propto \cos^4 \theta_c$$

and (4.11)

$$D^0 \to "\pi^0", \ "\eta"; \quad D^+ \to "\pi^+"; \quad F^+ \to "K^+" \propto \cos^2 \theta_c \sin^2 \theta_c \ .$$

To estimate the total non-leptonic decay rate we again use the Gaillard, Lee and Rosner[21] estimate that the rate is given essentially by the elementary process $c \to \lambda + p + \bar{n}$ shown in Fig. 23b, which yields

$$\Gamma_{total}(\text{charm} \to \text{hadrons}) \approx \frac{1}{196\pi^3} (G_F A \cos^2 \theta_c)^2 \, m_c^5 \tag{4.12}$$

where A is a typical factor for the enhancement of non-leptonic decays, estimated from strange particle decays to be $O(\sin^{-1} \theta_c)$. We then get for $m_c \approx 1.5$ GeV

$$\Gamma_{total}(\text{charm}\to\text{hadrons}) \approx 20 \left(\frac{m_c}{m_\mu}\right)^5 \Gamma(\mu\to e\nu\bar{\nu}) = O(10^{13})\,\text{sec}^{-1}. \tag{4.13}$$

Estimates of the specific $D^0 \to K^-\pi^+$ mode are consistent with this inclusive estimate, giving at most a few x 10^{12} sec^{-1} for this mode. Comparing the inclusive estimates (4.8) and (4.13) we get an estimate

$$\frac{\Gamma(\text{leptonic})}{\Gamma(\text{nonleptonic})} \approx O\left(\frac{1}{20}\right) .$$

This is very uncertain - there seems to be a general impression that it may be an overestimate - but a leptonic branching ratio of the order of 1 to 10% seems reasonable.

We used above a non-leptonic enhancement factor A; phenomenologically strange non-leptonic decays with $|\Delta I| = \frac{1}{2}$ are enhanced relative to $|\Delta I| = \frac{3}{2}$ transitions and leptonic decays by a factor of O(20) in the rate.

Data on SU(3) weak decays are consistent with a general octet dominance rule, and the question arises of finding the appropriate generalization of this rule to SU(4) decays. Asymptotically free gauge theories of the strong interactions give such a prescription[21,51] based on the differences in the anomalous dimensions of SU(4) tensor operators. The conclusion is that the dominant piece of L_{eff} for SU(4) decays should belong to a $\underline{20}$ representation of SU(4). Decomposed into SU(3) representations this has a $\Delta C = 0$ octet part, and $\Delta C = +1, -1$ parts belonging to $\underline{6}$ and $\underline{\bar{6}}$ representations of SU(3). The suppressed $\Delta C = +1, -1$ parts belong to $\underline{\overline{15}}$ and $\underline{15}$ representations of SU(3).

This scheme then makes predictions about the relative abundances of different decays of D^0, D^+, and F^+. For the two-pseudoscalar $\cos^4\theta_c$ decays they are[50,51]

$$\Gamma(D^0 \to \pi^+ K^-) = 2\,\Gamma(D^0 \to \bar{K}^0 \pi^0)$$
$$= 6\,\Gamma(D^0 \to \bar{K}^0 \eta) = \Gamma(F^+ \to \bar{K}^0 K^+) = \frac{3}{2}\,(F^+ \to \pi^+ \eta) . \tag{4.14}$$

Other decay modes are suppressed either because they are proportional to $\cos^2\theta_c \sin^2\theta_c$ or $\sin^4\theta_c$, or because they go via the $\underline{15}$ and $\overline{\underline{15}}$ SU(3) representations. We note that there are no enhanced two-body $D^\pm$ decay modes: hence their important two-body modes may come from $\cos^2\theta_c \sin^2\theta_c$ terms in L_{eff} or from $\underline{15}$ and $\overline{\underline{15}}$ contributions. For the $\cos^2\theta_c \sin^2\theta_c$ modes we have[50,51]

$$\Gamma(D^+ \to \bar{K}^o K^+) = \frac{3}{2}\,\Gamma(D^+ \to \eta\pi^+) \qquad (4.15)$$

which may be competitive with the unenhanced mode $D^+ \to \bar{K}^o\pi^+$. Note however there are still enhanced $\cos^4\theta_c$ multibody decay modes which may well be dominant.

It is unclear how many of the D^o and F^+ decay modes should also be multibody. Gaillard, Lee and Rosner[21] used a simple algebra/phase-space model to estimate that for D^o's and F^+'s with masses $\sim$ 2 GeV

$$\Gamma(2\text{ body}) : \Gamma(3\text{ body}) : \Gamma(4\text{ body}) : \Gamma(5\text{ body}) \simeq 51\,\% : 38\,\% : 9\,\% : 1\,\% \quad . \qquad (4.16)$$

However, this may well overestimate the two-body decays. It seems to be a phenomenological fact that no matter how you prepare a hadron "blob" with a given mass, final-state interactions have a universal effect, and the multiplicity of final decay products is about the same. For example, the charged multiplicities in $e^+e^- \to$ hadrons and $p\bar{p} \to$ mesons are essentially identical, and $\langle n_{ch}\rangle \simeq 3$ for systems with mass $\sim$ 2 GeV. Thus we might expect for non-leptonic decays of D and F with masses 2 GeV $\langle n_{ch}\rangle$ also ~ 3 and $\langle n_{total}\rangle \sim 4$ to 5. Because of this large multiplicity, the total K/π ratio in charmed particle decays might not be greater than 1/4. In this case the dominant decays would be messy multibody ones, and it might be difficult to see significant enhancements in mass plots for experimentally simpler channels such as $K^{\pm}\pi^{\mp}$. However, these decays would be very spectacular in emulsions: very short tracks producing many-pointed stars.

4.2. The production of charmed particles

In this section we discuss how charmed particles may be produced in various different types of experimental situations. The discussion of the previous section suggests that while it may be easy to see that something is going on, it may not be easy to verify that it is charm. As we shall see, there are plenty of suggestions that something is happening!

4.2.1. e^+e^- annihilation

According to the charm scheme, essentially all the increase above $\sqrt{q^2} = 4$ GeV of

$$R = \frac{\sigma(e^+e^- \to \text{hadrons})}{\sigma(e^+e^- \to \mu^+\mu^-)}$$

above the naive quark value 2 is due to the production of charmed particles. Because of the near degeneracy of the masses $m_D \approx m_F \approx m_{D^*} \gtrsim m_{F^*} \approx m_{D_s} \gtrsim m_{F_s}$, etc., we can expect a very dramatic threshold effect due to the opening up of many channels. Also, naive extrapolation of $m^2_{3.7} - m^2_{3.1}$ yields a new state (radial excitation in the quark language or lower daugther in the dual language) with a mass $\sim$ 4.1 to 4.2, which will further enhance the threshold effect. This state should have a typical hadronic decay width and decay mainly into the available charmed meson pairs$^+$.

The data are not inconsistent with this possibility, but the storage rings have not yet seen any sign of a charmed particle. Rumour[7] has it that in the data at $\sqrt{q^2}$ = 4.8 GeV the following channels have been examined for suspicious bumps in effective mass plots:

$K^o\ \pi^\pm$, $\quad \pi^+\ \pi^-$, $\quad K^o\ \pi^+\ \pi^-$ $\quad$ and $\quad \pi^+\ \pi^+\ \pi^-$

and that nothing has been seen with a mass around 2 GeV. How serious would this be for charm if the rumour were correct? Reference back to Section 4.1 shows that neither of the above two-body channels is expected to be populated by the dominant enhanced charmed meson decays. On the other hand, the $(K^o\pi^+\pi^-)$ and $(\pi^+\pi^+\pi^-)$ channels should be populated by D^o and F^+ decays, respectively, but with unknown branching ratios which could well be small. What is the branching ratio of $\bar{n}p \to \pi^+\pi^+\pi^-$ near 2 GeV? The most distinctive two-body decay favoured by the arguments of Section 4.2. is $D^o \to K^-\pi^+$. Unfortunately, a simple kinematical calculation shows that if a pair of charmed mesons with masses 2 to 2.1 GeV each are produced in an e^+e^- collision, then at a centre-of-mass energy of 4.8 GeV, 80 to 100% of the $K^\pm$ produced in their two-body decays have

---

$^+$ Remember, however, the cautionary footnote at the beginning of section 2.

momenta > 600 MeV, beyond the centre-of-mass momenta for which the SPEAR magnetic detector can discriminate $K^{\pm}$ from $\pi^{\pm}$. Therefore the prospects of seeing lots of charm in the near future may be little better than the amount seen so far.

There is one problem, namely the apparent excess of neutral energy or "energy crisis"[52], which seems to have no natural resolution in the charm scheme. Radiative transitions between charmed mesons (e.g. $D^* \to D\gamma$) should only emit relatively low energy γ-rays if the mass formulae of Lecture 2 are to be believed. Above charm threshold the quark diagram rules suggest that relatively few "hidden charmed" states should be produced - so no high-energy γ-rays from that source. Simple estimates with the preferred two-body decay modes (4.14) and (4.15) reveal little tendency for the neutral energy to be much more than the canonical 1/3. Also, as we said in Lecture 3, charm probably has to go back to the drawing board if R continues to rise above 5 for $\sqrt{q^2} > 5$ GeV.

4.2.2. Neutrino production

This is one place that a large production of charmed particles might be expected; the motivation for introducing charm required that their weak couplings be comparable to those of uncharmed particles. Here we will concentrate on deep inelastic experiments. Gaillard, Lee and Rosner[21] give a discussion of quasi-elastic experiments. Let us recall the main formulae of the quark parton model for deep inelastic scattering. The nucleons are supposed to contain constituents with momentum distributions $p(x)$, $n(x)$, $\bar{\lambda}(x)$, etc., where $x = |q^2|/2\nu$ is the fraction of the nucleons' longitudinal momentum carried by the quark partons p, n, $\bar{\lambda}$, etc. Scattering of leptons off antipartons or antileptons off partons is proportional to $(1 - y)^2$, where $y = E_h/E_\nu$, and scattering of leptons off partons or antileptons off antipartons is independent of y. Assuming the Weinberg-Salam model with charm (4.1) and setting $\theta_c = 0$, we get double differential cross-sections off heavy nuclei:

$$\frac{d^2\sigma^\nu}{dxdy} \approx \frac{G_F^2 m_N E_\nu}{\pi}[(n + \lambda) + (1-y)^2 (\bar{p} + \bar{c})]$$

and (4.17)

$$\frac{d^2\sigma^{\bar{\nu}}}{dxdy} \approx \frac{G_F^2 m_N E_\nu}{\pi}[(\bar{n} + \bar{\lambda}) + (1-y)^2 (p + c)]$$

where we have (optimistically) included a contribution from charmed partons. The constituents are divided into two classes: "valence" partons (p and n for the proton and neutron), which are expected to be copious; and "sea" partons (λ, c, $\bar{p}$, $\bar{n}$, $\bar{\lambda}$, and $\bar{c}$ for the proton and neutron), which are expected to be less important and to cluster around $x \approx 0$. It is expected that $p/\bar{p}$, $n/\bar{n}$, $\lambda/\bar{\lambda}$, and $c/\bar{c} \to 1$ as $x \to 0$, and SU(3) or even SU(4) symmetry may be restored there.

We see from equation (4.17)[+] that at high energies

[+] Since $\theta_c \neq 0$, there are contributions $\propto \sin^2 \theta_c \approx 0.05$ to the production of charmed particles arising from ν scattering off "valence" n partons or $\bar{\nu}$ scattering off "sea" $\bar{n}$ partons. These may be difficult to see, except that the former may be a source of dimuon events (see the second footnote at the end of 4.2.2.).

$$\frac{\sigma^{\bar{\nu}}_{total}}{\sigma^{\nu}_{total}} \simeq \frac{Q/3 + \bar{Q}}{Q + \bar{Q}/3} \tag{4.18}$$

where Q and $\bar{Q}$ are the x integrals of the parton and antiparton distributions. Experimentally the ratio (4.18) is close to 1/3, and suggests $\bar{Q}/Q < 10\%$. Gargamelle data[48] taken literally would yield $\bar{Q}/Q = 6 \pm 2\%$ and suggest that the antiparton distributions are very small above $x \approx 0.1$.

Presumably the charm degree of freedom is not being excited in the Gargamelle experiments, and the contributions of λ, $\bar{\lambda}$, c, and $\bar{c}$ are negligible at these low energies. Also, in scattering off heavy nuclei with almost equal numbers of neutrons $n \simeq p$. Hence for Gargamelle

$$\frac{d^2\sigma^{\nu}}{dxdy} \approx \frac{G_F^2 m_N E_\nu}{\pi} [p + (1-y)^2 \bar{p}]$$

and (4.19

$$\frac{d^2\sigma^{\bar{\nu}}}{dxdy} \approx \frac{G_F^2 m_N E_\nu}{\pi} [\bar{p} + (1-y)^2 p] \quad .$$

Hence the Gargamelle data on $\bar{Q}/Q$ suggest

$$\int dx\ \bar{p}(x) \simeq \frac{6}{100} \int dx\ p(x) \quad . \tag{4.20}$$

Now we proceed to FNAL energies. Comparing (4.17) to (4.19) we see that the increases in the cross-sections due to exciting the charm degree of freedom are proportional to

$$\begin{aligned} &\lambda(x) + (1-y)^2 \ \bar{c}(x) && \text{for} \ \nu \\ &\bar{\lambda}(x) + (1-y)^2 \ c(x) && \text{for} \ \bar{\nu} \ . \end{aligned} \tag{4.21}$$

Hence the absolute increases in the ν and $\bar{\nu}$ cross-sections are the same, but the fractional increase in the $\bar{\nu}$ cross-section is almost three times larger, just because $\sigma_{\bar{\nu}}/\sigma_{\nu} \sim 1/3$ at low energies. If we suppose that SU(3) symmetry is good for "sea" parton distributions, then

$$\lambda(x) = \bar{\lambda}(x) \simeq \bar{p}(x) \qquad \text{etc.}$$

Let us also assume SU(4) symmetry is badly violated so that $c(x) \ll \lambda(x)$, etc. Then the increases in both the $\bar{\nu}$ and ν cross-sections are essentially independent of y, and

$$\frac{\Delta\sigma^{\bar{\nu}}}{\sigma^{\bar{\nu}}} \cong \frac{\int dx \ \lambda(x)}{\int dx \,[\bar{p}(x) + \frac{1}{3}p(x)]} \simeq \frac{0.06}{0.39} \simeq O(15\%)$$

whereas (4.22)

$$\frac{\Delta\sigma^{\nu}}{\sigma^{\nu}} \simeq \frac{\int dx \ \bar{\lambda}(x)}{\int dx \,[p(x) + \frac{1}{3}\bar{p}(x)]} \simeq \frac{0.06}{1.02} \simeq O(6\ \%) \ .$$

What does this increase in the cross-section correspond to? According to the Weinberg-Salam model with $\theta_c \approx 0$, the charged current interacting with the λ or $\bar{\lambda}$ quark produces a c or $\bar{c}$ quark in the final state, and hence charmed particles in the final state. These particles presumably travel predominantly in the forward direction and have a branching ratio into μ + anything of the order of (1-10)%.

The Harvard-Pennsylvania-Wisconsin-Fermilab group has recently published two papers[53,54] on deep inelastic neutrino scattering which claim to see such effects. In one of them[53] they find an increase in the inclusive differential $\bar{\nu}$ cross-section $d^2\sigma^{\bar{\nu}}/dxdy$ above $E_{\bar{\nu}}$ = 30 GeV which is confined to $x \lesssim 0.1$, essentially independent of y and of the order of 14%. They also say that a similar effect in the ν data of the order of 5% cannot be excluded. It is possible to have questions about cuts on the experimental data and so on, but it may not be coincidental that the theoretical and experimental figures are so similar. There is, however, a problem: at small x the formula (4.17) indicates that the ν and $\bar{\nu}$ cross-sections should be equal, whereas experimentally their ratio seems to be $\simeq 2$[55].

The other paper[54] by the same group reports the observation of events of the type

$$\nu \text{ or } \bar{\nu} + \text{nucleus} \rightarrow \mu^+\mu^- + \text{anything} \tag{4.23}$$

with a partial cross-section of the order of a few parts per thousand. This observation is of the type expected on the basis of charm - the forward-moving charmed hadron should decay into a lepton of the opposite charge from that of the muon associated with the incoming ν or $\bar{\nu}$. However, it may still be wondered whether the data could be contaminated by the decay products of fast forward-moving pions or kaons. Since the ratio of π^+/π^- in the forward direction in ν scattering is experimentally much greater than 1 (and similarly for π^-/π^+ in $\bar{\nu}$ scattering), this contamination would also produce pairs of muons of opposite signs⁺. For the moment we do not regard either of the H-P-W-Fermilab papers as strong evidence for charm - but they may be indicative⁺⁺.

4.2.3. Electroproduction

If the nucleon contains some charmed partons, then they presumably would not contribute to the inclusive electroproduction cross-section at low energies, but would show up later, probably at Fermilab energies. As discussed in Lecture 3, any charmed partons should be lurking at low $x \lesssim 0.1$, and their distribution is probably less than that of uncharmed "sea" partons. If we suppose $c(x)/p(x) \gtrsim r_c$ as $x \to 0$, then the corresponding fractional rise in the electroproduction cross-section is

$$\frac{\frac{4}{9}\,(c+\bar{c})}{\frac{4}{9}(p+\bar{p})+\frac{1}{9}\,(n+\bar{n}+\lambda+\bar{\lambda})} = \frac{2}{3}\,r_c \qquad (4.24)$$

⁺ The CERN Theory Boson Workshop[16] estimated the contamination of $(\mu^+\mu^-)$ pairs due to weak production of the 3.1: it was very sensitive to the Weinberg angle, but $\lesssim 10^{-4}$ and hence not significant.

⁺⁺ In a seminar at CERN, B. Barish has reported the observation of events like (4.23) in the Caltech experiment at Fermilab, at least some of which are very difficult to explain as being due to π or K decays. Some of these have $x \gtrsim 0.1$, and may be due to the mechanism discussed in the footnote on p. 40.

It appears[47] that in an experiment at Fermilab the deep inelastic muon electroproduction scales when 56 GeV and 150 GeV data are compared, but that for $x \lesssim 0.1$ the cross-section at both energies is $\sim$ 20% higher than that extrapolated from the SLAC cross-section. There could of course be other explanations for this - for example, the "non-charmed" part of the cross-section may well not have scaled near $x \approx 0$ at SLAC energies - but it is not impossible that this increase may reflect charmed particle effects.

4.2.4. Photoproduction

We discussed in Lecture 1 the photoproduction of the 3.1 state, and how this corresponded to a total cross-section $\sigma_{tot}(3.1\ N) \approx 1$ mb. In Lecture 2 a model was suggested according to which this cross-section would correspond to the "conventional" dual Pomeron graph of Fig. 15, in which it is apparent from cutting down the dotted line that this cross-section comes mainly from the production of charmed particles in the intermediate state+. If we now make the usual model in which

$$\sigma_{total}(\gamma N) = \sum_{\substack{\text{vector}\\\text{mesons } V}} \frac{|\langle\gamma|V\rangle|^2}{m_V^4}\, \sigma_{total}(VN) \qquad (4.25)$$

then the contribution of $\phi_c = 3.1$ to this total is approximately

---

+ This model for the Pomeron is probably hopelessly naive, at infinite energies charmed particle yields in γN and πN collisions should be similar. But also in multiperipheral models one would expect $\sigma(\gamma N \to 3.1 N)$ to be the shadow of charmed particle production in the photon fragmentation region, and the bulk of $\sigma_{tot}(3.1N)$ to contain charmed particles.

$$\frac{\sigma_{total}(\phi_c N)}{\frac{4}{3}\sigma_{total}(\rho N)} \cdot \frac{\frac{1}{\gamma^2_{\phi_c}}}{\frac{1}{\gamma^2_\rho}} \approx (\frac{4}{3} \cdot 20)^{-1} \cdot \frac{2}{9} \approx \frac{1}{120} \quad . \tag{4.26}$$

Hence O(1 μb) of the asymptotic photoproduction cross-section could be due to charmed particles. A similar estimate gives a fraction of about 1/12 for associated production of strange particles, which seems to be consistent with experiment[56].

4.2.5. Hadronic collisions

The first task is to estimate the total inclusive cross-section for the production of charmed particles. The inclusive charged π cross-section is O(300) mb at ISR energies (recall that each inelastic event produces of the order of 10 charged pions): the charged kaon inclusive cross-section is one or two orders of magnitude less, and in qualitative agreement with the prediction of the thermodynamical model that the production rate of a particle of mass m is proportional to $e^{-m/T}$ with $T \approx m_\pi$. This model also predicts a p_T distribution at high energies of the form $e^{-p_T/T}$, which is known to be valid for $p_T \lesssim 1$ GeV, but to break down for $p_T > 1$ GeV, and is probably replaced by some sort of power law. Let us assume without justification that the thermodynamical predictions for large p_T and heavy mass production are violated in the same way. We then get

$$\frac{\sigma_{inclusive}(m)}{\sigma_{inclusive}(\pi)} \approx \frac{\sigma^\pi_{inclusive}(p_T)}{\sigma^\pi_{inclusive}(p_T=0)} \quad . \tag{4.27}$$

In the case of charm it is unclear whether we should take $p_T = m_D$ or $2m_D$ on the right-hand side because charmed particles have to be produced in pairs.

However, the cross-section for $p + p \rightarrow 3.1 +$ anything seen at Fermilab seems to be $O(1/10)$ μb corresponding to choosing $p_T \gtrsim 3$ in the formula (4.27), even though according to Lecture 2 (see Fig. 13) the majority of the final states in 3.1 production should also contain charmed-anticharmed particle pairs. Hence $O(1/10)$ μb should be a lower bound on the charmed particle cross-section. Taking (4.27) with $p_T \simeq m_D \simeq 2$ GeV and using ISR inclusive data[57] on large p_T phenomena we get

$$\frac{\sigma_{\text{inclusive}}(m=2)}{\sigma_{\text{inclusive}}(\pi)} \gtrsim 10^{-4\frac{1}{2}} \tag{4.28}$$

corresponding to

$$\sigma_{\text{inclusive}}(m = 2) \gtrsim 10 \ \mu\text{b} \quad . \tag{4.29}$$

This may well be an overestimate, but charmed particle cross-sections of $O(1)$ μb in hadronic collisions seem not impossible.

What are the prospects of enhancing the proportion of charmed particle events by using a selective trigger on the multiparticle events? One way would be to trigger on a 3.1 particle, as mentioned above. A second possibility[21] is to look for events with a μ or e coming from the decay of one charmed particle and a K coming from the decay of the other one. However, if the preferred decay modes of charmed particles are multi-body with a K/π ratio $\simeq$ 1/4, these events may not be very distinctive. Another possibility[21] is to look for $(\mu - e)$ coincidences generated by leptonic decays of both the charmed particles. This may also be difficult if the leptonic branching ratio is only O(1)%. A better chance may come from looking at large p_T events. The cross-section there is very low, but the D/π ratio may be O(1), as is known to be the case for the K/π ratio.

There are already indications that new things are happening at large p_T: the inclusive electron and muon distributions at large p_T have $e/\pi \approx \mu/\pi \approx 10^{-4}$ [58]. Is it possible they could come from the decays of charmed particles? If we use[59] the formula

$$E \frac{d\sigma}{d^3p} \propto (m^2 + p_T^2)^{-4} \tag{4.30}$$

which is qualitatively valid for familiar hadrons in the central region then the decays $D \to K\ell\nu$ + give

$$\left(\frac{L}{\pi}\right) \approx B\left(\frac{D}{\pi}\right) \left(\frac{1}{300} \leftrightarrow \frac{1}{120}\right) \tag{4.31}$$

where B is the branching ratio. With $(D/\pi) = O(1)$ and B $\sim$ few %, the formula (4.31) can just succeed in reproducing the observed lepton spectrum. A priori, there are other possible sources. For example, the decays $3.1 \to \to \ell^+\ell^-$ give

$$\left(\frac{L}{\pi}\right) \approx B\left(\frac{3.1}{\pi}\right) \left(\frac{1}{10} \leftrightarrow \frac{1}{6}\right)$$

+ $D_{\ell 2}$ and $F_{\ell 2}$ decays should have too small a branching ratio to be relevant, and would not produce electrons.

In this case B is known to be 7 x 10^{-2}, so that a ratio $(3.1/\pi) \sim O(1)\%$ would suffice to give the lepton spectrum. With the assumed p_T distribution (4.30) there is no problem with a peak in the p_T distribution of the produced leptons. However, if the 3.1 were the source of the leptons, the ISR-CCRS group believe they should have seen more (e^+e^-) pairs in their experiment. Furthermore, the familiar ϕ cannot be the source of the large (ℓ/π) ratio as data on $(K\bar{K})$ pairs at large p_T indicate that $(\phi/\pi) \lesssim$ $\lesssim$ 20%.

Hence it seems that at the moment no possible source of high p_T leptons is known which is less implausible than charmed particles.

4.3. Conclusions

We have seen how the new narrow resonances and other structures observed in $e^+e^- \to$ hadrons could be understood in terms of charm. We have also discussed the properties of charmed particles, how they should decay, and how they should be produced in various different reactions. As was remarked in the introduction, the author's prejudice is that some form of "charm" is the least unreasonable model for the new resonances and other phenomena. However, it is not at all obvious that the conventional SU(4) charm model is the right one, and it has various experimental problems, such as:

Why have no charmed particles yet been seen at SPEAR or DORIS? If they have a large two-body branching rate, why are they not seen in mass plots, and why does the ratio of strange particle production not increase very much? If their dominant decay modes are multibody, why does $<n_{ch}>$ not rise abruptly at $\sqrt{q^2} = 4$ GeV? What about the "energy crisis"? What happens to R for $\sqrt{q^2} >$ > 5 GeV?

Is charm really consistent with the Fermilab deep inelastic neutrinoproduction data? The ν and $\bar{\nu}$ cross-sections at small x seem unequal, contrary to simple charmed parton model expectations from Eq. (4.21), and this is related to an apparent charge symmetry violation[60]. One may also wonder whether charmed particles would give quite so many dimuon events, particularly at x not near 0 as reported by the Caltech Group at Fermilab.

The experimental indications discussed above have all been indirectly related to the possible existence of charmed particles. Now there are some more direct experimental claims[61] for their existence. Only time will tell whether these swallows will turn into a summer.

REFERENCES

1. J. J. Aubert et al., Phys. Rev. Letters 33, 1404 (1974); and preprint submitted to Nuclear Phys. B.
2. J.-E. Augustin et al., Phys. Rev. Letters 33, 1406 (1974).
3. C. Bacci et al., Phys. Rev. Letters 33, 1408 (1974).
4. G. S. Abrams et al., Phys. Rev. Letters 33, 1453 (1974).

5. L. Criegee et al., Phys. Letters 53B, 480 (1975).

6. J.-E. Augustin et al., Phys. Rev. Letters 34, 764 (1975).

7. This and much other information is derived from seminars at CERN, reports of seminars at SLAC, letters to Israel from H. Harari, newspaper reports and rumours. The most complete written source so far seems to be: F. J. Gilman, SLAC-PUB-1537-Talk presented at Coral Gables (1975).

8. W. Braunschweig et al., Phys. Letters 53B, 491 (1975). R. L. Ford et al., Phys. Rev. Letters 34, 604 (1975).

9. J. J. Aubert et al., Phys. Rev. Letters 33, 1624 (1974).

10. L. M. Lederman, unpublished note of 11.30 a.m., 6 December, 1974.

11. B. Knapp et al., Photoproduction of narrow resonances, Fermilab preprint (1975).

12. D. E. Andrews et al., Phys. Rev. Letters 34, 233 (1975).

13. J. F. Martin et al., Phys. Rev. Letters 34, 288 (1975).

14. B. Knapp et al., Dimuon production by neutrons, Fermilab preprint (1975).

15. Rumour confirmed by F. J. Gilman, Ref. 7.

16. For reviews see: H. Harari, SLAC-PUB-1514 (1974); CERN Theory Boson Workshop, CERN TH 1964 (1974); I. Bars et al., SLAC-PUB-1522 (1974); J. D. Jackson, unpublished notes. Similar studies have been made by many other individuals and groups. My apologies to everyone not refereed to, but it is impossible to be fair and complete.

17. R. E. Marshak and R. N. Mohapatra, Phys. Rev. Lett. 34, 426 (1975).

18. Seminar at CERN by C. Rubbia, December 1974.

19. M. Y. Han and Y. Nambu, Phys. Rev. B 139, 1006 (1965). Y. Nambu and M. Y. Han, Phys. Rev. D 10, 674 (1974).

20. T. Tati, Progr. Theor. Phys. 35, 973 (1966).

21. For a review and references see, M. K. Gaillard, B. W. Lee and J. L. Rosner, Fermilab preprint 74/86-THY (1974).

22. See many papers: for example, I. Bars and R. D. Peccei, Stanford University preprint ITP-484 (1975). I thank David Broadhurst and Roger Cashmore for talking and writing to me about colour.

23. We will return to this point in later lectures.

24. J. Ellis, CERN TH 1880 (1974); also Ref. 16.

25. Many calculations of this have been made. It could, however, be either a virtual bound state just below threshold, or a very impure ($c\bar{c}$) state decaying into ordinary hadrons.

26. The origin of these is lost in the mists of time.

27. M. Gell-Mann, R. J. Oakes and B. Renner, Phys. Rev. 175, 2195 (1968).

28. R. J. Crewther, Phys. Rev. Letters 28, 1421 (1972). M. S. Chanowitz and J. Ellis, Phys. Letters 40B, 387 (1972), and Phys. Rev. D 7, 2490 (1973).

29. S. Orito et al., Phys. Letters 48B, 380 (1974).

30. T. Appelquist and H.D. Politzer, Phys. Rev. Letters 34, 43 (1975). R. J. Harrington, S. Y. Park and

A. Yildiz, Phys. Rev. Letters 34, 168 (1975).

31. E. Eichten et al., Phys. Rev. Letters 34, 369 (1975).

32. T. Appelquist, A. Rujula, H. D. Politzer and S. L. Glashow, Phys. Rev. Letters 34, 365 (1975).

33. Similar ideas have been discussed by: S. Kitakado, S. Orito and T. F. Walsh, DESY preprint 74/54 (1974). C. Montonen, M. Roos and N. Törnquist, Helsinki preprint 30-74 (1974); and by M. Gourdin, G. Preparata and others.

34. R. Carlitz, M. B. Green and A. Zee, Phys. Rev. Letters 26, 1515 (1971) and Phys. Rev. D 4, 3439 (1971). For the recent application, see, for example, R. C. Brower and J. Primack, U. C. Santa Cruz preprint 74/103 (1974) and C. E. Carlson and P.G.O. Freund, Enrico Fermi Institute preprint 74/60 (1974).

35. A. Litke et al., Phys. Rev. Letters 30, 1189 (1973). G. Tarnopolsky et al., Phys. Rev. Letters 32, 432 (1974).

36. B. Richter, Proc. 17th Internat. Conf. on High-Energy Physics, London, 1974 (Rutherford Laboratory, UK, 1974) p. IV-37; J.-E. Augustin et al., Ref. 6.

37. See, for example, the results quoted by F. J. Gilman, Ref. 7, from a talk by G. Salvini.

38. References can be traced from M. A. Gonzàles and J. H. Weis, Phys. Letters 49B, 351 (1974).

39. M. S. Chanowitz and S. D. Drell, Phys. Rev. Letters 30, 807 (1973) and Phys. Rev. D 9, 2078 (1974). M.Pavković, Phys. Letters 46B, 435 (1973). G. B. West and P.Zerwas, SLAC-PUB-1420 (1974).

40. J. Kogut, Phys. Letters 51B, 383 (1974).

41. See, for example, J. Engels, K. Schilling and H. Satz, Nuovo Cimento 17A, 535 (1973).

42. See, for example, F. Cooper, G. Frye and E. Schonberg, Phys. Rev. Letters 32, 862 (1974), and references therein.

43. J. Erwin et al., Phys. Rev. Letters 27, 1534 (1971).

44. See, for example, C. H. Llewellyn Smith and D. V. Nanopoulos, Nuclear Phys. B878, 205 (1974).

45. T. Goldman and P. Vinciarelli, Phys. Rev. Letters 33, 246 (1974).

46. J. C. Pati and A. Salam, ICTP preprint IC/74/81 (1974), and references therein.

47. D. J. Fox et al., Phys. Rev. Letters 33, 1504 (1974). K. W. Chen, talk given at the Fermilab Muon Physics Workshop, January 1975.

48. S. Weinberg, Phys. Rev. Letters 19, 1264 (1967). A. Salam, Proc. 8th Nobel Symposium, Stockholm, 1968 (Amquist and Wiksell, Stockholm, 1968), p. 367.

49. See, for example, H. Georgi and S. L. Glashow, Phys. Rev. Letters 32, 438 (1974).

50. R. L. Kingsley, S. B. Treiman, F. Wilczek and A. Zee, Weak decays of charmed hadrons, Princeton University preprint (1975).

51. G. Altarelli, N. Cabibbo and L. Maiani, Nucl. Phys. B88, 285 (1975).

52. C. H. Llewellyn Smith, Proc. 9th Rencontre de Moriond, (C.N.R.S., Orsay, 1974), p. 161.

53. A. Benvenuti et al., Phys. Rev. Letters 34, 597 (1975).

54. A. Benvenuti et al., Phys. Rev. Letters 34, 419 (1975).

55. V. Barger, T. Weiler and R.J.N. Phillips, Wisconsin preprint COO-881-441 (1975).

56. H. Meyer, Proc. 6th Internat. Symposium on Electron and Photon Interactions at High Energies, Bonn, 1973 (eds. H. Rollnik and W. Pfeil) (North-Holland Publ. Co., Amsterdam, 1974), p. 175.

57. M. Frisch, Proc. 17th Internat. Conf. on High-Energy Physics, London 1974 (Rutherford Laboratory, UK, 1974), p. V-8.

58. Ibid., p. V-41.

59. J. M. Gaillard, M. K. Gaillard and R. Savit, private communication.

60. B. Aubert et al., Phys. Rev. Letters 33, 984 (1974).

61. K. Niu, E. Mikumo and Y. Maeda, Prog. Theor. Phys. 46, 1644 (1971); K. Hoshino et al., Nagoya Univ. preprint DPNU-3 (1975); E.G. Cazzoli et al., BNL preprint NG-308 (1975).

FIGURE CAPTIONS

Fig. 1: Coupling of 3.1 to hadrons via the photon.

Fig. 2: Direct coupling of 3.1 to hadrons.

Fig. 3: Diffractive diagram supposed responsible for the photoproduction of the 3.1.

Fig. 4: Disallowed quark diagram for $\phi \to 3\pi$ decay.

Fig. 5: The Han-Nambu colour scheme. Marked on the figure are the quark charges and the colours (Red, Yellow and Blue) of primary coloured quarks.

Fig. 6: Disallowed quark diagram for 3.7 $\rightarrow$ 3.1 $\pi\pi$ decay. In the colour scheme III $q = \lambda$; in the charm scheme $q = c$.

Fig. 7: Quantum number diagram for the conventional charm scheme. The numbers are the quark charges.

Fig. 8: Quantum number diagram for the pseudoscalar mesons in the usual charm scheme. Similar $\underline{16}$-plets are expected for mesons with other spin parities. The I = 0, Y = 0, C = 0 states can in general mix strongly.

Fig. 9: The "charmonium" picture for (a) the hadronic decays, (b) the photonic coupling of the 3.1 state, and (c) the hadronic decays of η_c "paracharmonium".

Fig. 10: (a) The single-loop diagram allowed for $\phi_c \rightarrow$ $\rightarrow K^+K^-$ decay. (b) The same diagram twisted.

Fig. 11: The single-loop diagram after a duality transformation.

Fig. 12: Diagrams expected to give vector meson mixing (a) in a quark gluon picture, (b) in a phenomenological hadronic approach.

Fig. 13: Typical (a) disallowed and (b) allowed quark diagrams for ϕ_c production in nucleon-nucleon collisions.

Fig. 14: The diagrams (a) for $\phi_c \rightarrow X\gamma$, (b) for $\phi_c' \rightarrow$

$\rightarrow \phi_c\varepsilon$ decay, including mass mixing for the pseudoscalar X and the scalar ε.

Fig. 15: (a) A dual Pomeron contribution to $\phi_c p$ scattering.
(b) The same diagram after a double duality transformation.

Fig. 16: (a) The simple parton model for $\sigma_{tot}(e^+e^- \rightarrow \rightarrow \gamma \rightarrow \text{hadrons})$.
(b) Modifications to this model found in renormalization group models including asymptotic freedom.

Fig. 17: The parton model for the final state in e^+e^- annihilation.

Fig. 18: Heavy lepton pair production via one-photon exchange where (a) both happen to decay semileptonically, (b) both happen to decay purely leptonically.

Fig. 19: The two-photon mechanism for producing hadrons.

Fig. 20: Comparison of the data on

$$R = \frac{\sigma_{tot}(e^+e^- \rightarrow \text{hadrons})}{\sigma(e^+e^- \rightarrow \mu^+\mu^-)}$$

with quark models and asymptotic freedom.

Fig. 21: A simple dual current model. For e^+e^- annihilation we calculate graphs like this with no external hadrons.

Fig. 22: No-photon models for $e^+e^- \rightarrow$ hadrons.

Fig. 23: Diagrams for (a) inclusive semileptonic decays, (b) inclusive non-leptonic decays of charmed mesons.

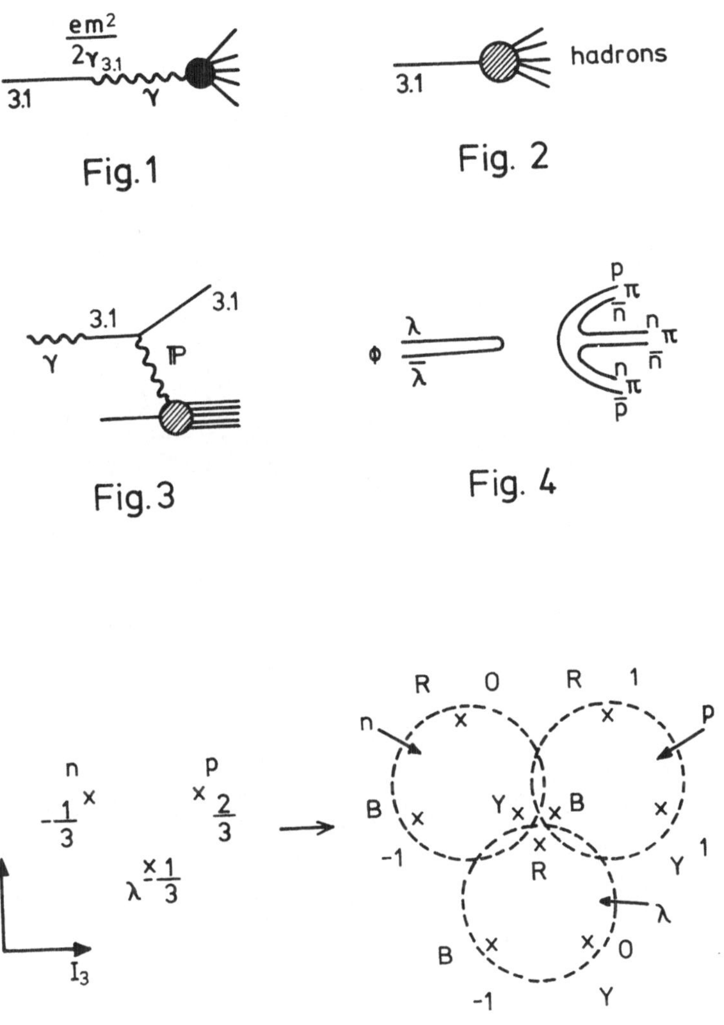

Fig. 1

Fig. 2

Fig. 3

Fig. 4

Fig. 5

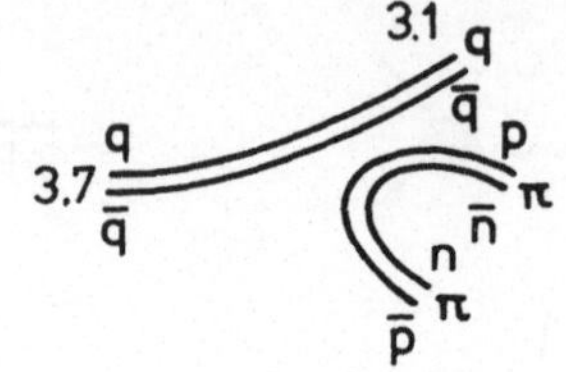

Fig. 6

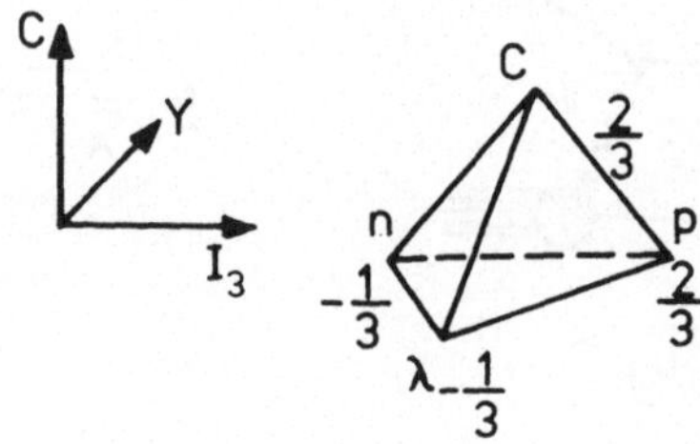

Fig. 7

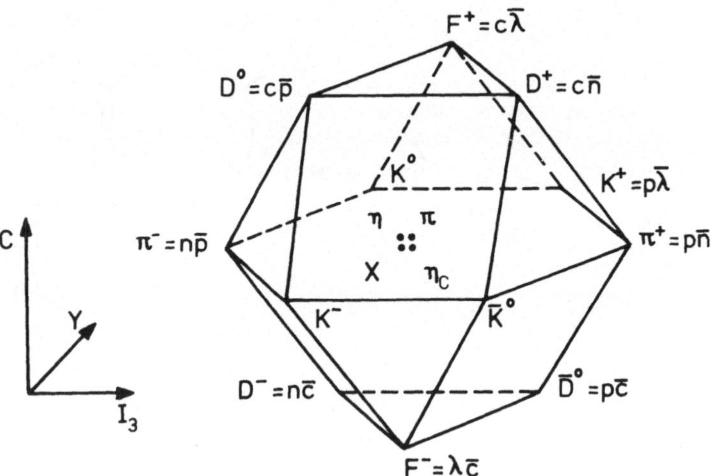

Fig. 8

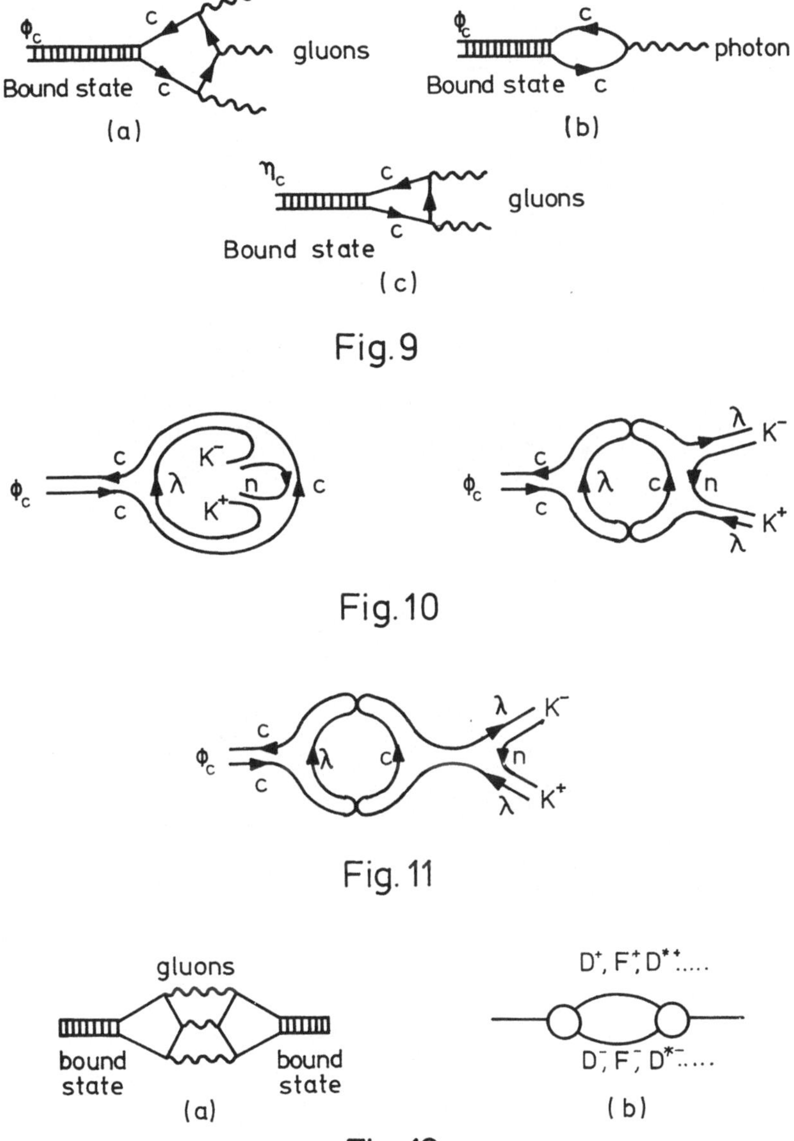

Fig. 9

Fig. 10

Fig. 11

Fig. 12

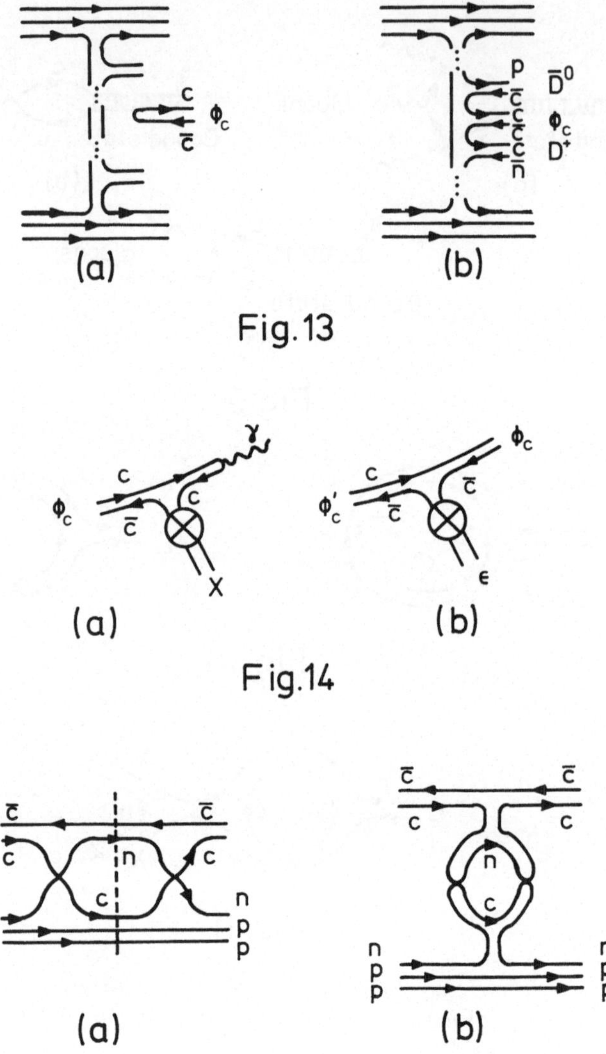

Fig. 13

Fig. 14

Fig. 15

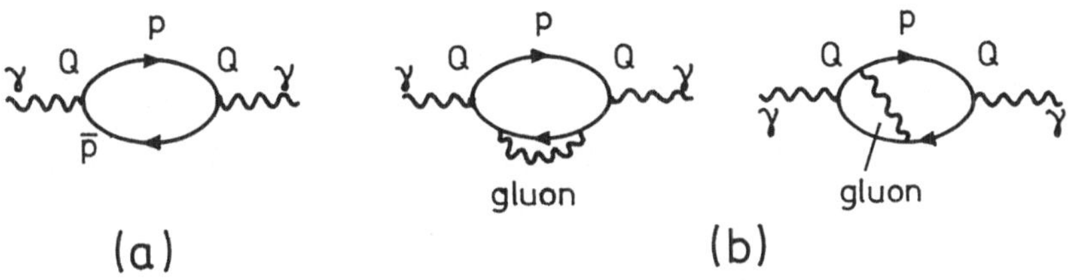

(a) (b)

Fig. 16

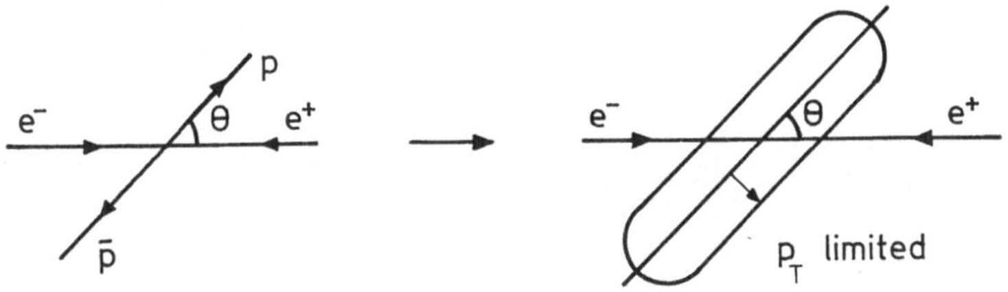

Fig. 17

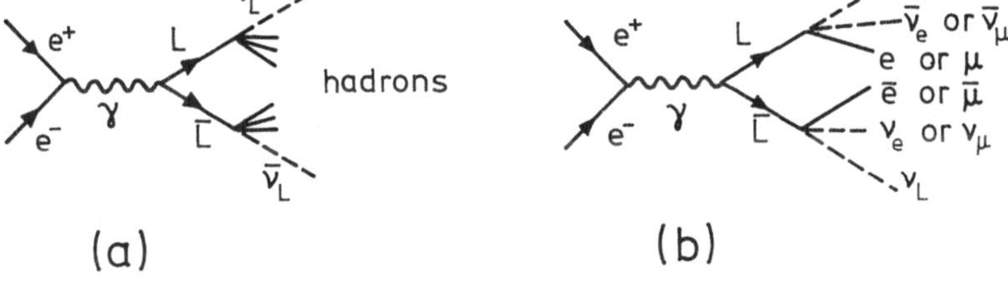

Fig. 18

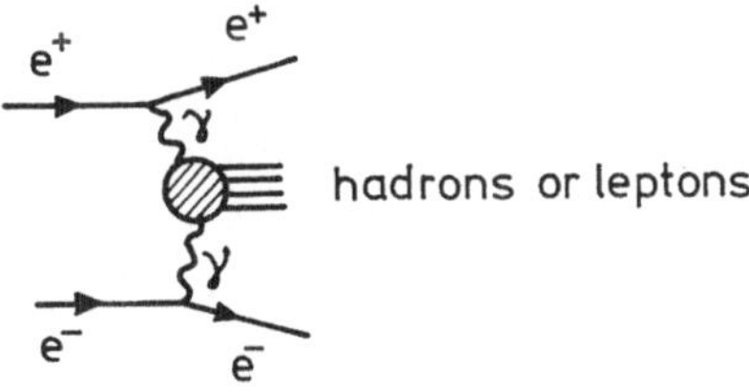

Fig. 19

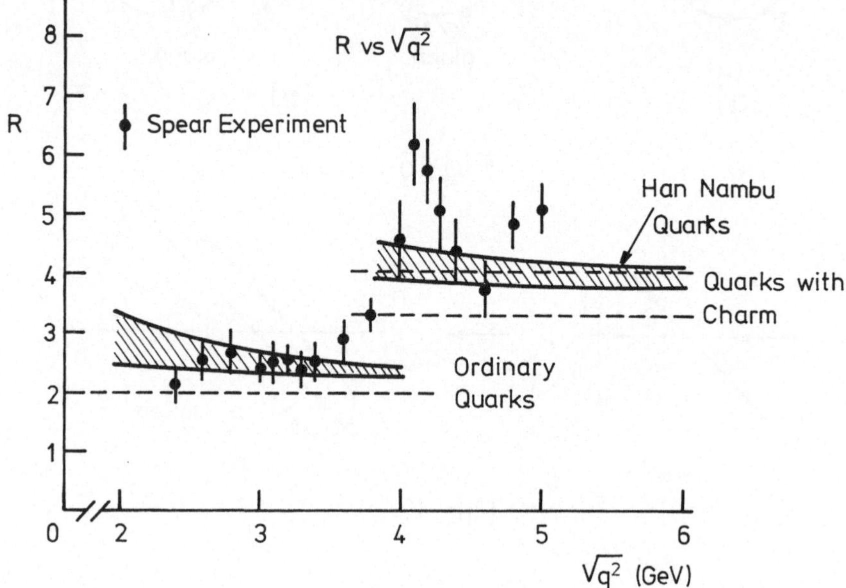
R vs $\sqrt{q^2}$
R
Spear Experiment
Han Nambu
Quarks
Quarks with
Charm
Ordinary
Quarks
$\sqrt{q^2}$ (GeV)
0
2
3
4
5
6
1
2
3
4
5
6
7
8

Fig. 20

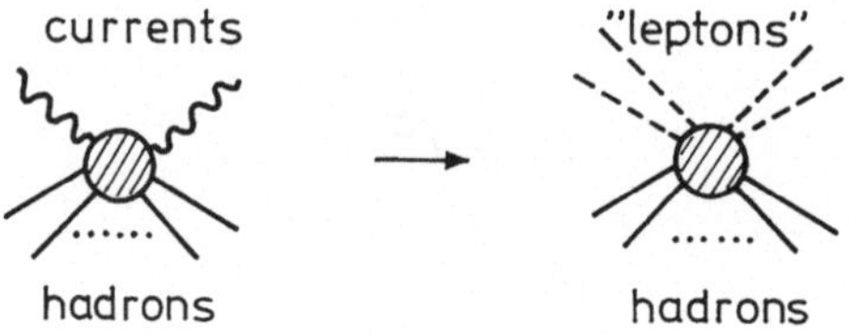
currents
"leptons"
hadrons
hadrons

Fig. 21

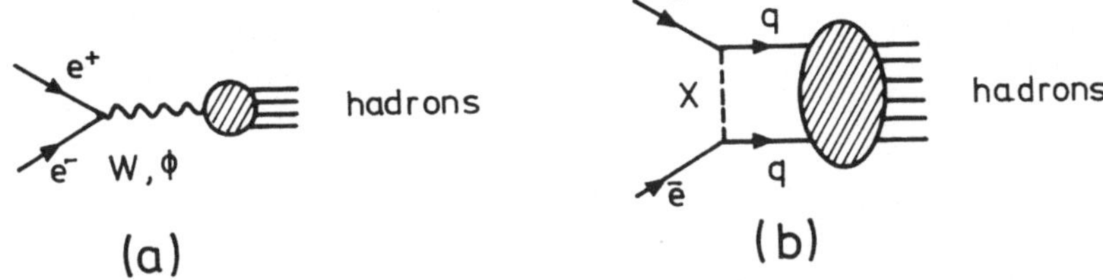
e+
e−
W, φ
hadrons
(a)
e
q
X
ē
q
hadrons
(b)

Fig. 22

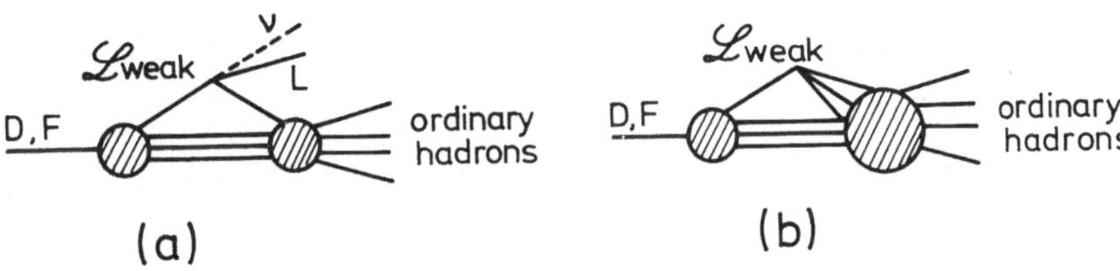
ℒweak
ν
L
D, F
ordinary
hadrons
(a)
ℒweak
D, F
ordinary
hadrons
(b)

Fig. 23

Acta Physica Austriaca, Suppl. XIV, 233 — 306 (1975)

PHYSICS WITH e^+e^--STORAGE RINGS+

by

H. SCHOPPER
DESY, Hamburg

Electron storage rings offer excellent possibilities to study the electromagnetic and strong interactions. If at a later stage center of mass energies in the region of 30 to 40 GeV became available also interference effects between the electromagnetic and weak interaction could be investigated.

Test of Quantum Electrodynamics

The graphs which contribute to various processes are shown in fig. 1. To separate the graphs it is important to study all the processes $e^+e^- \rightarrow e^+e^-$, $e^-e^- \rightarrow e^-e^-$, $e^+e^- \rightarrow$ $\rightarrow \mu^+\mu^-$, $e^+e^- \rightarrow \gamma\gamma$. Recently at SPEAR $e^+e^- \rightarrow e^+e^-$ and $e^+e^- \rightarrow$ $\rightarrow \mu^+\mu^-$ were measured at total energies between 3 and 4.8 GeV which allowed to obtain lower limits for the cut-off parameter of the photon propagator. Depending on whether one assumes that the cut-off parameter for spacelike and timelike photons are different or equal one obtains somewhat different limits. The results which are obtained for

---

+ Lecture given at XIV.Internationale Universitätswochen für Kernphysik,Schladming,Austria,February 24-March 7,1975.

the two assumptions of the metric are shown in fig. 2. The lowest two lines indicate the results which one obtains by introducing independent cut-off parameters for the electron and muon formfactors.

In conclusion one might say that the experiments are compatible with cut-off parameters larger than about 20 GeV which corresponds to a distance of 10^{-15} cm. This implies that down to these distances the electron and muon behave like a pointlike particle.

Hadronic Interactions

e^+e^--annihilation is an excellent tool to study the hadronic interactions. The reason is that in all the processes studied so far the annihilation leads to a virtual intermediate photon which is then converted to a hadronic state. The implication is that the intermediate state has the quantum numbers of the photon and is, therefore, very well defined independent of the total energy. This is in contrast to processes with two incident hadrons where higher and higher angular momenta contribute with increasing energy.

a) Two-Body Reactions

The virtual photon can decay into $\pi^+\pi^-$, K^+K^-, etc. The photon-hadron vertex (see fig. 3) is determined by the formfactor of the hadrons and hence this formfactor can be deduced from such experiments. Of course, the formfactor in the timelike region is obtained in this way. It is quite remarkable that by such experiments the formfactor of very short lived particles could be determined. Unfortunately, I have no time to discuss these very interesting experiments.

b) Resonance Production

If the energy of the virtual photon matches the mass of a resonance the photon can be converted to this resonance which subsequently decays into hadrons. Particles with the quantum numbers of the photon are the ω, ρ, Φ. Indeed all these vector mesons have been observed in e^+e^--annihilation and interesting results were obtained. As I shall discuss in more detail later there is strong evidence that the newly discovered particles also have the quantum numbers of the photon and, therefore, can be produced in a similar way (compare fig. 3).

c) Multi-Hadron Production

The decay of a resonance, of course, can produce a multi-hadron final state. However, multi-hadrons can also be produced by the decay of the virtual gamma of a resonance (fig. 3). For the theoretical interpretation it is easier to consider the ratio R which is defined by the cross section for hadron production at a certain energy devided by a pointlike cross section at the same energy e.g. $e^+e^- \rightarrow \mu^+\mu^-$. If one is away from the resonance region at high energies one would expect that this ratio is smaller than one since the coupling at the photon-hadron vertex is again determined by hadron formfactors. However, the ratio A can be about 1 if the hadron contains pointlike constituents with a formfactor F = 1.

In terms of the quark model one might assume that the virtual photon is converted into a quark-antiquark pair and the two quarks subsequently get "dressed" and appear as jets of hadrons. As usual the coupling at the photon-quark vertex is determined by the square of the charges and hence it is predicted that at very high energies the ratio R should be given by $R = \sum Q_q^2$, where Q_q are the charges of the quarks. For the ordinary quark model with three quarks one expects that R is approaching the value of 2/3.

Hence it came as a big surprise when measurements at CEA at Cambridge indicated that R assumed values between 4 and 6 at energies between 4 and 5 GeV. This was confirmed about a year ago by detailed measurements at SPEAR and the most recent results are shown in fig. 4. As one notices it seems that R is rising from 3 to 5 GeV and in addition there is a broad peak at around 4.2 GeV. Of course, it is very important to extend the measurements of R to larger energies in order to see if there is further structure or whether R settles down at some asymptotic value which is predicted by an extended quark model. Two of such predictions are shown in fig. 4.

In parenthesis I should like to mention here that the newly discovered resonances which I shall discuss in detail can hardly be drawn in a figure like fig. 4. With the energy scale of that figure the experimental widths of the resonances would be much less than the thickness of the lines indicating the error bars. The extremely narrow widths of these particles is indeed the outstanding and puzzling property.

The remaining part of this lecture will be devoted to a discussion of these particles.

Discovery of the New Particles

S. Ting who had developed the art of detecting e^+e^--pairs during many years at DESY performed with his group an experiment at BNL where they bombarded a Be-target with 30

GeV protons and searched for e^+e^--pairs. In summer 1974 they found a sharp peak at an invariant mass of the electron pair at 3.1 GeV. Fig. 5 shows the now already classical picture of the mass distribution. No other resonance could be detected with this experiment. In particular, no enhancement at an energy of 3.7 GeV was observed. After the discovery of the 3.7-resonance this could be at least qualitatively understood in terms of decreasing cross sections and branching ratios.

From the experimental mass resolution an upper limit for the width of the 3.1 GeV-resonance, which this group called J-particle, could be deduced as $\Gamma < 20$ MeV.

Since most elementary particle physicists thought that at energies of 3 to 4 GeV the particle spectrum is practically a continuum the first storage ring experiments were done with energy steps of the order of 100 MeV. Only when the energy scanning was done in steps of the order of 1 MeV the 3.1 GeV-resonance could clearly be established for the first time at SPEAR in November 1974 at an electron storage ring. Within a few days this discovery was confirmed by groups working at ADONE in Frascati and DORIS at Hamburg. A second resonance was soon found around 3.7 GeV at SPEAR and confirmed at DORIS. SPEAR introduced the nomenclature ψ and ψ' for the two resonances. Fig. 6 shows as an example the observation of the two resonances with the superconducting spectrometer PLUTO at DORIS in multihadron events. The position of the two resonances is somewhat lower than the energies first published by SPEAR. This is due to a small calibration error at SPEAR which in the meantime has been corrected.

At SPEAR a very systematic energy scan has been performed covering the region from 3.1 to 6 GeV c.m. energies.

No other sharp resonance has been detected (fig. 7). However, the existence of wide resonances (widths of the order of 100 MeV) can not yet be excluded by these measurements.

Total and Leptonic Widths of New Particles

The cross section for the reaction $e^+e^- \rightarrow$ hadrons is presented for the ψ- and the ψ'-particle in figs. 8 and 9. These results were obtained with the magnetic detector at SPEAR. Results for the decay of these resonances into lepton pairs were reported from the SPEAR magnetic detector and the double arm spectrometer DASP at DORIS. These two instruments are in certain respect complementary. DASP has the possibility to identify different particles in a very clean way whereas the magnetic SPEAR detector has a larger acceptance and hence better statistics. Results for the decay of the ψ-particle into electron pairs as obtained from DASP are shown in fig. 10 and the SPEAR results are shown for the muon decay in fig. 11.

The width of the resonance curves does not correspond to the natural line width but rather to the momentum spread in the colliding beams. The true widths of the resonance can be derived assuming that the resonance can be fitted by a Breit-Wigner curve. Neglecting all finer details the procedure is outlined in fig. 12. The ratio of the hadronic to the leptonic cross section at the resonance peak is equal to the ratio of the corresponding widths. This ratio should be independent of the resolution since the finite experimental resolution should smear out both resonance curves in the same way. If it is further assumed that this smearing out does change the shape of the resonance curve but not its area it follows (see fig. 12) that the integral over the hadronic cross section yields the width for the electron decay and if this is inserted in the ratios one obtains the hadronic width. Strangly enough, the observation of the decay of the resonance into hadrons gives the leptonic width whereas a decay into leptons allows a determination of the hadronic partial width. The directly measured cross sections have, of course, to be corrected for radiative corrections.

The results for the 3.1 GeV-resonance are shown in fig. 13. Some recent unpublished results as obtained from DASP at DORIS are given in fig. 14. The surprising result is the very small total width (relative width $2.5 \cdot 10^{-5}$). The partial widths for electron and muon agree within the experimental error and, therefore, electron-muon universality seems to hold.

From these results one can also get some information on the decay mechanism of the ψ-particle. It may decay directly into hadrons (graph designated by I in fig. 13) or it may decay via a virtual photon (designated by II in the lower part of fig. 13). The information on the coupling of the virtual photon to the final hadron state can be obtained from a measurement of the total cross section outside the resonance and if multiplied by the width determining the coupling of the virtual photon to the ψ (which is equal to $\Gamma_{\mu\mu}$) one obtains for the partial width corresponding to graph II a value of 13 keV. If the widths for the electron and muon decay (which certainly go via a virtual photon) are added one obtains a partial width for all decays going via a virtual photon of about 25 keV. If this is compared to the total width of 77 keV a width of about 52 keV remains for processes due to a direct coupling. Hence about two thirds of the decays have to be ascribed to a direct coupling mechanism.

The available results for the second resonance and the broad peak around 4.2 GeV are summarized in fig. 15.

The Nature of the New Particles

In view of the strange properties of the new particles and in particular their narrow decay widths the question was asked what their nature could be and to what kind of couplings they are subjected. Various possibilities have been discussed.

a) Electromagnetic Coupling

If the partial decay widths of the new particles into electrons are compared with corresponding widths of other particles, e.g. vector mesons, one notices that these branching ratios are all of the order of a few keV (see fig.16). Hence the situation seems to be quite normal as far as the electromagnetic coupling is concerned.

b) Weak Bosons

The small widths could be understood if the new particles decay by weak interactions only. From the diagram given in fig. 16 and the measured lepton width one can deduce the coupling constant g and indeed one finds that it is of the same order of magnitude as the Fermi coupling constant. As far as the interaction strength is concerned the new particles might indeed be weak bosons. However, a number of reasons are against this. First is not quite clear why there should be two such bosons, further one expects that the mass of the intermediate bosons should be larger than about 40 GeV. Weak interactions violate parity and hence asymmetric angular distributions are expected. No such asymmetries have been observed. The fact that the 3.7 GeV-particle decays mainly via the 3.1 GeV-particle can also not be understood in terms of weak interactions.

c) Strong Interaction

For a strongly interacting particle with a mass of about 3 GeV one expects a typical decay width between 40 and 200 MeV. This follows from the ordinary coupling strength and the large phase space available for the decay of such heavy particles. The observed decay widths are about a factor 10^{-3} too small compared to this estimate. This is the big puzzle which so far has not yet been fully understood.

A number of models have been proposed in which by the introduction of a new quantum number and a corresponding selection rule the decay of the particles is strongly inhibited. As we will discuss below some of these models are in disagreement with results on particular decays. Hence it is still possible that a completely unknown effect is the origin of the long lifetime of these particles.

The first task, of course, is to determine the quantum numbers of the new states. If it is true that the new particles are produced like ordinary vector mesons by the conversion of a virtual gamma into a resonance then one would expect that the new particles have the quantum numbers of the photon i.e. $J^P = 1^-$. Several experiments indicate that this might indeed be true.

If the resonance decays into leptons one expects an angular distribution of the decay particle with respect to the incident particles which is isotropic for a spin = 0 particle whereas one expects $1 + \cos^2\theta$ for a spin = 1 particle. This angular distribution has been measured at DASP both for electrons and muons. For the electron decay two graphs can interfere (compare fig. 1) and from Q.E.D. one expects an angular distribution as shown in fig. 17a. In the resonance peak a distribution as shown in fig.17b was observed. The difference between the measured results and the Q.E.D. expectation is shown in fig. 17c. This angular distribution is compatible with $1 + \cos^2\theta$ but does not exclude an isotropic decay. In fig. 18 the angular distributions of the muons is shown which again is compatible with a spin = 1 particle but does not exclude completely a spin = 0 particle. This distribution is also a strong argument against a weak boson. Because of parity violation one would expect a forward-backward asymmetry of about 54 %. No such asymmetry is observed with an upper limit of about 11 %. The clean identification of muons as in DASP is quite essential for such measurements.

The inner detector of DASP (see appendix) is very well suited to detect neutral particles, in particular gamma rays. Hence a search was made for the decay of the J-particle into a collinear gamma pair. No such events were found, with a sensitivity of about 8 % relative to the decay into electrons (see fig. 19a). This excludes a J = 0 particle.

A very strong argument for the ψ-particle to have the photon quantum numbers comes from the shape of the resonance curve as observed in the SPEAR magnetic detector with high statistics. As fig. 11 shows there is a destructive interference at the lower side of the resonance and constructive interference at the higher side. Apparently the resonance amplitude is interfering with the Q.E.D.-background amplitude indicating that both have the same quantum numbers.

A Possible Interpretation of the New Particles: The Charm Model

In order to have a better basis to discuss the meaning of further experiments I should like to discuss possible interpretations of the new particles. Since experiments very strongly indicate that these particles are hadrons and not weak bosons one has to explain their properties in terms of strong interaction physics. Their extremely narrow decay width indicates that a new selection is at work which implies the introduction of a new quantum number. Most of the models which have been proposed during the recent months are in one way or the other an extension of the well known quark model. In addition to the three known quarks other quarks are introduced with a new quantum number, e.g. charm, colour, heaviness. The total numbers of quarks in these models are 4, 9, 6; but other models with different numbers of quarks have been proposed. As an example which is chosen more or less by personal prejudice I should like to discuss the charm model. However, it might turn out that the final interpretation might be completely different.

In the usual quark model the three quarks u, d and s can be represented by three arrows in a I_3-Y-plane where I_3 is the third component of isospin and Y is the hypercharge (see fig. 20). In the charm model a fourth quark is added which has a new quantum number charm c = +1. This quark can be represented by an arrow which is perpendicular to the I_3-Y-plane. The properties of this charmed quark are given in fig. 20.

In the frame of the quark model all mesons are obtained by combining a quark with an antiquark. Hence a meson is a bound $q\bar{q}$-state. The baryons are states composed of three quarks. In the following I shall discuss only the mesons. All possible combinations of quarks and antiquarks are obtained if the three pairs representing the antiquarks are superimposed on the quark triangle (fig. 21). One obtains the well known nonet with three particles sitting at the center. Since these particles have the same quantum numbers the physical particles are a mixture of the corresponding wave functions. In terms of a nonrelativistic quark model the spins of the two quarks can either be antiparallel (paraquarkonium) or parallel (orthoquarkonium). Hence a nonet with $J^P = 0^-$ and another one with 1^- is expected. Indeed the pseudoscalar and vector mesons found in nature exactly fit into that scheme (fig. 22).

If one goes through the same procedure with the four quarks one obtains the structure shown in fig. 23 which corresponds to a graphical representation of SU(4). In addition to the well known nonet one obtains three more particles consisting of a charm quark and one of the three ordinary quarks. For these mesons the nomenclature D and F is used. Of course, there are the corresponding antiparticles consisting of a charmed antiquark and one of the ordinary quarks. These triplets are one of the essential predictions of the charm scheme.

One further particle is obtained by combining a charmed quark with its antiquark. The location of this particle is at the center of the ordinary nonet and hence, in principle, this particle can mix with the other central particles. The $c\bar{c}$-particle is identified with the 3.1 GeV-resonance and it can be shown that this particle is essentially a $c\bar{c}$-state with small admixtures only from other wave functions. This is similar to the Φ which is almost a pure $s\bar{s}$-state. Since the total charme charge of the $c\bar{c}$-particle is zero the J-particle is referred to as having hidden charm whereas the D- and F-mesons are real charmed mesons. As explained above each of these states can be formed with the quark spins either parallel or antiparallel, corresponding to the formation of ortho- or para-charmonium.

Like a hydrogen molecule the state formed by a charmed quark and its antiquark can be excited to radial oscillations or rotational excitations. Let us discuss as a next step, therefore, the level scheme of the bound quark-antiquark system. Neglecting for a moment the level splitting due to hyperfine and LS interactions one obtains for various potentials the level scheme shown schematically in fig. 24. For a Coulomb potential the 2s and 1p states are degenerate whereas for the harmonic oscillator the 1p state is halfway between the 1s and 2s state and the latter is degenerate with the 1d level. Theoretical ideas and phenomenological fits indicate that the potential between two quarks is at small distances weaker than the harmonic oscillator but stronger than the Coulomb potential. Among other possibilities a modified Coulomb potential or a linear potential have been tried. The level scheme then is somewhere in between the Coulomb level scheme and the harmonic oscillator levels as indicated in fig. 24.

If we now introduce a level splitting we obtain a level scheme as shown schematically in fig. 25. Each of the s-levels breaks up into the ortho- and paracharmonium. The $2s_1$-state is identified with the 3.7 GeV-resonance and may be the $3s_1$-level can be associated with the broad resonance at 4.2 GeV. From a phenomenological fit to the known masses of particles one finds that the hyperfine structure splitting should be of the order of 50 to 60 MeV and hence the 0^--states should lie close to but somewhat below the 1^--states. As we shall discuss below no such state has been seen so far.

The p-states with S = 1 can be split by LS or tensor forces. The expectation is that the center of gravity of the ^{3}p-states coincides with the ^{0}p-state. It is further expected that the LS splitting is rather small. In addition to that each level now is split by SU(3)-breaking mechanism and in conclusion a rather complicated level scheme is obtained. This is simpler for charmed or hidden charmed particles since in this case one has SU(3)-triplets or even singlets compared to octets or singlets for normal particles with charm C = 0. A further advantage of the charmonium system compared to bound states of ordinary quarks is the fact that from the phenomenological fit one finds that the mass of the charmed quark is much higher than that for ordinary quarks (1.6 GeV compared to 0.5 or 0.35 GeV). As a consequence a nonrelativistic quark model might yield more reliable results for charmonium and hence the charmonium system would play a keyrole in the understanding of the dynamics of the quark system.

It might finally be remarked that the 1^3d_1-state is depressed and gets very close to the $2s_1$-state. As a consequence the ψ'-particle might be a mixture of these two states.

In fig. 26 the total level scheme for the quark-antiquark system is displayed. The group of low lying levels corresponds to the well known particles. The other group of levels corresponds to the charm-anticharm quark system and in the gap between these two groups there are the three levels with charm quantum number different from zero.

For the purpose of showing an alternative possibility of interpretation in fig. 27 the level scheme is shown which one obtains from the assumption of having in total nine so-called coloured quarks. The main distinction between the two schemes is that in the coloured quark level scheme there is a wide gap between the low lying and high lying particles which is not filled as in the charmed quark scheme.

In both schemes as displayed in figs. 26 and 27 electromagnetic transitions are possible between various levels if the corresponding selection rules are fulfilled. Some of these transitions are shown in fig. 27 for the charmonium states. The implication is that one expects to see monoenergetic gamma lines. None have been seen so far; but perhaps it is too early to draw a definite conclusion.

The main reason for introducing a new quantum number was to explain the narrow widths of the new particles in terms of a new selection rule. How this can be achieved is demonstrated for charm in fig. 29 using duality diagrams. Each line represents a quark or an antiquark. A particle with hidden charm can decay according to fig. 29a into two mesons with "open" charm. Since the mass of the charmed quark is about 1.6 GeV and that of an ordinary quark around 0.4 GeV the total energy required is about 4 GeV. Since the masses of ψ and ψ' are lower these particles cannot decay into charmed mesons. However, they could decay into ordinary mesons according to fig. 29c. From the decay of ordinary hadrons, however, a rule has been established which is usually connected with the name of Zweig and which says that decays corresponding to duality diagrams with unconnected quark groups are strongly hindered. According to that rule the decays corresponding to diagrams 29b and c are not completely forbidden but strongly inhibited. This would explain the narrow widths of the ψ- and ψ'-particle. However, a quantitative estimate of the decay widths applying Zweig's rule shows that even with this inhibition one expects wider decay widths than observed experimentally. Therefore, although explained at least qualitatively the very narrow widths of the new particles are still somewhat puzzling. If the peak observed at 4.2 GeV is associated to the second radial excitation of the quarkonium system its rather large widths could be associated to the decays into charmed mesons. However, no such mesons have been found so far.

Before concluding this chapter it should be mentioned that charm has been introduced before the detection of the new particles for a completely different reason. Weak interactions can be described by the coupling of pairs of particles. With the old quark model having three quarks it was difficult to group them in a natural way in pairs. With four quarks, however, this can easily be done as is shown in fig. 30. There is complete symmetry between the two lepton pairs and the two quark pairs also in the sense that the two particles in one pair differ by one unit of electric charge. The main difference is that the leptons have integer charges whereas the quarks have broken charges. A further complication which cannot be discussed here is that the d-quark has a small admixture of the s-quark and vice versa. The angle determining this admixture is the well known Cabibbo angle. The big success of this scheme is that it explains the absence of neutral currents in processes with a change of strangeness, like the decay $K^0 \to \mu^+\mu^-$.

Hadronic Decay Channels of the New Particles

During the last months a considerable amount of data on decay channels of the new particles have been obtained as well at SPEAR as at DORIS. For the 3.1 GeV-resonance the data are summarized in fig. 31.

In DASP non-collinear gamma-ray events have been studied. Such events could originate from a decay into $\pi^o\gamma$ or $\eta_C\gamma$ where η_C is the 0^--paracharmonium state somewhat below the 3.1 GeV-resonance. The η_C would decay into two gammas which are emitted almost in opposite directions. No $\pi^o\gamma$ event has been seen; but the limit is still somewhat too high to be conclusive. An improvement by a factor of 10 or even 100 would provide a very critical test for various models. The fact that the decay into $\eta_C\gamma$ has not been seen could be explained by a small branching ratio for $\eta_C \to \gamma\gamma$. Otherwise the non-existence of the η_C would be embarrassing for the charm model.

The absence of the decay into $\pi^+\pi^-$ and K^+K^- is not surprising since these decay channels would be forbidden for a I = 0 state.

The decay into proton-antiproton pairs has been seen although it is rather rare. The decay into lambda-antilambda has also been seen and it is comparable to the $p\bar{p}$ decay. The approximate equality of these two decay rates is an indication that the decay does not go via a virtual gamma since in this case the branching ratio should be 1/9. The smallness of the $\Gamma(p\bar{p})$ and $\Gamma(\Lambda\bar{\Lambda})$ is also an indication that the interpretation of the ψ-particles as bound baryon-antibaryon states is hardly acceptable.

Very recently DASP succeeeded to observe three gamma events. From the invariant mass of two of the gamma rays three events could be identified as $\psi \rightarrow \eta\gamma$. The limits for the decay width which can be deduced from this result are indicated in fig. 31. This is a very interesting result since in the coloured quark scheme one expects a width of the order of 1 MeV. In the charmed quark scheme a width of about 2 keV is expected. This experimental result, therefore, seems to be in disagreement with the colour prediction and is still perhaps a little bit low for the charm model.

Decays into many pions have been studied at SPEAR. The decay channels which have been seen are given in fig. 31 and examples are shown in fig. 32 and 33.

For the 3.7 GeV-resonance the very interesting result was found that it decays mainly via a cascade through the 3.1 GeV-particle. This was seen at SPEAR by looking at missing mass plots (fig. 34) and at DORIS with DASP which has very good muon identification and could trigger on muon pairs with a mass of 3.1 GeV. In DASP the detection of neutral particles was also possible. The results for the various branching ratios are shown in figs. 35 and 36. These results can be summarized in the following way: About 60% of the decay of the ψ'-particle go via the ψ-cascade. This cascade in turn goes mainly via the channel $\psi' \rightarrow \psi + \pi\pi$ (more than 80%). The fraction going into two π^o agrees very nicely with the expectation for the two pions being in an isospin 0 state. About 10 % of the cascade go via the channel $\psi' \rightarrow \psi + \eta$ which has been clearly seen at DASP.

The fact that ψ' mainly decays via the ψ-particle indicates that the two particles are very similar in nature and, therefore, the interpretation that the ψ' is a radial excitation of the ψ is supported by experiment. On the other hand phase space would allow many other direct decay channels into mesons not going through ψ and a dynamical principle which is not yet understood must suppress such decays. The branching ratio $\psi' \to \psi + \eta$ seems to be too small in terms of the colour model. The decay $\psi' \to p\bar{p}$ is more frequent than the corresponding decay of the ψ-particle, a fact which still needs an explanation.

Inclusive Spectra

Single particle inclusive spectra have been measured with DASP at the 3.1 GeV-resonance. This means that in the process $e^+e^- \to h^\pm X$ the momentum spectrum of the single hadron h is measured.

In fig. 37 the momentum distribution of charged hadrons without identifying different kinds of particles is shown. For comparison the previously at SPEAR measured spectra of the resonance are also included in this figure. As one notices there is no difference between the two spectra.

By using time-of-flight and range measurements DASP is able to identify different particles up to momenta of about 1.5 GeV/c. The results for π, K and $\bar{p}$ are shown in fig. 38. In this figure the invariant cross section is plotted against the hadron energy. In this kind of plot the spectra for the different particles almost coincide.

In fig. 39 the ratios of kaons and antiprotons relative to pions are plotted. For comparison some older SLAC data are again included which were taken off the resonance. Unfortunately the overlap of the two measurements is rather small. However, it seems that the trend of the data is the same. In particular there is no indication for more abundant kaon production at the resonance. This would have been expected in the frame of the charm model. In fig. 40 finally the particle ratios are plotted as a function of the Feynman scaling variables. Within these still rather large errors the data are compatible with a flat distribution.

Photoproduction of the New Particles

Recently in an experiment at FNAL the photoproduction of muon pairs was studied. The invariant mass distribution of muon pairs which was obtained is shown in fig. 41. Above the QED distribution a prominent peak at 3.1 GeV is seen. In fig. 42 the t-distribution for the production of this particle is shown. The very steep rise at its smallest t-value can be attributed to coherent production off the nucleons in the Be-nucleus. The distribution at higher t-values has a slope of about 4 $(\mathrm{GeV}/c)^{-2}$. This indicates that the particle is produced diffractively. These results are rather important since they indicate strongly that the particle is a hadron. In terms of the vector dominance model one can derive the total cross section for the scattering of the ψ-particle on protons. The related expressions are given in fig. 43. As a result one finds a total cross section of about 1 mb which is low but still typical for a hadronic interaction.

It should also be mentioned that a fit of the dynamical non-relativistic quark model to the level scheme as explained above led to the conclusion that the geometric size of the charmonium system is about a factor 2 smaller than for ordinary particles. This is corroborated by the t-distribution shown in fig. 42 since the slope is considerably smaller than for ordinary particles.

In the meantime the photoproduction of the ψ-particle has also been observed at CORNELL and SLAC at lower energies. The preliminary results are summarized in fig. 44. Some people speculate that the data might hint at the existence of a threshold which could be associated to the opening-up of new channels like the production of charmed particles.

In an experiment at FNAL the 3.1 GeV-resonance could also be produced by neutrons hitting Be-nuclei. The mass distribution of the observed muon pairs is shown in fig. 45. A comparison of these results with the cross sections obtained in e^+e^- collisions tells us that the ψ-particle is directly produced on a nucleon without an intermediate photon. The details cannot be discussed here.

APPENDIX

DORIS and its Detectors

Some information on the electron-positron storage ring in Hamburg and its detectors might be of interest.

The double-ring system DORIS came into operation at the beginning of 1974 and the experimental programs started earlier than originally scheduled in October 1974. DORIS is the only storage ringe in existence with two separated rings which allows not only to collide positrons on electrons but also electrons on electrons and electrons on protons. One has to pay for this advantage by having a very complicated machine with beams crossing at a finite angle (24 mrad in our case). Presently the storage ring is operated at energies between 1.5 to 3.4 GeV per beam. After increasing the RF power and improving the injection system it should be possible to reach energies up to 4.2 GeV per beam at the beginning of 1976. By improving further the rf system and adding power supplies to the magnet it is hoped that energies up to 5 GeV per beam could be reached toward the end of 1976.

Fig. 46 shows the schematic lay-out of DORIS and its injection system. Fig. 47 shows a plan of DORIS with its main parameters. Typical beam parameters are given in fig. 48.

In order to observe the new particles the request from the experimenters has been to make the energy spread in each beam as small as possible. An energy spread of about 500 keV could be obtained by special measures. Under these conditions experimental runs are performed with electron and positron currents between 200 and 300 mA. Yielding luminosities of about 10^{30} cm^{-2} sec^{-1} at 2 GeV per beam. Beam live times of many hours are achieved. Without the boundary on the momentum spread in the beam currents between 1/2 and 1 A have been obtained.

Fig. 49 shows a top view of the double arm spectrometer DASP. On each side of the detection region a magnet can be recognized (total weight 500 tons). In front and behind the magnet spark chambers are placed to measure the direction of the particles before and after magnetic deflection. The spark chambers are followed by time-of-flight counters, shower counters and steel absorbers. The measurement of momentum, time-of-flight, shower production and range in steel allows an excellent discrimination of various particles. Apart from particle identification very good momentum measurements can be achieved (less than 1 %). On the other hand the solid angle is limited to about 1 sr. Inside the magnets surrounding the intersection region the so-called inner detector is installed (fig. 50). It consists of layers of scintillation counters, proportional tubes and shower counters. It covers almost 90 % of 4π and allows the detection of charged and neutral particles.

In fig. 51 the spectrometer PLUTO is shown. It consists of a superconducting solenoid about 1.5 m long and with the same diameter. This spectrometer has a moderate momentum resolution but covers a very big solid angle and, therefore, is complementary to DASP. Inside the solenoid there is a number of cylindrical proportional chambers which can be sandwiched with a lead absorber in order to distinguish neutral particles. The iron yoke for the magnetic return flux has slots in which chambers can be placed in order to identify muons.

In fig. 52 an installation is shown which consists essentially of sodium iodid crystals and lead-glass counters. The intersection region is surrounded by cylindrical drift chambers and scintillation counters. A converter, consisting of mercury, can easily be removed and put back in place. The main interest will be, of course, to investigate photon spectra. This equipment is presently installed at one of the intersection regions and is taking data.

Electron-Electron Interaction at high energies

1. Quantum electrodynamics

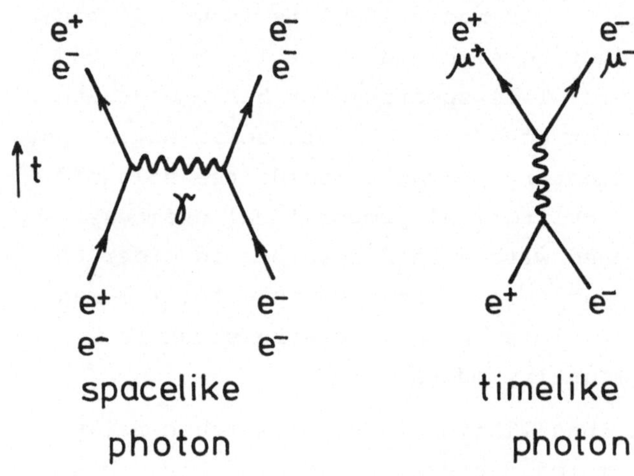

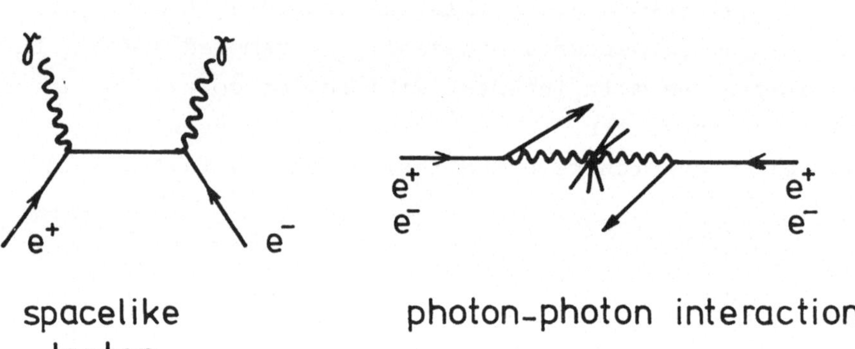

to separate graphs compare $e^+ e^-$
$e^- e^-$
$e^+ e^- \longrightarrow \mu^+ \mu^-$

Fig. 1

Validity of QED

$e^+e^- \to e^+e^-$, $e^+e^- \to \mu^+\mu^-$ (SLAC)

E = 3.0, 3.8, 4.8 GeV

	Assumption	Λ_+ (GeV)	Λ_- (GeV)
e^+e^-	Λ_S	> 15	> 19
	Λ_T independent	> 13	> 16
	$\Lambda_S = \Lambda_T$	> 15	> 19
e^+e^-	Λ_S	> 21	> 23
	Λ_T independent	> 33	> 36
and	$\Lambda_S = \Lambda_T$	> 35	> 47
$\mu^+\mu^-$	Λ_e	> 21	> 19
	Λ_μ independent	> 27	> 16

$$F(Q^2) \simeq 1 \pm \frac{Q^2}{\Lambda_\pm^2}$$

<u>e, μ pointlike</u>

Λ = 20 GeV $\hat{=}$ r = 10^{-15} cm

Fig. 2

Hadronic Interaction

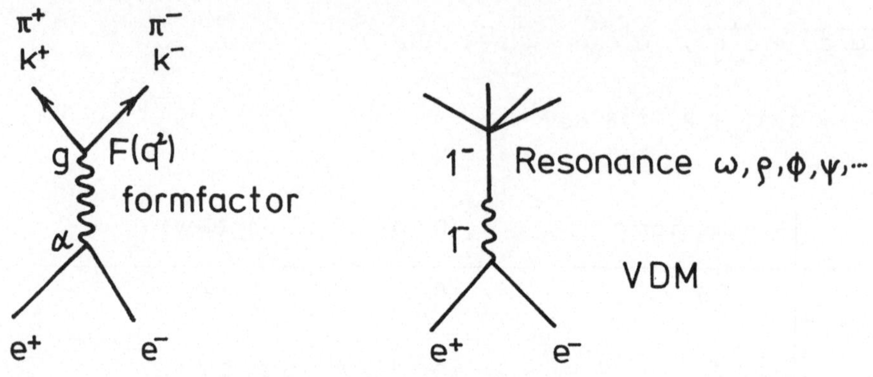

Multihadron production

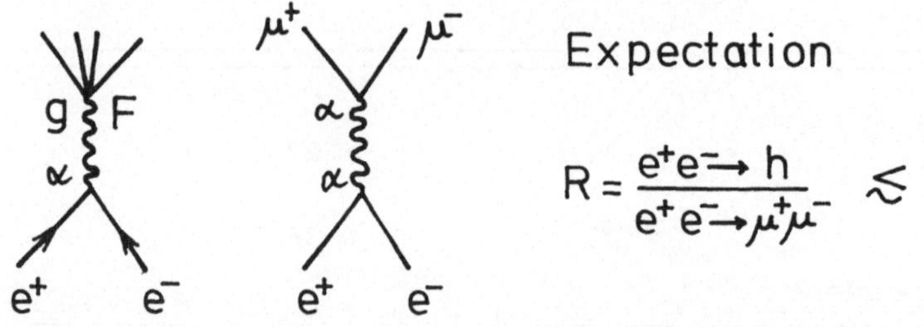

Expectation

$$R = \frac{e^+e^- \to h}{e^+e^- \to \mu^+\mu^-} \lesssim 1$$

However, if hadrons contain pointlike constituents (F = 1)

asymtotic ($E^* \to \infty$)

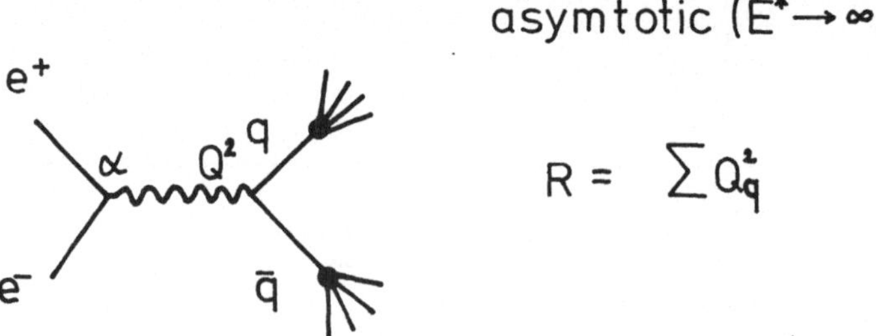

$$R = \sum Q_q^2$$

Fig. 3

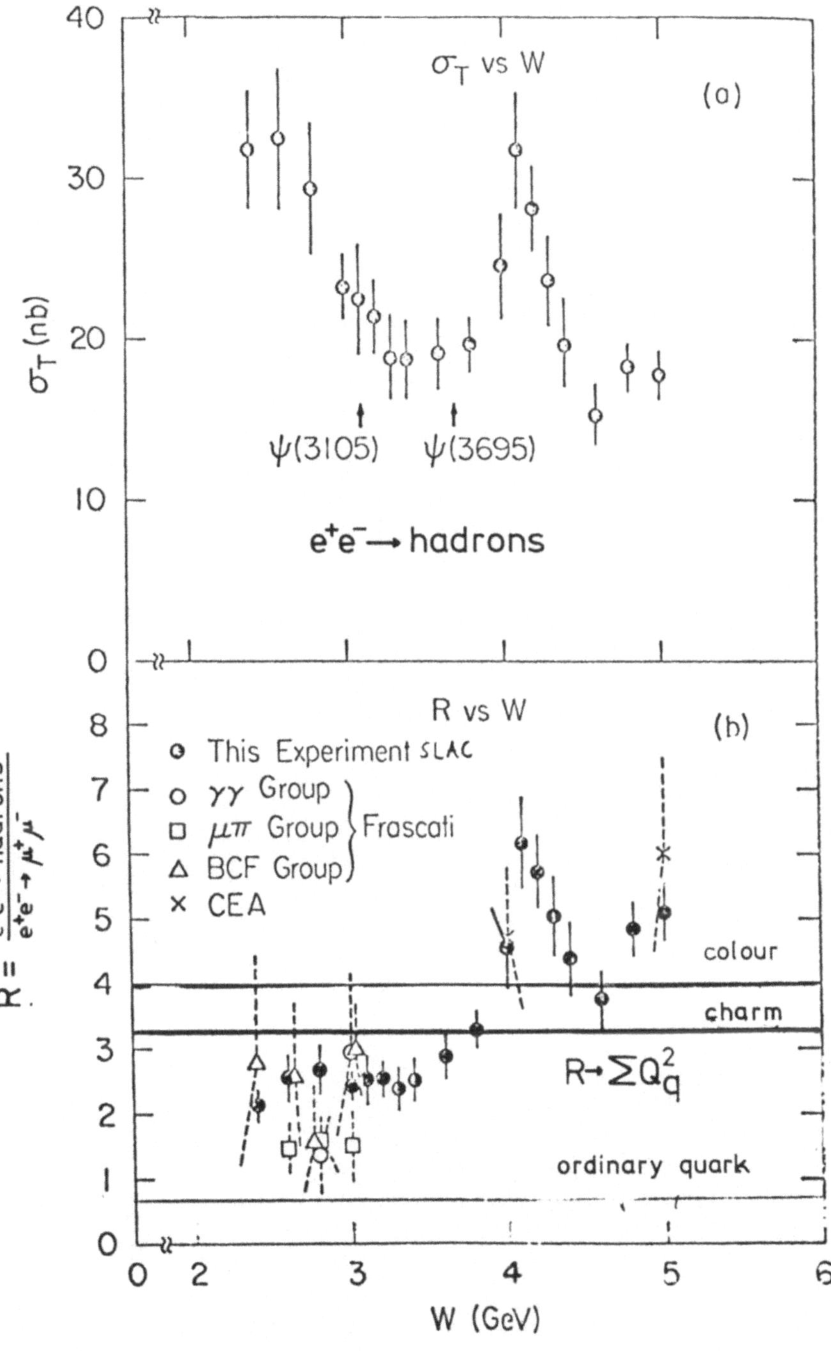

σ_T vs W
(a)
σ_T (nb)
ψ(3105)
ψ(3695)
$e^+e^- \to$ hadrons
R vs W
(b)
This Experiment SLAC
$\gamma\gamma$ Group
$\mu\pi$ Group
BCF Group
Frascati
CEA
$R = \frac{e^+e^- \to \text{hadrons}}{e^+e^- \to \mu^+\mu^-}$
colour
charm
$R \to \Sigma Q_q^2$
ordinary quark
W (GeV)

Fig. 4

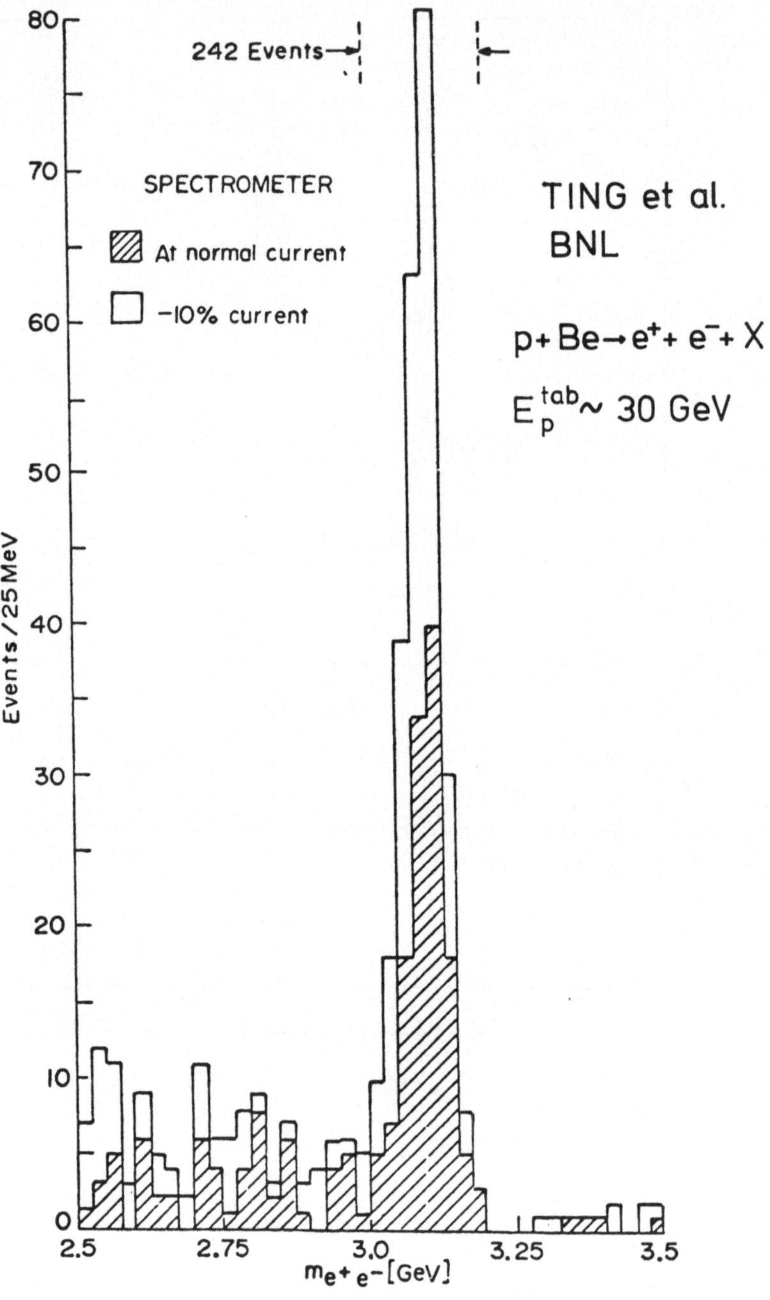

80
70
60
50
40
30
20
10
0
Events / 25 MeV
242 Events
SPECTROMETER
At normal current
-10% current
TING et al.
BNL
p+ Be→e⁺+ e⁻+ X
E_p^tab ~ 30 GeV
2.5
2.75
3.0
3.25
3.5
m_e+ e- [GeV]

Fig. 5

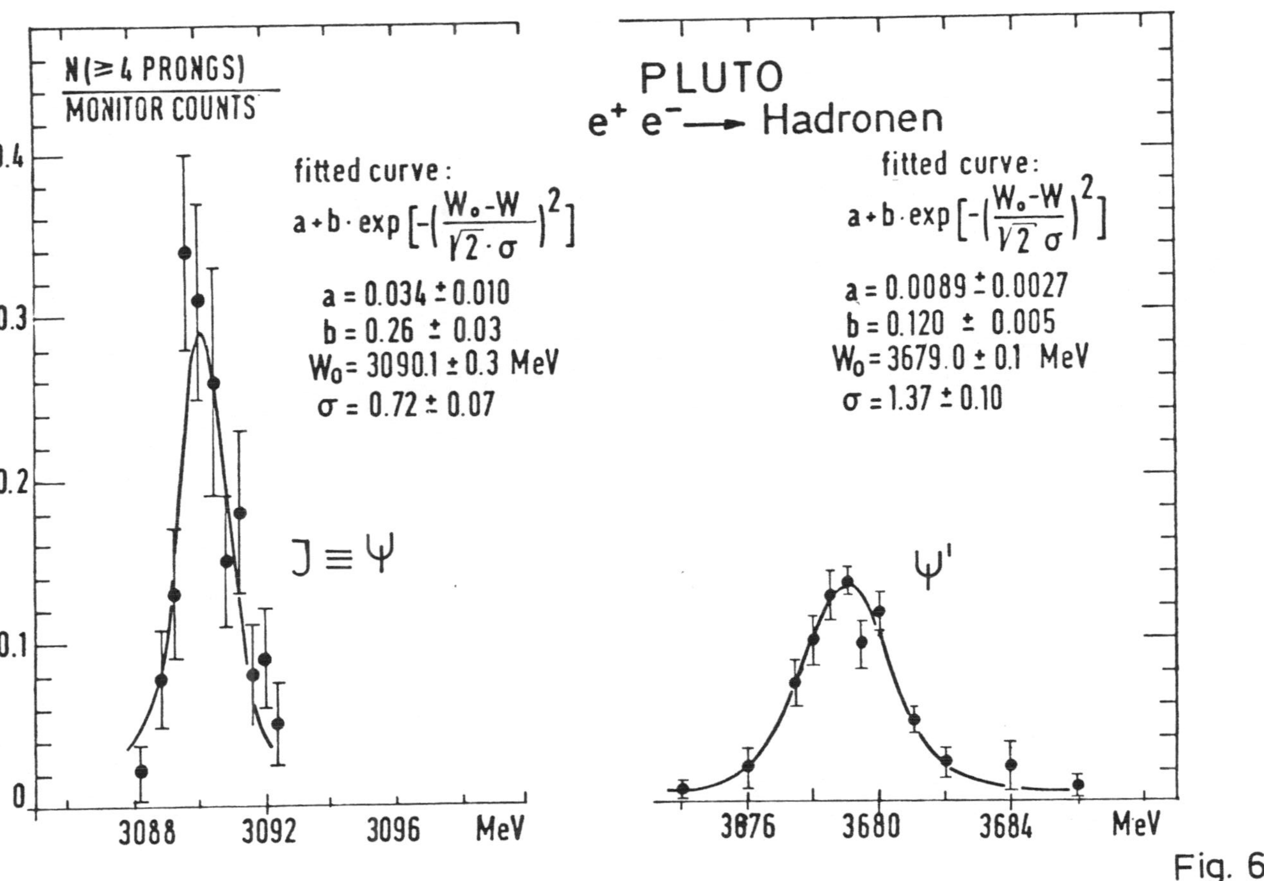

N(≥ 4 PRONGS)
MONITOR COUNTS
PLUTO
e+ e- ⟶ Hadronen
fitted curve:
a + b · exp[-(W₀-W / √2 · σ)²]
a = 0.034 ± 0.010
b = 0.26 ± 0.03
W₀ = 3090.1 ± 0.3 MeV
σ = 0.72 ± 0.07
J ≡ ψ
fitted curve:
a + b · exp[-(W₀-W / √2 σ)²]
a = 0.0089 ± 0.0027
b = 0.120 ± 0.005
W₀ = 3679.0 ± 0.1 MeV
σ = 1.37 ± 0.10
ψ'
0.4
0.3
0.2
0.1
0
3088
3092
3096
MeV
3876
3680
3684
MeV

Fig. 6

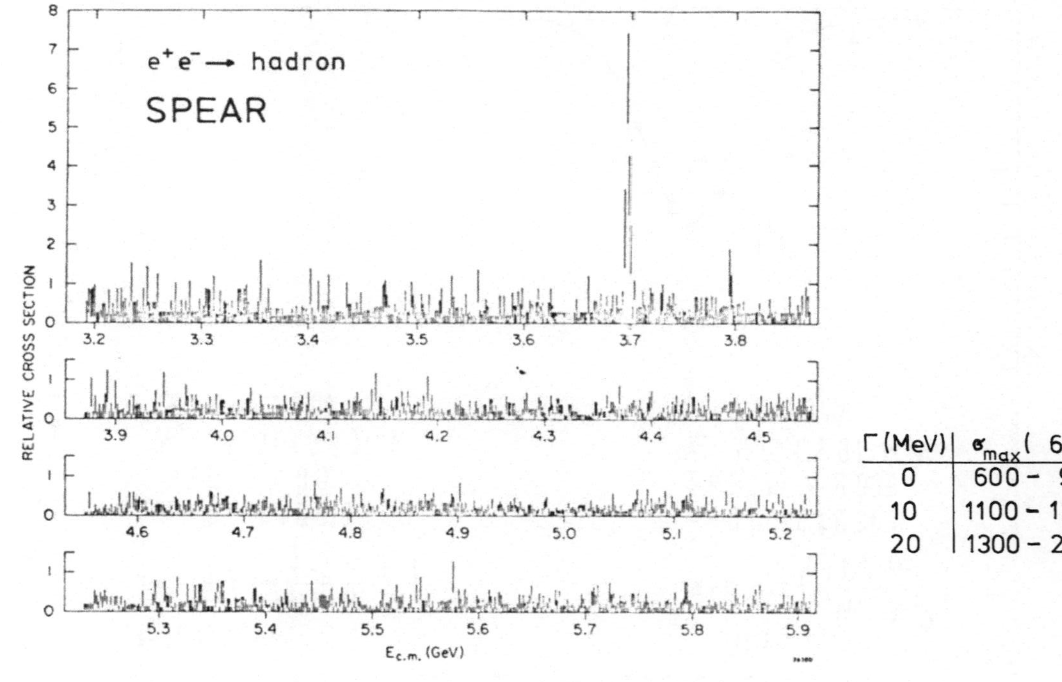

Γ (MeV)	σ_{max} (6MeV)
0	600 - 900
10	1100 - 1700
20	1300 - 2200

Fig. 7

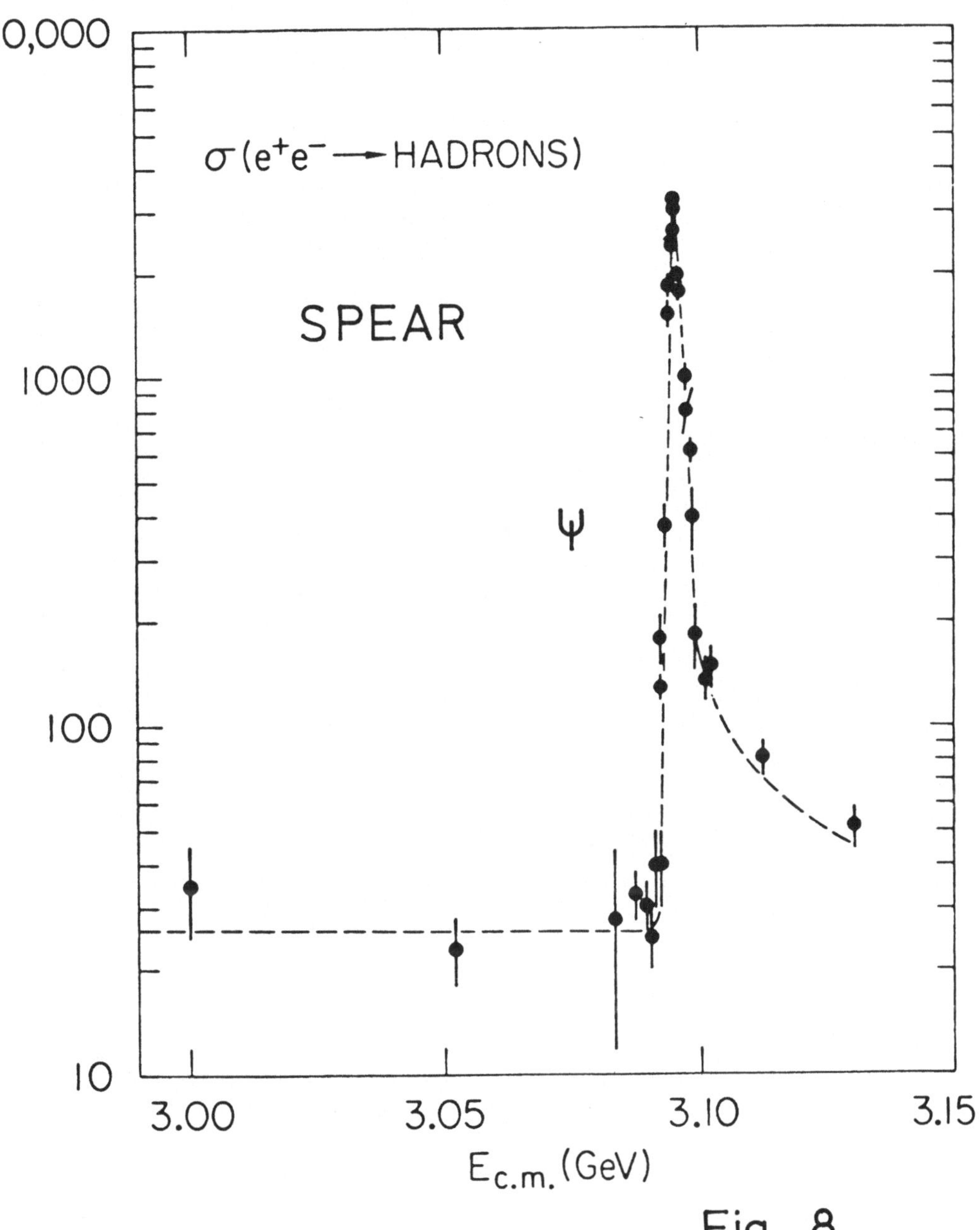

Fig. 8

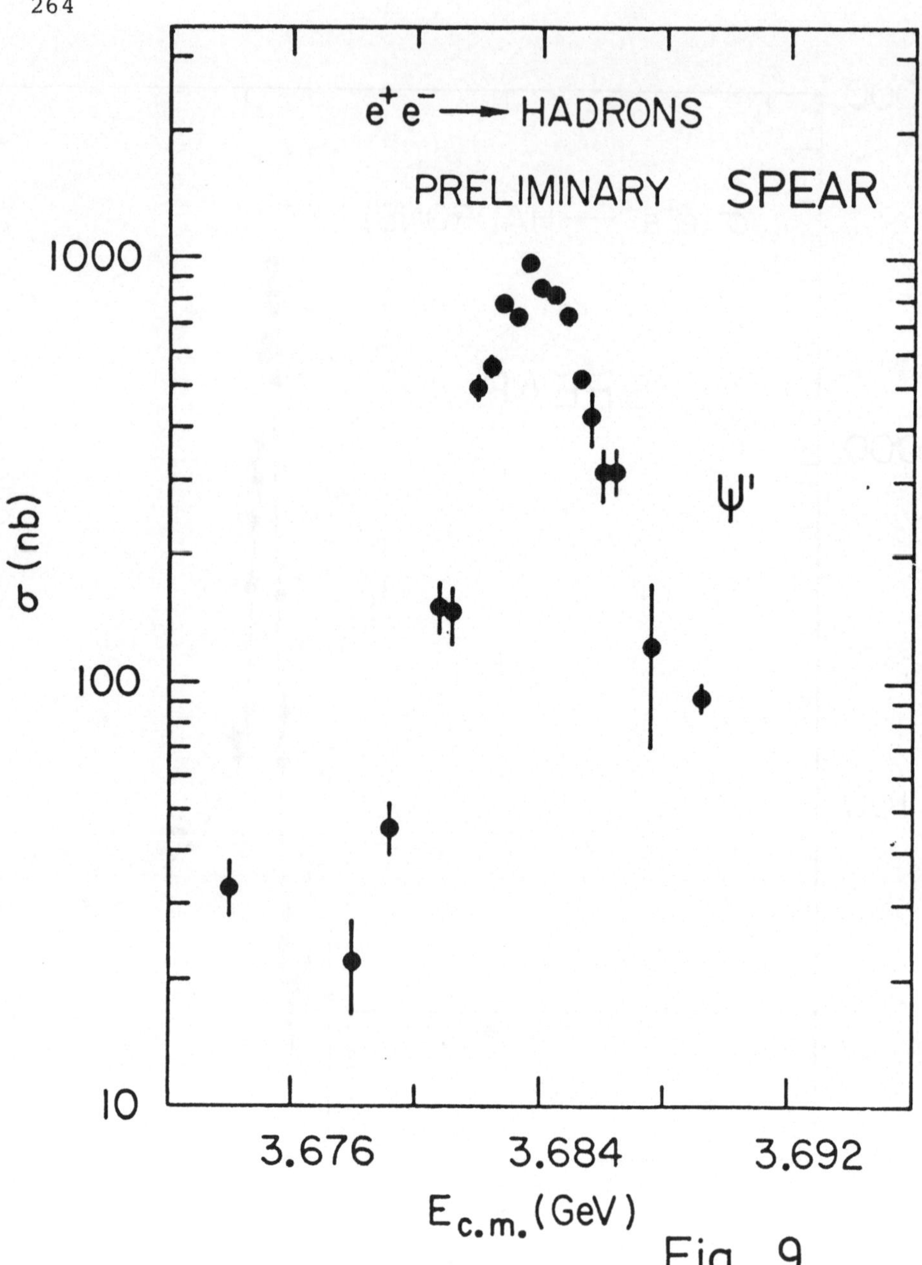

e⁺e⁻ ⟶ HADRONS
PRELIMINARY
SPEAR
ψ'
1000
100
10
σ (nb)
3.676
3.684
3.692
E c.m. (GeV)

Fig. 9

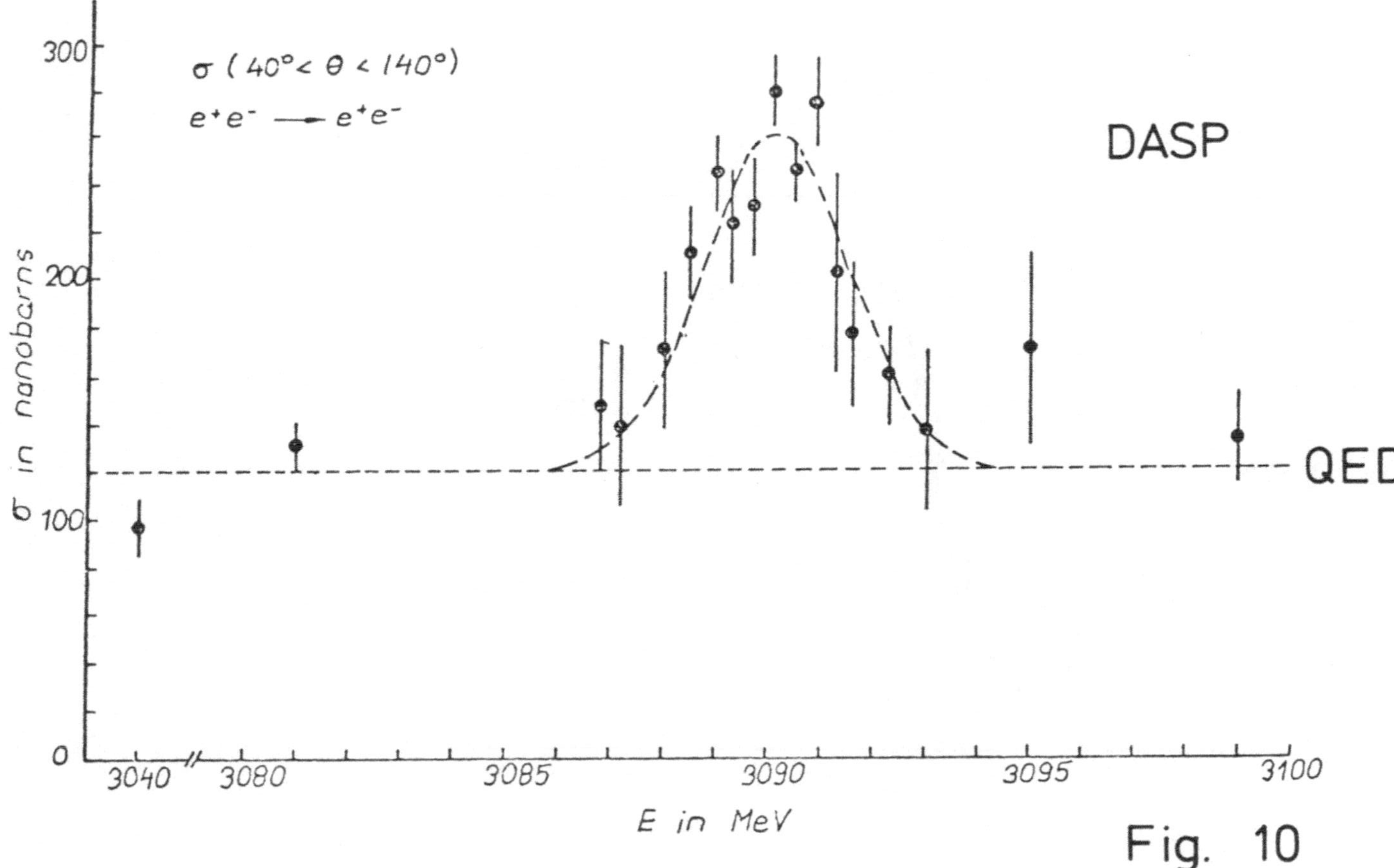
σ (40° < θ < 140°)
e⁺e⁻ → e⁺e⁻
DASP
QED
300
200
100
0
σ in nanobarns
3040
3080
3085
3090
3095
3100
E in MeV

Fig. 10

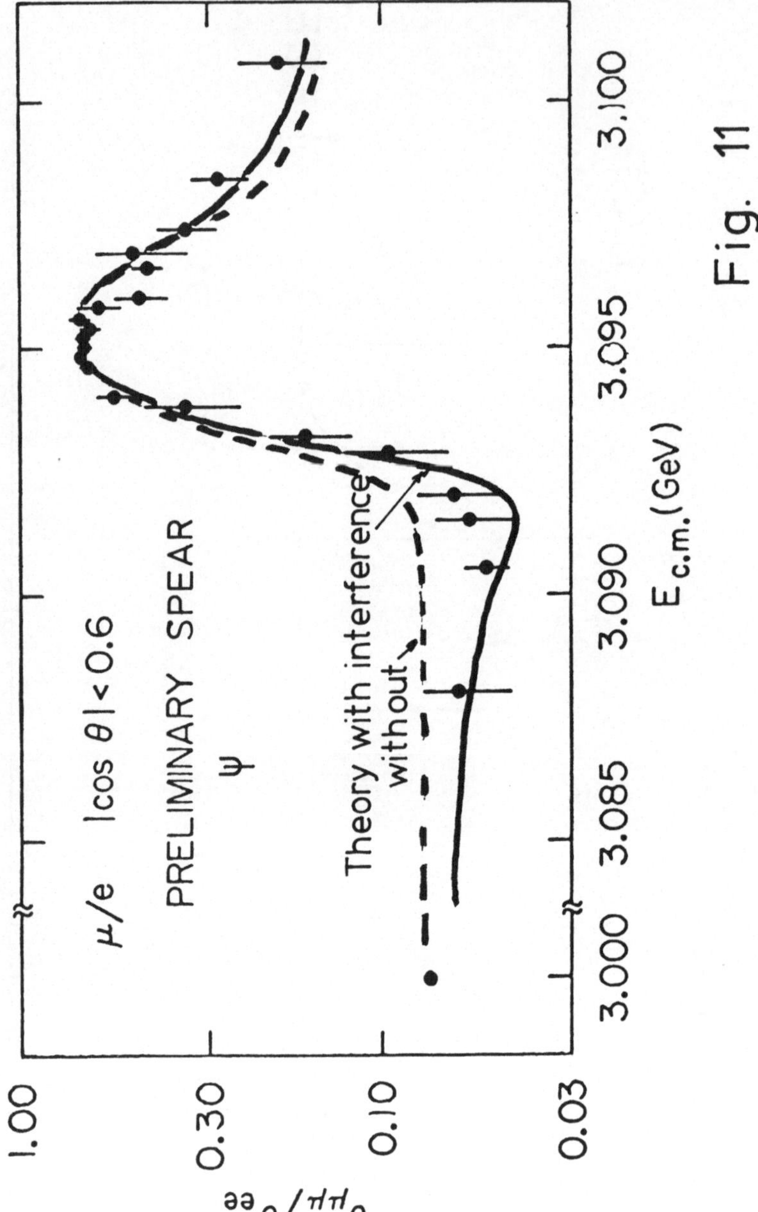

μ/e |cos θ|<0.6
PRELIMINARY SPEAR
ψ
Theory with interference
without
$\sigma_{\mu\mu}/\sigma_{ee}$
1.00
0.30
0.10
0.03
3.000
3.085
3.090
3.095
3.100
$E_{c.m.}$ (GeV)

Fig. 11

Breit-Wigner Resonance 1^-

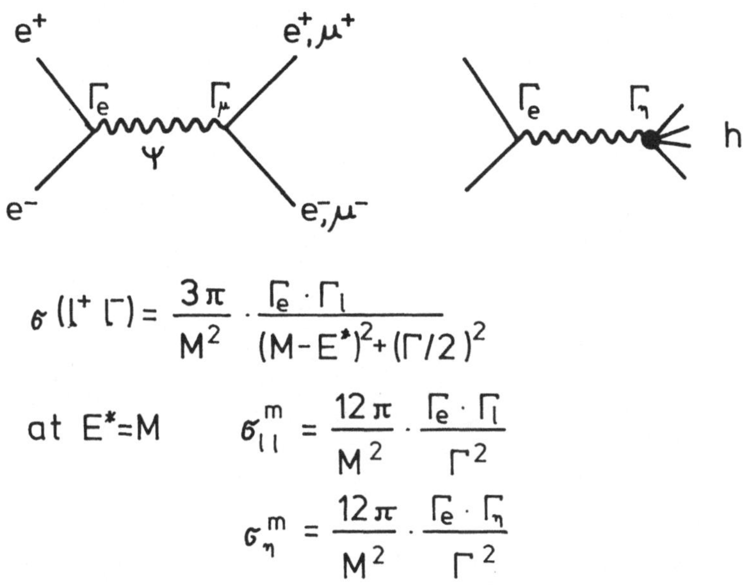

e^+
e^+, μ^+
Γ_e
Γ_μ
Ψ
e^-
e^-, μ^-
Γ_e
Γ_η
h
$\sigma(l^+ l^-) = \frac{3\pi}{M^2} \cdot \frac{\Gamma_e \cdot \Gamma_l}{(M-E^*)^2 + (\Gamma/2)^2}$
at $E^* = M$
$\sigma_{ll}^m = \frac{12\pi}{M^2} \cdot \frac{\Gamma_e \cdot \Gamma_l}{\Gamma^2}$
$\sigma_\eta^m = \frac{12\pi}{M^2} \cdot \frac{\Gamma_e \cdot \Gamma_\eta}{\Gamma^2}$

$$\frac{\sigma_\eta^m}{\sigma_e^m} = \frac{\Gamma_\eta}{\Gamma_e} \quad \text{independent of resolution}$$

$$\int \sigma_h \cdot dE^* \approx \sigma_\eta^m \cdot \Gamma = \frac{12\pi}{M^2} \cdot \frac{\Gamma_e \cdot \Gamma_\eta \cdot \cancel{\Gamma}}{\Gamma^{\cancel{2}}}$$

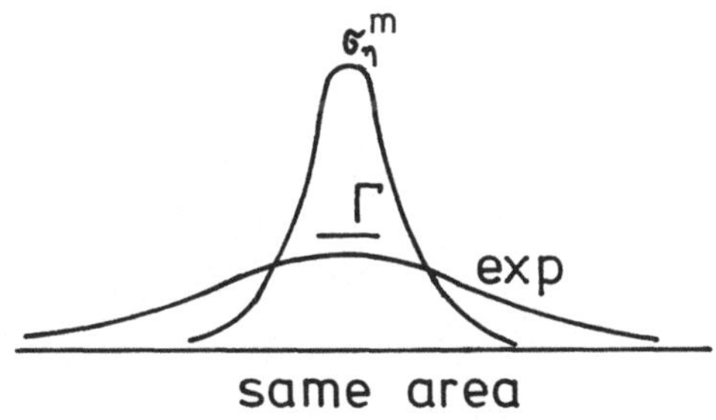

$$\Gamma_e = \frac{M^2}{12\pi} \cdot \underbrace{\frac{\Gamma}{\Gamma_\eta}}_{\sim 1} \cdot \int \sigma_\eta \cdot dE^*$$

Fig. 12

Results for Ψ (3.1)

$\underline{\underline{\Psi}}$	M = 3095 MeV ±5	(S)	SPEAR: (S)
	3090 MeV	(D)	DORIS: (D)
	3101 MeV	ADONE	

$\int \sigma_{tot}\, dM = 10800$ nb MeV ± 25 % (S)
(radiative corrections)

$\Gamma_{tot} = 77$ keV ± 20 (S)

$\Gamma_{ee} = 5.5 \pm 0.4$ keV (D), 5.2 keV (S)

$\Gamma_{\mu\mu} = 6.9 \pm 1.8$ keV (D)

---

indication
for spin 1^-: $e^+e^- \to \mu^+\mu^-$ interference with QED (S)

$1+\cos^2\Theta$ for $e^+e^- \to e^+e^-$ (D)

$e^+e^- \to \mu^+\mu^-$ (D)

$\Psi \to \gamma\gamma$ forbidden, $\Gamma_{\gamma\gamma} < 0.4$ keV (D)

Decay: direct: indirect:

Ψ I Ψ γ II

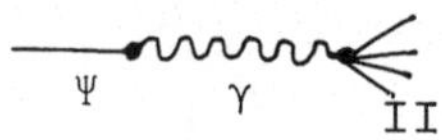

$$\Gamma_{\eta}^{II} = \Gamma_{\mu\mu} \cdot \underbrace{\left\{\frac{\sigma(e^+e^- \to h)}{\sigma(e^+e^- \to \mu^+\mu^-)}\right\}}_{R=2.4\pm 0.3} \text{ outside resonance}$$

$$\Gamma_{\eta}^{II} = 2.4 \times 5.5 = 13 \text{ keV}$$

Γ^{II}:	η	13 keV
	ee	5.5keV
	$\mu\mu$	6.5keV
		25 keV
Γ^{I}		52 keV

Fig. 13

Ψ(3.1 GeV) DASP-Results

$\int \sigma(e^+e^-)\, dE = 965 \pm 140$ nb MeV

$\Gamma^2_{ee}/\Gamma = 0.41 \pm 0.06$ keV

$\int \sigma(\mu^+\mu^-)\, dE = 1240 \pm 230$ nb MeV

$\Gamma_{\mu\mu}\, \Gamma_{ee}/\Gamma = 0.51 \pm 0.09$

$\Gamma_{ee}/\Gamma_{\mu\mu} = 1.24 \pm 0.28$ universality

$\Gamma_{\gamma\gamma} < 0.4$ keV

Fig. 14

Ψ' $M = 3684$ MeV (SPEAR)

3680 MeV (DORIS)

$\int \sigma_{tot}\, dM = 3700$ nb MeV (corrected) (S)

$\Gamma_{tot} = 200 - 800$ keV (S)

$\Gamma_{ee} = 2.2$ keV $\pm$ 0.5 keV (S)

4.15 GeV Resonance, threshold?

$\int \sigma_{tot}\, dM = 3600$ nb MeV (S)

$\Gamma_{tot} = 250 - 300$ MeV

$\Gamma_{ee} = 4.0 \pm ?$ keV

Fig. 15

Nature of particles?

Compare:

1) <u>electromagnetic:</u> $\Gamma(\rho \rightarrow e^+e^-) = 6.5$ keV
$\Gamma(\Phi \rightarrow e^+e^-) = 1.3$ keV
comparable to
$\Gamma(\Psi \rightarrow e^+e^-)$
$\Gamma(\Psi' \rightarrow e^+e^-)$

2) <u>weak boson:</u>

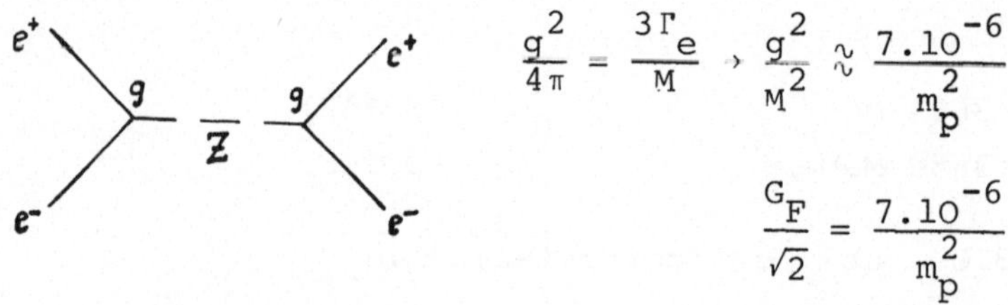

$$\frac{g^2}{4\pi} = \frac{3\Gamma_e}{M} , \quad \frac{g^2}{M^2} \approx \frac{7.10^{-6}}{m_p^2}$$

$$\frac{G_F}{\sqrt{2}} = \frac{7.10^{-6}}{m_p^2}$$

but:
- why Ψ, Ψ' ?
- why m so small ? $\qquad m(Z) \gtrsim 40$ GeV
- no μ asymmetry
- $\Psi(3.7) \rightarrow \Psi(3.1) + \pi^+\pi^-$ cannot be explained.

3) <u>strong interaction:</u>

typically $\Gamma \approx 40 \ldots\ldots 200$ MeV at $m \approx 3$ GeV

$\Gamma_{tot}(\Psi) = 77$ keV about 10^{-3} too small

explanation: i) new quantum number (approximate selection rule)
ii) unknown effect

Fig. 16

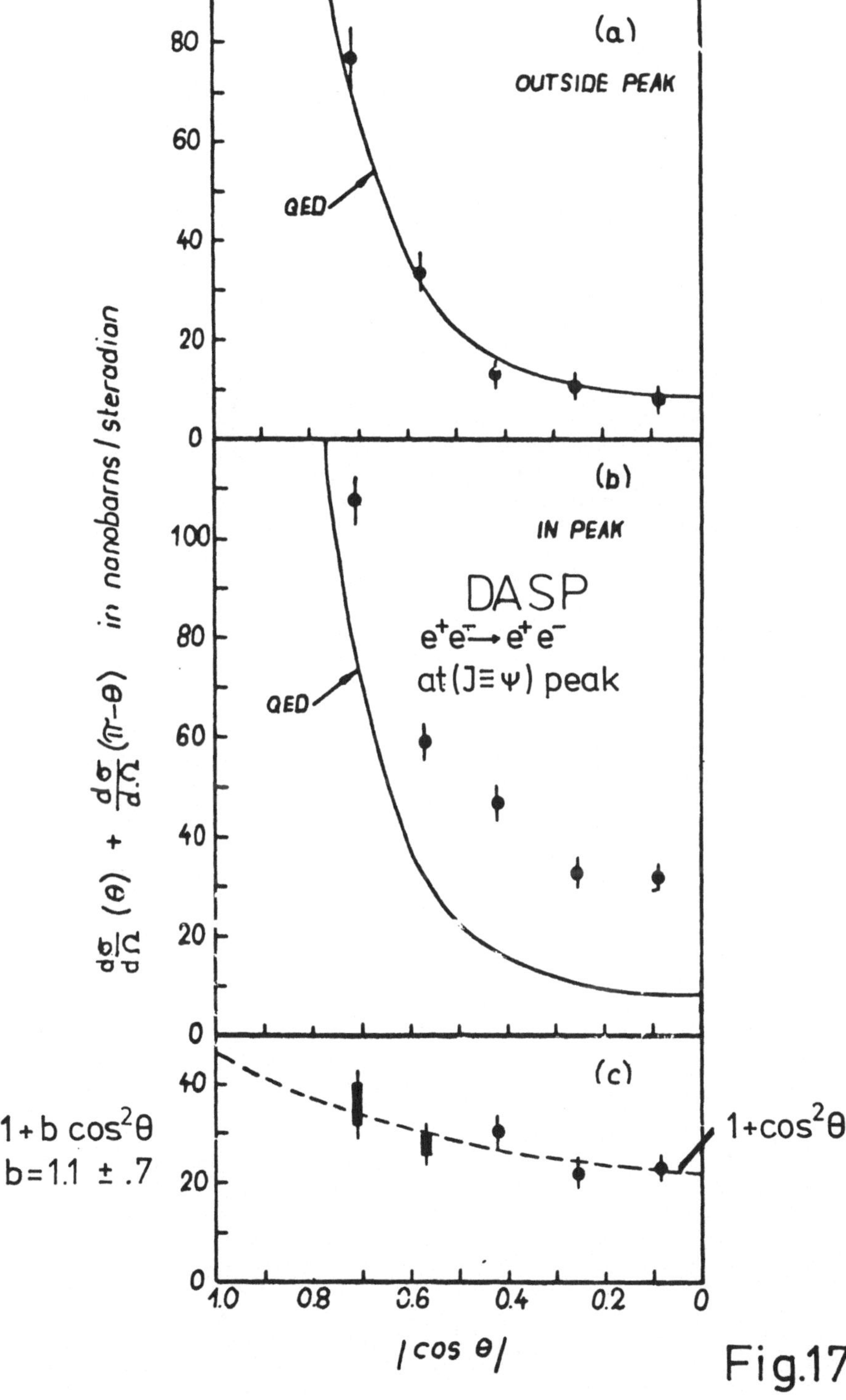
(a)
OUTSIDE PEAK
QED
(b)
IN PEAK
DASP
$e^+e^- \to e^+e^-$
at (J ≡ ψ) peak
QED
(c)
1+b cos²θ
b=1.1 ± .7
1+cos²θ
$\frac{d\sigma}{d\Omega}(\theta) + \frac{d\sigma}{d\Omega}(\pi-\theta)$ in nanobarns / steradian
|cos θ|

Fig.17

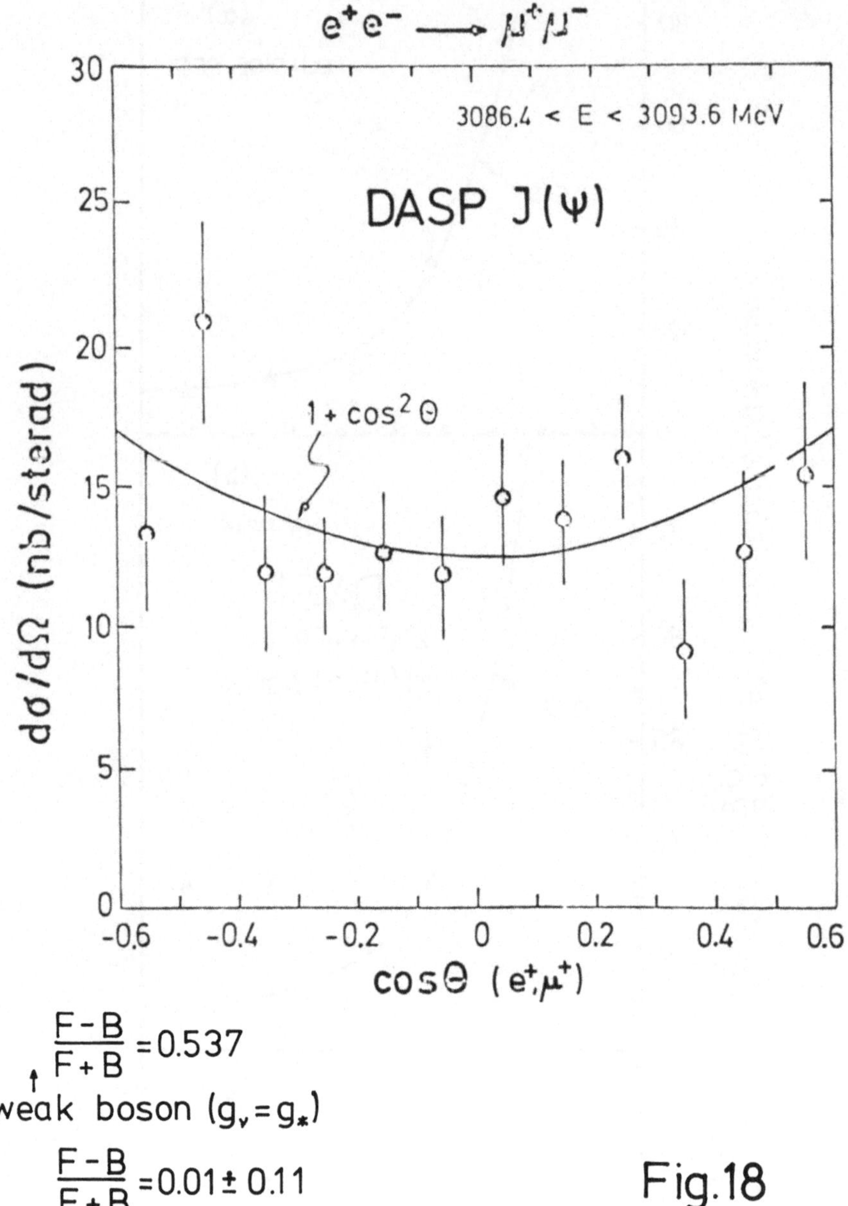
$e^+e^- \longrightarrow \mu^+\mu^-$
3086.4 < E < 3093.6 MeV
DASP J(ψ)
$1 + \cos^2\Theta$
$d\sigma/d\Omega$ (nb/sterad)
$\cos\Theta$ (e^+,μ^+)
0
5
10
15
20
25
30
-0.6
-0.4
-0.2
0
0.2
0.4
0.6
$\frac{F-B}{F+B} = 0.537$
weak boson $(g_v = g_*)$
$\frac{F-B}{F+B} = 0.01 \pm 0.11$

Fig.18

Fig.19

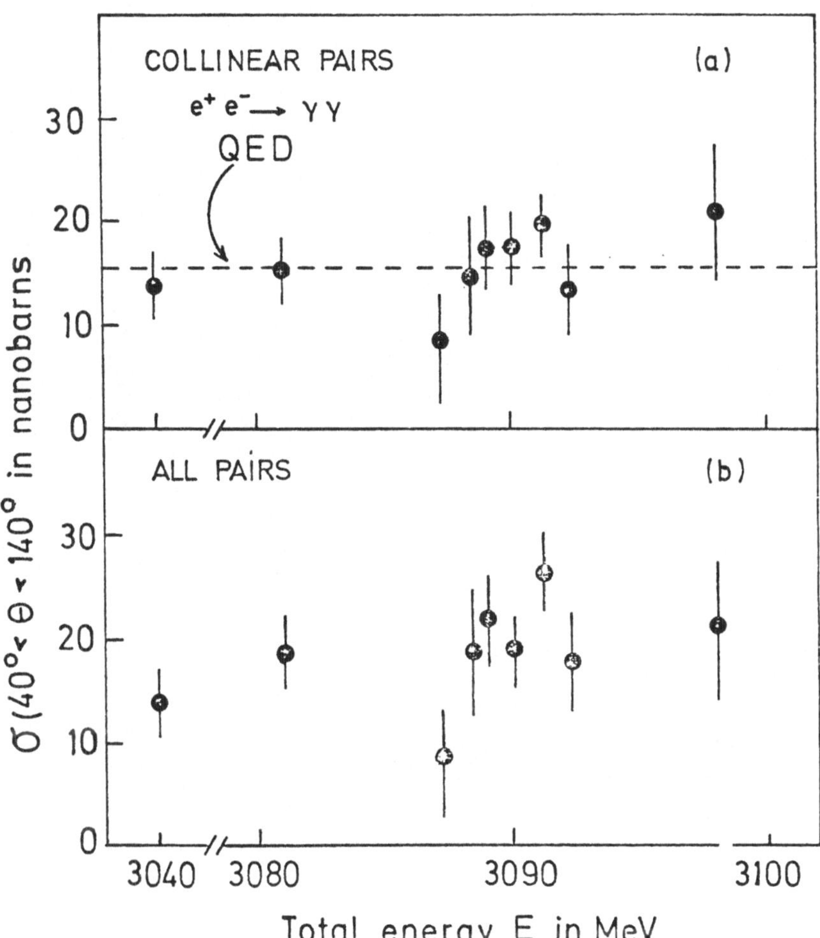

$\Gamma_{\gamma\gamma} / \Gamma_{ee} < 0.08$ (90%C.L)

$\Gamma_{\pi^0\gamma} / \Gamma_{ee} < 0.08$

$\Gamma_{\chi\gamma} / \Gamma_{ee} < 0.2$

$M_\chi > 2.6$ GeV, $\chi \to \gamma\gamma$

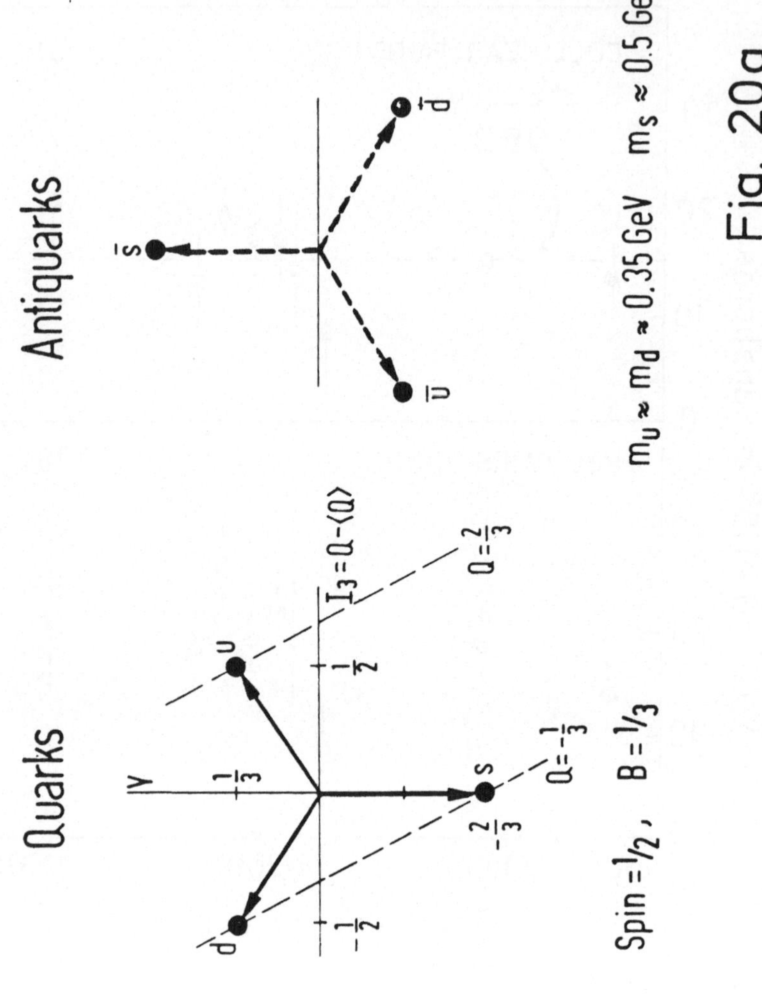
Quarks
Antiquarks
Y
$\frac{1}{3}$
$-\frac{2}{3}$
$-\frac{1}{2}$
$\frac{1}{2}$
$I_3 = Q - \langle Q \rangle$
$Q = \frac{2}{3}$
$Q = -\frac{1}{3}$
d
u
s
$\bar{u}$
$\bar{d}$
$\bar{s}$
Spin = $1/2$, B = $1/3$
$m_u \approx m_d \approx 0.35$ GeV $m_s \approx 0.5$ GeV

Fig. 20a

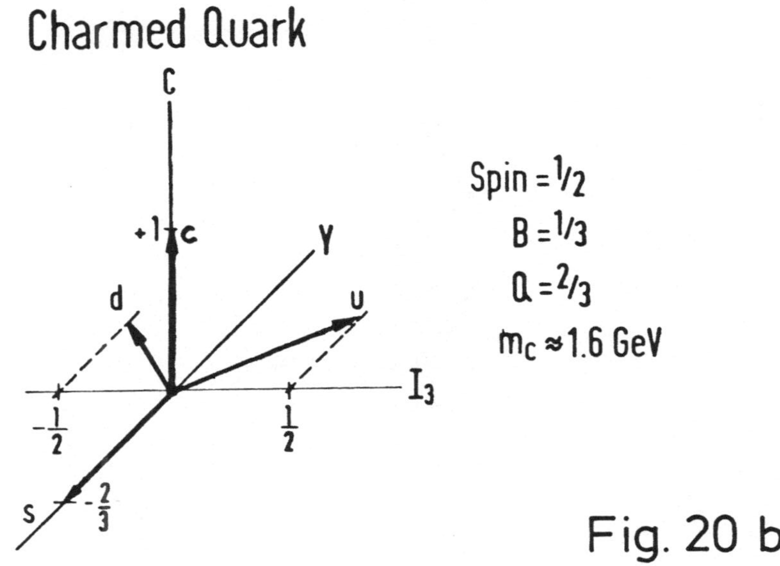

Fig. 20 b

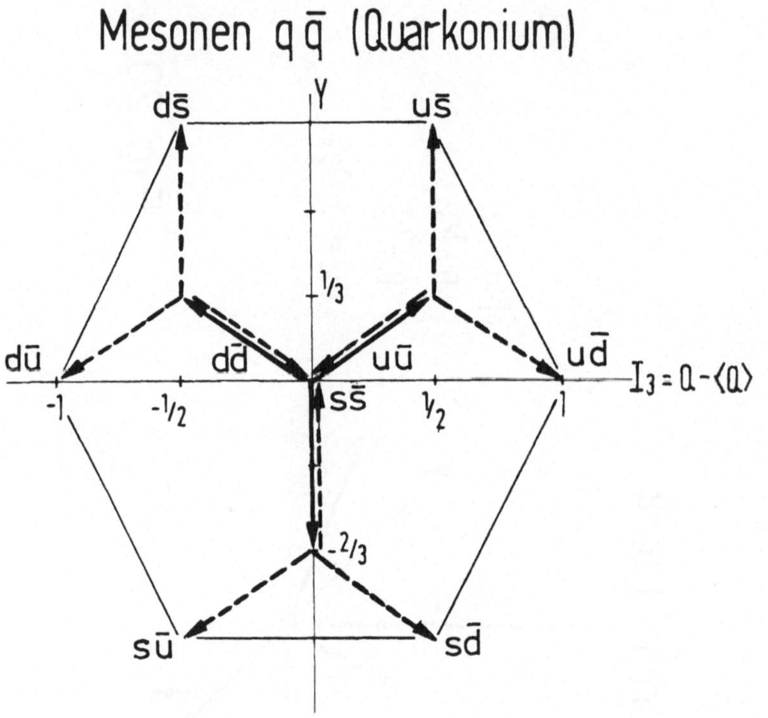

Mesonen $q\bar{q}$ (Quarkonium)

Y

$d\bar{s}$ $u\bar{s}$

$1/3$

$d\bar{u}$ $d\bar{d}$ $u\bar{u}$ $u\bar{d}$ $I_3 = Q - \langle Q \rangle$

-1 $-1/2$ $s\bar{s}$ $1/2$ 1

$-2/3$

$s\bar{u}$ $s\bar{d}$

Fig.21

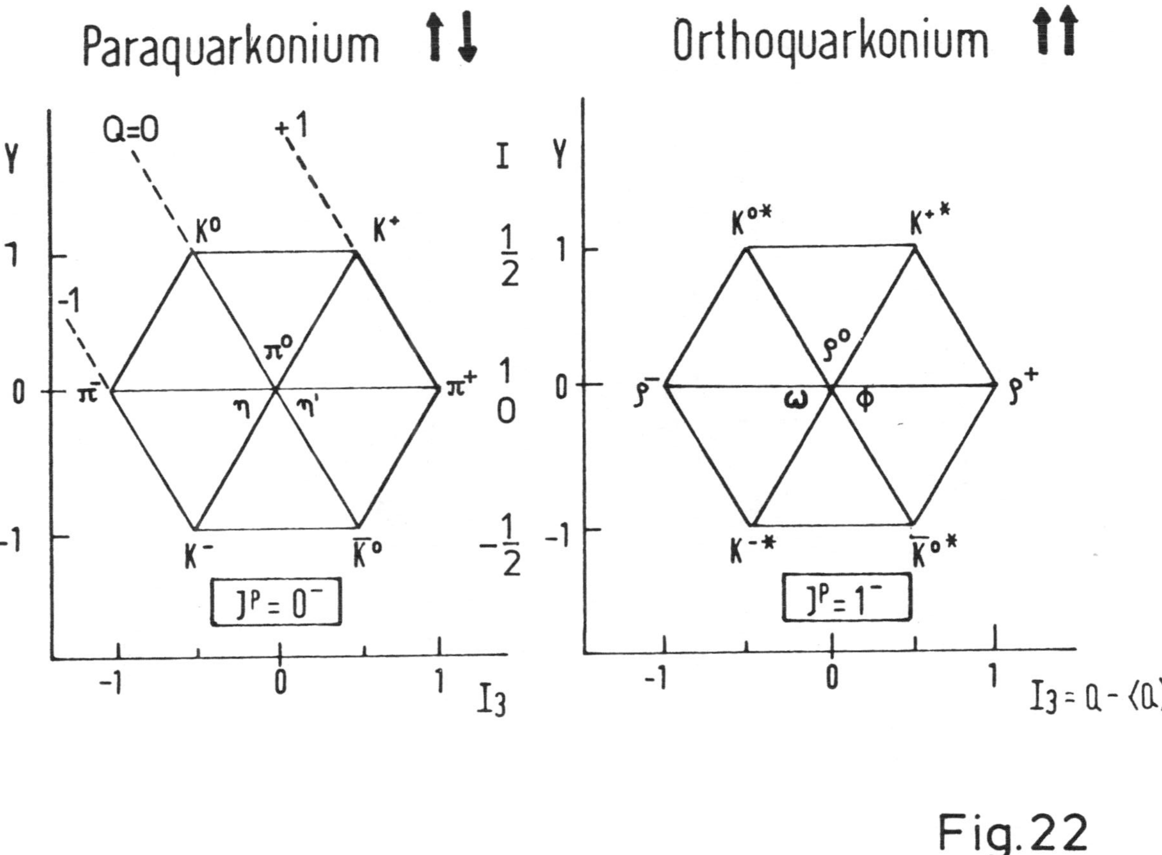

Paraquarkonium ↑↓
Orthoquarkonium ↑↑
Q=0
+1
-1
Y
I
1
0
-1
1/2
1
0
-1/2
K^0
K^+
π^-
π^0
π^+
η
η'
K^-
$\bar{K}^0$
$J^P = 0^-$
I_3
K^{0*}
K^{+*}
ρ^-
ρ^0
ρ^+
ω
ϕ
K^{-*}
$\bar{K}^{0*}$
$J^P = 1^-$
$I_3 = Q - \langle Q \rangle$

Fig.22

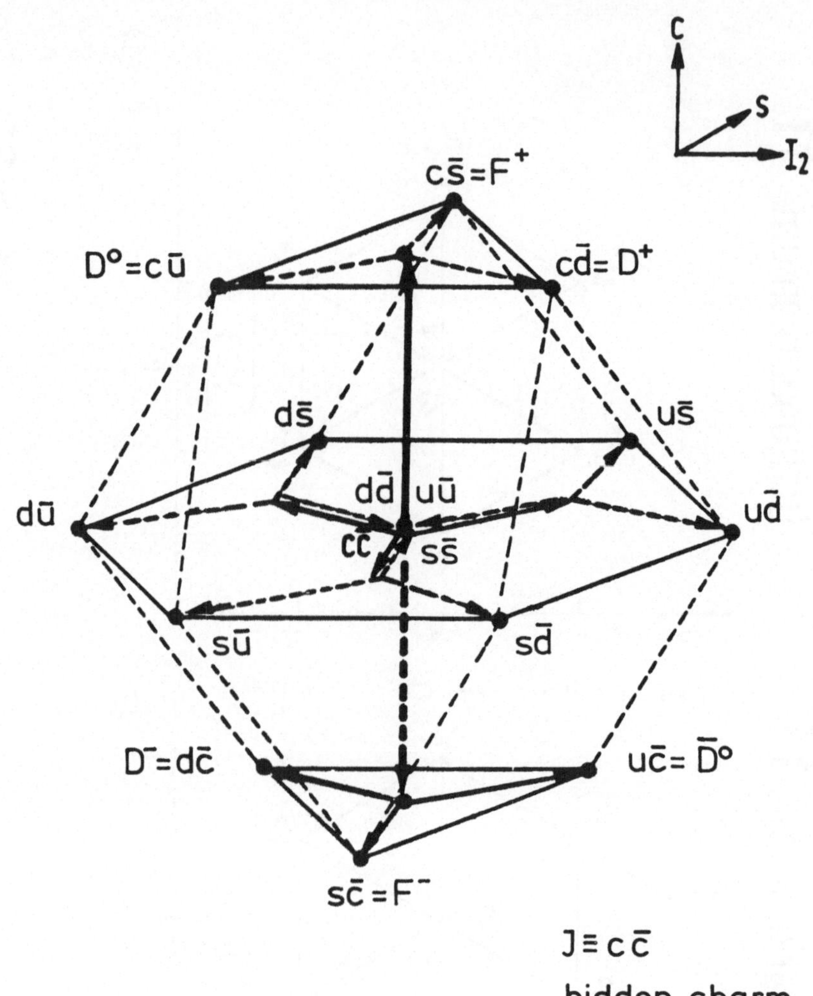
c
s
I_2
$c\bar{s}=F^+$
$D^o=c\bar{u}$
$c\bar{d}=D^+$
$d\bar{s}$
$u\bar{s}$
$d\bar{u}$
$d\bar{d}$ $u\bar{u}$
$u\bar{d}$
$c\bar{c}$
$s\bar{s}$
$s\bar{u}$
$s\bar{d}$
$D^-=d\bar{c}$
$u\bar{c}=\bar{D}^o$
$s\bar{c}=F^-$
$J\equiv c\bar{c}$
hidden charm

Fig. 23

Level scheme for various potentials

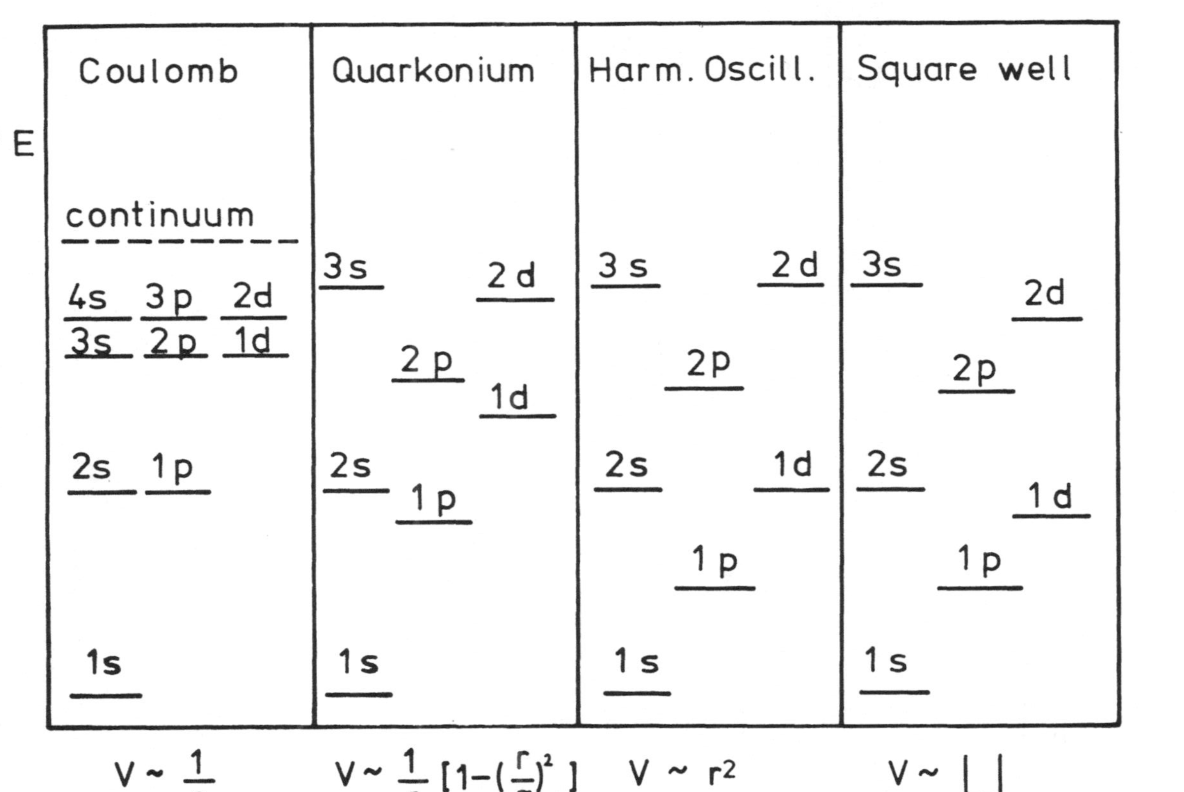

Fig. 24

Level splitting for quarkonium

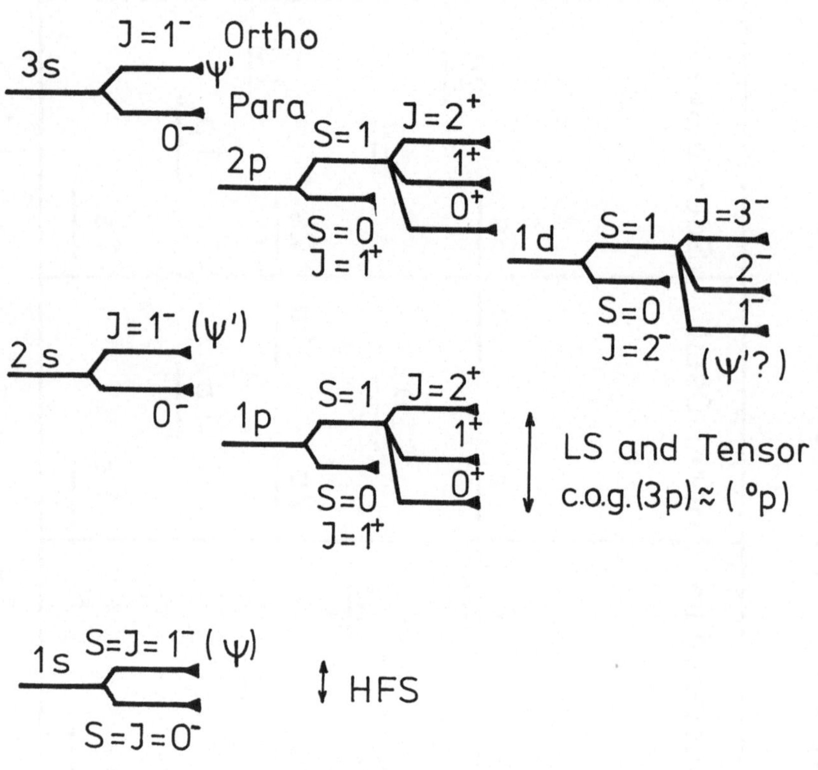

SU(3)-splitting ⑧+ ① for C=0

③+③̄ +1 C=±1

Fig. 25

Quarkonium level scheme
Charm modell

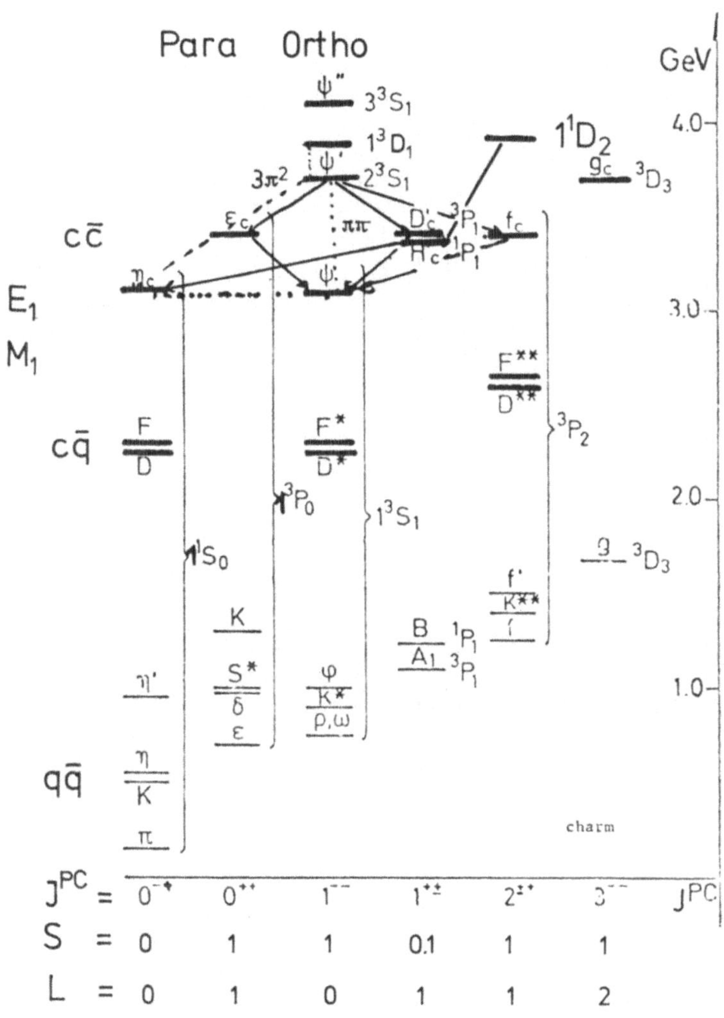

Fig. 26

Quarkonium levels coloured quarks

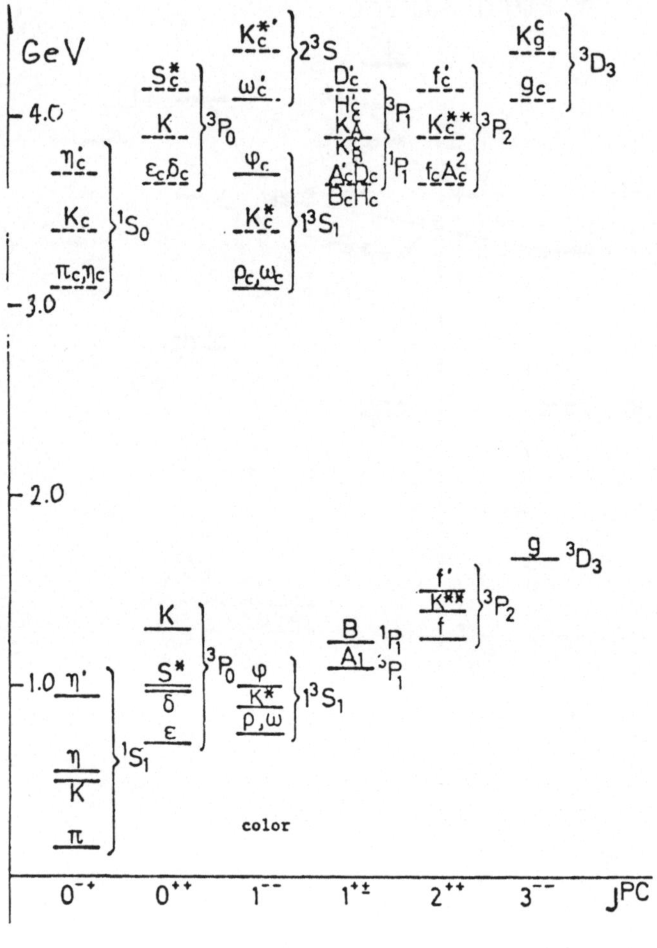

GeV

4.0 3.0 2.0 1.0

η_c' K_c π_c, η_c $\}^1S_0$

S_c^* K $\varepsilon_c \delta_c$ $\}^3P_0$

$K_c^{*\prime}$ ω_c' $\}2^3S$

φ_c K_c^* ρ_c, ω_c $\}1^3S_1$

D_c' H_c' K_A^c $\}^3P_1$ K_B^c $A_c' D_c$ $B_c H_c$ $\}^1P_1$

f_c' K_c^{**} $f_c A_c^2$ $\}^3P_2$

K_g^c g_c $\}^3D_3$

η' η K π $\}^1S_1$

K S^* δ ε $\}^3P_0$

φ K^* ρ, ω $\}1^3S_1$

B $^1P_1$ A_1 $^3P_1$

f' K^{**} f $\}^3P_2$

g $^3D_3$

color

0^{-+} 0^{++} 1^{--} $1^{+\pm}$ 2^{++} 3^{--} J^{PC}

Fig. 27

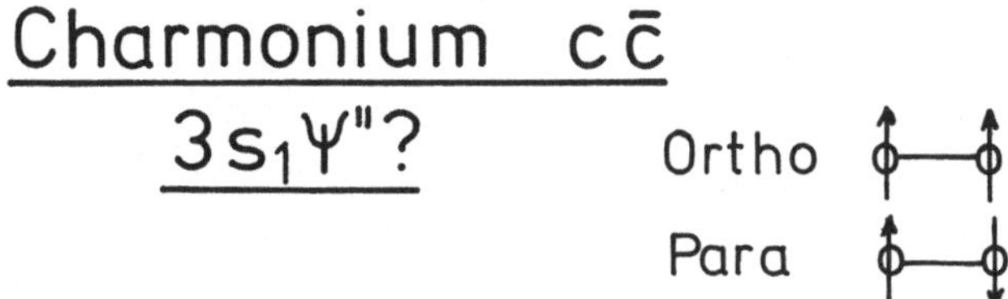
Charmonium $c\bar{c}$
$3s_1 \psi''$?
Ortho
Para

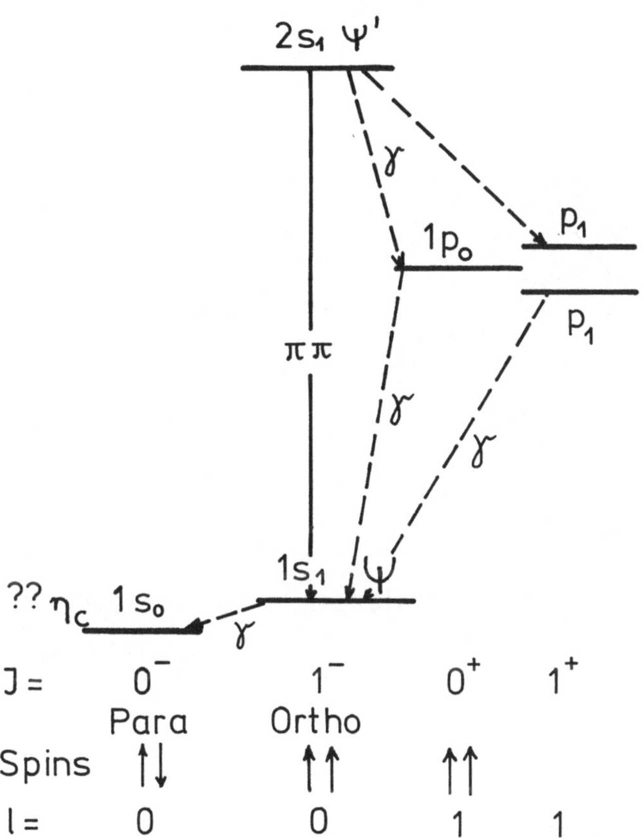
$2s_1$ ψ'
γ
$1p_0$
p_1
p_1
$\pi\pi$
γ
γ
$1s_1$ ψ
?? η_c $1s_0$
γ
J = 0^- 1^- 0^+ 1^+
Para Ortho
Spins ↑↓ ↑↑ ↑↑
l = 0 0 1 1

Fig. 28

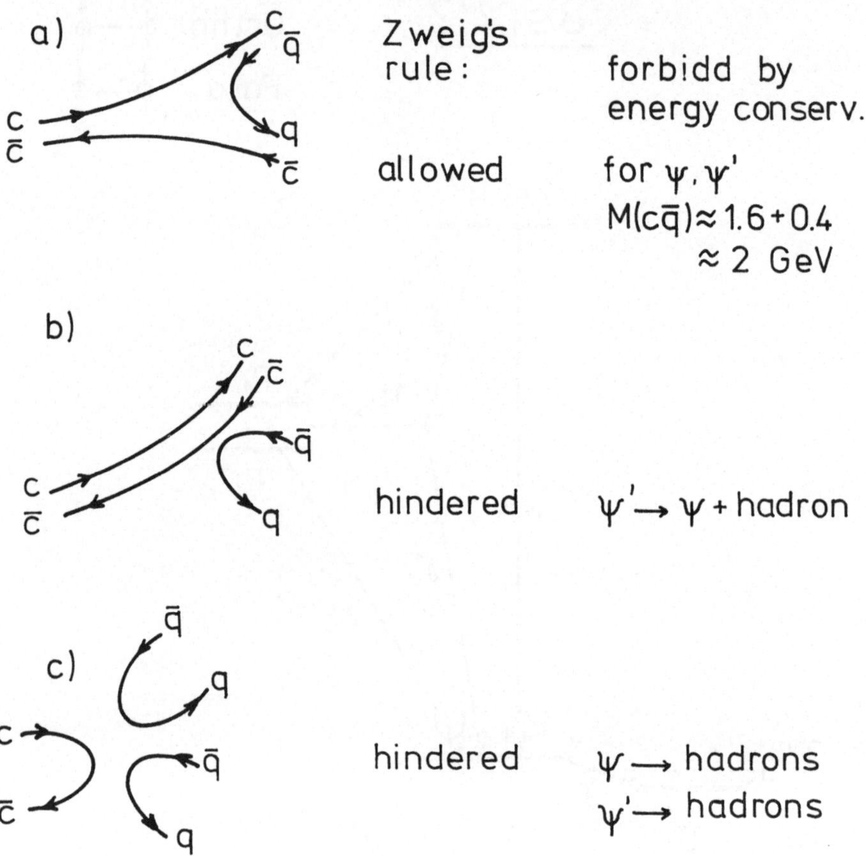
Decay of ψ-particles
a)
c
q̄
c
c̄
q
c̄
Zweig's
rule:
forbidd by
energy conserv.
allowed
for ψ, ψ'
M(cq̄)≈1.6+0.4
≈2 GeV
b)
c
c̄
q̄
c
c̄
q
hindered
ψ'→ ψ + hadron
c)
q̄
q
c
q̄
c̄
q
hindered
ψ → hadrons
ψ'→ hadrons

Fig. 29

Weak interaction with charm

Leptons | Quarks

$$\begin{matrix} Q \\ -1 \\ 0 \end{matrix} \quad \begin{pmatrix} e^- \\ \nu_e \end{pmatrix} \quad \begin{pmatrix} \mu^- \\ \nu_\mu \end{pmatrix} \quad \overset{\longleftrightarrow}{\text{symmetry}} \quad \begin{pmatrix} d.\cos\Theta + s.\sin\Theta \\ u \end{pmatrix} \quad \begin{pmatrix} s.\cos\Theta - d.\sin\Theta \\ c \end{pmatrix} \quad \begin{matrix} Q \\ -\frac{1}{3} \\ +\frac{2}{3} \end{matrix}$$

Fig. 30

Ψ'(3.7 GeV) DASP

$$\frac{\Psi' \to p\bar{p}}{\Psi \to p\bar{p}} \approx 2$$

$\Psi' \to \eta_c \gamma$ not seen (< few %)

$\Psi' \to \Psi \eta$ observed, Γ too small for colour

$\frac{K}{\pi}$ similar as for Ψ(3.1)

Conclusion:

Ψ' decays mainly via Ψ ($\sim$ 6o %)

(phase space would favour many other channels: dynamical principle?)

$\Psi' \to \Psi + X$ goes mainly as $\Psi' \to \Psi + \pi\pi$
($\gtrsim$ 80 %)
$\Psi' \to \Psi + \eta$ seen

Fig. 36

Ψ(3.1 GeV)

hadronic decay channels

DASP results

non collinear $\gamma\gamma$: $e^+e^- \rightarrow \Psi \rightarrow \gamma\gamma + X$

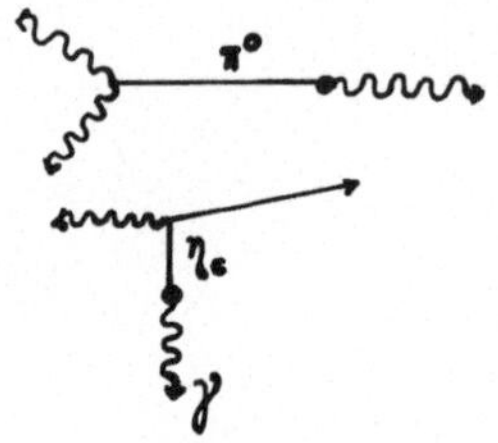

$\Psi \rightarrow \pi^o \gamma$ $\quad \Gamma(\pi^o\gamma) < 0.4$ keV

$\Psi \rightarrow \eta_c \gamma$ $\quad \Gamma(\eta_c\gamma) < 1.1$ keV ($\eta_c \rightarrow \gamma\gamma$)

if $m(\eta_c) > 2.6$ GeV

$\Psi \rightarrow \pi^+\pi^-$ $\quad \Gamma(\pi^+\pi^-) < 130$ eV — I = 0 (less than 5 events

$\Psi \rightarrow K^+K^-$ $\quad \Gamma(K^+K^-) < 130$ eV

$\Psi \rightarrow p\bar{p}$ $\quad \Gamma(p\bar{p}) = 180^{+130}_{-70}$ eV

$\Psi \rightarrow \Lambda\bar{\Lambda}$ $\quad \Gamma(\Lambda\bar{\Lambda}) \approx \Gamma(p\bar{p})$

(12 pairs)
no bound baryon
direct decay
(via virtual γ,
$\Gamma(p\bar{p})/\Gamma(\Lambda\bar{\Lambda}) = 1/9$)

$\Psi \rightarrow \eta\gamma$ ($\eta \rightarrow \gamma\gamma$) $\quad 0.1 \text{ keV} \lesssim \Gamma(\eta\gamma) \lesssim 2\text{keV}$ (3 events)

too small
for colour:
$\Gamma \approx 1$ MeV
for charm $\Gamma \sim 2$keV

SPEAR

$\Psi \rightarrow \pi^\pm\rho^\mp$ seen

$\Psi \rightarrow \pi^+\pi^-\omega$ $\quad \Gamma \sim 400$ eV

G = - 1

$\Psi \rightarrow K^+\pi^-K^*$ (830) seen

Fig. 31

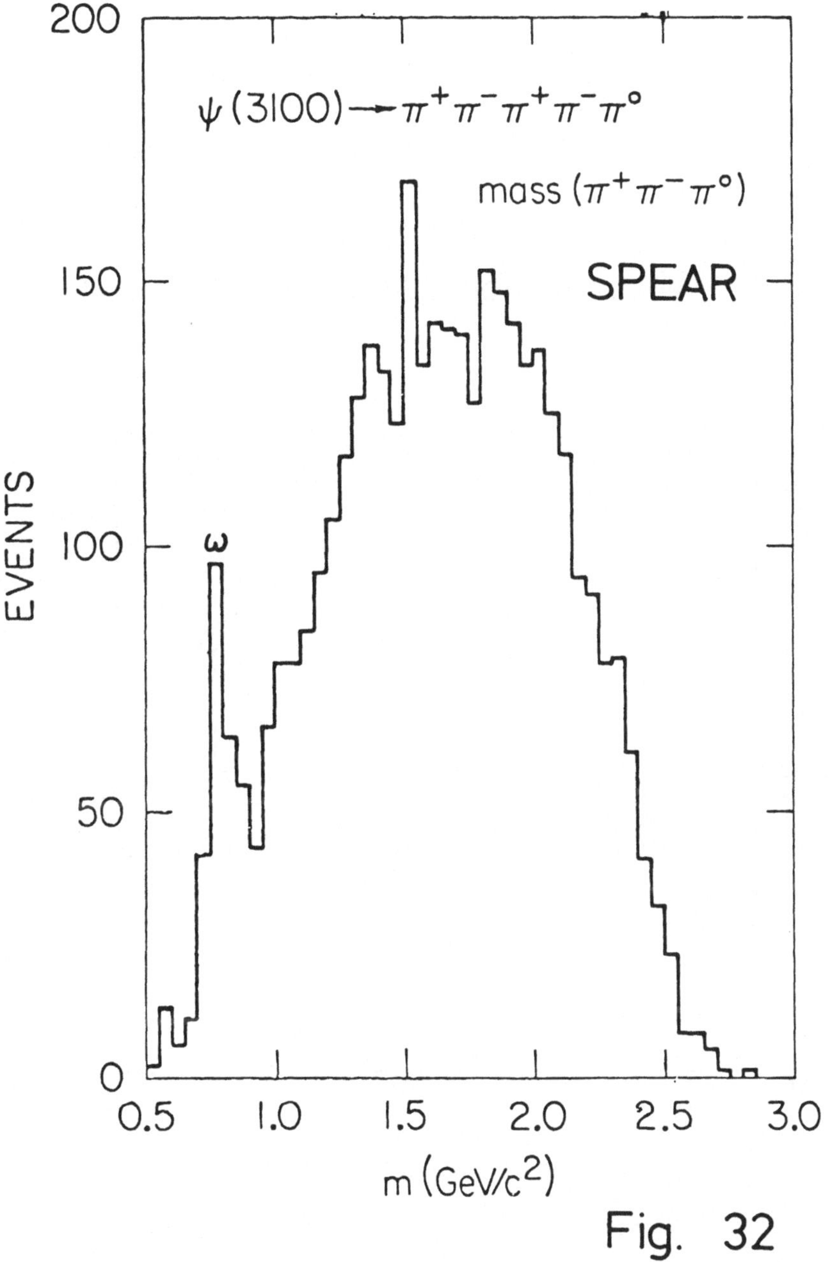

200
150
100
50
0
EVENTS
ψ(3100) → π⁺π⁻π⁺π⁻π°
mass (π⁺π⁻π°)
SPEAR
ω
0.5
1.0
1.5
2.0
2.5
3.0
m (GeV/c²)

Fig. 32

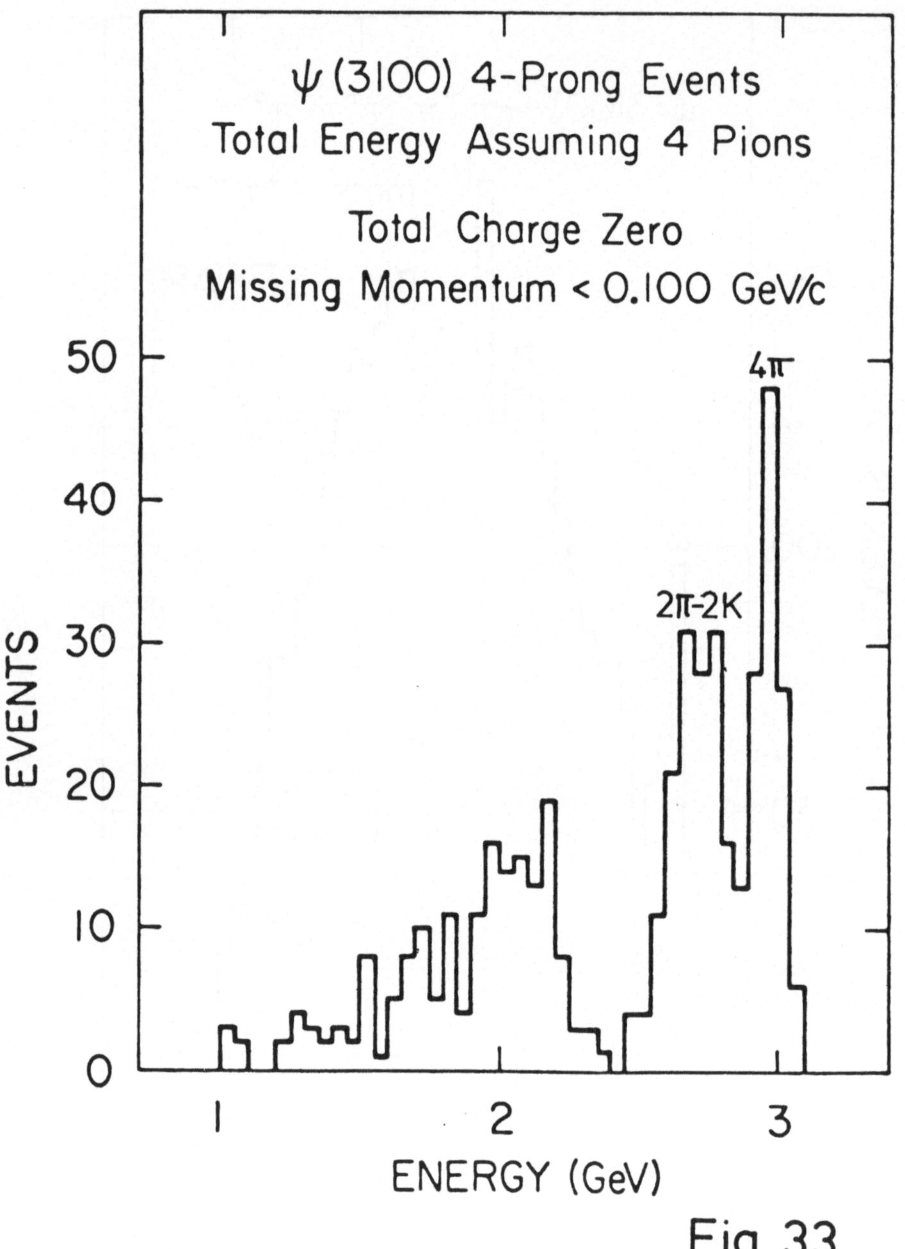

ψ (3100) 4-Prong Events
Total Energy Assuming 4 Pions
Total Charge Zero
Missing Momentum < 0.100 GeV/c
4π
2π-2K
EVENTS
0
10
20
30
40
50
1
2
3
ENERGY (GeV)

Fig. 33

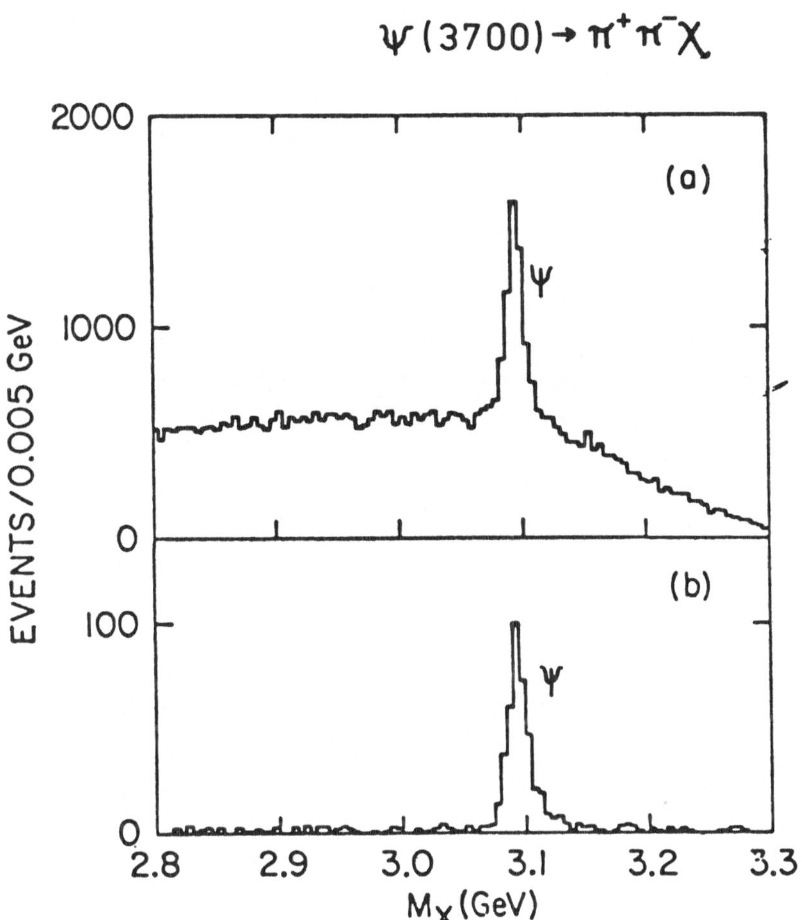
ψ(3700) → π⁺π⁻X
2000
1000
0
100
0
(a)
(b)
ψ
ψ
EVENTS/0.005 GeV
2.8
2.9
3.0
3.1
3.2
3.3
Mx (GeV)

Fig. 34

Ψ' (3.7 GeV) cascades

$\Psi' \to \Psi + X$

SPEAR: $\mu^+\mu^-(3.1)$, $\pi^+\pi^-$

$$\frac{\Psi' \to \Psi + X}{\Psi' \to \text{all}} = 0.57 \pm 0.08$$

$$\frac{\Psi' \to \Psi + \pi^+ + \pi^-}{\Psi' \to \text{all}} = 0.32 \pm 0.04$$

Theory:

$$\frac{\Psi' \to \Psi + \text{neutrals}}{\Psi' \to \Psi + X} = 0.44 \pm 0.03 \leftrightarrow \frac{1}{3} \text{ for } I_{\pi\pi} = 0$$

0	1
2/3	2

$$\frac{\Psi' \to \Psi + \pi^+ + \pi^-}{\Psi' \to \Psi + X} = 0.56 \pm 0.1$$

$0.44 \leftrightarrow 0.33$
other decays into $\pi^0\pi^0$!

DASP: good μ identification, detection of γ; $\mu^+\mu^-(3.1)$ bigger

$$\frac{\Psi' \to \Psi + \pi^+ + \pi^-}{\Psi' \to \Psi + X} = 0.53 \pm 0.1$$

Theory:

$$\frac{\Psi' \to \Psi + \pi^0\pi^0}{\Psi' \to \Psi + \pi^+\pi^-} \approx 0.5 \qquad \frac{1}{2} \text{ for } I_{\pi\pi} = 0$$

0	1
1	2

$\Psi' \to \Psi + \eta$ seen in $\Psi' \to \Psi + \pi^+\pi^-\pi^0$
$\Psi + \gamma\gamma \qquad m(\gamma\gamma) \approx m(\eta)$

$$\frac{\Psi' \to \Psi + \eta}{\Psi' \to \Psi + X} \approx 0.1$$

Fig. 35

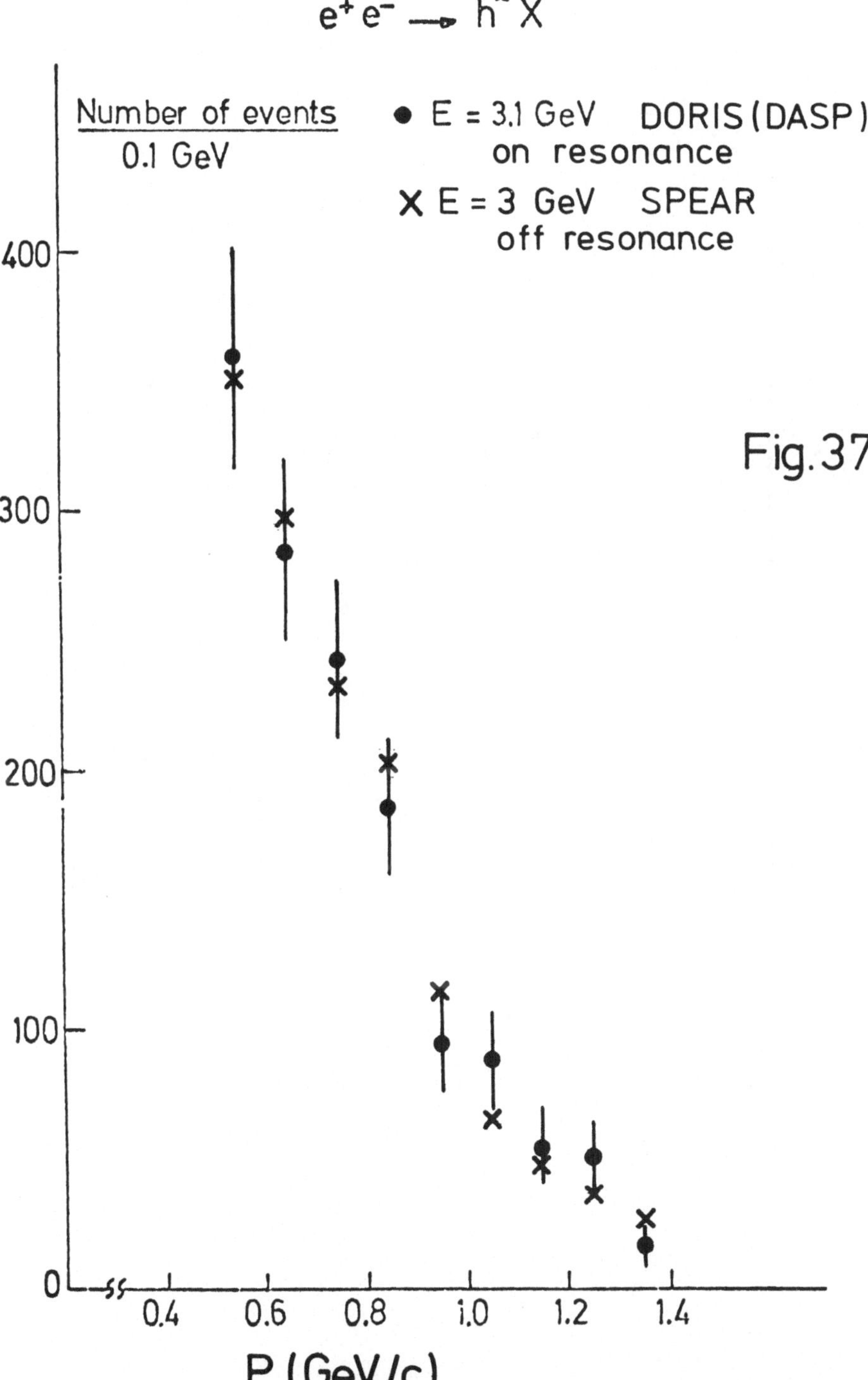
$e^+e^- \to h^\pm X$
Number of events / 0.1 GeV
● E = 3.1 GeV DORIS (DASP) on resonance
× E = 3 GeV SPEAR off resonance
400
300
200
100
0
0.4
0.6
0.8
1.0
1.2
1.4
P (GeV/c)

Fig. 37

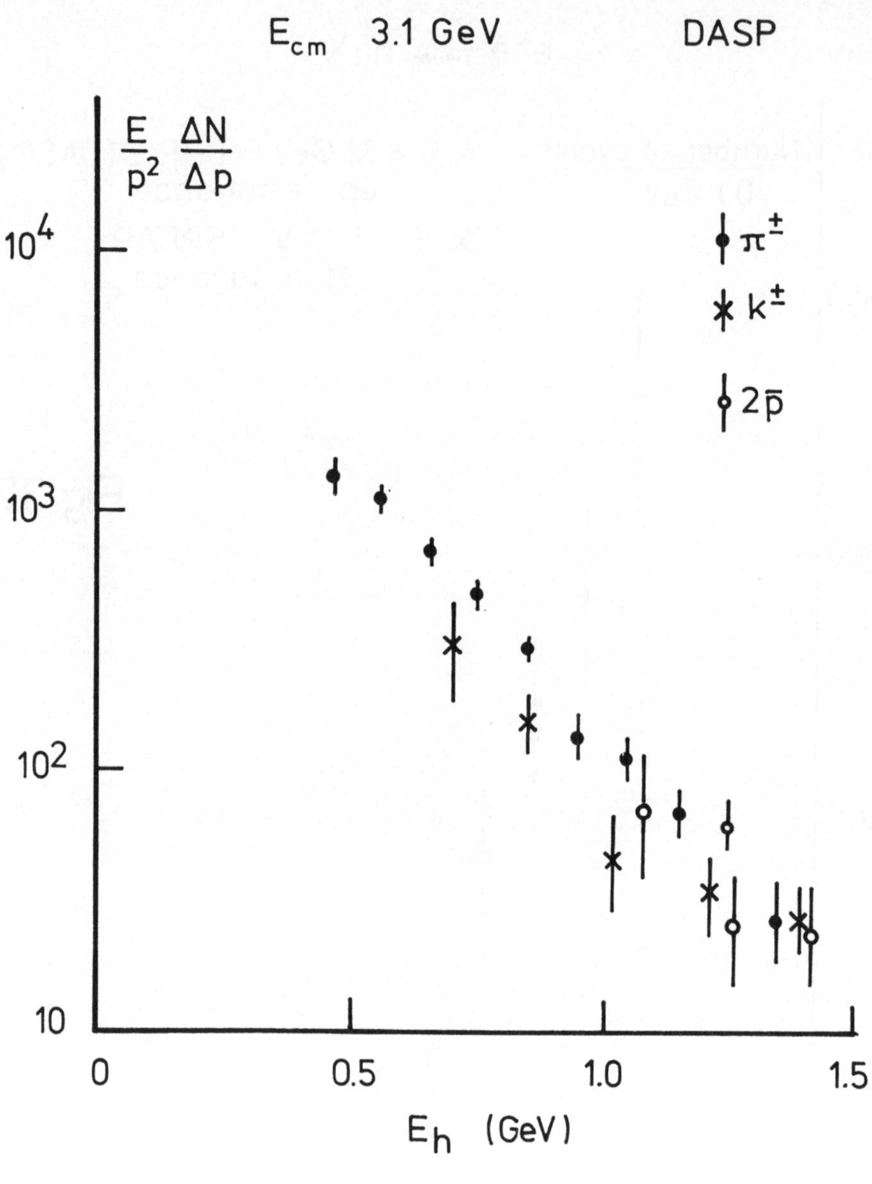

Fig. 38

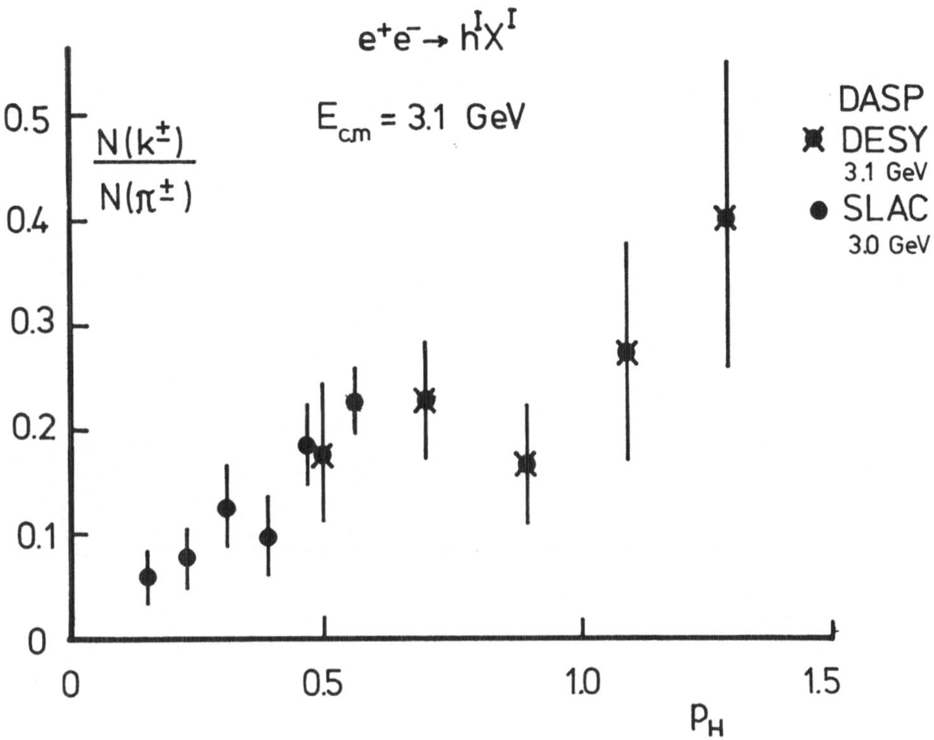
$e^+e^- \to h^I X^I$
$E_{cm} = 3.1$ GeV
$\frac{N(k^\pm)}{N(\pi^\pm)}$
DASP
DESY
3.1 GeV
SLAC
3.0 GeV
0.5
0.4
0.3
0.2
0.1
0
0
0.5
1.0
1.5
p_H

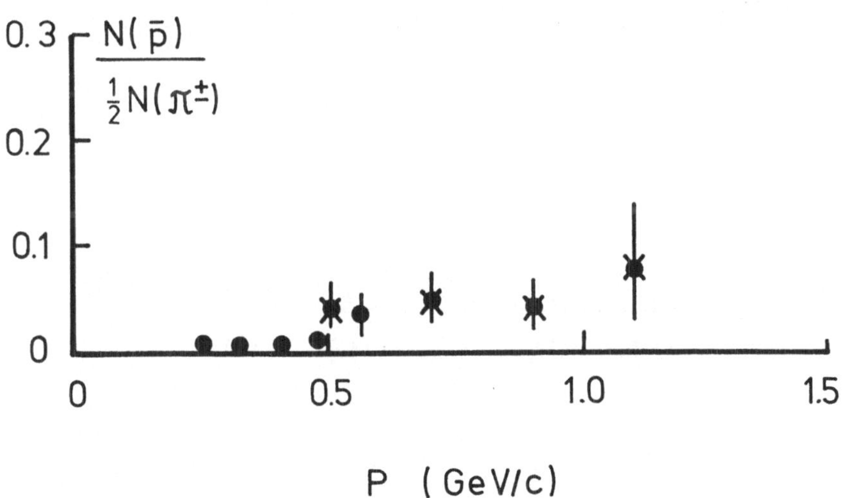
$\frac{N(\bar{p})}{\frac{1}{2}N(\pi^\pm)}$
0.3
0.2
0.1
0
0
0.5
1.0
1.5
P (GeV/c)

Fig. 39

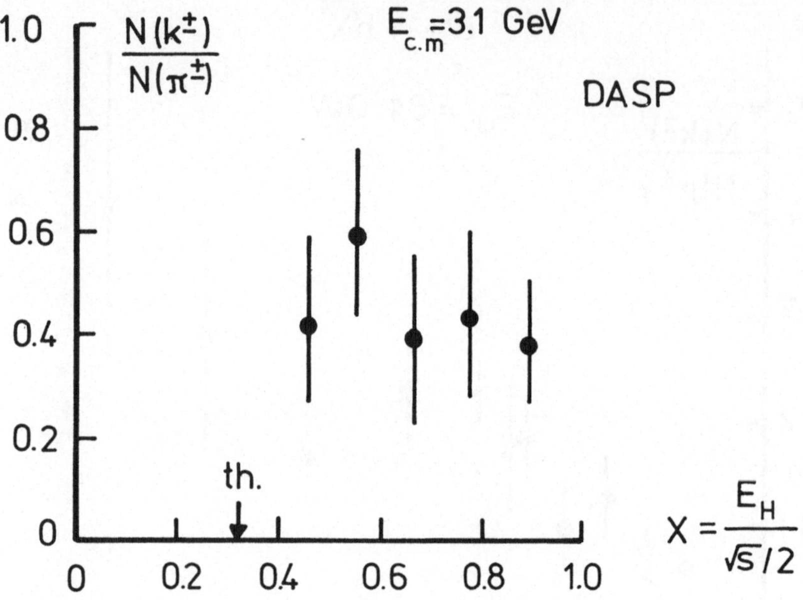

1.0
0.8
0.6
0.4
0.2
0
$\frac{N(k^\pm)}{N(\pi^\pm)}$
$E_{c.m}$=3.1 GeV
DASP
th.
0
0.2
0.4
0.6
0.8
1.0
$X = \frac{E_H}{\sqrt{s}/2}$

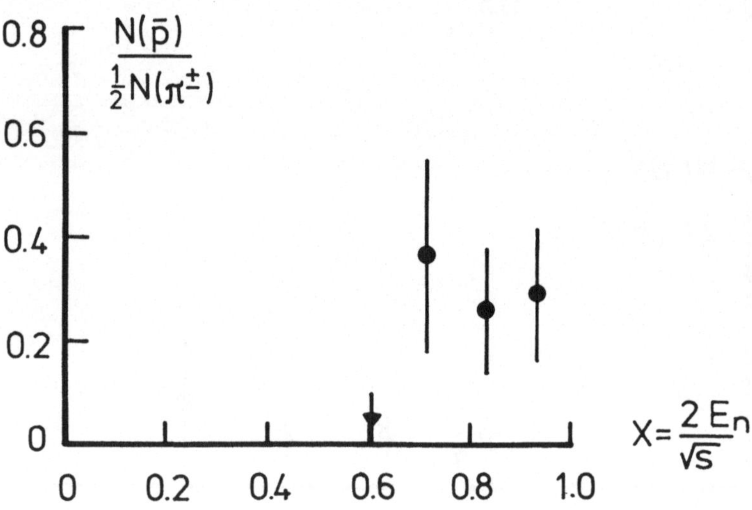

0.8
0.6
0.4
0.2
0
$\frac{N(\bar{p})}{\frac{1}{2}N(\pi^\pm)}$
0
0.2
0.4
0.6
0.8
1.0
$X = \frac{2E_n}{\sqrt{s}}$

Fig. 40

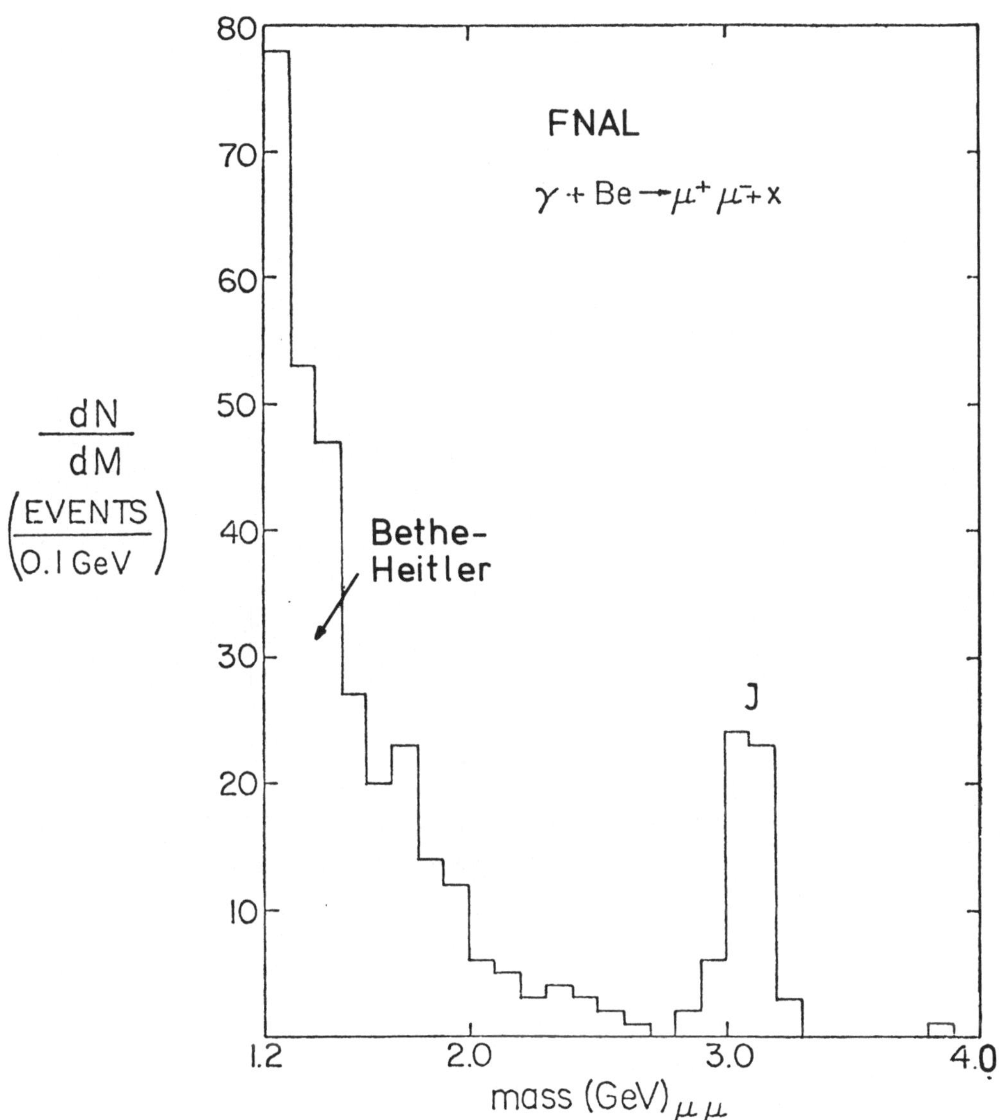
FNAL
$\gamma + Be \to \mu^+ \mu^- + x$
Bethe-
Heitler
J
$\frac{dN}{dM}$
$\left(\frac{EVENTS}{0.1 GeV}\right)$
80
70
60
50
40
30
20
10
1.2
2.0
3.0
4.0
mass (GeV)$_{\mu\mu}$

Fig. 41

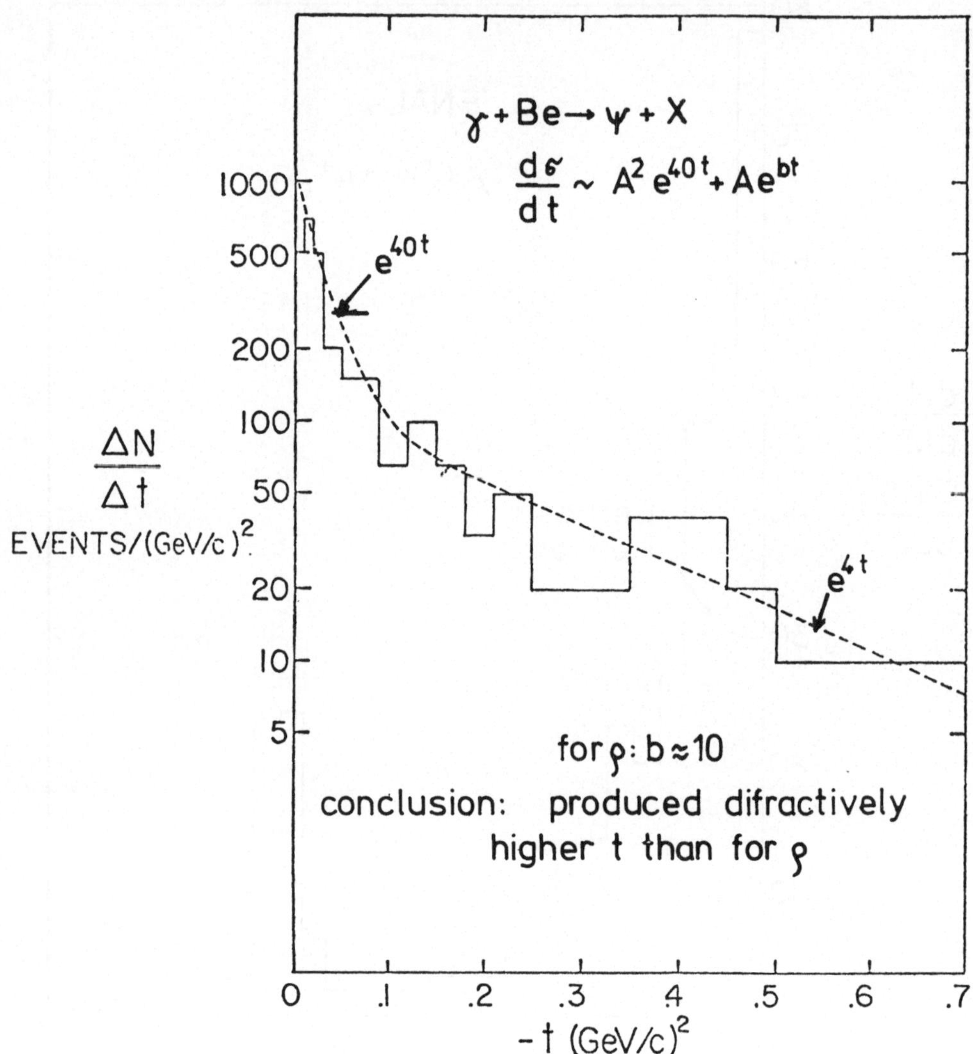

$\gamma + Be \rightarrow \psi + X$
$\frac{d\sigma}{dt} \sim A^2 e^{40t} + A e^{bt}$
1000
500
200
100
50
20
10
5
$\frac{\Delta N}{\Delta t}$
EVENTS/(GeV/c)^2
e^{40t}
e^{4t}
for ρ: b ≈ 10
conclusion: produced difractively
higher t than for ρ
0
.1
.2
.3
.4
.5
.6
.7
$-t$ (GeV/c)^2

Fig. 42

$$\sigma(\gamma Be \to \psi X)\ \ (\psi \to \mu^+\mu^-) = 16 \pm 5 \text{ nb/nucleus}$$

with $\Gamma_{\mu\mu}/\Gamma_{tot} \simeq 5.6/77 \simeq 0.072$

$$\sigma(\gamma\ Be \to \psi X) = 220 \pm 70 \text{ nb/nucleus}$$

$$\frac{d\sigma}{dt}(\gamma p \to \psi p)_{t=0} = \frac{\int_0^{0.5} \frac{d\sigma}{dt}(\gamma\ Be \to \psi X)dt}{\int_0^{0.5} (A^2\ e^{40t} + A\ e^{4t})dt}$$

VDM:

optical theorem

Re t = 0

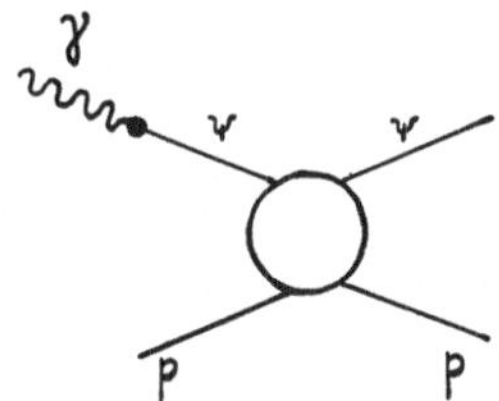

$$\sigma_{tot}(\psi + p) = \left\{ \frac{\gamma^2_{\psi\gamma}/4\pi}{\alpha/64} \cdot \frac{d\sigma}{dt}(\gamma p \to \psi p)_{t=0} \right\}^{\frac{1}{2}}$$

$$\gamma^2_{\psi\gamma}/4\pi = \frac{\alpha^2 M_\psi}{12\Gamma_{ee}} = 2.7$$

$$\sigma_{tot}(\psi + p) \simeq 1 \text{ mb } (\pm 0.3)$$

Ψ is a hadron!

Fig. 43

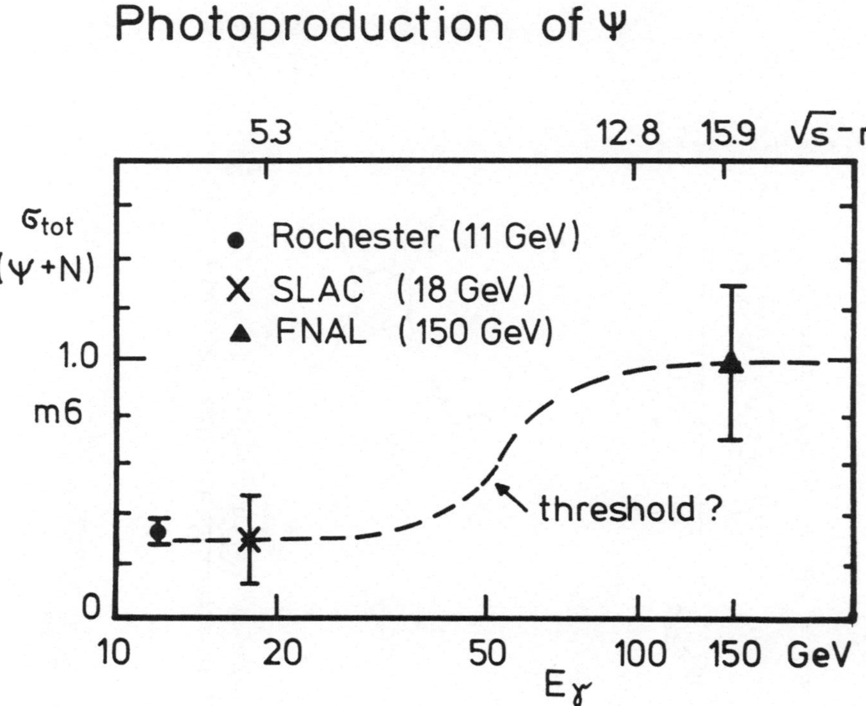
Photoproduction of Ψ
5.3
12.8
15.9
√s - m_p
σ_tot
(Ψ+N)
1.0
mb
0
● Rochester (11 GeV)
X SLAC (18 GeV)
▲ FNAL (150 GeV)
threshold ?
10
20
50
100
150
GeV
E_γ

Fig. 44

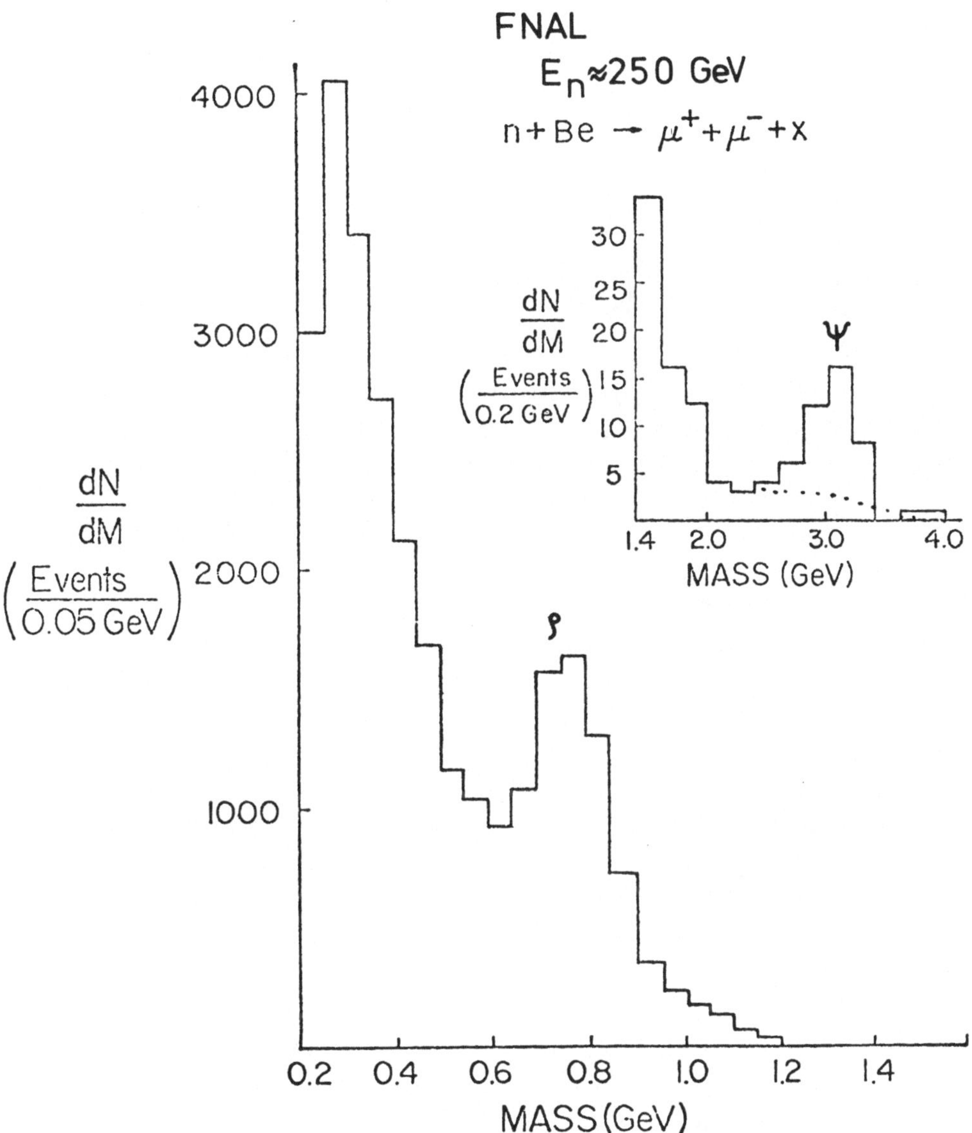
FNAL
$E_n \approx 250$ GeV
$n + Be \to \mu^+ + \mu^- + x$
dN/dM (Events/0.05 GeV)
4000
3000
2000
1000
ρ
0.2
0.4
0.6
0.8
1.0
1.2
1.4
MASS (GeV)
dN/dM (Events/0.2 GeV)
30
25
20
15
10
5
Ψ
1.4
2.0
3.0
4.0
MASS (GeV)

Fig. 45

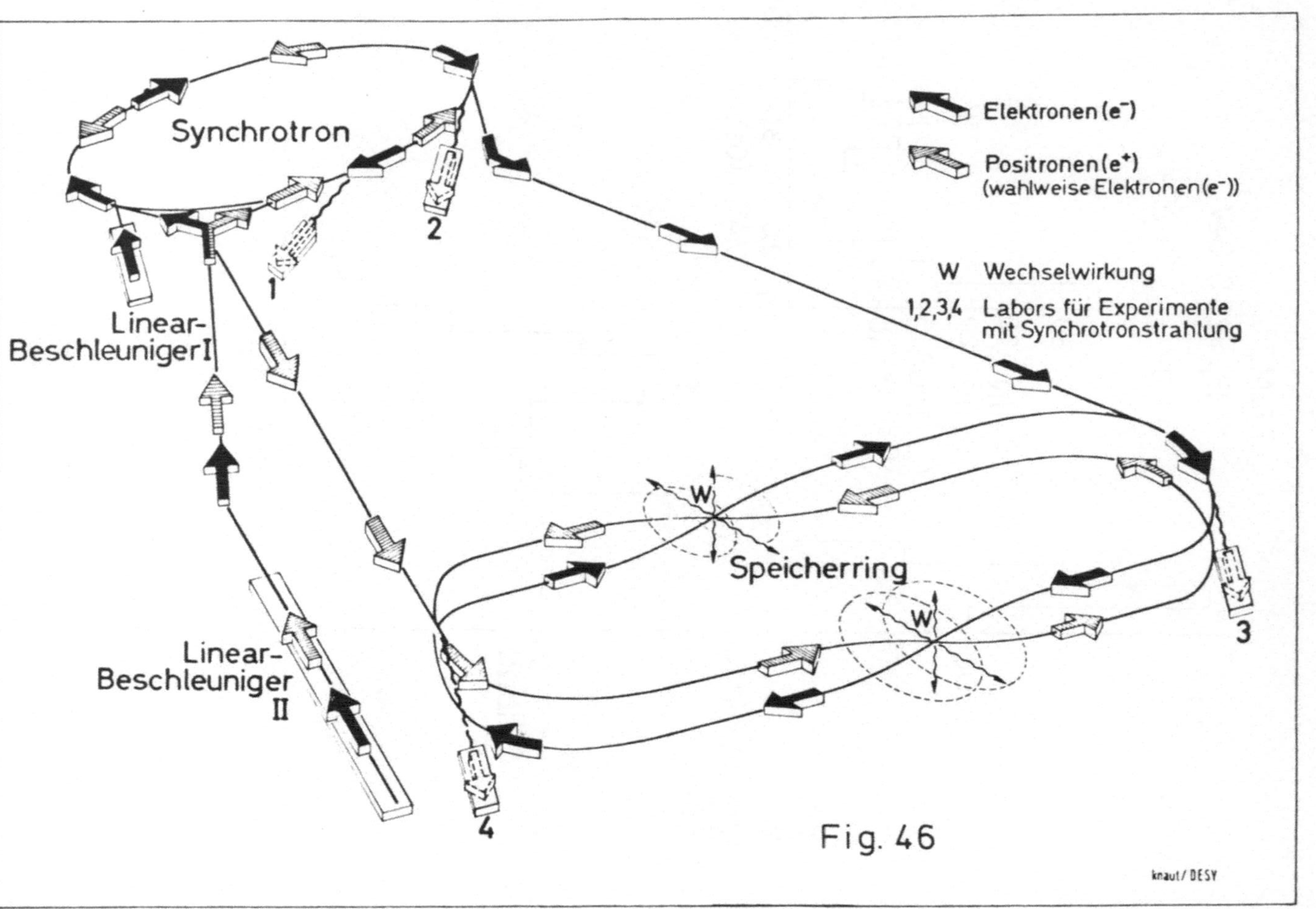
Synchrotron
Elektronen (e⁻)
Positronen (e⁺)
(wahlweise Elektronen (e⁻))
W Wechselwirkung
1,2,3,4 Labors für Experimente mit Synchrotronstrahlung
1
2
3
4
Linear-Beschleuniger I
Linear-Beschleuniger II
W
Speicherring
W
knaut/DESY

Fig. 46

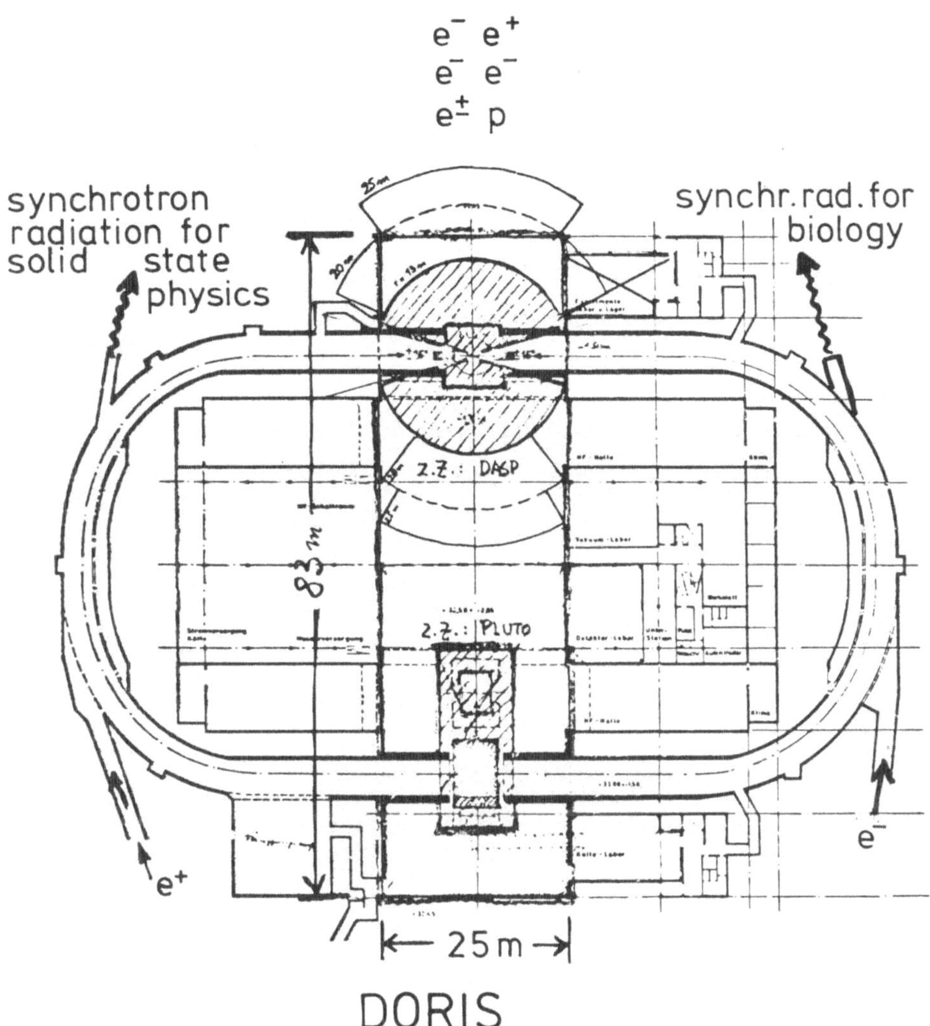

e⁻ e⁺
e⁻ e⁻
e± p
synchrotron radiation for solid state physics
synchr. rad. for biology
Z.Z.: DASP
Z.Z.: PLUTO
83 m
e⁺
e⁻
25 m
DORIS

Fig. 47

DORIS $\begin{matrix} e^+e^- \\ e^-e^- \end{matrix}$

Energy: now 2 x 1.5 GeV ..... 2 x 3.4 GeV.

beginning 76 up to ...2 x 4.2 GeV

(better injection)
(more rf end 76 2 x 5 GeV)

crossing angle: 24 mrad

(complications: L reduced by bunch lengthing, beam-beam instabilities)

beam heigth at I.P.:	0.2 mm	
width	0.5 mm	
bunch length :	$\sim$ 2 cm	(at 1.5 GeV)
	$\sim$ 3 cm	(at 2.0 GeV)

Energy spread/beam: $\sim$ 500 keV

(if coherent synchrotron oscillations damped and $I \lesssim 200$ mA)

with $I_{e^+} \approx I_{e^-} \approx 200$ mA $\rightarrow L \approx 10^{30}$ cm^{-2} sec^{-1} at 2 GeV/ beam

average L about 30% less
beam life time: 4 .... 7 hrs.

If $\Delta p/p >$ 500 MeV tolerable: $I_{e^-} \simeq 1$ A

$I_{e^+} \simeq 0.4$ A

Fig. 48

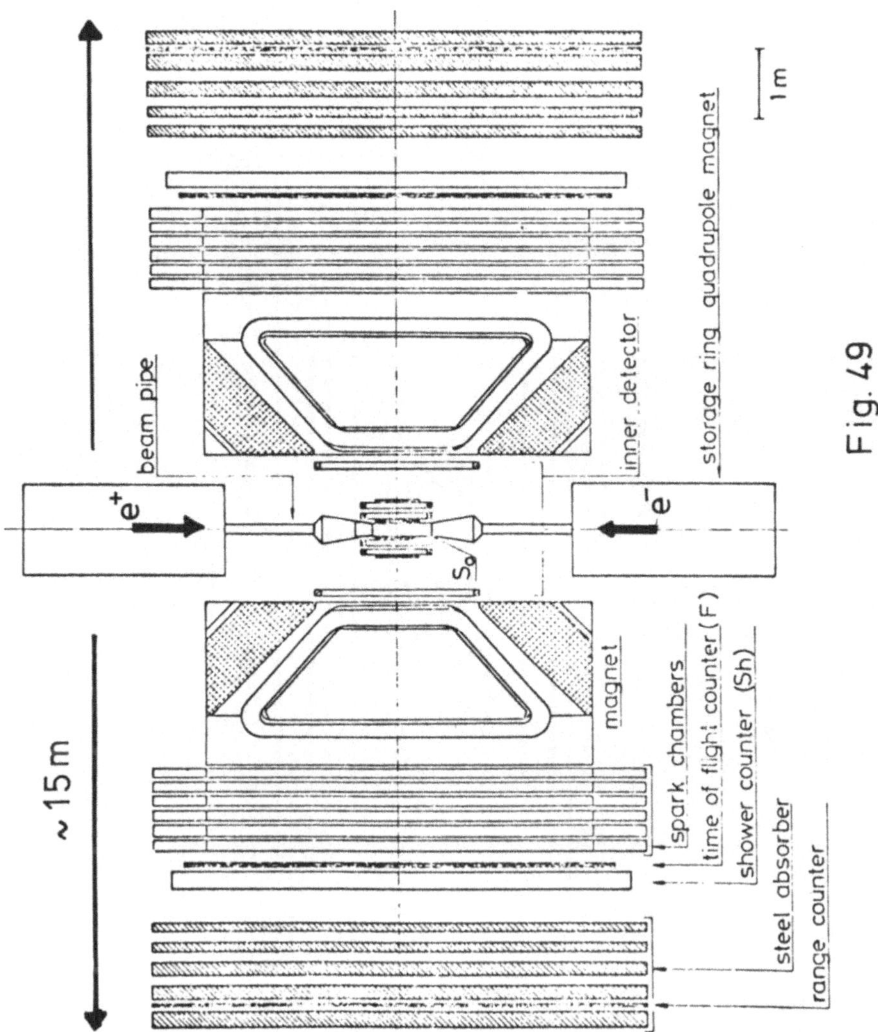
~15 m
e+
e-
beam pipe
inner detector
storage ring quadrupole magnet
1 m
S0
magnet
spark chambers
time of flight counter (F)
shower counter (Sh)
steel absorber
range counter

Fig. 49

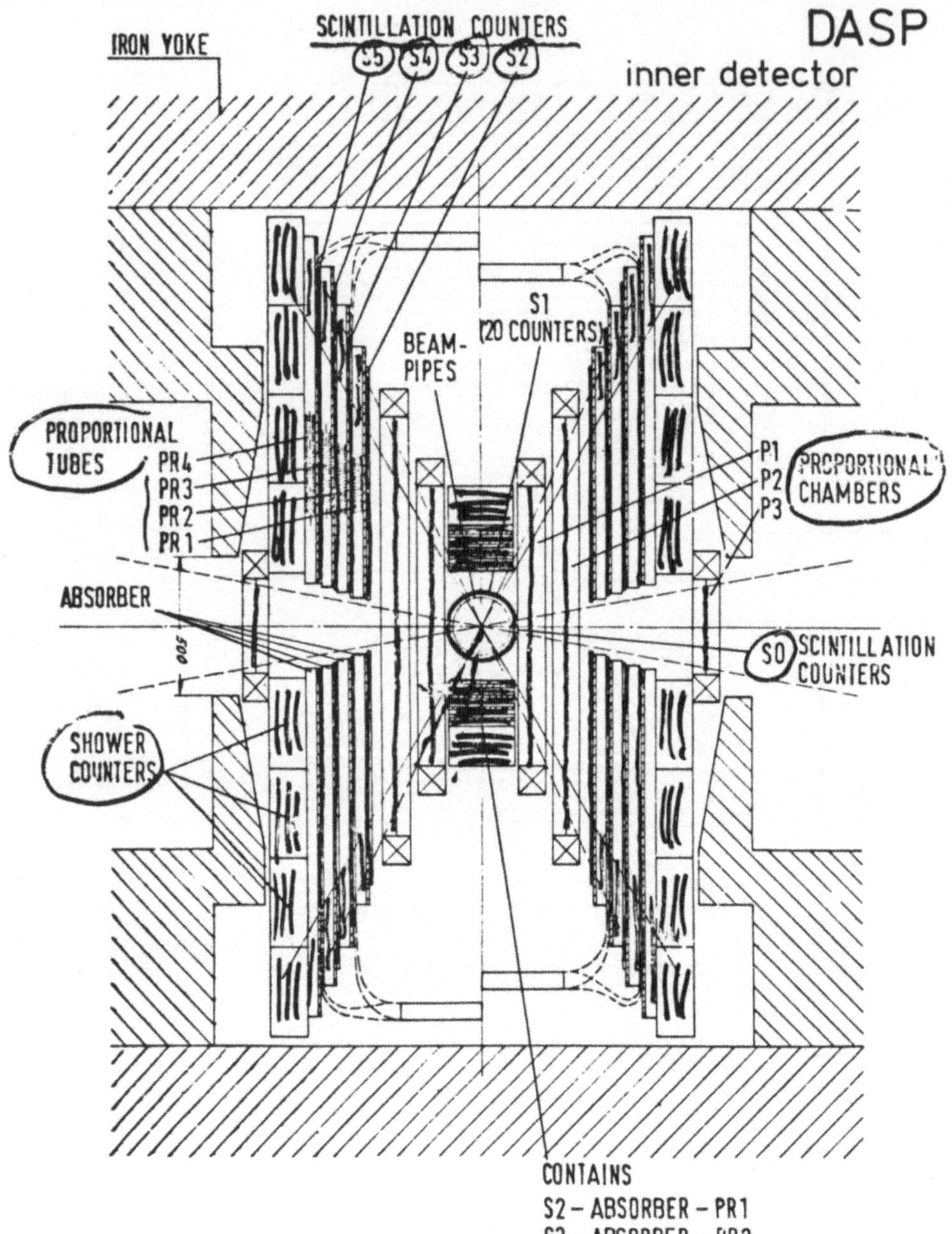

DASP
inner detector
IRON YOKE
SCINTILLATION COUNTERS
S5
S4
S3
S2
S1
(20 COUNTERS)
BEAM-PIPES
PROPORTIONAL TUBES
PR4
PR3
PR2
PR1
P1
P2
P3
PROPORTIONAL CHAMBERS
ABSORBER
500
S0
SCINTILLATION COUNTERS
SHOWER COUNTERS
CONTAINS
S2 – ABSORBER – PR1
S3 – ABSORBER – PR2
S4 – ABSORBER – PR3
S5 – ABSORBER – PR4
SHOWER COUNTERS

Fig. 50

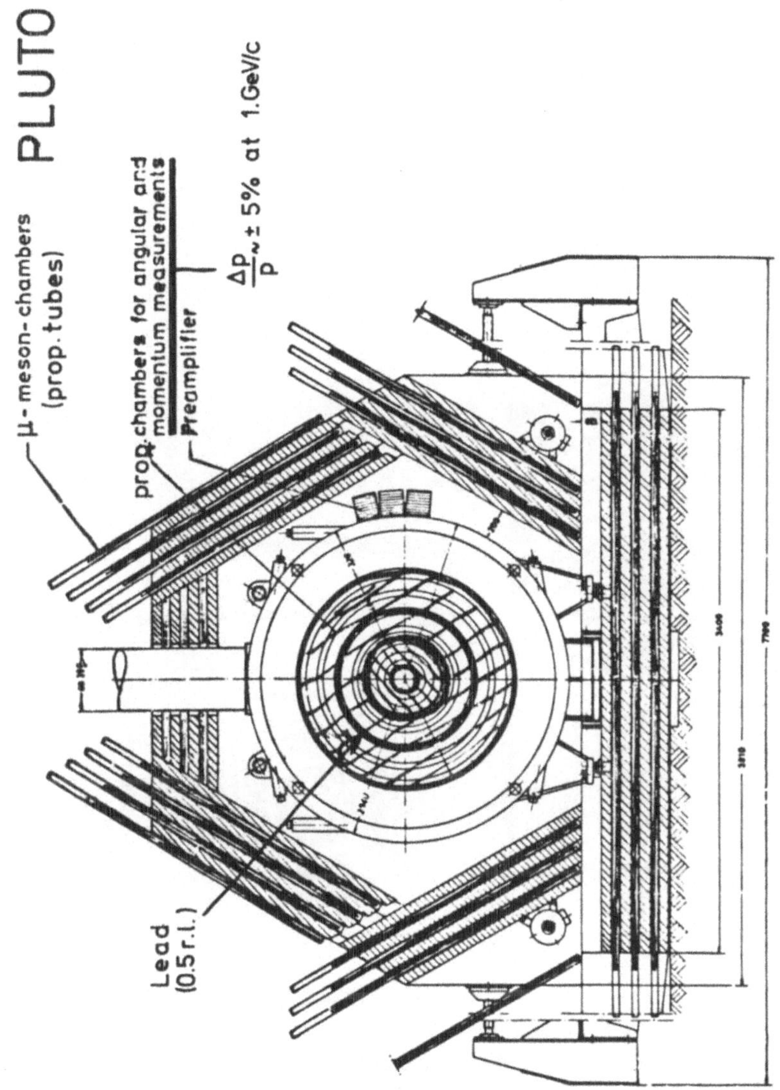
PLUTO
μ - meson - chambers
(prop. tubes)
prop. chambers for angular and
momentum measurements
Preamplifier
Δp/p ≈ ± 5% at 1.GeV/c
Lead
(0.5 r.l.)

Fig.51

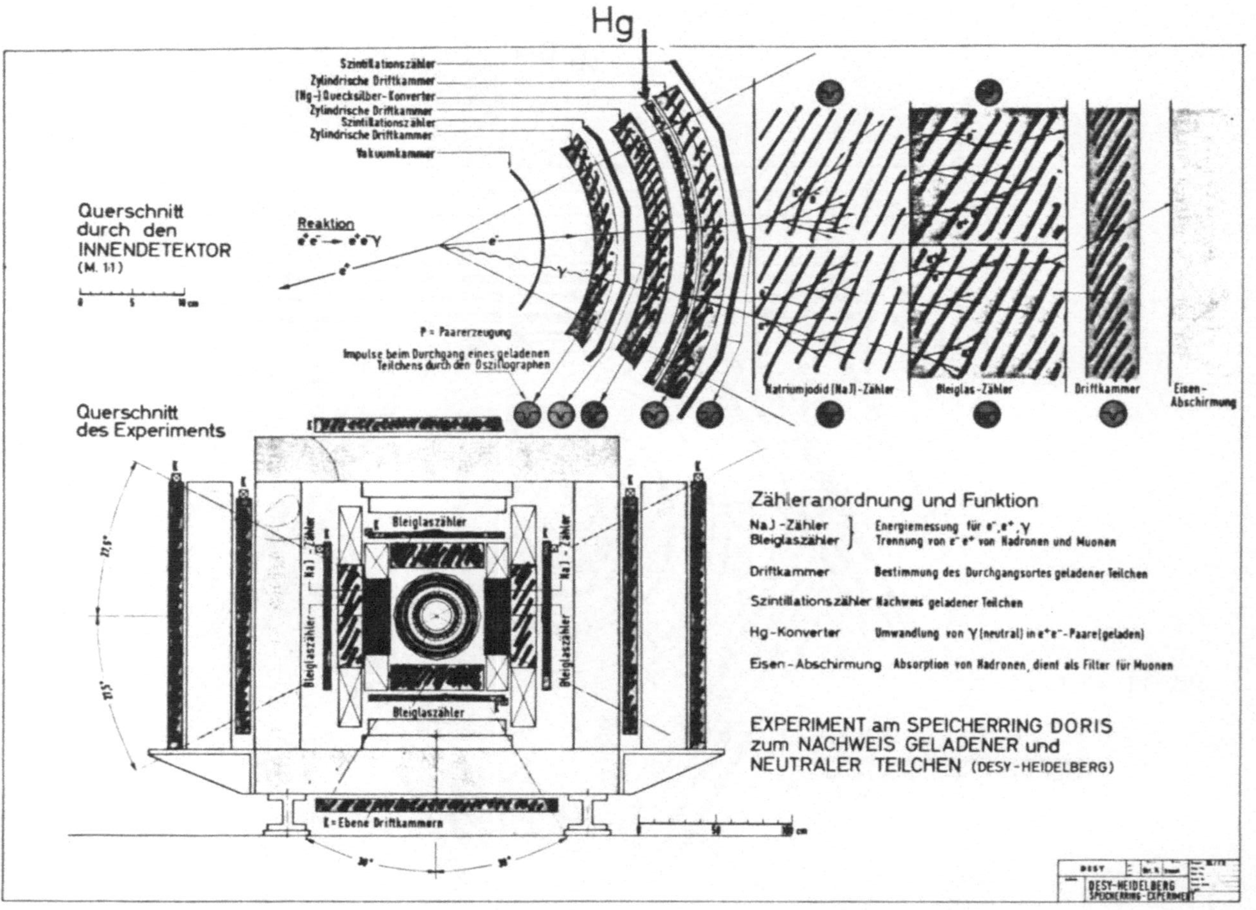

Hg
Szintillationszähler
Zylindrische Driftkammer
(Hg-) Quecksilber-Konverter
Zylindrische Driftkammer
Szintillationszähler
Zylindrische Driftkammer
Vakuumkammer
Querschnitt durch den INNENDETEKTOR (M. 1:1)
Reaktion
$e^+e^- \rightarrow e^+e^-\gamma$
P = Paarerzeugung
Impulse beim Durchgang eines geladenen Teilchens durch den Oszillographen
Natriumjodid (NaJ)-Zähler
Bleiglas-Zähler
Driftkammer
Eisen-Abschirmung
Querschnitt des Experiments
Bleiglaszähler
NaJ-Zähler
E = Ebene Driftkammern
Zähleranordnung und Funktion
NaJ-Zähler, Bleiglaszähler: Energiemessung für e^-, e^+, γ; Trennung von e^-e^+ von Hadronen und Muonen
Driftkammer: Bestimmung des Durchgangsortes geladener Teilchen
Szintillationszähler: Nachweis geladener Teilchen
Hg-Konverter: Umwandlung von γ (neutral) in e^+e^--Paare (geladen)
Eisen-Abschirmung: Absorption von Hadronen, dient als Filter für Muonen
EXPERIMENT am SPEICHERRING DORIS zum NACHWEIS GELADENER und NEUTRALER TEILCHEN (DESY-HEIDELBERG)
DESY-HEIDELBERG
SPEICHERRING-EXPERIMENT

Fig. 52

Acta Physica Austriaca, Suppl. XIV, 307 — 395 (1975)

PHENOMENOLOGY OF WEAK CURRENTS+

by

M. GOURDIN
Laboratoire de Physique Théorique
et Hautes Energies, Paris

INTRODUCTION

The discovery of leptonic and hadronic weak neutral currents is of fundamental importance for our understanding of weak interactions. From a naive and rational point of view the absence of such currents was a difficult puzzle to solve and only tricky models were able to explain - or better to incorporate - this feature. Because of the insisting calls of gauge theory people and thanks to the improvement of techniques the experimentalists made considerable and expensive efforts in that direction and we now have five independent evidences for the existence of weak neutral currents [1] - [5].

We obviously are at the beginning of the neutral current era if the results are of fundamental importance they still remain dissiminated and they carry large uncertainties. Therefore the temptation is extremely com-

---

+ Lecture given at XIV. Internationale Universitätswochen für Kernphysik,Schladming,Austria, February 24-March 7, 1975.

pelling for experimentalists to translate their results in terms of fashionable models. The aim of this series of lectures is to react against such an attitude and to present an analysis of the data as model independent as possible in order to determine the main properties of the weak neutral current. Our approach proceeds in the same phenomenological way as that used by Pais and Treiman [6] in their pioneering work and more recently by Rajasekaran and Sarma [7]. We are first interested in general features and peculiar models are only considered as illustrations of more general properties.

With the limited time allowed to me I am obliged to restrict this study to inclusive reactions induced by neutrinos and antineutrinos. Other processes will be approached in subsequent publications.

The lectures are organized as follows:

LECTURE 1: We present the general method on the two well-known cases of inclusive reactions involving the weak charged current and the electromagnetic current. The relation between them by means of the conserved vector current hypothesis is discussed. An illustration of a calculation of hadronic matrix elements of the currents is presented in the framework of the quark parton model. The link with the previous analysis of the experimental data is given.

LECTURE 2: We then turn to weak processes involving neutral currents on a nuclear target. The general framework is discussed and the experimental data coming from three different groups are presented. We determine very general features of these reactions and we use an assumption for the isovector part of the weak neutral current. Consequences of the positivity are then discussed and constraints on the parameters obtained from the experiments. A second assumption is proposed for the vector part of the current and we elaborate again on the consequences of positivity.

LECTURE 3: Putting together the two assumptions we look at their implications. A particular class of neutral currents without isoscalar axial vector component is studied in some detail and three particular realizations are presented and compared with experiment. The Weinberg model is a special case of this class of neutral currents and the mixing angle θ_W is determined from the data.

LECTURE 4: A third assumption is made for the isoscalar axial vector component of the current. Again consequences of positivity are obtained and the hadronic matrix elements are computed in the quark parton model based on U(3) symmetry. Three particular models are studied as an illustration of this class of neutral currents. As a special case an extension of the Beg and Zee gauge theory is compared with experiment. Finally we introduce new components for the isoscalar current in particular components having the quantum numbers of the baryonic current. Two interesting proposals are described and analyzed: the first one is an extended Sakurai model involving only baryonic currents; the second one a quark parton model based on U(4) symmetry and which constitutes a consistent extension to hadrons of the Salam-Weinberg gauge theory.

LECTURE 1

ELECTROPRODUCTION AND WEAK CHARGED CURRENTS

1. Kinematics and Scaling

1.) Inclusive reactions induced by leptons

$$\ell + p \to \ell' + \text{HADRONS} \qquad \text{(R)}$$

are described to lowest order by the diagram of Fig. 1. where the kinematical notation is indicated. We introduce, as usual, the scalar variables q^2, W^2 and ν defined by:

$$q^2 = (k-k')^2 \qquad W^2 = -(p+q)^2 \qquad M\nu = -p.q$$

with the relation $W^2 = M^2 + 2M\nu - q^2$, M being the nucleon mass. In the laboratory frame the lepton variables are

E incident lepton energy E' final lepton energy

θ scattering angle between leptons

and they can be expressed in terms of the Lorentz invariants q^2 and ν by

$$q^2 = 4\ EE' \sin\frac{\theta}{2} \qquad \nu = E - E' .$$

2.) As a consequence of the spin one exchange between leptons and hadrons the transition matrix element for reactions (R) is factorized into the product of two matrix elements of the electromagnetic or weak current, one for leptons, and one for hadrons.

When the nucleon target is unpolarized and when

the lepton mass can be neglected the inclusive reactions (R) are described by three structure functions $\sigma_\lambda(q^2,W^2)$ one can interpret as the total cross sections for the processes:

$\gamma + p \rightarrow$ HADRONS $\qquad \sigma_\lambda^Q\ (q^2,\ W^2)$

$W^\pm + p \rightarrow$ HADRONS $\qquad \sigma_\lambda^\pm\ (q^2,\ W^2)$

λ being the helicity of the virtual intermediate boson ($\lambda = \pm 1, 0$).

The double differential cross section for processes induced by neutrinos and antineutrinos can be written in the form

$$d^2\sigma^\pm = d^2\sigma_o^{WEAK}\ \{\sigma_T^\pm + \varepsilon\sigma_L^\pm \pm \frac{\sqrt{1-\varepsilon^2}}{2}[\sigma_{-1}^\pm - \sigma_{+1}^\pm]\} \tag{1}$$

with

$$\frac{d^2\sigma_o^{WEAK}}{dq^2dW^2} = \frac{G^2}{4\pi^2}\ \frac{q^2\sqrt{\nu^2+q^2}}{2ME^2}\ \frac{1}{1-\varepsilon} \tag{2}$$

$$\varepsilon^{-1} = 1 + 2\ tg^2\ \frac{\theta}{2}\ \frac{\nu^2+q^2}{q^2}\ . \tag{3}$$

The +(-) sign corresponds to reactions initiated by neutrinos (antineutrinos). In the limit where the squared intermediate vector boson mass is large as compared to the involved value of q^2, the local approximation is valid and G is the Fermi weak coupling constant.

We obtain analogous results for inclusive electroproduction

$$d^2\sigma = d^2\sigma_o^{EM} \{\sigma_T^Q + \varepsilon\sigma_L^Q\} \tag{4}$$

with

$$\frac{d^2\sigma_o^{EM}}{dq^2 dW^2} = \frac{2\alpha^2}{q^2} \frac{\sqrt{\nu^2+q^2}}{ME^2} \frac{1}{1-\varepsilon} . \tag{5}$$

Here the photon propagator plays a dominant role in the cross section.

$$d^2\sigma_o^{WEAK} = X\, d^2\sigma_o^{EM} \qquad \text{with} \qquad X = 2\,\frac{G^2 q^4}{e^4} \tag{6}$$

and as usual α is the fine structure constant:

$$\alpha = \frac{e^2}{4\pi} \simeq \frac{1}{137} .$$

3.) Using the dimensionless variables:

$$\rho = \frac{p.k'}{p.k} = \frac{E'}{E} \quad \text{and} \quad \xi = \frac{q^2}{2M\nu} \tag{7}$$

we define the scaling limit as the high-energy limit $E \to \infty$ with ρ and ξ fixed. It follows, in particular, that q^2, W^2 and ν become all large with ξ fixed and Bjorken's conjecture for the structure functions can be written as:

$$\lim_{\substack{q^2, W^2 \to \infty \\ \xi \text{ fixed}}} \frac{M\sqrt{\nu^2+q^2}}{\pi} \sigma_\lambda(q^2, W^2) = F_\lambda(\xi). \tag{8}$$

In the scaling limit we have:

$$\varepsilon \to \frac{2\rho}{1+\rho^2} \tag{9}$$

and the weak cross sections (1) and (2) take the simple forms:

$$\begin{aligned} \frac{d^2\sigma^\nu}{d\rho\, d\xi} &\to \frac{G^2 ME}{\pi} \xi [\rho^2 F_+^\nu(\xi) + F_-^\nu(\xi) + 2\rho F_0^\nu(\xi)] \\ \frac{d^2\sigma^{\bar{\nu}}}{d\rho\, d\xi} &\to \frac{G^2 ME}{\pi} \xi [\rho^2 F_-^{\bar{\nu}}(\xi) + F_-^{\bar{\nu}}(\xi) + 2\rho F_0^{\bar{\nu}}(\xi)] \end{aligned} \tag{10}$$

<u>These differential cross sections increase linearly with the incident energy E and their ρ dependence is simply a second order polynomial.</u>

4.) We now compute the total cross sections by integrating over the variables q^2, W^2 or equivalently ρ and ξ. The assumption generally made is that the leading term of the total cross section is obtained by using everywhere the form of $d^2\sigma/d\rho d\xi$ valid in the scaling region. The result is simply a linear rising with energy E of the total neutrino and antineutrino cross sections.

$$\lim_{E\to\infty} \sigma_{TOT} = \frac{G^2 ME}{\pi} A \tag{11}$$

where the coefficients A are given by:

$$\begin{aligned} A^{\nu} &= \frac{1}{3} I_{+}^{\nu} + I_{-}^{\nu} + I_{o}^{\nu} \\ A^{\bar{\nu}} &= \frac{1}{3} I_{-}^{\bar{\nu}} + I_{+}^{\bar{\nu}} + I_{o}^{\bar{\nu}} \end{aligned} \tag{12}$$

and the integrals I_{λ} by:

$$I_{\lambda} = \int_{o}^{1} \xi F_{\lambda}(\xi) d\xi \quad . \tag{13}$$

5.) The structure functions $\sigma_{\lambda}(q^2, W^2)$ being total cross sections are positive functions of q^2 and W^2 in the physical region.

$$q^2 \geq 0 \qquad\qquad W^2 \geq M^2$$

Therefore the scaling functions $F_{\lambda}(\xi)$ and the first moment I_λ integrals are also positive:

$$F_\lambda(\xi) \geq 0 \qquad\qquad 0 \leq \xi \leq 1$$

$$I_\lambda \geq 0 .$$

2. Neutrino and Antineutrino Reactions with Charged Currents

1.) We consider only total cross sections in the scaling region and we discuss the general relations between the slope parameters defined in the previous section. An extension of these considerations to differential cross sections is given in appendix A.

2.) The weak charged is written in the form:

$$J_\mu^W = \cos\theta_C \; J_\mu^W(\Delta s = 0) + \sin\theta_C \; J_\mu^W(\Delta s = 1) \tag{14}$$

where θ_C is the Cabibbo angle.

The strangeness conserving part is, as usual, the V - A isotopic spin current.

$$J_\mu^W(\Delta s = 0) = V_\mu^{1+i2} + A_\mu^{1+i2} \tag{15}$$

and the strangeness violating component is the V - A current associated to V spin in the SU(3) language.

$$J^W_\mu(\Delta s = 0) = V_\mu^{4+i5} + A_\mu^{4+i5} \quad . \tag{16}$$

It follows that inclusive reaction on a nucleon target will lead to two classes of final states

those of zero strangeness with a slope parameter $B \cos^2 \theta_C$

those of strangeness ±1 with a slope parameter $C \sin^2 \theta_C$

when these states are not separated the total slope is

$$A = B \cos^2 \theta_C + C \sin^2 \theta_C \; . \tag{17}$$

3.) Let us restrict now to the strangeness conserving transition which strongly dominate because of the smallness of the Cabibbo angle. The weak current being a superposition of vector and axial vector components the slope parameters on a nucleon target can be written as follows:[2]

$$\begin{aligned}
B^{\nu p}_{cc} &= V^p_3 + A^p_3 + I^p_3 \\
B^{\nu n}_{cc} &= V^n_3 + A^n_3 + I^n_3 \\
B^{\bar{\nu} p}_{cc} &= V^n_3 + A^n_3 - I^n_3 \\
B^{\bar{\nu} n}_{cc} &= V^p_3 + A^p_3 - I^p_3 \; .
\end{aligned} \tag{18}$$

It would be convenient, in what follows, to introduce the more symmetrical quantities:

$$\Sigma_c = \frac{1}{4}[B^{\nu p}_{cc} + B^{\nu n}_{cc} + B^{\bar{\nu} p}_{cc} + B^{\bar{\nu} n}_{cc}] = \frac{1}{2}(V^p_3 + V^n_3 + A^p_3 + A^n_3)$$

$$\Delta_c = \frac{1}{4}[B^{\nu p}_{cc} + B^{\nu n}_{cc} - B^{\bar{\nu} p}_{cc} - B^{\bar{\nu} n}_{cc}] = \frac{1}{2}(I^p_3 + I^n_3) \tag{19}$$

$$D_c = \frac{1}{4}[B^{\nu p}_{cc} - B^{\nu n}_{cc} + B^{\bar{\nu} p}_{cc} - B^{\bar{\nu} n}_{cc}] = \frac{1}{2}(I^p_3 - I^n_3)$$

$$S_c = \frac{1}{4}[B^{\nu p}_{cc} - B^{\nu n}_{cc} - B^{\bar{\nu} p}_{cc} + B^{\bar{\nu} n}_{cc}] = \frac{1}{2}(V^p_3 - V^n_3 + A^p_3 - A^n_3) \ .$$

Defining:

$$V^p_3 + V^n_3 = 2V^c_3 \qquad A^p_3 + A^n_3 = 2A^c_3 \qquad I^p_3 + I^n_3 = 2I^c_3 \tag{20}$$

we get:

$$\Sigma_c = V^c_3 + A^c_3$$

$$\Delta_c = I^c_3 \ . \tag{21}$$

4.) In the performed experiments either with the Gargamelle bubble chamber at CERN or with counters at FERMILAB proton and neutron cross sections have not yet been separated. We are then interested in the quantities Σ_c and Δ_c for which experimental data exist and we shall use the notation N for an average over proton and neutron $N = \frac{p+n}{2}$.

Let us first separate the vector and the axial vector contributions by defining an angle Φ_c

$$V_3^c = \Sigma_c \cos^2 \Phi_c \qquad\qquad A_3^c = \Sigma_c \sin^2 \Phi_c \tag{22}$$

with: $\quad 0 \leq \Phi_c \leq \frac{\pi}{2}$.

The interference term I_3^c has its magnitude restricted by a Schwartz inequality

$$(I_3)^2 \leq V_3 \, A_3 \tag{23}$$

which is now written using equations (21) and (22) as:

$$4 \, \Delta_c^2 \leq \Sigma_c^2 \sin^2 2\Phi_c \quad . \tag{24}$$

The reality of Φ_c implies bounds for the cross sections

$$4 \, [B_{cc}^{\nu N} - B_{cc}^{\bar{\nu} N}]^2 \leq [B_{cc}^{\nu N} + B_{cc}^{\bar{\nu} N}]^2 \tag{25}$$

which can be equivalently written as:

$$\frac{1}{3} \leq r_c \leq 3 \tag{26}$$

where r_c is the ratio of antineutrino to neutrino total

cross sections averaged over nucleons.

$$r_c = \frac{\sigma_{cc}^{\bar{\nu}N}}{\sigma_{cc}^{\nu N}} \quad . \tag{27}$$

It is convenient to parametrize the experimental data with an angle δ_c, $0 \leq \delta_c \leq \frac{\pi}{4}$, defined by

$$\sin 2\delta_c = 2\,\frac{|\Delta_c|}{\Sigma_c} = 2\,\frac{|1-r_c|}{1+r_c} \quad . \tag{28}$$

The Schwartz inequality (24) is then equivalently written as:

$$\delta_c \leq \Phi_c \leq \frac{\pi}{2} - \delta_c \quad . \tag{29}$$

For a discussion of the experimental situation we refer to review talks at the last conferences and to specialized papers for the analysis of these data. Using the CERN Gargamelle resuls[1]:

$$2\,\frac{\Delta_c}{\Sigma_c} = 0.932 \pm 0.042 \tag{30}$$

we obtain

$$\delta_c = 34^\circ \,{}^{+4^\circ}_{-3^\circ} \tag{31}$$

giving severe limitations on Φ_c.

$$[34^{\circ}\,{}^{+4^{\circ}}_{-3^{\circ}}] \leq \Phi_C \leq [56^{\circ}\,{}^{+3^{\circ}}_{-4^{\circ}}] \quad .$$

3. Electroproduction

1.) The electromagnetic current is a pure vector superposition of an isoscalar part and of an isovector part

$$J^Q_\mu = V^3_\mu + V^0_\mu \quad . \tag{32}$$

In order to compare the electroproduction data with the neutrino and antineutrino total cross section we shall use quantities analogous to the slopes B by defining "total cross sections" after removing the photon propagator.

$$\sigma^Q = \int\int X\, d^2\, \sigma^{EM} \qquad \text{with} \qquad X = 2\,\frac{G^2 q^4}{e^4}$$

and in the scaling region we simply have

$$\sigma^Q \to \frac{G^2\, EM}{\pi}\, B^Q \quad .$$

2.) The quantities B^Q associated to a nucleon target can be written as follows:

$$B^{ep} = V^Q_3 + V^Q_0 + J^Q_V$$

$$B^{en} = V_3^Q + V_0^Q - J_V^Q \quad . \tag{33}$$

We now introduce the sum and the difference of these quantities (b)

$$\Sigma_Q = \frac{1}{2}(B^{ep} + B^{en}) = V_3^Q + V_0^Q \tag{34}$$

$$D_Q = \frac{1}{2}(B^{ep} - B^{en}) = J_V^Q \; . \tag{35}$$

We separate the isovector and the isoscalar parts by introducing an angle Φ_Q.

$$V_3^Q = \Sigma_Q \cos^2 \Phi_Q \qquad\qquad V_0^Q = \Sigma_Q \sin^2 \Phi_Q \tag{36}$$

with $\qquad 0 \leq \Phi_Q \leq \frac{\pi}{2}$.

The interference term I_{30}^Q is limited by a Schwartz inequality:

$$(J_V^Q)^2 \leq 4 \; V_3^Q \; V_0^Q \tag{37}$$

which can be written, using equation (35) and (36) as:

$$D_Q^2 \leq \Sigma_Q^2 \sin^2 2\Phi_Q \quad . \tag{38}$$

The reality of Φ_Q is simply insured by the positivity of the two cross-sections σ^{ep} and σ^{en}.

We represent the data with an angle δ_Q, $0 \leq \delta_Q \leq \frac{\pi}{4}$, defined by:

$$\sin 2\delta_Q = \frac{|D_Q|}{\Sigma_Q} \tag{39}$$

and the inequality (38) is equivalently written as:

$$\delta_Q \leq \Phi_Q \leq \frac{\pi}{2} - \delta_Q \quad . \tag{40}$$

Another way to write the same constraint is:

$$\Sigma_Q - \sqrt{\Sigma_Q^2 - D_Q^2} \leq 2V_3^Q \leq \Sigma_Q + \sqrt{\Sigma_Q^2 - D_Q^2} \quad . \tag{41}$$

The angle δ_Q is now computed using the SLAC data

$$\frac{D_Q}{\Sigma_Q} = 0.133 \pm 0.066 \tag{42}$$

and we obtain, for δ_Q, a small value

$$\delta_Q = 4^\circ \pm 2^\circ \; . \tag{43}$$

4. Conserved vector current hypothesis

1.) The so-called conserved vector current hypothesis states that the vector currents

$$V_\mu^{1+i2}, \quad V_\mu^{3}, \quad V_\mu^{1-i2}$$

previously introduced for the weak charged current and the electromagnetic current, belong to the same isotopic spin triplet.

The direct consequence is an equality between squared matrix elements.

$$V_3^C = 2\, V_3^Q \ . \tag{44}$$

We then have a relation between the two parametrization angles Φ_C and Φ_Q which is given by

$$\Sigma_C \cos^2 \Phi_C = 2\, \Sigma_Q \cos^2 \Phi_Q \quad . \tag{45}$$

2.) The constraint (40) on Φ_Q is easily translated in the language of Φ_C using equations (41) and (45)

$$\frac{\Sigma_Q-\sqrt{\Sigma_Q^2-D_Q^2}}{\Sigma_C} \leq \cos^2 \Phi_C \leq \frac{\Sigma_Q+\sqrt{\Sigma_Q^2-D_Q^2}}{\Sigma_C} \ . \tag{46}$$

Because of the smallness of the ratio D_Q/Σ_Q only the upper limit in equation (46) will be useful. We now introduce an angle δ_{Qc}, $0 \leq \delta_{Qc} \leq \frac{\pi}{2}$, relating electroproduction and weak process data

$$\cos^2 \delta_{Qc} = \frac{\Sigma_Q + \sqrt{\Sigma_Q^2 - D_Q^2}}{\Sigma_c} \quad . \tag{47}$$

The intersection of the domains (29) and (46) for Φ_c turns out to be:

$$\delta_{Qc} \leq \Phi_c \leq \frac{\pi}{2} - \delta_c \quad . \tag{48}$$

We use the SLAC and the CERN-Gargamelle data in order to compute the angle δ_{Qc}. With the value

$$\frac{\Sigma_Q}{\Sigma_c} = 0.298 \pm 0.045 \tag{49}$$

we obtain:

$$\delta_{Qc} = 40^\circ \pm 5^\circ \ . \tag{50}$$

3.) Let us remark that the value $\Phi_c = 45^\circ$ belongs to the domain allowed by experiments. This feature is the reason why <u>the quark parton model based on U(3) symmetry is compatible with experiment at this level of the first moment of the scaling function.</u>

Moreover the corresponding bound

$$\frac{\Sigma_c}{2} \leq \Sigma_Q + \sqrt{\Sigma_Q^2 - D_Q^2}$$

is practically saturated by experiment and the value of $\delta_c = 45^o$ is very close to its lower limit δ_{Qc}. This situation makes the quark parton model simple and predictive.

5. Quark parton model

1.) In the parton model the composite structure of the hadrons is described by a set of positive functions $D_j(\xi)$ which are the distribution functions in the hadron, of the parton of type j carrying the fraction ξ of the hadron momentum.

The scaling functions $F_\lambda(\xi)$ are simply linear combinations of the functions $D_j(\xi)$ with coefficients describing the pointlike coupling of the parton j with the current. It follows that for total cross sections the slope parameters are linear combinations of the first moments d_j of the distributions $D_j(\xi)$

$$d_j = \int_o^1 \xi D_j(\xi) d\xi \quad .$$

2.) In the quark parton model based on U(3) symmetry there are six such distributions (c), three for quarks (j = 1, 2, 3) and three for antiquarks (j = -1, -2, -3). The six cross sections involved in electroproduction and charged current reactions are simply written[8]:

$$\Sigma_Q = \frac{5}{27} (d_1 + d_{-1} + d_2 + d_{-2}) + \frac{2}{27}(d_3 + d_{-3})$$

$$D_Q = \frac{1}{9} \ (d_1 + d_{-1} - d_2 - d_{-2})$$

$$\Sigma_c = \frac{2}{3}(d_1 + d_{-1} + d_2 + d_{-2})$$

$$\Delta_c = \frac{1}{3}(d_1 - d_{-1} + d_2 - d_{-2})$$

$$D_c = -\frac{1}{3}(d_1 + d_{-1} - d_2 - d_{-2}) \tag{51}$$

$$S_c = -\frac{2}{3}(d_1 - d_{-1} - d_2 + d_{-2}) \ .$$

Complete experimental information will determine five moments, d_1, d_2, d_{-1}, d_{-2} and the sum $d_3 + d_{-3}$. In addition the quark parton model predicts the relation:

$$D_c + 3D_Q = 0 \tag{52}$$

and by positivity of the d_j's we get the Llewellyn-Smith inequality:

$$\Sigma_Q \geq \frac{5}{18}\Sigma_c \ . \tag{53}$$

3.) In order to separate the strange quark and anti-quark moments we must observe the $|\Delta S| = 1$ transitions for weak processes (d). Using the upper index (1) for the corresponding cross sections we get the relation

$$\Delta_c^{(1)} = -\frac{1}{6}(d_1 + d_2 - d_{-1} - d_{-2}) + (d_3 - d_{-3})$$

and three constraints

$$\Sigma_c^{(1)} = \frac{27}{2}\Sigma_Q - \frac{7}{2}\Sigma_c$$

$$D_c^{(1)} = -\frac{1}{2} D_c$$

$$S_c^{(1)} = \frac{1}{4} S_c \quad . \tag{54}$$

The first of these relations leads, by positivity, to the inequality

$$\Sigma_c^{(1)} \geq \frac{1}{4} \Sigma_c \quad . \tag{55}$$

4.) A phenomenological analysis of the SLAC and CERN-Gargamelle data, taking into account the $|\Delta S| = 1$ currents gives the following numerical estimates[8]:

$$V_3^c = A_3^c = \frac{1}{3} (d_1 + d_2 + d_{-1} + d_{-2}) = 0.168 \pm 0.018$$

$$I_3^c = \frac{1}{3} (d_1 + d_2 - d_{-1} - d_{-2}) = 0.156 \pm 0.028$$

$$V_0^Q = \frac{1}{54} (d_1 + d_2 + d_{-1} + d_{-2}) + \frac{2}{27} (d_3 + d_{-3}) = 0.016 \begin{smallmatrix} +0.011 \\ -0.008 \end{smallmatrix}$$

$$3 J_V^Q = \frac{1}{3} (d_1 - d_2 + d_{-1} - d_{-2}) = 0.04 \pm 0.02 \quad . \tag{56}$$

LECTURE 2

WEAK NEUTRAL CURRENTS

Assumptions

1. Generalities

1.) Let us begin with two general properties of the hadronic weak neutral current. By analogy with the electromagnetic and weak charged currents we take for granted the usual Lorentz covariance of the currents.

(A) the weak neutral current is a superposition of vector and axial vector components.

Experimentally the strangeness violating neutral current is strongly suppressed and we shall consider here only the strangeness conserving part. With respect to isotopic spin rotations we assume the following property.

(B) the weak neutral current is a superposition of isovector and isoscalar components.

We then use the following decomposition:

$$J_\mu^Z = V_\mu^{3,N} + A_\mu^{3,N} + V_\mu^{O,N} + A_\mu^{O,N} \ . \tag{57}$$

2.) By analyzing the various cross sections we wish to isolate the various hadronic matrix elements of each component of the current. In the case of an unpolarized target we obtain such quantities:

i) the diagonal terms V_3^N, A_3^N, V_O^N, A_O^N,
ii) the vector axial vector interference terms I_3^N, I_O^N

iii) the isovector isoscalar interference terms J_V^N, J_A^N

iv) the mixed interference terms K_{3o}^{VA}, K_{3o}^{AV}.

In inclusive experiments on a nucleon unpolarized target we can measure four cross sections and it is therefore impossible to determine from experiment the ten quantities without supplementary assumptions. Let us now give the expression of the slope parameters B of the total inclusive cross sections, assuming scaling.

$$B_{Nc}^{\nu p} = [V_3^N+A_3^N+V_o^N+A_o^N] + [I_3^N+I_o^N] + [J_V^N+J_A^N] + [K_{3o}^{VA}+K_{3o}^{AV}]$$

$$B_{Nc}^{\nu n} = [V_3^N+A_3^N+V_o^N+A_o^N] + [I_3^N+I_o^N] - [J_V^N+J_A^N] - [K_{3o}^{VA}+K_{3o}^{AV}]$$

$$B_{Nc}^{\bar{\nu} p} = [V_3^N+A_3^N+V_o^N+A_o^N] - [I_3^N+I_o^N] + [J_V^N+J_N^A] - [K_{3o}^{VA}+K_{3o}^{AV}]$$

$$B_{Nc}^{\bar{\nu} n} = [V_3^N+A_3^N+V_o^N+A_o^N] - [I_3^N+I_o^N] - [J_V^N+J_N^A] + [K_{3o}^{VA}+K_{3o}^{AV}] \quad .$$

It will be convenient, for practical purpose, to introduce four linear combinations of these cross-sections.

$$\Sigma_N = \frac{1}{4} [B_{Nc}^{\nu p}+B_{Nc}^{\nu n}+B_{Nc}^{\bar{\nu} p}+B_{Nc}^{\bar{\nu} n}] = V_3^N + A_3^N + V_o^N + A_o^N$$

$$\Delta_N = \frac{1}{4} [B_{Nc}^{\nu p}+B_{Nc}^{\nu n}-B_{Nc}^{\bar{\nu} p}-B_{Nc}^{\bar{\nu} n}] = I_3^N + I_o^N \tag{58}$$

$$D_N = \frac{1}{4}\,[B_{Nc}^{\nu p} - B_{Nc}^{\nu n} + B_{Nc}^{\bar{\nu} p} - B_{Nc}^{\bar{\nu} n}] = J_V^N + J_A^N$$

$$S_N = \frac{1}{4}\,[B_{Nc}^{\nu p} - B_{Nc}^{\nu n} - B_{Nc}^{\bar{\nu} p} + B_{Nc}^{\bar{\nu} n}] = K_{3o}^{VA} + K_{3o}^{AV} \quad .$$

3.) In the actual situation, inclusive data on neutron and proton are not available and only averaged quantities like Σ_N and Δ_N have been measured. Six parameters are now involved in the analysis V_3^N, A_3^N, V_o^N, A_o^N and the two interference terms I_3^N and I_o^N. Again the problem is undetermined without supplementary assumptions.

A systematic research of neutral currents has been carried out at CERN with neutrino and antineutrino beams entering the Gargamelle chamber filled out with freon. The results for relative rates of neutral current events to charge current events for interaction with hadron energy release larger than 1 GeV are as follows (a)[2]:

$$\left(\frac{Nc}{cc}\right)^{\nu} = 0.217 \pm 0.026 \tag{59}$$

$$\left(\frac{Nc}{cc}\right)^{\bar{\nu}} = 0.43 \pm 0.12 \quad . \tag{60}$$

We emphasize that these quantities refer to numbers of events with identical cuts and not to total cross sections. In the same situation the ratio for antineutrino to neutrino for charged currents has been found to be

$$\frac{(cc)^{\bar{\nu}}}{(cc)^{\nu}} = 0.26 \pm 0.03 \quad . \tag{61}$$

A good way to minimize the effect due to the hadron cut is to consider the ratios $\frac{\Sigma_N}{\Sigma_c}$ and $\frac{\Delta_N}{\Delta_c}$. From equations (59), (60) and (61) we get:

$$\frac{\Sigma_N}{\Sigma_c} = 0.26 \pm 0.03 \tag{62}$$

$$\frac{\Delta_N}{\Delta_c} = 0.14 \pm 0.06 \ . \tag{63}$$

Two different experiments have been performed at the FERMI-LAB by:

- the Harvard, Pennsylvania-Wisconsin group (HPW)
- the Caltech group.

Data obtained by the HPW group correspond to a mean incident energy of 50 GeV and only events carrying a total hadronic energy larger than 4 GeV have been retained. The average of successive experiments gives[3]

$$\left(\frac{Nc}{cc}\right)^{\nu} = 0.11 \pm 0.05 \tag{64}$$

$$\left(\frac{Nc}{cc}\right)^{\bar{\nu}} = 0.32 \pm 0.09 \tag{65}$$

from which we deduce:

$$\frac{\Sigma_N}{\Sigma_c} = 0.16 \pm 0.04 \tag{66}$$

$$\frac{\Delta_N}{\Delta_c} = 0.005 \pm 0.086 \ . \tag{67}$$

Positive evidences for hadronic neutral currents are also claimed by the Caltech group. They give the following estimates without errors[4]

$$\left(\frac{Nc}{cc}\right)^{\nu} = 0.22 \qquad \left(\frac{Nc}{cc}\right)^{\bar{\nu}} = 0.33 \tag{68}$$

For the ratios Σ_N/Σ_c and Δ_N/Δ_c

$$\frac{\Sigma_N}{\Sigma_c} = 0.25 \qquad \frac{\Delta_N}{\Delta_c} = 0.16 \tag{69}$$

the agreement with the CERN data is very nice.

4.) Let us first separate the vector and axial vector parts by defining an angle Φ_N

$$V_3^N + V_o^N = \Sigma_N \cos^2 \Phi_N \tag{70}$$

$$A_3^N + A_o^N = \Sigma_N \sin^2 \Phi_N \tag{71}$$

with

$$0 \leq \Phi_N \leq \frac{\pi}{2} \quad .$$

The interference part $\Delta_N = I_3^N + I_o^N$ is restricted by a Schwartz inequality analogous to equation (24)

$$4\,\Delta_N^2 \leq \Sigma_N^2 \sin^2 2\,\Phi_N \quad . \tag{72}$$

We then get bounds for the total cross sections as a consequence of the reality of Φ_N

$$4\ [B_{Nc}^{\nu N} - B_{Nc}^{\bar{\nu}N}]^2 \leq [B_{Nc}^{\nu N} + B_{Nc}^{\bar{\nu}N}]^2 \tag{73}$$

or, using the ratio r_N defined by

$$r_N = \frac{\sigma_{Nc}^{\bar{\nu}N}}{\sigma_{Nc}^{\nu N}} \tag{74}$$

we get the positivity constraint

$$\frac{1}{3} \leq r_N \leq 3 \quad . \tag{75}$$

It is convenient to parametrize the experimental data with an angle δ_N, $0 \leq \delta_N \leq \frac{\pi}{4}$, defined by

$$\sin 2\ \delta_N = 2\ \frac{|\Delta_N|}{\Sigma_N} = 2\ \frac{|1-r_N|}{1+r_N} \quad . \tag{76}$$

The Schwartz inequality (72) is then equivalently written

$$\delta_N \leq \Phi_N \leq \frac{\pi}{2} - \delta_N \ . \tag{77}$$

From the CERN-Gargamelle data we have

$$r_N = 0.6 \pm 0.2 \tag{78}$$

and the angle δ_N is badly determined

$$\sin 2\,\delta_N = 0.5 \pm 0.3$$
$$\delta_N = 15^\circ \begin{array}{l} +\,12^\circ \\ -\,10^\circ \end{array} . \tag{79}$$

From the result of the H.P.W. collaboration of FERMILAB $r_N = 1 \pm 0.2$ and the value of δ_N is compatible with zero.

$$\sin 2\,\delta_N = 0 \pm 0.2$$
$$\delta_N = 0^\circ - 6^\circ . \tag{80}$$

5.) We proceed in the same way for the isovector and isoscalar parts introducing an angle Φ_Z

$$V_3^N + A_3^N = \Sigma_N \cos^2 \Phi_Z \tag{81}$$

$$V_o^N + A_o^N = \Sigma_N \sin^2 \Phi_Z \tag{82}$$

with

$$0 \leq \Phi_Z \leq \frac{\pi}{2} .$$

The interference part $D_N = J_V^N + J_A^N$ is restricted by a Schwartz inequality analogous to equation (38)

$$D_N^2 \leq \Sigma_N^2 \sin^2 2\,\Phi_Z \quad . \tag{83}$$

The reality of Φ_Z is a simple consequence of the positivity of the cross sections and does not give new bounds.

We now introduce an angle δ_Z, $0 \leq \delta_Z \leq \frac{\pi}{4}$, by the equality

$$\sin 2\,\delta_Z = \frac{|D_N|}{\Sigma_N} \tag{84}$$

and the positivity requirement (83) is simply

$$\delta_Z \leq \Phi_Z \leq \frac{\pi}{2} - \delta_Z \quad . \tag{85}$$

There are no available experimental data on D_N and therefore we cannot compute δ_Z.

2. Assumption for the isovector part

1.) We make the following assumption:

The isovector neutral currents belong to the same isotopic spin triplets as the isovector charged currents (I)

$$V_\mu^{3,N} = \alpha\, V_\mu^3 \qquad\qquad A_\mu^{3,N} = \beta A_\mu^3 \quad . \tag{86}$$

It follows that V_3^N, A_3^N and I_3^N are related to charged current quantities averaged over proton and neutron

$$V_3^N = \frac{1}{2}\alpha^2 V_3^c \qquad A_3^N = \frac{1}{2}\beta^2 A_3^c \qquad I_3^N = \frac{1}{2}\alpha\beta I_3^c . \tag{87}$$

We now study, in the α,β plane, the domain allowed by experiments on weak charged and neutral currents.

2.) We isolate the isoscalar contributions:

$$\Sigma_o = V_o^N + A_o^N = \Sigma_N - \frac{1}{2}(\alpha^2 V_3^c + \beta^2 A_3^c) \tag{88}$$

$$\Delta_o = I_o^N = \Delta_N - \frac{1}{2}\alpha\beta I_3^c = \Delta_N - \frac{1}{2}\alpha\beta\Delta_c . \tag{89}$$

It is convenient to use for V_3^c and A_3^c the parametrization with an angle Φ_c already introduced in the previous lecture

$$V_3^c = \Sigma_c \cos^2\Phi_c \qquad A_3^c = \Sigma_c \sin^2\Phi_c . \tag{90}$$

The positivity of Σ_o leads to the inequality

$$\alpha^2 \cos^2\Phi_c + \beta^2 \sin^2\Phi_c \leq 2\frac{\Sigma_N}{\Sigma_c} \tag{91}$$

which represents, in the α,β plane, the interior of an ellipse $E(\Phi_c)$ of axis $\alpha = 0$, $\beta = 0$, and passing through four fixed points

$$\alpha^2 = \beta^2 = 2\frac{\Sigma_N}{\Sigma_c} .$$

From electroproduction and charged current experiments the angle Φ_c is restricted to be in the range

$$\delta_{Qc} \leq \Phi_c \leq \frac{\pi}{2} - \delta_c \quad . \tag{92}$$

The domain in the α,β plane associated to the condition $\Sigma_o \geq 0$ is then the union of the interiors of the ellipse $E(\Phi_c)$ and it turns out to be also the union of the interiors of the limiting ellipses

$$E(\delta_{Qc}) \quad \cup \quad E(\frac{\pi}{2} - \delta_c) \ .$$

3.) The vector axial vector interference term I_o^N is restricted by a Schwartz inequality

$$(I_o^N)^2 \leq V_o^N \, A_o^N \tag{93}$$

from which we deduce the constraint

$$2 \, |\Delta_o| \leq \Sigma_o \quad . \tag{94}$$

Using the previous expressions (88) and (89) for Σ_o and Δ_o we get two inequalities

$$\alpha^2 V_3^c + \beta^2 A_3^c \pm 2\alpha\beta\Delta_c \leq 2(\Sigma_N \pm 2\Delta_N) \tag{95}$$

depending on the parameter Φ_c via the relations (90).

The right-hand side of equation (95) is positive because of the relation (72)

$$2\ |\Delta_N| \leq \Sigma_N \quad . \tag{96}$$

The left-hand side of equation (95) is a bilinear form in α,β whose discriminant is negative or zero because of the positivity constraint (24) for charged currents. It follows that equation (95) represents the interiors of two ellipses $E^{\pm}(\Phi_c)$.

The allowed domain $D_1(\Phi_c)$ in the α,β plane is the intersection of the interiors of these two ellipses

$$D_1(\Phi_c) = E^+(\Phi_c) \cap E^-(\Phi_c) \quad . \tag{97}$$

Some remarks are now in order:

i) the ellipses $E^{\pm}(\Phi_c)$ have four fixed points independent of Φ_c located on the lines $\alpha \pm \beta = 0$

ii) the equations $E^{\pm}(\Phi_c) = 0$ correspond to the equalities $\Sigma_o \pm 2\ \Delta_o = 0$

iii) the common points of the two ellipses $E^+(\Phi_c)$ and $E^-(\Phi_c)$ which are the corners of the domain $D_1(\Phi_c)$ are associated to the vanishing of the isoscalar contributions - pure isovector weak neutral current - $\Sigma_o = 0$ $\Delta_o = 0$. They are also located at the intersection of the ellipse $E(\Phi_c)$ corresponding to $\Sigma_o = 0$ with the rectangular hyberbola H associated to $\Delta_o = 0$ and of equation

$$\alpha\beta = 2\ \frac{\Delta_N}{\Delta_c} \quad . \tag{98}$$

We have represented on Fig. 2 the domain D_1 for the particular case

$$\Phi_c = \frac{\pi}{4} \qquad \text{where} \qquad V_3^c = A_3^c = \frac{1}{2} \Sigma_c \quad .$$

From the CERN-Gargamelle experiments the coordinates of the corners P and Q are given by:

$$\begin{array}{lll} P & \alpha = \varepsilon[0.98 \pm 0.11] & \beta = \varepsilon[0.29 \pm 0.11] \\ Q & \alpha = \varepsilon[0.29 \pm 0.11] & \beta = \varepsilon[0.98 \pm 0.11] \end{array} \tag{99}$$

with $\varepsilon = \pm 1$.

For $\Phi_c = \pi/4$ the domain D_1 looks very close to a rectangle whose dimensions are given by the fixed points M and N. Putting

$$g_L = \frac{\alpha+\beta}{2} \qquad g_R = \frac{\alpha-\beta}{2} \tag{100}$$

we obtain, using the CERN-Gargamelle data

$$\text{M} \qquad |g_L| \leq 0.64 \pm 0.05 \tag{101}$$

$$\text{N} \qquad |g_R| \leq 0.36 \pm 0.09 \quad . \tag{102}$$

4.) The total domain allowed by experiments in the α, β plane is the union of the $D_1(\Phi_c)$'s with Φ_c varying in its physical range (92). Let us notice that in the limiting case $\Phi_c = \frac{\pi}{2} - \delta_c$

the domain D_1 reduces to a parallelogram. A graphical representation of the complete domain

$$D_1(\delta_{Qc}) \cup D_1(\frac{\pi}{2} - \delta_c)$$

is given on Fig. 3.

3. Assumption for the vector part

1.) We now make the following assumption

> The vector neutral currents are proportional to the isotopic spin components of the electromagnetic current (II)

$$V_\mu^{3,N} = \alpha V_\mu^3 \qquad\qquad V_\mu^{o,N} = \gamma V_\mu^o \; .$$

It follows that V_3^N, V_o^N and J_V^N are related to quantities measured in electroproduction.

$$V_3^N = \alpha^2 \, V_3^Q \qquad\qquad V_o^N = \gamma^2 \, V_o^Q \qquad\qquad J_V^N = \alpha\gamma J_V^Q \; . \qquad (103)$$

We now study, in the α,γ plane the domain allowed by experiments on weak neutral and electromagnetic currents.

2.) We isolate the axial vector contributions

$$\Sigma_A = A_3^N + A_o^N = \Sigma_N - (\alpha^2 V_3^Q + \gamma^2 V_o^Q) \tag{104}$$

$$D_A = J_A^N = D_N - \alpha\gamma J_V^Q = D_N - \alpha\gamma D_Q \tag{105}$$

and we use for V_3^Q and V_o^Q the parametrization introduced in electroproduction with the angle Φ_Q

$$V_3^Q = \Sigma_Q \cos^2 \Phi_Q \qquad V_o^Q = \Sigma_Q \sin^2 \Phi_Q \; . \tag{106}$$

The positivity of Σ_A leads to the inequality

$$\alpha^2 \cos^2 \Phi_Q + \gamma^2 \sin^2 \Phi_Q \leq \frac{\Sigma_N}{\Sigma_Q} \tag{107}$$

which represents, in the α,γ plane, the interior of an ellipse $F(\Phi_Q)$ of axis $\alpha = 0$, $\gamma = 0$ and passing through four fixed points independent of Φ_Q

$$\alpha^2 = \gamma^2 = \frac{\Sigma_N}{\Sigma_Q} \quad .$$

3.) The isoscalar isovector interference term J_A^N is restricted by a Schwartz inequality

$$(J_A^N)^2 \leq 4\, A_3^N\, A_o^N \tag{108}$$

from which we deduce the constraint

$$|D_A| \leq \Sigma_A \quad . \tag{109}$$

Using the expressions (104) and (105) for Σ_A and D_A we get two inequalities

$$\alpha^2\, V_3^Q + \gamma^2\, V_o^Q \pm \alpha\gamma D_Q \leq \Sigma_N \pm D_N \quad . \tag{110}$$

Taking into account the constraints due to positivity

$$|D_N| \leq \Sigma_N \qquad\qquad (J_V^Q)^2 \leq 4\, V_3^Q\, V_o^Q$$

we easily see that equation (110) represents the interiors of two ellipses $F^{\pm}(\Phi_Q)$. The allowed domain $\bar{D}_2(\Phi_Q)$ in the α,γ plane is the intersection of the interiors of these two ellipses.

$$\bar{D}_2(\Phi_Q) = F^+(\Phi_Q) \cap F^-(\Phi_Q) \quad .$$

The discussion of $\bar{D}_2(\Phi_Q)$ can proceed as that of $D_1(\Phi_C)$ in the previous section. In particular the four corners of $\bar{D}_2$ are associated to the vanishing of the axial vector contributions - pure vector weak neutral current -

$$\Sigma_A = 0 \qquad\qquad D_A = 0$$

and they are located at the intersection of the ellipse $F(\Phi_Q)$ with the rectangular hyperbola $\bar{H}$ of equation:

$$\alpha\gamma = \frac{D_N}{D_Q} \quad .$$

All the considerations of this paragraph imply the knowledge of D_N, e.g. the separation of proton and neutron neutral current cross sections which has not yet been experimentally achieved. Therefore numerical estimates of $\bar{D}_2$ cannot be done.

LECTURE III

WEAK NEUTRAL CURRENT

Models

1. Consequences of the assumptions I and II

1.) We now analyse the weak neutral current data with the two assumptions I and II which, by the way, imply the conserved vector current hypothesis between electroproduction and charged current. Using the simplified notation

$$V_3^Q = V_3 \qquad V_o^Q = V_o \qquad J_V^Q = J_V$$

$$V_3^c = 2V_3 \qquad A_3^c = 2A_3 \qquad I_3^c = 2I_3 \tag{111}$$

we get

$$V_3^N = \alpha^2 V_3 \qquad V_o^N = \gamma^2 V_o \qquad A_3^N = \beta^2 A_3$$
$$I_3^N = \alpha\beta I_3 \qquad J_V^N = \alpha\gamma J_V \tag{112}$$

and we study, in the α,β,γ space, the domain allowed by experiments.

2.) We first isolate the isoscalar axial vector contribution

$$A_o^N = \Sigma_N - (\alpha^2 V_3 + \beta^2 A_3 + \gamma^2 V_o) \tag{113}$$

and we use for V_3, A_3 and V_o the one parameter representation (c)

$$V_3 = \frac{1}{2}\Sigma_c \cos^2\Phi_c \qquad A_3 = \frac{1}{2}\Sigma_c \sin^2\Phi_c \qquad V_o = \Sigma_Q - \frac{1}{2}\Sigma_c \cos^2\Phi_c \; . \tag{114}$$

The positivity of A_o^N leads to the inequality

$$\alpha^2 V_3 + \beta^2 A_3 + \gamma^2 V_o \leq \Sigma_N \tag{115}$$

or equivalently:

$$\alpha^2\cos^2\Phi_c+\beta^2\sin^2\Phi_c + \gamma^2[2\,\frac{\Sigma_Q}{\Sigma_c} - \cos^2\Phi_c] \le 2\,\frac{\Sigma_N}{\Sigma_c} \tag{116}$$

which represents, in the α,β,γ space, the interior of an ellipsoid $E(\Phi_c)$ of axis the three coordinate axis and passing through a fixed curve independent of Φ_c and located on the intersection of the cone

$$\alpha^2 = \beta^2 + \gamma^2$$

with the cylinder

$$\beta^2 + 2\,\gamma^2\,\frac{\Sigma_Q}{\Sigma_c} = 2\,\frac{\Sigma_N}{\Sigma_c} \quad . \tag{117}$$

3.) It is now possible to extract the full content of the Schwartz inequality (93). We obtain

$$(\Delta_N-\tfrac{1}{2}\alpha\beta\Delta_c)^2 \le \gamma^2V_o[\Sigma_N - (\alpha^2V_3 + \beta^2A_3 + \gamma^2V_o)] \tag{118}$$

which represents in the α,β,γ space a volume interior to the ellipsoid $E(\Phi_c)$ and limited by a fourth order surface $S_1(\Phi_c)$. The ellipsoid and S_1 are tangent along a curve associated to the vanishing of the axial vector isoscalar contribution $A_o^N = 0$. This curve is also the intersection of the ellipsoid $E(\Phi_c)$ with the cylinder $\Delta_o = 0$ of equation (99)

$$\alpha\beta = 2\,\frac{\Delta_N}{\Delta_c} \quad .$$

We can solve the inequality (118) in γ^2 for fixed values of α and β and we get

$$\gamma_m(\alpha,\beta) \le |\gamma| \le \gamma_M(\alpha,\beta) \tag{119}$$

with

$$\gamma_m(\alpha,\beta) = \frac{1}{2}\left\{\sqrt{\frac{\Sigma_o+2\Delta_o}{V_o}} - \sqrt{\frac{\Sigma_o-2\Delta_o}{V_o}}\right\}$$

(120)

$$\gamma_M(\alpha,\beta) = \frac{1}{2}\left\{\sqrt{\frac{\Sigma_o+2\Delta_o}{V_o}} + \sqrt{\frac{\Sigma_o-2\Delta_o}{V_o}}\right\}$$

Σ_o and Δ_o are functions of α,β given by the definitions (90) and (91). The reality of γ_m and γ_M is insured when the point α,β belongs to the domain $D_1(\Phi_c)$ which is precisely defined by $\Sigma_o \ge 2|\Delta_o|$. On the frontier of that domain γ^2 is known

$$V_o^N = A_o^N = \frac{1}{2}\Sigma_o \qquad \text{and} \qquad \gamma^2 = \frac{\Sigma_o}{2V_o} \;. \tag{121}$$

This value vanishes at the four corners of $D_1(\Phi_c)$ as previously noticed for a pure isovector weak neutral current.

In the opposite situation of a pure isoscalar weak neutral current, $\alpha = 0$, $\beta = 0$, the limits of $|\gamma|$ become

$$\gamma_m(0,0) = \frac{1}{2}\left\{\sqrt{\frac{\Sigma_N+2\Delta_N}{V_o}} - \sqrt{\frac{\Sigma_N-2\Delta_N}{V_o}}\right\}$$

$$\gamma_M(o,o) = \frac{1}{2}\{\sqrt{\frac{\Sigma_N+2\Delta_N}{V_o}} + \sqrt{\frac{\Sigma_N-2\Delta_N}{V_o}}\} \tag{122}$$

4.) It may also be useful to consider equation (118) as a second order inequality in β for fixed values of α and γ. The result is:

$$\beta_-(\alpha,\gamma) \leq \beta \leq \beta_+(\alpha,\gamma) \tag{123}$$

were

$$\beta_\pm(\alpha,\gamma) = \frac{2\Delta_C\Delta_N \pm \sqrt{4\gamma^2 V_o \bar{Y}}}{\alpha^2\Delta_C^2+4\gamma^2 V_o A_3} \tag{124}$$

with

$$\bar{Y} = [\alpha^2\Delta_C^2+4\gamma^2 V_o A_3][\Sigma_N \dot{-} (\alpha^2 V_3+\gamma^2 V_o)]-4A_3\Delta_N^2 \quad . \tag{125}$$

The reality of the solutions (124) corresponds to the positivity of the discriminant $\bar{Y}(\alpha,\gamma) \geq 0$. In the α,γ plane this condition defines a domain $\bar{D}_1(\Phi_C)$ interor to the ellipse $F(\Phi_Q)$ - associated to $\Sigma_A = 0$ - and limited by a fourth order curve $\bar{Y}(\alpha,\gamma) = 0$. This domain is nothing but the orthogonal projection of $S_1(\Phi_C)$ on the α,γ plane.

We have represented on Fig. 4 the domain $\bar{D}_1$ for the particular case $\Phi_C = \frac{\pi}{4}$ where $V_3 = A_3 = \frac{1}{4}\Sigma_C$, $V_o = \Sigma_Q - \frac{1}{4}\Sigma_C$.

Combining now the results in the α,γ plane we easily see that solutions in β, for fixed α,γ can exist if and only if belongs to a domain $\bar{D}$ intersection of $\bar{D}_1$ and $\bar{D}_2$.

$$\bar{D}(\Phi_c) = \bar{D}_1(\Phi_c) \cap \bar{D}_2(\Phi_Q) \quad .$$

5.) In an analogous way the Schwartz inequality (1o8) can be written as:

$$(D_N - \alpha\gamma D_Q)^2 \leq 4\beta^2 A_3[\Sigma_N - (\alpha^2 V_3 + \beta^2 A_3 + \gamma^2 V_o)] \tag{126}$$

which represents in the α,β,γ space a volume interior to the ellipsoid $E(\Phi_c)$ and limited by a 4th order surface $S_2(\Phi_c)$. The ellipsoid and S_2 are tangent along a curve associated to the vanishing of the axial vector isoscalar contribution $A_o^N = 0$. This curve is also the intersection of the ellipsoid $E(\Phi_Q)$ with the cylinder $D_A = 0$ of the equation.

$$\alpha\gamma = \frac{D_N}{D_Q} \quad .$$

The inequality (126) can be considered as a second order inequality in β^2 for fixed values of α and γ. The solutions can be discussed as in the previous section and we get:

$$\beta_m(\alpha,\gamma) \leq |\beta| \leq \beta_M(\alpha,\gamma) \tag{127}$$

with

$$\beta_m(\alpha,\gamma) = \frac{\sqrt{\Sigma_A+D_A} - \sqrt{\Sigma_A-D_A}}{2\sqrt{A_3}}$$

(128)

$$\beta_M(\alpha,\gamma) = \frac{\sqrt{\Sigma_A+D_A} + \sqrt{\Sigma_A-D_A}}{2\sqrt{A_3}} \quad .$$

The reality of these solutions is obtained when the point α,γ belongs to the domain $\bar{D}_2(\Phi_Q)$ which is defined by $\Sigma_A \geq |D_A|$. On the frontier of this domain β^2 and A_o^N are known:

$$A_3^N = A_o^N = \frac{\Sigma_A}{2} \qquad \text{and} \qquad \beta^2 = \frac{\Sigma_A}{2A_3} \tag{129}$$

and this value vanishes at the four corners of D_2 as previously noticed for a pure vector weak neutral current.

In the opposite situation of a pure axial vector weak neutral current, $\alpha = 0$, $\gamma = 0$, and the solutions for β are limited by

$$\beta_m(0,0) = \frac{1}{2}\left\{\sqrt{\frac{\Sigma_N+D_N}{A_3}} - \sqrt{\frac{\Sigma_N-D_N}{A_3}}\right\}$$

(130)

$$\beta_M(0,0) = \frac{1}{2}\left\{\sqrt{\frac{\Sigma_N+D_N}{A_3}} + \sqrt{\frac{\Sigma_N-D_N}{A_3}}\right\} \quad .$$

In order to compare these constraints with those of the domain D_1 we consider equation (126) as a second order inequality in γ for fixed values of α and β . The result is:

$$\gamma_-(\alpha,\beta) \leq \gamma \leq \gamma_+(\alpha,\beta) \qquad (131)$$

where

$$\gamma_\pm(\alpha,\beta) = \frac{\alpha D_Q D_N \pm \sqrt{4\beta^2 A_3 Y}}{\alpha^2 D_Q^2 + 4\beta^2 A_3 V_o} \qquad (132)$$

with

$$Y = [\alpha^2 D_Q^2 + 4\beta^2 A_3 V_o][\Sigma_N - (\alpha^2 V_3 + \beta^2 A_3)] - V_o D_N^2 \,. \qquad (133)$$

The reality of the solutions implies the positivity of the discriminant $Y \geq 0$. In the α,β plane this last inequality defines a domain $D_2(\Phi_c)$ interior to the ellipse $E(\Phi_c)$ - associated to $\Sigma_o = 0$ - and limited by a 4th order curve $Y = 0$. This domain is the orthogonal projection of $S_2(\Phi_c)$ on the α,β plane.

We see that solutions in γ can exist if and only if α,β belongs to a domain $D(\Phi_c)$ intersection of $D_1(\Phi_c)$ and $D_2(\Phi_c)$

$$D(\Phi_c) = D_1(\Phi_c) \cap D_2(\Phi_c)$$

2. Models without axial vector isoscalar current

1.) We discuss in this section a class of models where the axial vector isoscalar current is absent. For the other components we use assumptions I and II as explicited by equations (111) and (112).

In the α, β, γ space the physical points are at the intersection of the ellipsoid $E(\Phi_C)$

$$\Sigma_N = \alpha^2 V_3 + \beta^2 A_3 + \gamma^2 V_o$$

with the two cylinders:

$$\alpha\beta = 2\frac{\Delta_N}{\Delta_C} \qquad \alpha\gamma = \frac{D_N}{D_Q} .$$

We then obtain four solutions symmetrical two by two with respect to the origin we shall call two sets of solutions depending on the parameter Φ_C.

$$\alpha^2 = \frac{\Sigma_N}{2V_3}\{1 \pm [1 - \frac{16V_3A_3\Delta_N^2}{\Delta_C^2\ \Sigma_N^2} - \frac{4V_3V_oD_N^2}{D_Q^2\ \Sigma_N^2}]^{1/2}\} \qquad (134)$$

$$\beta = 2\frac{\Delta_N}{\Delta_C}\frac{1}{\alpha} \qquad \gamma = \frac{D_N}{D_Q}\frac{1}{\alpha} .$$

The compatibility of our hypothesis A_o^N with experiment implies the positivity of the discriminant in equation

(134) and we obtain the inequality:

$$\frac{16V_3A_3\Delta_N^2}{\Delta_c^2\ \Sigma_N^2} + \frac{4V_3V_oD_N^2}{D_Q^2\ \Sigma_N^2} \leq 1$$

which is equivalently written, using the parametrization (114) as

$$\left(\frac{D_N}{D_Q}\right)^2 \leq \frac{\left(\frac{\Sigma_N}{\Sigma_c}\right)^2 - \sin^2 2\Phi_c \left(\frac{\Delta_N}{\Delta_c}\right)^2}{2\cos^2\Phi_c\left[\frac{\Sigma_Q}{\Sigma_c} - \frac{1}{2}\cos^2\Phi_c\right]} . \tag{135}$$

The equation (135) restricts the unmeasured quantity D_N as a function of Φ_c. In the particular case $\Phi_c = \frac{\pi}{4}$ we get

$$\left(\frac{D_N}{D_Q}\right)^2 \leq \frac{\left(\frac{\Sigma_N}{\Sigma_c}\right)^2 - \left(\frac{\Delta_N}{\Delta_c}\right)^2}{\frac{\Sigma_Q}{\Sigma_c} - \frac{1}{4}} . \tag{136}$$

whose numerical value, using the CERN-Gargamelle and SLAC data is

$$\left|\frac{D_N}{D_Q}\right| \leq 1.07 \pm .15 . \tag{137}$$

Conversely when D_N will experimentally be known the inequality (135) could be considered as a constraint on the angle Φ_C. The absence of data on D_N makes unuseful more considerations on this point.

Let us finally notice that the interference term K_{3o}^{AV} will be determined by a measurement of S_N

$$K_{3o}^{AV} = \alpha^2 \frac{S_N}{2 \frac{\Delta_N}{\Delta_c} \frac{D_N}{D_Q}} .$$

The interference term K_{3o}^{AV} must obviously obey a Schwartz inequality

$$(K_{3o}^{AV})^2 \leq A_3 V_o$$

and this constraint gives a limitation on S_N that we do not discuss because of the lack of experimental information.

2.) Model I $\gamma = 0$

The weak neutral current is purely isovector and we first have the two predictions

$$D_N = 0 \qquad S_N = 0$$

corresponding to the equality of proton and neutron cross

sections.

The parameters α and β can be determined from experiment using the relations

$$\Sigma_N = \alpha^2 V_3 + \beta^2 A_3$$

$$\Delta_N = \alpha\beta\, I_3$$

and the solutions are

$$\alpha = \frac{1}{2}\left\{\varepsilon_+\sqrt{\frac{\Sigma_N + 2\Delta_N \frac{\sqrt{V_3 A_3}}{I_3}}{V_3}} + \varepsilon_-\sqrt{\frac{\Sigma_N - 2\Delta_N \frac{\sqrt{V_3 A_3}}{I_3}}{V_3}}\right\} \tag{138}$$

$$\beta = \frac{1}{2}\left\{\varepsilon_+\sqrt{\frac{\Sigma_N + 2\Delta_N \frac{\sqrt{V_3 A_3}}{I_3}}{A_3}} - \varepsilon_-\sqrt{\frac{\Sigma_N - 2\Delta_N \frac{\sqrt{V_3 A_3}}{I_3}}{A_3}}\right\} . \tag{139}$$

In the particular case $\Phi_c = \frac{\pi}{4}$ these equalities take a simpler form

$$(\alpha \pm \beta)^2 = 4\left[\frac{\Sigma_N}{\Sigma_Q} \pm \frac{\Delta_N}{\Delta_c}\right] . \tag{140}$$

Using the CERN-Gargamelle data (62) and (63) we deduce the following values for the left hand $g_L = \frac{\alpha+\beta}{2}$ and right hand $g_R = \frac{\alpha-\beta}{2}$ coupling constants

$$|g_L| = 0.63 \pm 0.05$$
$$|g_R| = 0.34 \pm 0.09 \tag{141}$$

corresponding for α and β to the points P and Q of Fig. 2

$$\alpha = \pm [0.98 \pm 0.11] \qquad \beta = \pm [0.29 \pm 0.11]$$
$$\alpha = \pm [0.29 \pm 0.11] \qquad \beta = \pm [0.98 \pm 0.11] \quad . \tag{142}$$

3.) Modell II $\qquad \alpha = \beta + \gamma$

The weak neutral current is a linear combination of the V - A isovector current and of electromagnetic current.

The neutral current cross sections depend on two parameters α and β.

$$\Sigma_N = \alpha^2 V_3 + \beta^2 A_3 + (\alpha-\beta)^2 V_o \tag{143}$$

$$\Delta_N = \frac{1}{2}\alpha\beta\Delta_c \tag{144}$$

$$D_N = \alpha(\alpha-\beta)\ D_Q \tag{145}$$

$$S_N = \beta(\alpha-\beta)\ K_{3o}^{AV} \quad . \tag{146}$$

The solution for α and β are obtained by solving the system of equations (143) and (144). The result can be written

$$\alpha = \frac{1}{2}\,\frac{A_+ + A_-}{\sqrt{V_3 + V_o}} \qquad\qquad \beta = \frac{1}{2}\,\frac{A_+ - A_-}{\sqrt{\Lambda_3 + V_o}} \tag{147}$$

where

$$A_\pm = \varepsilon_\pm \left\{\Sigma_N + 4\,\frac{\Delta_N}{\Delta_c}\left[V_o \pm \sqrt{(V_3 + V_o)(\Lambda_3 + V_o)}\right]\right\}^{\frac{1}{2}} \tag{148}$$

with $\varepsilon_+^2 = \varepsilon_-^2 = 1$.

The parameter γ is given by the difference $\gamma = \alpha - \beta$ and from equation (145) we obtain the prediction for D_N

$$\frac{D_N}{D_Q} = \alpha(\alpha-\beta) = \alpha^2 - 2\,\frac{\Delta_N}{\Delta_c} \quad . \tag{149}$$

The expression (148) takes a simpler form in the particular case $\Phi_c = \frac{\pi}{4}$ and we obtain for α and β the two equations

$$(\alpha-\beta)^2 = \frac{\Sigma_N}{\Sigma_Q} - \frac{\Sigma_c}{\Sigma_Q}\,\frac{\Delta_N}{\Delta_c}$$

$$(\alpha+\beta)^2 = \frac{\Sigma_N}{\Sigma_Q} + \left(\delta - \frac{\Sigma_c}{\Sigma_Q}\right)\frac{\Delta_N}{\Delta_c} \quad .$$

Using the CERN-Gargamelle and SLAC data we deduce the following numerical estimates for the left hand $g_L = \frac{\alpha+\beta}{2}$ and right hand $g_R = \frac{\alpha-\beta}{2}$ coupling constants

$$|g_L| = 0.62 \pm 0.05$$
$$|g_R| = 0.32 \pm 0.09 \quad . \tag{150}$$

This corresponds to two sets of solutions in α, β, γ given by

Set 1.
$$\alpha = \pm [0.93 \pm 0.10]$$
$$\beta = \pm [0.30 \pm 0.10] \tag{151}$$
$$\gamma = \pm [0.63 \pm 0.18]$$

with the prediction

$$\frac{D_N}{D_Q} = 0.59 \pm 0.23 \tag{152}$$

Set 2.
$$\alpha = \pm [0.30 \pm 0.10]$$
$$\beta = \pm [0.93 \pm 0.10] \tag{153}$$
$$\gamma = \mp [0.63 \pm 0.18]$$

with the prediction

$$\frac{D_N}{D_Q} = - 0.19 \pm 0.03 \quad . \tag{154}$$

It must be noticed that the set 2 solution with the upper

sign is compatible with the Weinberg model where

$$\alpha = 1-2 \sin^2 \theta_W , \beta = 1 , \quad \gamma = -2 \sin^2 \theta_W \tag{155}$$

and the value of the mixing angle θ_W is given by

$$\sin^2 \theta_W = 0,35 \pm 0,05 \quad . \tag{156}$$

4.) Weinberg weak neutral current[9]

The most naive extension to non strange hadrons of the Weinberg-Salam gauge theory leads to a weak neutral current which is a particular case of Model II with the additional constraints

$$\beta = 1 \qquad\qquad -2 \leq \gamma \leq 0 \quad . \tag{157}$$

The mixing angle θ_W is determined from the relation $\alpha\beta = 2 \frac{\Delta_N}{\Delta_C}$ with $\beta = 1$ and we get

$$\sin^2 \theta_W = \frac{1}{2} - \frac{\Delta_N}{\Delta_C} \quad . \tag{158}$$

This model depending only on the mixing angle parameter θ_W it is possible to compute, from experiment, the angle Φ_C. An elementary algebraic calculation gives:

$$\cos^2\Phi_c = \frac{1}{2}\,\frac{1-2\,\frac{\Sigma_N}{\Sigma_c}}{1-2\,\frac{\Delta_N}{\Delta_c}} + \left(1 - 2\,\frac{\Delta_N}{\Delta_c}\right)\frac{\Sigma_Q}{\Sigma_c} \quad . \tag{159}$$

From the CERN-Gargamelle and SLAC data we obtain

$$\sin^2\,\theta_W = 0{,}36 \pm 0{,}06 \tag{160}$$

$$\cos^2\,\Phi_c = 0{,}55 \pm 0{,}06 \tag{161}$$

or

$$\Phi_c = 42^\circ \pm 4^\circ \quad . \tag{162}$$

The value of Φ_c is obviously compatible with $\frac{\pi}{4}$ as pointed out in the previous paragraph. Such a property is also reflected in the fact that the two values (156) and (160) of $\sin^2\,\theta_W$ are the same inside the errors.

The prediction for D_N is now

$$\frac{D_N}{D_Q} = -\,0.20 \pm 0.05 \quad . \tag{163}$$

5.) Model III $\qquad \beta = 0$

The weak neutral current is purely vector and we first have the general relation

$$d\sigma^{\nu}_{Nc} = d\sigma^{\bar{\nu}}_{Nc} \tag{164}$$

for any arbitrary target. In particular, for total cross sections in inclusive reactions, the predictions are

$$\Delta_N = 0 \qquad\qquad S_N = 0 \quad . \tag{165}$$

The first equality is equivalently written $r_N = 1$ and we recall the experimental results for that quantity

CERN-Gargamelle	$r_N = 0.6 \pm 0.2$
H.P.W.	$r_N = \ \ 1 \pm 0.2$.

The experimental uncertainties make difficult to reject now a parity conserving model and we must wait for improved data before settling that question.

The non vanishing cross sections Σ_N and D_N are given in terms of two parameters α and γ.

$$\Sigma_N = \alpha^2 V_3 + \gamma^2 V_o$$

$$D_N = \alpha\gamma J_V \quad .$$

The solutions are:

$$\alpha = \frac{1}{2}\left\{\varepsilon_+\sqrt{\frac{\Sigma_N+2D_N\frac{\sqrt{V_3V_o}}{J_V}}{V_3}} + \varepsilon_-\sqrt{\frac{\Sigma_N-2D_N\frac{\sqrt{V_3V_o}}{J_V}}{V_3}}\right\} \tag{166}$$

$$\gamma = \frac{1}{2}\{\varepsilon_+\sqrt{\frac{\Sigma_N+2D_N\frac{\sqrt{V_3V_o}}{J_V}}{V_o}}-\varepsilon_-\sqrt{\frac{\Sigma_N-2D_N\frac{\sqrt{V_3V_o}}{J_V}}{V_o}}\} \tag{167}$$

and their reality imply a constraint on $D_N^{(f)}$

$$\left|\frac{D_N}{D_Q}\right| \leq \frac{1}{2}\frac{\Sigma_N}{\sqrt{V_3(\Sigma_Q-V_3)}} \tag{168}$$

which, in the case $\Phi_c = \frac{\pi}{4}$ where $V_3 = \frac{1}{4}\Sigma_c$ takes the numerical value (f)

$$\left|\frac{D_N}{D_Q}\right| \leq 1.20 \pm 0.15 \tag{169}$$

LECTURE IV

I WEAK NEUTRAL CURRENT

1. Assumption for the isoscalar axial vector part

1.) We take for granted the assumption I concerning the isovector part and we extend the assumption II to the axial vector part. The combined hypothesis can be expressed as follows:

<u>The vector and axial vector weak neutral currents have the same internal quantum numbers as the electromagnetic current</u> (III)

$$V_\mu^{3,N} = \alpha V_\mu^3 \qquad A_\mu^{3,N} = \beta A_\mu^3$$
$$V_\mu^{o,N} = \gamma V_\mu^o \qquad A_\mu^{o,N} = \delta A_\mu^o \quad . \tag{170}$$

Unfortunately the hadronic matrix elements of A_μ^o have not been measured and this piece of the weak neutral current does not enter either in electroproduction or in weak reactions involving charged currents without supplementary assumptions concerning the internal symmetry structure of the currents. It follows that the new parameter δ cannot be determined by reference to another measured reaction and we shall use specific models, like parton models, in order to estimate the new quantities A_o, I_o, J_A, K_{3o}^{VA}.

In such a framework the experimental knowledge of Σ_N, Δ_N, D_N and S_N will determine the four parameters α, β, γ and δ. This information being missing we shall not discuss here the general problem but rather look into particular models satisfying assumption (III).

2.) Quark parton model based on U(3) symmetry

The computation of the hadronic matrix elements of the current components is straightforward and the result is simply

$$V_3 = A_3 = \frac{1}{6}\,(d_1 + d_{-1} + d_2 + d_{-2})$$

$$I_3 = \frac{1}{6} (d_1 - d_{-1} + d_2 - d_{-2})$$

$$V_o = A_o = \frac{1}{54} (d_1 + d_{-1} + d_2 + d_{-2}) + \frac{2}{27} (d_3 + d_{-3})$$

$$I_o = \frac{1}{54} (d_1 - d_{-1} + d_2 - d_{-2}) + \frac{2}{27} (d_3 - d_{-3}) \tag{171}$$

$$J_V = J_A = \frac{1}{9} (d_1 + d_{-1} - d_2 - d_{-2})$$

$$K^{VA}_{3o} = K^{AV}_{3o} = \frac{1}{18} (d_1 - d_{-1} - d_2 + d_{-2}) \quad .$$

Using the expression obtained in equation (51) for the electroproduction and charged current reactions we obtain the following equalities

$$V_3 = A_3 = \frac{1}{4} \Sigma_c$$

$$I_3 = \frac{1}{2} \Delta_c$$

$$V_o = A_o = \Sigma_Q - \frac{1}{4} \Sigma_c$$

$$I_o = \frac{5}{54} \Delta_c + \frac{2}{27} \Delta_c^{(1)} \tag{172}$$

$$J_V = J_A = D_Q = - \frac{1}{3} D_c$$

$$K^{VA}_{3o} = K^{AV}_{3o} = - \frac{1}{12} S_c \quad .$$

3.) Model IV $\alpha = 0$ $\beta = 0$

The weak neutral current is purely isoscalar and we first have the two predictions

$$D_N = 0 \qquad\qquad S_N = 0 \quad . \tag{173}$$

The parameters γ and δ can be determined from experiment using the relations

$$\Sigma_N = \gamma^2 V_o + \delta^2 A_o$$

$$\Delta_N = \gamma\delta \quad J_o$$

and the solutions are

$$\gamma = \frac{1}{2}\{\varepsilon_+ \sqrt{\frac{\Sigma_N + 2\Delta_N \frac{\sqrt{V_o A_o}}{I_o}}{V_o}} + \varepsilon_- \sqrt{\frac{\Sigma_N - 2\Delta_N \frac{\sqrt{V_o A_o}}{I_o}}{V_o}}\} \tag{174}$$

$$\delta = \frac{1}{2}\{\varepsilon_+ \sqrt{\frac{\Sigma_N + 2\Delta_N \frac{\sqrt{V_o A_o}}{I_o}}{A_o}} - \varepsilon_- \sqrt{\frac{\Sigma_N - 2\Delta_N \frac{\sqrt{V_o A_o}}{I_o}}{A_o}}\} \quad . \tag{175}$$

Of course, the equality (174) is compatible with the limits (122) because of the Schwartz inequality $I_o^2 \leq V_o A_o$.

Using the quark parton model equalities (172) the CERN-Gargamelle and the SLAC data, we obtain two sets of solutions

$$\gamma = \pm [1.95 \begin{smallmatrix} + 0.5 \\ - 0.3 \end{smallmatrix}] \qquad \delta = \pm [1.3 \begin{smallmatrix} + 0.3 \\ - 0.5 \end{smallmatrix}]$$

(176)

$$\gamma = \pm [1.3 \begin{smallmatrix} + 0.3 \\ - 0.5 \end{smallmatrix}] \qquad \delta = \pm [1.95 \begin{smallmatrix} + 0.5 \\ - 0.3 \end{smallmatrix}] \quad .$$

4.) Model V $\qquad \alpha = \gamma \qquad \beta = \delta$

The weak neutral current is a linear combination of the electromagnetic vector and axial vector currents[10].

The neutral current cross sections are given by:

$$\Sigma_N = \alpha^2 (V_3 + V_o) + \beta^2 (A_3 + A_o)$$

$$\Delta_N = \alpha\beta (I_3 + I_o)$$

(177)

$$D_N = \alpha^2 \, J_V + \beta^2 \, J_A$$

$$S_N = \alpha\beta (K_{3o}^{VA} + K_{3o}^{AV})$$

and using the quark parton model equalities (172) they can be expressed in terms of electroproduction and charged current weak reactions as follows

$$\Sigma_N = (\alpha^2 + \beta^2) \; \Sigma_Q \qquad (178)$$

$$\Delta_N = \frac{2}{27}\,\alpha\beta[8\Delta_c + \Delta_c^{(1)}] \tag{179}$$

$$D_N = (\alpha^2 + \beta^2)\,D_Q \tag{180}$$

$$S_N = -\frac{1}{6}\,\alpha\beta\quad S_c\quad . \tag{181}$$

Eliminating α and β we first obtain two predictions

$$\frac{D_N}{D_Q} = \frac{\Sigma_N}{\Sigma_Q} \tag{182}$$

$$\frac{S_N}{S_c} = -\frac{9}{4}\,\frac{\Delta_N}{8\Delta_c + \Delta_c^{(1)}} \tag{183}$$

and by solving equations (178) and (179) we can compute the two parameters α and β. The result is

$$(\alpha \pm \beta)^2 = \frac{\Sigma_N}{\Sigma_Q} \pm \frac{27\;\Delta_N}{8\Delta_c + \Delta_c^{(1)}}\quad .$$

Using the CERN-Gargamelle and SLAC data we obtain the two sets of values (g)

$$\begin{aligned} &\alpha = \pm\,[0.89 \pm 0.10] &\qquad &\beta = \pm\,[0.29 \pm 0.10] \\ &\alpha = \pm\,[0.29 \pm 0.10] &\qquad &\beta = \pm\,[0.89 \pm 0.10] \end{aligned} \tag{184}$$

or, using the left-hand and right-hand couplings and $g_L = \frac{\alpha+\beta}{2}$ and $g_R = \frac{\alpha-\beta}{2}$ (g)

$$|g_L| = 0.59 \pm 0.05$$
$$|g_R| = 0.30 \pm 0.09 \quad . \tag{185}$$

For the predictions (182) and (183) the numerical estimates are (g)

$$\frac{D_N}{D_Q} = 0.88 \pm 0.11 \tag{186}$$

$$\frac{S_N}{S_c} = -\,0.043 \pm 0.017 \quad . \tag{187}$$

5.) Model VI $\qquad \alpha = 0 \qquad \gamma = 0$

The weak neutral current is purely axial vector and we first have the two predictions $\Delta_N = 0$ and $S_N = 0$. The discussion of the constraint $\Delta_N = 0$ has already been made for model of a pure vector weak neutral current.

The non vanishing cross sections Σ_N and D_N are given in terms of the parameters β and δ:

$$\Sigma_N = \beta^2 A_3 + \delta^2 A_o$$

$$D_N = \beta\delta\, J_A \quad .$$

The solutions are:

$$\beta = \frac{1}{2}\{\varepsilon_+ \sqrt{\frac{\Sigma_N+2\Delta_N \frac{\sqrt{A_3A_o}}{J_A}}{A_3}} + \varepsilon_- \sqrt{\frac{\Sigma_N-2\Delta_N \frac{\sqrt{A_3A_o}}{J_A}}{A_3}}\} \qquad (188)$$

$$\delta = \frac{1}{2}\{\varepsilon_+ \sqrt{\frac{\Sigma_N+2\Delta_N \frac{\sqrt{A_3A_o}}{J_A}}{A_o}} - \varepsilon_- \sqrt{\frac{\Sigma_N-2\Delta_N \frac{\sqrt{A_3A_o}}{J_A}}{A_o}}\} . \qquad (189)$$

Of course the equality (188) is compatible with the limits (130) because of the Schwartz inequality $J_A^2 \leq A_3A_o$.

In the quark parton model where

$$A_3 = V_3 \qquad A_o = V_o \qquad J_A = J_V$$

the expressions (188) and (189) are identical to those obtained for α (166) and γ (167) in the case of Model III and the consistency constraint on D_N has the same numerical value (169).

II THE BARYONIC CURRENT

1.) If assumption I concerning the isovector weak neutral current looks extremely reasonable, the assumptions II and III for the isoscalar part are very restrictive. In fact the most general form for an isoscalar weak neutral current will be two linear combinations - one for vector and one for axial vector - of isoscalar currents constructed in the framework of the internal symmetry group of hadrons. Such combinations have no a priori reasons to be identical to that occurring in the isoscalar electromagnetic current. But an analysis of available data on neutral currents is difficult without a few assumptions and that was the motivation of the previous section.

We shall now discuss two particular models where assumptions II and III are not satisfied. The first one is a simple generalization of Sakurai's proposal and the weak neutral current is a linear combination of the vector and axial vector baryonic currents. In the second one the isoscalar weak neutral current is a linear combination of the isoscalar electromagnetic current and of the baryonic current for both vector and axial vector parts. We restrict ourselves in this last case to the model associated to an SU(4) symmetry of hadrons.

2.) We now discuss a model where the weak neutral current is a linear combination of the vector and axial vector baryonic currents[11]

$$J^Z_\mu = \eta_V V^B_\mu + \eta_A A^B_\mu \quad . \tag{190}$$

As for any pure isoscalar neutral current we have two predictions

$$D_N = 0 \qquad\qquad S_N = 0 \quad . \tag{191}$$

The two other cross sections Σ_N and Δ_N depend on three quantities: B_V, B_A and B_{VA}

$$\Sigma_N = \eta_V^2 B_V + \eta_A^2 B_A \tag{192}$$

$$\Delta_N = \eta_V \eta_A B_{VA} \tag{193}$$

restricted by a Schwartz inequality

$$[B_{VA}]^2 \leq B_V B_A \quad . \tag{194}$$

Assuming the nucleon matrix elements B_V, B_A and B_{VA} to be known we can solve equations (192) and (193) in η_V and η_A and the result is

$$\eta_V = \frac{A_+ + A_-}{2\sqrt{B_V}} \qquad \eta_A = \frac{A_+ - A_-}{2\sqrt{B_A}} \tag{195}$$

with

$$A_\pm = \varepsilon_\pm \{ \Sigma_N \pm 2\Delta_N \frac{\sqrt{B_V B_A}}{B_{VA}} \}^{\frac{1}{2}} \tag{196}$$

$$\varepsilon_+^2 = \varepsilon_-^2 = 1 \ .$$

In general the baryonic charge operator B belongs to a singulet representation of the internal symmetry group and in terms of quark fields the baryonic currents have the form

$$V_\mu^B = \frac{1}{r}\, i \sum_j \bar{q}_j \gamma_\mu q_j \tag{197}$$

$$A_\mu^B = \frac{1}{r}\, i \sum_j \bar{q}_j \gamma_\mu \gamma_5 q_j \tag{198}$$

where $\frac{1}{r}$ is the quark baryonic number. In models where the nucleon is made of three constituent quarks $r = 3$.

It is now straightforward to compute B_V, B_A, B_{VA} and the result is:

$$B_V = B_A = \frac{2}{3r^2} \sum_j (d_j + d_{-j}) \tag{199}$$

$$B_{VA} = \frac{2}{3r^2} \sum_j (d_j - d_{-j}) \quad . \tag{200}$$

In all these equalities the summation over j extends to quarks only and the d_j's are the first moments of the quark and antiquark distribution functions previously introduced. The inequality (194) is then a trivial consequence of the positivity of the d_j's.

The parton sums (199) and (200) are easily computed in the quark parton model based on U(3) symmetry.

$$\sum_j (d_j + d_{-j}) = \frac{9}{4} [6\Sigma_Q - \Sigma_c]$$

(201)

$$\sum_j (d_j - d_{-j}) = \frac{1}{2} [7\Delta_c + 2\Delta_c^{(1)}] \quad .$$

We obtain numerical estimates for the coupling constants using the CERN-Gargamelle and SLAC data. The result is (g)

$$|h_L| = 0.90 \pm 0.08$$
$$|h_R| = 0.42 \pm 0.17 \qquad (202)$$

where

$$h_L = \frac{\eta_V + \eta_A}{2} \qquad\qquad h_R = \frac{\eta_V - \eta_A}{2}$$

III. THE SU(4) CURRENTS[12]

1.) In gauge theories unifying weak and electromagnetic interactions the simplest way of obtaining weak neutral currents is to use the gauge group SU(2) x U(1) with a mixing angle parameter θ_W. The weak hadronic currents are then simply expressed in terms of the weak isotopic spin current J^k_μ (k = 1, 2, 3) and of the electromagnetic current J^Q_μ

$$J^W_\mu = c\ [J^1_\mu + i\ J^2_\mu]$$
$$J^Z_\mu = d[J^3_\mu - \sin^2 \theta_W\ J^Q_\mu] \quad . \qquad (203)$$

The normalization coefficients c and d are model dependent.

The extension of these theories to hadrons is not as straightforward as expected and the main difficulty is due to the experimental suppression of the strangeness violating neutral current. For instance the quark parton model based on U(3) symmetry will predict such a current as the result of the commutation of the Cabibbo current with its hermitian conjugate. Therefore the unwanted part must be canceled with a contribution coming from additional quarks. A larger symmetry for hadrons will emerge out of these schemes.

2.) The Salam-Weinberg model is an economical realization of this class of gauge theories. In this case $c = d = 2$. One possible solution to cure the difficulty with $|\Delta S| = 1$ neutral currents is to introduce four quarks as the basic constituents of an SU(4) unitary symmetry and the new additive quantum number is usually called "charm". Such a solution is known as the Glashow-Iliopoulos-Maiani mechanism and we now study in some detail the explicit form of the currents.

The weak charged current is made of two pieces.

i) the Cabibbo current with a strangeness conserving component and a strangeness violating one;

ii) a charm changing current with also a strangeness conserving component and a strangeness violating one.

In terms of quark, we have

$$J_\mu^W = \frac{i}{2}\,[\bar{p}\gamma_\mu(1 + \gamma_5)n_c + \bar{p}'\gamma_\mu(1 + \gamma_5)\lambda_c] \qquad (204)$$

where

$$n_c = \cos\theta_c\, n + \sin\theta_c\, \lambda$$
$$\lambda_c = -\sin\theta_c\, n + \cos\theta_c\, \lambda\ . \tag{205}$$

The Gell-Mann-Nishijima formula for the electric charge operator is modified as follows

$$Q = I_3 + \frac{1}{2}\,(B + S + c)$$

and the electromagnetic current with fractionally charged quarks is simply written as

$$J_\mu^Q = i\{\ \frac{2}{3}[\bar{p}\gamma_\mu p + \bar{p}'\gamma_\mu p'] - \frac{1}{3}[\bar{n}\gamma_\mu n + \bar{\lambda}\gamma_\mu\lambda]\}. \tag{206}$$

The weak neutral current is now immediately computed from equation (203) and the result is

$$J_\mu^Z = \frac{i}{2}\,\{\bar{p}\gamma_\mu(1+\gamma_5)p + \bar{p}'\gamma_\mu(1+\gamma_5)p' - \bar{n}\gamma_\mu(1+\gamma_5)n - \bar{\lambda}\gamma_\mu(1+\gamma_5)\lambda\} - 2\sin^2\theta_W\, J_\mu^Q\ . \tag{207}$$

We notice that the electromagnetic and weak neutral currents are both strangeness and charm conserving. Therefore strange and charmed particles must be produced in pairs when neutral currents are involved.

3.) The cross sections for inelastic scattering of charged leptons ($e^{\pm}$, $\mu^{\pm}$) of nucleons are given, in terms of quark and anti-quark momenta by

$$\Sigma_Q = \frac{5}{27} (d_1 + d_{-1} + d_2 + d_{-2}) + \frac{2}{27}(d_3+d_{-3})+\frac{8}{27}(d_4+d_{-4}) \quad (208)$$

$$D_Q = \frac{1}{9} (d_1 + d_{-1} - d_2 - d_{-2}) \quad . \quad (209)$$

For neutrino and antineutrino cross sections involving charged currents the result is

$$\Sigma_C = \frac{2}{3}(d_1 + d_{-1} + d_2 + d_{-2}) + (d_3 + d_{-3}) + \frac{1}{3}(d_4 + d_{-4}) \quad (210)$$

$$\Delta_C = \frac{1}{3}(d_1 - d_{-1} + d_2 - d_{-2}) + (d_3 - d_{-3}) - \frac{1}{3}(d_4 - d_{-4}) \quad (211)$$

$$D_C = - \frac{1}{3} (d_1 + d_{-1} - d_2 - d_{-2}) \quad (212)$$

$$S_C = - \frac{2}{3} (d_1 - d_{-1} - d_2 + d_{-2}) \quad . \quad (213)$$

The relation (52)

$$D_C + 3 D_Q = 0$$

remains valid and in order to measure the eight momenta d_j's and to check the consistency of this model with experiment it is necessary to separate the final states associated to weak transitions $|\Delta S| = 0,1$ $|\Delta C| = 0,1$. The theoretical expressions have been given in Appendix B.

Let us notice two interesting relations independent of such a separation: the first one gives a measure of the amount of gluons through the parameter ε defined by

$$\varepsilon = 1 - \sum_j (d_j + d_{-j}) \qquad 0 \leq \varepsilon \leq 1$$

and it is written

$$\sum_j (d_j + d_{-j}) = \frac{9}{11} (3\Sigma_Q + \Sigma_c) \tag{214}$$

the second one generalizes the inequality (53) and compares the distributions for strange and charmed quarks

$$(d_4 + d_{-4}) - (d_3 + d_{-3}) = \frac{3}{11} (18\Sigma_Q - 5\Sigma_c) \ . \tag{215}$$

We do not know if charmed particles have been produced in charged current experiments. The high energy total cross sections are in remarkable agreement with a linear extrapolation of the low energy ones so that it is difficult, within errors, to detect any change in the values of the slope parameters $A^{\nu N}$ and $A^{\bar{\nu} N}$. Nevertheless in order to have numerical estimates for equations (214) and (215) we use the Gargamelle and SLAC data even if such a procedure is poorly justified. The results are:

$$\sum_j (d_j + d_{-j}) = 0.513 \pm 0.027$$

or $$\varepsilon = 0.49 \pm 0.03$$

and

$$(d_4 + d_{-4}) - (d_3 + d_{-3}) = 0.045 \pm 0.049$$

showing that the strange quark and charmed quark distributions cannot be very different.

4.) The weak neutral current based on SU(4) symmetry has the general structure

$$J^Z_\mu = (V^3_\mu + A^3_\mu) + \frac{1}{2}(V^S_\mu + A^S_\mu + V^C_\mu + A^C_\mu) - 2 \sin^2 \theta_W \, J^Q_\mu \quad . \tag{216}$$

Assumption I is obviously fulfilled with

$$\alpha = 1 - 2 \sin^2 \theta_W \qquad \beta = 1 \tag{217}$$

and the isoscalar components of the neutral current are linear combinations of the isoscalar electromagnetic current and of the baryonic current given by

$$V^{o,N}_\mu = (1 - 2 \sin^2 \theta_W) \, V^Q_\mu - \frac{1}{2} V^B_\mu \tag{218}$$

$$A^{o,N}_\mu = A^Q_\mu - \frac{1}{2} A^B_\mu \quad . \tag{219}$$

The various matrix elements of the current can be computed using a quark parton model and they are then related to analogous quantities measurable in electroproduction and in weak processes involving charged currents. The result is:

$$\Sigma_N = \frac{3}{11}\,(3\Sigma_Q + \Sigma_c) - \kappa\,\Sigma_c^{(o,o)} + 4\kappa^2\,\Sigma_Q \tag{220}$$

$$\Delta_N = \frac{1}{6}\,(8\Delta_c^{(o,o)} + 4\Delta_c^{(1,o)} - 3\Delta_c) - \kappa\Delta_c^{(o,o)} \tag{221}$$

$$D_N = -\,2\kappa(1 - 2\kappa)\,D_Q \tag{222}$$

$$S_N = \frac{1}{6}\,\kappa\,S_c \tag{223}$$

where $\kappa = \sin^2\theta_W$.

The relations (222) and (223) were already valid in the simple Weinberg model. The same feature occurs also for equation (221) in particular quark parton models where $d_3 = d_{-3}$; $d_4 = d_{-4}$

$$\Delta_N = (\frac{1}{2} - \kappa)\,\Delta_c \quad . \tag{224}$$

Finally from equation (220) and using the positivity of momenta

$$\Sigma_c^{(o,o)} \leq \Sigma_c$$

we obtain the inequality

$$\frac{\Sigma_N}{\Sigma_c} \geq [\frac{9}{11} + 4\kappa^2]\,\frac{\Sigma_Q}{\Sigma_c} + \frac{3}{11} - \kappa \tag{225}$$

which in the absence of information about the various charged current cross sections may be useful to test the model.

We have represented in Fig. 5 the lower bound in the one standard deviation limit using the charged current and electromagnetic results from CERN-Gargamelle and SLAC. The CERN-Gargamelle point for neutral currents is consistent with such a constraint.

CONCLUSION

The analysis of inclusive reactions induced by neutrinos and antineutrinos can give important information about the properties of the hadronic weak neutral current. We have attempted such an analysis of the existing data in order to present a model independent systematics of treating the experimental results. As previously emphasized by many people the situation concerning the structure of the weak neutral current is still open and the eight models considered in these lectures as particular illustrations of general assumptions can fit the data in spite of their big differences. A great improvement would take place when neutrino and antineutrino data on proton and neutron separately - both for charged and neutral currents - will become available[13].

Exclusive reactions are a second important source of information and among them the elastic scattering

$$\nu(\bar{\nu}) + N \rightarrow \nu(\bar{\nu}) + N \qquad (E)$$

and the one pseudoscalar (or vector) meson production

$$\nu(\bar{\nu}) + N \rightarrow \nu(\bar{\nu}) + M + N' . \qquad \text{(M)}$$

Reaction of type M have been experimentally observed[5] and a quantitative analysis of the results can be undertaken following the same method as that presented in these lectures.

For the elastic processes (E) we only have upper limits[1] for the cross sections but measurements on a proton target are expected in a very near future[14]. The theoretical analysis of these reactions involves a comparison of electromagnetic, weak charged and weak neutral elastic form factors. We then have a powerful means to investigate the properties of the weak neutral current. Of course new assumptions are needed as, for instance, a universal q^2 dependence of all the elastic form factors which is certainly consistent with actual data to a reasonable accuracy.

We also expect manifestations of the weak hadronic neutral current in processes where neutrinos and antineutrinos are not involved as for instance the electroproduction[15] or the electron-positron annihilation into hadrons[16]. For values of $|q^2|$ small as compared with the squared neutral vector boson mass the dominant correction to the one photon exchange cross-section - apart from many photon exchange contributions - will be an interference between electromagnetic and weak neutral amplitudes. A careful study of these corrections will teach us very useful information about the properties of the weak neutral current. In most of the gauge theories the neutral vector boson mass is expected very large and values of $|q^2|$ of 100-200 GeV^2 are necessary in order to make possible such measurements.

APPENDIX A

DIFFERENTIAL CROSS SECTION BOUNDS

Let us consider the total cross sections $\sigma_\lambda(q^2,W^2)$ for weak processes. The transverse and longitudinal parts are the sum of a vector and an axial vector contribution

$$\sigma_{T,L} = \sigma^V_{T,L} + \sigma^A_{T,L}$$

and the difference of the helicity $\lambda = \pm 1$ terms is due to the vector axial vector interference.

$$\sigma_{-1} - \sigma_{+1} = 4\ \tau^{VA} \quad .$$

Following equation (1) we define three differential quantities

$$d\ V^c_3 = d\sigma_o\ [\sigma^V_T + \varepsilon\sigma^V_L]$$

$$d\ A^c_3 = d\sigma_o\ [\sigma^A_T + \varepsilon\sigma^A_L]$$

$$d\ I^c_3 = d\sigma_o\ \ [2\sqrt{1-\varepsilon^2}\ \tau^{VA}]$$

We now restrict our considerations to the strangeness conserving transitions for which charge symmetry holds. Using

$$\sigma_\lambda^{\bar{\nu}p} = \sigma_\lambda^{\nu p} \qquad \sigma_\lambda^{\bar{\nu}n} = \sigma_\lambda^{\nu n}$$

we obtain

$$d\sigma^{\nu p} = dV_3^{c,p} + dA_3^{c,p} + dI_3^{c,p}$$

$$d\sigma^{\nu n} = dV_3^{c,n} + dA_3^{c,n} + dI_3^{c,n}$$

$$d\sigma^{\bar{\nu}p} = dV_3^{c,n} + dA_3^{c,n} - dI_3^{c,n}$$

$$d\sigma^{\bar{\nu}n} = dV_3^{c,p} + dA_3^{c,p} - dI_3^{c,p} \quad .$$

The Schwartz inequality

$$[\tau^{VA}]^2 \leq \sigma_T^V \sigma_T^A$$

takes the simple form for any arbitrary target

$$(dI_3^c)^2 \leq 4(1-\varepsilon^2) dV_3^c \, dA_3^c \quad .$$

Using charge symmetry we obtain the two constraints between differential cross sections

$$[d\sigma^{\nu p} - d\sigma^{\bar{\nu}n}]^2 \leq (1-\varepsilon^2) \, [d\sigma^{\nu p} + d\sigma^{\bar{\nu}n}]^2$$

$$[d\sigma^{\nu n} - d\sigma^{\bar{\nu}p}]^2 \leq (1-\varepsilon^2)\ [d\sigma^{\nu n} + d\sigma^{\bar{\nu}p}]^2 \quad .$$

As a consequence of these relations we also get

$$[d\sigma^{\nu p}+d\sigma^{\nu n}-d\sigma^{\bar{\nu}p}-d\sigma^{\bar{\nu}n}]^2 \leq (1-\varepsilon^2)\ [d\sigma^{\nu p}+d\sigma^{\nu n}+d\sigma^{\bar{\nu}p}+d\sigma^{\bar{\nu}n}]^2$$

$$[d\sigma^{\nu p}-d\sigma^{\nu n}+d\sigma^{\bar{\nu}p}-d\sigma^{\bar{\nu}n}]^2 \leq (1-\varepsilon^2)\ [d\sigma^{\nu p}+d\sigma^{\nu n}+d\sigma^{\bar{\nu}p}+d\sigma^{\bar{\nu}n}]^2 \quad .$$

A parametrization of dV_3^c and dA_3^c with an angle Φ_c can be done as for the total cross sections. The discussion of the allowed domain of Φ_c will proceed in an analogous way.

APPENDIX B

The separation of the various final states for weak processes involving charged currents is made by specifying for the slope parameter A the values of $|\Delta S|$ and $|\Delta C|$.

$$A=\cos^2\theta_c\ A^{(o,o)}+\sin^2\theta_c A^{(1,o)}+\sin^2\theta_c A^{(o,1)}+\cos^2\theta_c A^{(1,1)} .$$

In the notation (17) used for the Cabibbo current

$$A^{(o,o)} = B \qquad\qquad A^{(1,o)} = C \quad .$$

The four linear combinations Σ_c, Δ_c, D_c and S_c are split into four parts. The result is:

$$\Sigma_c^{(o,o)} = \frac{2}{3}\ (d_1 + d_{-1} + d_2 + d_{-2})$$

$$\Sigma_c^{(1,o)} = \frac{1}{6} (d_1 + d_{-1} + d_2 + d_{-2}) + (d_3 + d_{-3})$$

$$\Sigma_c^{(o,1)} = \frac{1}{2} (d_1 + d_{-1} + d_2 + d_{-2}) + \frac{1}{3} (d_4 + d_{-4})$$

$$\Sigma_c^{(1,1)} = (d_3 + d_{-3}) + \frac{1}{3} (d_4 + d_{-4})$$

$$\Delta_c^{(o,o)} = \frac{1}{3} (d_1 - d_{-1} + d_2 - d_{-2})$$

$$\Delta_c^{(1,o)} = - \frac{1}{6} (d_1 - d_{-1} + d_2 - d_{-2}) + (d_3 - d_{-3})$$

$$\Delta_c^{(o,1)} = \frac{1}{2} (d_1 - d_{-1} + d_2 - d_{-2}) - \frac{1}{3} (d_4 - d_{-4})$$

$$\Delta_c^{(1,1)} = (d_3 - d_{-3}) - \frac{1}{3} (d_4 - d_{-4})$$

$$D_c^{(o,o)} = - \frac{1}{3} (d_1 + d_{-1} - d_2 - d_{-2})$$

$$D_c^{(1,o)} = - \frac{1}{2} D_c^{(o,o)}$$

$$D_c^{(o,1)} = \frac{3}{2} D_c^{(o,o)}$$

$$D_c^{(1,1)} = 0$$

$$S_c^{(o,o)} = - \frac{2}{3} (d_1 - d_{-1} - d_2 + d_{-2})$$

$$S_c^{(1,o)} = \frac{1}{4} S_c^{(o,o)}$$

$$S_c^{(o,1)} = \frac{3}{4} S_c^{(o,o)}$$

$$S_c^{(1,1)} = 0 \quad .$$

The quark parton model based on U(4) symmetry is described by eight distributions functions $D_j(\xi)$. In an ideal experiment we can measure 16 inclusive cross sections. Therefore the model predicts 8 constraints

Measurement	Constraint
4 Σ	1
4 Δ	1
4 D	3
4 S	3

Moreover the two electroproduction cross sections Σ_Q and D_Q are linear combinations of corresponding weak quantities. In addition to relation (52) for D_Q we obtain:

$$\Sigma_Q = \frac{2}{27} \Sigma_c^{(1,o)} + \frac{8}{9} \Sigma_c^{(o,1)} - \frac{11}{27} \Sigma_c^{(o,o)} \quad .$$

FOOTNOTES

(a) We use the subscript cc for charged current and Nc for neutral currents.

(b) The quantities Σ_Q and D_Q are proportional to the first moment of the transverse scaling function for electroproduction usually defined by:

$$I^e = \int_0^d \xi 2 \, [F_T(\xi) + F_L(\xi)] d\,\xi \quad .$$

The relations are

$$\Sigma_Q = \frac{1}{3} (I^{ep} + I^{en}) \qquad\qquad D_Q = \frac{1}{3} (I^{ep} - I^{en}) \; .$$

(c) We use the notation $j = 1, 2, 3$, for respectively the p, n, λ quarks and $j = -1, -2, -3$ for the $\bar{p}, \bar{n}, \bar{\lambda}$ antiquarks.

(d) Relations analogous to equation (17) give the complete total cross sections and the quantities with index (1) are constructed from the slope parameters C.

(e) The relation between the angles Φ_C and Φ_Q has previously been given in equation (45).

(f) In the numerical calculation we shall use the value of Σ_N/Σ_C despite the fact that Δ_N/Δ_C is not zero in the CERN Gargamelle experiment.

(g) In the absence of experimental information on $\Delta_C^{(1)}$ we assume the equality $d_3 = d_{-3}$ leading to the relation $\Delta_C^{(1)} = -\frac{1}{2} \Delta_C$.

REFERENCES

1. D. Cundy, Report at the London Conference (1974).

2. F. J. Hasert and al., Physical Letters 46B, 138 (1973); Nuclear Physics B73, 1 (1974); Paper presented at the London Conference 1974 n° 1013.

3. A. Benvenuti and al., Physical Review Letters 32, 800 (1974); B. Aubert and al., Physical Review Letters 32, 1454 (1974); 1457 (1974).

4. B. C. Barish and al., Paper presented at the London Conference 1974 N° 587.

5. S. J. Barish and al., Physical Review Letters 33, 448 (1974); D. C. Cundy and al., Paper presented at the London Conference 1974 N° 1024.

6. A. Pais and S. B. Treiman,Physical Review D6, 2700 (1972); Physical Review D9, 1459 (1974).

7. Rajasekaran and Sarma, Pramana 2, 62 (1974); Preprint (1974).

8. M. Gourdin, Lectures given at the International Summer Institute in Theoretical Physics (Bonn: 1974).

9. A. Salam, Elementary particle Theory (Almquist and Wiksell, Stockholm 1968); S. Weinberg, Physical Review Letters 19, 1264 (1967); Physical Review D5, 1412 (1972).

10. M. A.B. Bég and A. Zee, Physical Review Letters 30, 675 (1975); M. A. B. Bég, Physical Letters 49B, 361 (1974); C.M. Viallet, Thesis 1975.

11. J. J. Sakurai, Physical Review D9, 250 (1974); S. Pakvasa and S. F. Tuan, Physical Review D9, 2698 (1974); J. J. Sakurai and Urrutia UCLA/74 TEP/16.

12. R. Budny, Physical Letters 39B, 553 (1972); R. Budny and Scharbach, Physical Review D6, 3651 (1972); Riazuddin & Fayyazuddin, Physical Review D6, 2032 (1972); C. H. Albright, Nuclear Physics B70, 486 (1974); L. M. Sehgal, Nuclear Physics B65, 141 (1973).

13. A CERN-Gargamelle experiment with propane now in progress will provide such information.

14. D. Cundy, private communication.

15. A. Love, G. C. Ross and D. V. Nanopoulos, Nucl. Phys. B49, 519 (1972); E. Petronzio, preprint; E. Derman, Phys. Rev. D7, 2755 (1973); S. M. Berman and J. R. Primack, preprint; W. J. Wilson, preprint; C. P. Korthals-Altes, M. Perrottet and E. De Rafael, preprint 1974; C. H. Llewellyn-Smith and D. V. Nanopoulos, Nucl. Phys. B; M. Gourdin, Invited Talk at the IXth Balaton Symposium (1974).

16. J. Godive and A. Hankey, Phys. Rev. D6, 3301 (1972); A. Love, Lett. Nuovo Cimento 5, 113 (1972); V. K. Cung, A. K. Mann and E. A. Paschos, Phys. Letters 41B, 335 (1972); M. Gourdin, Lectures given at the Majorana School, Erice (1974).

TABLE 1

Resume of the results for the isovector part of the neutral current

MODELS	$\|g_L\|$	$\|g_R\|$	D_N/D_Q	S_N
Assumption I	$\leq 0.639 \pm 0.051$	$\leq 0.364 \pm 0.087$		
I pure I = 1	0.635 ± 0.051	0.345 ± 0.100	0	0
II	0.618 ± 0.052	0.316 ± 0.091	0.590 ± 0.231 -0.191 ± 0.033	
III pure V	$\leq 0.357 \pm 0.089$	$\leq 0.357 \pm 0.089$	$\|\|\leq 1.20 \pm 0.15$	0
IV pure I = 0	0	0	0	0
V	0.589 ± 0.048	0.303 ± 0.093	0.878 ± 0.107	
VI pure A	$\leq 0.357 \pm 0.089$	$\leq 0.357 \pm 0.089$	$\|\|\leq 1.20 \pm 0.15$	0

FIGURE CAPTION

Fig. 1. Kinematics of inclusive reactions.

Fig. 2. The allowed domain in the α, β plane for $\Phi_c = \frac{\pi}{4}$.

————	Domain	$D_1\ (\frac{\pi}{4})$
.-.-.-.-.	Circle	$E\ (\frac{\pi}{4})$
....	Ellipses	$E^{\pm}\ (\frac{\pi}{4})$
---------	Hyperbola	H

Fig. 3. The complete domain in the α, β plane for Φ_c in its physical range

————	Domain	$D_1(\delta_{Qc}) \cup D_1(\frac{\pi}{2} - \delta_c)$
-.-.-.-.-	Ellipses	$E(\delta_{Qc}), E(\frac{\pi}{2} - \delta_c)$
---------	Hyperbola	H

Fig. 4. The allowed domain in the α, γ plane for $\Phi_c = \frac{\pi}{4}$

————	Domain	$\bar{D}_1(\frac{\pi}{4})$
-.-.-.-.-	Ellipse	$F(\frac{\pi}{4})$
....	Projection of the intersection of H with $E(\frac{\pi}{4})$	

Fig. 5. The lower bound curve in the $\frac{\Sigma_N}{\Sigma_c}, \frac{\Delta_N}{\Delta_c}$ plane for the SU(4) model

————	Lower limit + 1 standard deviation
---------	Lower limit - 1 standard deviation
(error cross symbol)	CERN-Gargamelle Result.

N. B. For Fig. 2. 3. 4. errors are not represented for clarification. They must be taken into account in a numerical realistic calculation as in the text.

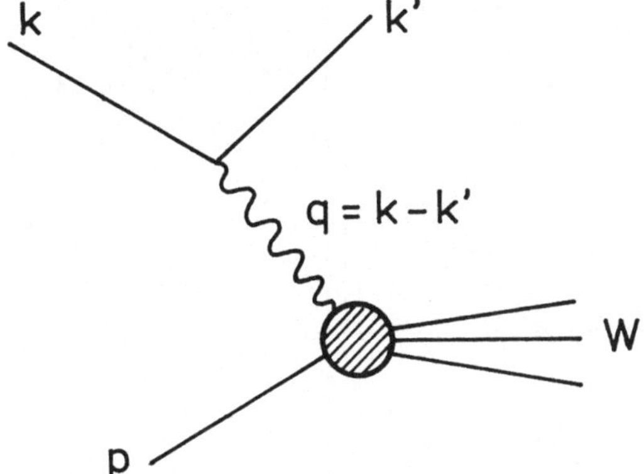

Fig.1

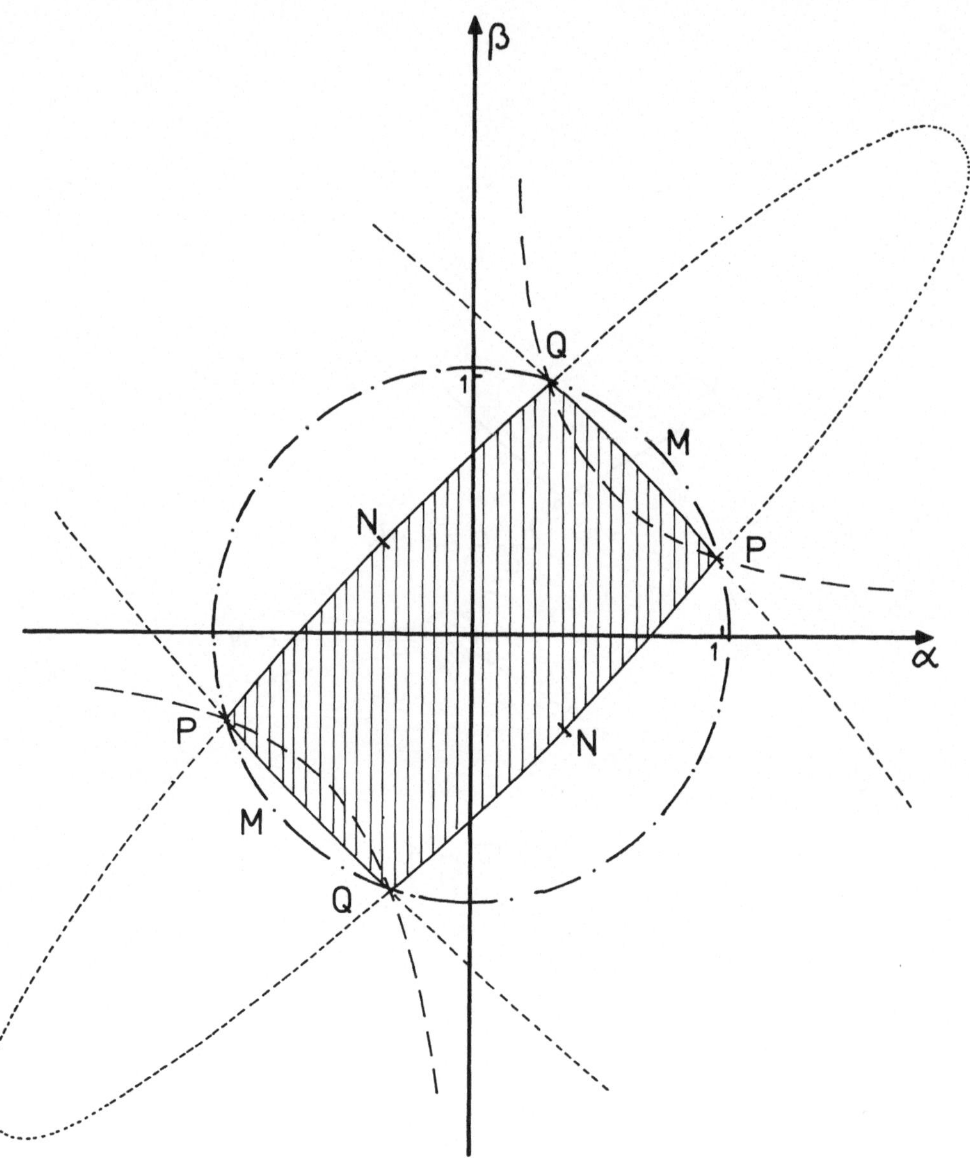

Fig. 2

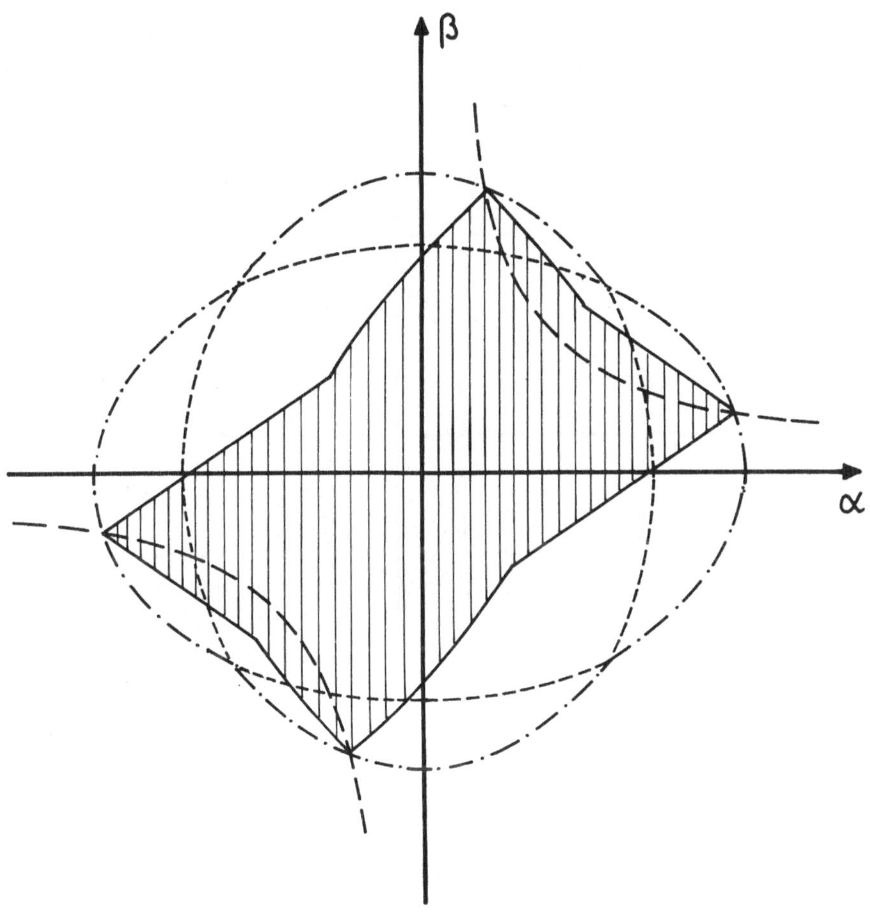

Fig. 3

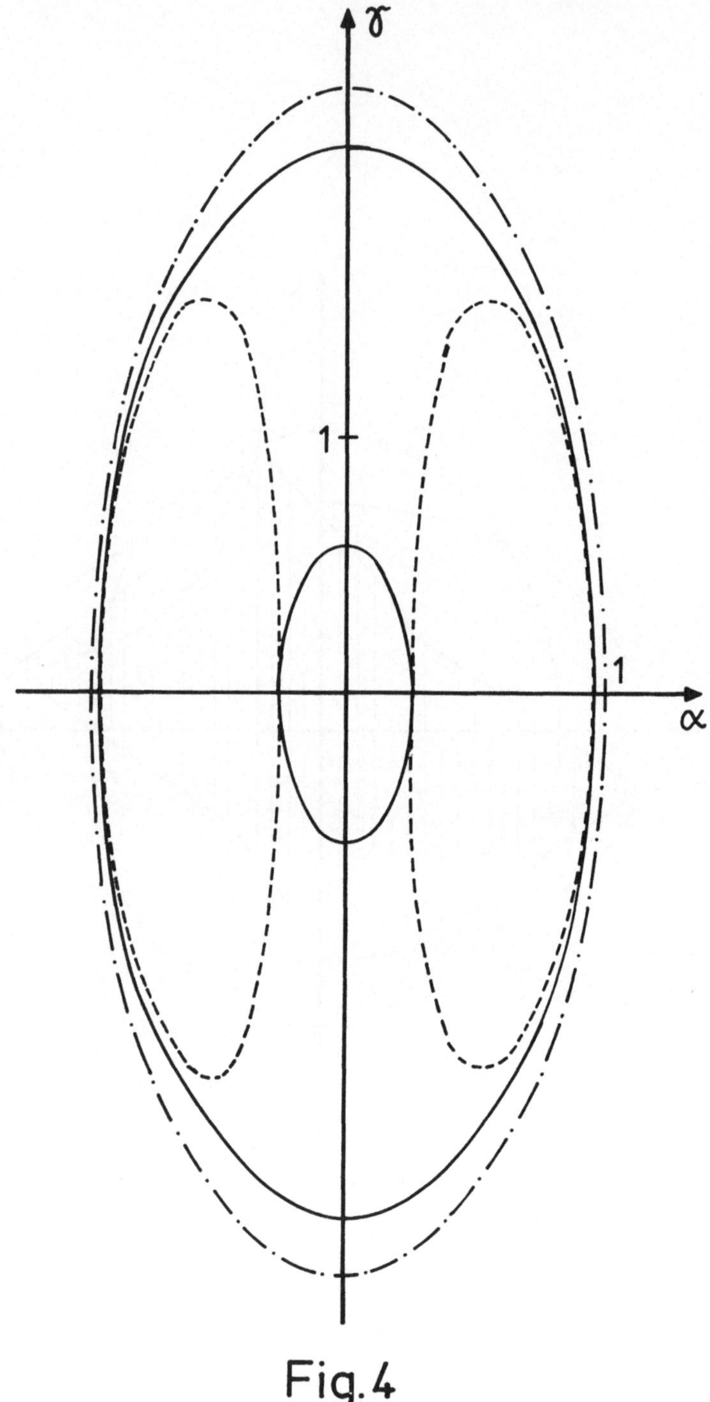

Fig. 4

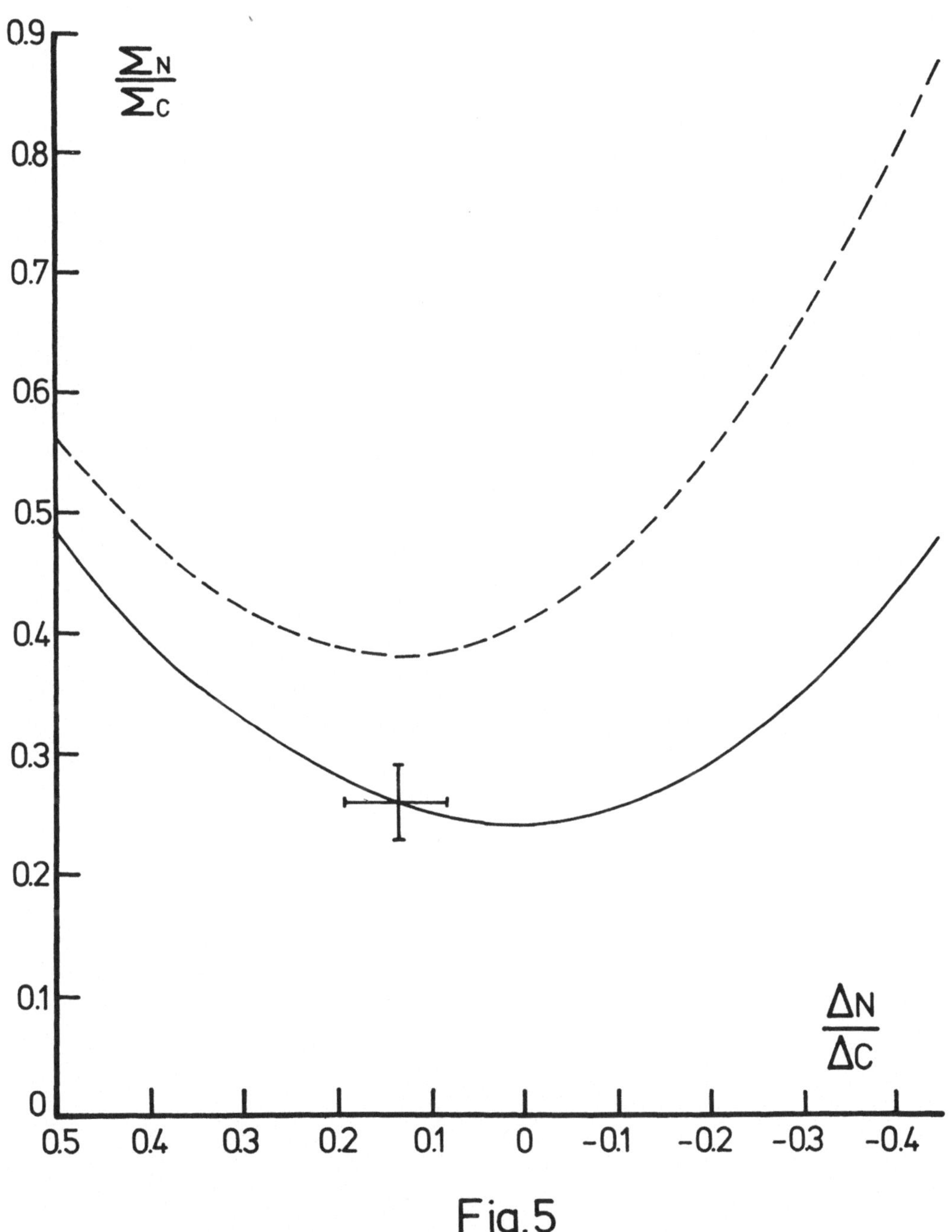

0.9
0.8
0.7
0.6
0.5
0.4
0.3
0.2
0.1
0
$\frac{\Sigma_N}{\Sigma_C}$
$\frac{\Delta N}{\Delta C}$
0.5
0.4
0.3
0.2
0.1
0
-0.1
-0.2
-0.3
-0.4

Fig.5

Acta Physica Austriaca, Suppl. XIV, 397 — 468 (1975)

QUANTUM ELECTRODYNAMICS IN LASER FIELDS[+]

by

H. MITTER
Institut für Theoretische Physik
Universität Tübingen

I. INTRODUCTION

The theory to be described here is a special case of quantum electrodynamics (QED) in presence of external (i.e. classical) fields. We shall therefore begin with a short review of this more general topic. The problem of interest is here the influence of given, classical electromagnetic fields on QED. These fields are supposed to be strong, i.e. the coupling of charged particles to the fields is not to be considered as a small perturbation. Some motivation for studying such a theory can be derived from the following reasoning. It is known that a formal solution of QED can be formulated in terms of the free Green's function in presence of an arbitrary external field. Any approximation to this function implies a corresponding one for QED. The study

[+] Lecture given at XIV. Internationale Universitätswochen für Kernphysik, Schladming, Austria, February 24-March 7, 1975.

of special external field problems might give some insight in the machinery, which could help in the general case. To learn what happens if an expansion parameter is no longer small should be easier in special cases than in general (nobody knows really how to sum perturbation theory for a quantum field theory with interaction).

Apart from these mathematical aspects, there are also some attractive physical features. In presence of an external field all the usual QED effects receive corrections which can be calculated once the wave function and Green's function in presence of the external field are known. If sufficiently strong fields can be produced experimentally, one can measure these effects and thus test QED at another end, namely for high field intensities. It is obvious that among all the effects those which reflect the particular properties of the ground state are the most interesting ones. Because of charge fluctuations the vacuum in presence of an external field acts like a nonlinear dielectric medium. These nonlinearities are perhaps the most specific aspect of pure QED. There are few cases, where they have been tested and one should look, whether strong fields allow for new possibilities. Also the one-particle states are influenced by the properties of the ground state: a particle in the field displays usually a (quasi-) level structure, which is changed by the emission and absorption of virtual photons. To some extent this change corresponds to a correction of the static particle properties (mass, magnetic moment) by field-dependent contributions. Also here a measurement would test renormalization concepts. Usually the effects turn out to be very small, but sometimes a resonance behavior is predicted, which may be helpful.

Thus far we know of three types of electromagnetic fields, which can be strong in the sense that a coupling parameter is not small.

a) Coulomb fields near heavy nuclei.

The expansion parameter is in this case $Z\alpha$, which is 0.6 for Pb. The Born approximation (expansion in powers of $Z\alpha$) is certainly not good for heavier nuclei. As far as the theory is concerned, one knows the exact wave function, but for the Green's function one has only rather complicated expressions. Almost all results, which go beyond the Born approximation, had to be derived by numerical methods (the only exception is the simplest vacuum polarization loop). Very accurate results are available for the energy levels of hydrogen and helium. For heavy atoms the complications due to the many-body structure prevent a clear separation of effects due to the field-theoretic vacuum from those due to relativistic many-body effects. For mesic atoms the bound-state problem can in good approximation be treated as a one-body-problem in the center of mass system. Vacuum polarization is a very important correction, but the effects due to the nuclear structure are also present and one controls in fact always a mixture of model assumptions on the nuclear charge structure plus field theory results. For some other effects (like e.g. Delbrück scattering or photon splitting) there are discrepancies between theory and experiment, which are probably due to the fact, that the strong field situation has not been treated correctly in the theory. For a pure Coulomb field the vacuum would become instable against spontaneous pair creation for $Z\alpha > 1$ (for the potential of a finite charge distribution the instability occurs at a higher value of $Z\alpha$). The question whether one could produce "super-nuclei" with $Z\alpha > 1$, which live long enough to do experiments, is open at present. It is also not quite clear how and whether the instability shows up (since the K-shell is filled, the Pauli principle forbids accommodation of an additional electron) and how higher radiative corrections affect the instability.

b) Strong constant magnetic fields.

Here an expansion parameter is B/B_{cr} where $B_{cr} = m^2c^3/eh$, which is $\sim 4.41 \times 10^{13}$ Gauss. Near the surface of pulsars fields of some 10^{12} Gauss are predicted. On earth one has reached 10^7 Gauss by magnetic flux compression using implosion techniques. There is some rumor that one could reach 10^9-10^{10} G, if the implosion is caused by a shock wave, which in turn is started by evaporation or small piece of matter with lasers. Even the one is still far from $B/B_{cr} = 1$.

c) Strong wave fields produced by lasers.

Here an expansion parameter is $\nu^2 = (ea/mc^2)^2$, where a is the amplitude of the vector potential. In technical units we have $\nu^2 \sim 7.5 \times 10^{-11} \lambda^2 S$, where λ is the wave length in cm and S is the illumination density in W/cm^2. According to the experts in the field one may obtain at present an energy of several kJ over a time of 0.1 Nanoseconds with Neodym-glass-lasers ($\lambda = 1\mu$) or CO_2-lasers ($\lambda = 10\mu$). For fusion experiments devices are in construction, which aim even at 10-100 kJ. If the beam could be focussed to a spot of dimension λ^2 (which is the theoretical limit), we would obtain $\nu^2 \sim 750$ for 1 kJ. In reality this limit can certainly not be obtained. For an illumination density of 10^{18} W/cm^2

(which is the goal of devices for laser fusion) we have $\nu^2 \sim 0.7$ for Neodym-glass-lasers and $\nu^2 = 75$ for CO_2-lasers. At present one may be off by an order of magnitude, but it is clear, that we should not expand in powers of ν^2.

Unfortunately ν^2 is not the only parameter, which controls the magnitude of effects.

If we consider particles in the field, another parameter is $\rho = 2(pk)/\kappa^2$, where p is the momentum of the particle, $\kappa = mc/\hbar$ and k is the wave vector of the laser wave. For a relativistic particle (v/c = 1) with energy E travelling in opposite direction to the laser wave we have

$$\rho \sim 2 \times 10^{-6} \frac{E(\mathrm{GeV})}{\lambda(\mathrm{cm})}$$

Note that $\rho \sim 1$ corresponds to E = 50 GeV for an infrared laser! In order to observe effects proportional to ρ at lower energies one would need high intensity X-ray- or γ-ray-lasers (GRASERS), which have not (yet) been constructed.

Having discussed the basic orders of magnitude, we shall now outline the theoretical framework (for the notation used here see Appendix 1). We shall start from the usual basic Lagrangian density

$$L = L_D(\psi) + L_M(A) + L_W(\psi, A, A^{ext}) \tag{1}$$

where the first two terms refer to the free Dirac-resp. Maxwell field and

$$L_W = -e(\bar{\psi}(x)\gamma^\mu\psi(x))\ (A_\mu(x) + A_\mu^{ext}(x)) \ . \tag{2}$$

Here A_μ is a quantized field describing real and virtual photons (in our case: non-laser quanta), whereas A_μ^{ext} is a given, classical field. The S-matrix is obtained after a transformation to the Furry picture in the form

$$S = T\ \exp(-ie\int :\ \bar{\psi}_e(x)\gamma^\mu\psi_e(x)A_\mu(x)d^4x:) \ . \tag{3}$$

Here ψ_e is the electron field in presence of A_μ^{ext}, i.e. the solution of the Dirac equation

$$(\gamma^\mu(i\ \partial_\mu - \varepsilon A_\mu^{ext}) - \kappa)\psi_e = 0 \ . \tag{4}$$

If we expand S in the usual perturbation series (which is an expansion in α), we may describe its matrix elements in terms of Feynman diagrams. We have however to insert other expressions as usual for the electron lines.

In order that the asymptotic conditions are correctly met, we must imagine A_μ^{ext} switched on/off in the remote past/future, so that $\psi_e \rightarrow \psi$ in the sense of a usual asymptotic condition. Practically one uses sometimes fields, for which the particles described by ψ_e are literally not free asymptotically (e.g. an infinitely extended homogeneous magnetic field or an infinitely extended plane laser wave). This is for mathematical reasons: the solutions of the Dirac equation are simple in these cases. If one calculates a process in a careless way the wrong asymptotic behavior shows up through the appearance of infinities (e.g. infinite phase factors). Nevertheless one may extract physically meaningful information. The only case, where consistency has been shown to some extent is the laser field. One can there see at least for some problems, that the solution with an infinitely extended external field is an approximation to the realistic case in which the particles are free asymptotically. We shall now specialize to this case.

II. DESCRIPTION OF THE LASER FIELD

We shall investigate moves with plane wave front characterized by the wave vector

$$k^{\mu} = \omega(1,\vec{n}),\ \vec{n}^2 = 1,\ \omega = \frac{2\pi}{\lambda} \quad . \tag{1}$$

The vector potential is

$$A_{\mu}^{ext} = a(e_{1,\mu}\ a_1(\xi) + e_{2,\mu}\ a_2(\xi)) \tag{2}$$

where a is the amplitude,

$$e_i^{\mu} = (0,\vec{e}_i),\ e_i^{\mu}e_{j,\mu} = -\delta_{ij},\ k^{\mu}e_{i,\mu} = 0 \quad i = 1,2 \tag{3}$$

are polarization vectors and $a_i(\xi)$ are arbitrary functions of the argument

$$\xi = k_{\mu}x^{\mu} \quad . \tag{4}$$

For a laser pulse of finite duration $a_i(\xi)$ differs from zero only inside a finite region in ξ, i.e. we have

$$a_i(\xi) = 0 \quad \text{for} \quad \xi > \xi_1 \quad \text{and} \quad \xi < \xi_o \tag{5}$$

with fixed ξ_o, ξ_1 corresponding to switching the laser on resp. off.

Note that the actual length of a laser pulse is quite large. For a pulse duration of 0.1 Nanoseconds the corresponding interval in ξ (which is a pure number) is $\xi_1 - \xi_o \sim 1.36 \times 10^5$ for $\lambda = 1\mu$. We shall therefore occasionally consider an infinitely extended plane wave as an approximation. For circular polarization we have

$$a_1 = \cos\xi, \; a_2 = -\sin\xi \quad \text{CPPW} \tag{6}$$

whereas for linear polarization

$$a_1 = \cos\xi, \; a_2 = 0 \; \text{LPPW} . \tag{7}$$

Before we go on, we have to consider whether this description comes close to reality. In the theory considered here ω and $\vec{n}$ are fixed. A generalization to superpositions in ω seems possible. Realistic lasers are, however, rather monochromatic, so that the restriction to one frequency is a good approximation. The restriction to a plane wave front ($\vec{n}$ fixed) seems more dangerous, since one would like to focus the laser to obtain a high intensity. It has not been possible so far to deal with non-plane waves in the theory. The plane wave is a good approximation, if the realistic field is "plane" at least over a volume of dimension λ^3. This is a large volume for particles, for which the unit of length is the Compton wave length.

One could also ask whether the external field picture is a realistic one. Actually the number of quanta in an intense laser beam is very large (for the Nd-glass laser mentioned in Chapter 1 some 10^{21} photons per pulse) and thus a classical description should be reasonable. The limits of the concept can be seen sometimes from the calculations. The concept of an "external" field implies the neglect of the recoil, which this field suffers in a given process, i.e. the field may absorb or emit an arbitrary amount of energy and momentum (i.e. an arbitrary number n of laser quanta with four-momentum $\hbar k^{\mu}$) without being changed. Therefore energy- and momentum-conservation in a process is fulfilled only up to nk^{μ} and all matrix elements and cross sections will have the form of an infinite sum on $n(-\infty \leq n \leq +\infty)$. In practice, only a finite number $n \leq N$ of quanta can be emitted or absorbed at all. Thus our theory will certainly be incorrect for high $|n|$. If we want the laser field to remain unchanged in a process, we should trust our results only for $|n| \ll N$. Practically the theory is certainly not applicable to processes, in which the laser beam is depleted considerably. A depletion is only to be expected, if the beam interacts with a relatively dense medium (whereby the density scale is set by the photon density of the beam). This may be the case for laser-plasma interactions, but will certainly not happen if the laser beam interacts with a particle beam from an accelerator.

The mathematical description is simplified considerably, if we use light-like coordinates. These coordinates are quite fashionable both in high-energy physics and gravitational theory. They are very natural for all processes dealing with mass zero particles and have occasionally been used for the description of light phenomena long before their use in high energy physics was proposed. We shall use a fixed Vierbein defined by the vectors

$$n^\mu = \frac{1}{\sqrt{2}\omega} k^\mu = \frac{1}{\sqrt{2}} (1,\vec{n}) , \quad \hat{n}^\mu = \frac{1}{\sqrt{2}}(1,-\vec{n}) , \quad e_i^\mu = (0,\vec{e}_i)$$

$$i = 1,2 . \tag{8}$$

These vectors form a complete, orthogonal set. The only abnormal feature is that n and $\hat{n}$ are light-like ($\hat{n}^2 = n^2 = 0$, $(n,\hat{n}) = 1$). We shall decompose any vector p^μ according to

$$p^\mu = n^\mu p_u + \hat{n}^\mu p_v + e_i^\mu p_i \tag{9}$$

in terms of the "components"

$$p_u = \hat{n}^\mu p_\mu , \quad p_v = n^\mu p_\mu , \quad p_i = -e_i^\mu p_\mu \quad . \tag{10}$$

If $\vec{n}$ is taken parallel to the z-axis, we have

$$p_v = \frac{1}{\sqrt{2}} (p^o - p_z) , \quad p_u = \frac{1}{\sqrt{2}}(p^o + p_z) , \quad p_i = (p_x, p_y) . \tag{11}$$

For the coordinate vector we use a special notation. We call

$$x_v = \frac{1}{\sqrt{2}\omega} \xi =: u , \quad x_u =: v . \tag{12}$$

Then (u,p_u) and (v,p_v) are pairs of canonically conjugate variables, if p^μ is the four-momentum. A collection of formulae for light-like components (including the light-like Dirac algebra) is given in Appendix 2 and 3.

It is easily noted that the external field depends only on u. Therefore translational invariance is only violated in the u-direction. As a consequence, the three components P_v, P_i of the total energy-momentum vector P^μ will be conserved in all processes (in ordinary coordinates only the conservation of P_i is manifest), whereas P_u will change. Note that the wave vector has the components

$$k_u = \sqrt{2}\omega, \; k_v = k_i = 0 \tag{13}$$

so that a change of P^μ can occur only by a multiple of k^μ.

III. THE VOLKOV SOLUTION

It has been noted long time ago [1] that the Dirac equation in presence of an external plane wave field can be solved exactly. The original paper is hard to read, but it is rather easy to derive the solution in light-like coordinates. We note that the Dirac operator reads

$$\not{\pi} - \kappa = i\gamma_v\partial_u + i\gamma_u\partial_v + \gamma_i(i\partial_i + \varepsilon a a_i(u)) - \kappa \; . \tag{1}$$

We shall seek a solution of

$$(\not{\pi} - \kappa)\; \phi = 0 \tag{2}$$

of the form

$$\phi \equiv \phi_p \ (x) = \Phi(u|p) \ \Phi_F(x|p) \tag{3}$$

where p_μ is a fixed vector labelling the solution and

$$\Phi_F(x|p) = u(p)e^{-i(px)} = u(p)\exp -i(up_u^f + vp_v - x_i p_i) \tag{4}$$

is the usual solution of the free Dirac equation with p on the mass shell, i.e. we have

$$(\not{p}-\kappa)u(p) = 0, \ p^2 = \kappa^2 \ \text{viz.} \ p_u^f = \frac{1}{2p_v}(\kappa^2 + p_i p_i) . \tag{5}$$

The initial condition $a_i(u_o) = 0$ requires

$$\Phi(u_o|p) = 1 . \tag{6}$$

Note that our ansatz (3) corresponds to the Bloch ansatz for the wave function of an electron in a periodic potential. There is also a "Bloch theorem" in our case: if a_i has periodicity properties, then we may write

$$\Phi(u|p) = \Psi \ (u|p) \ e^{i\frac{uC}{2p_v}}$$

where Ψ is periodic with the same period as a_i. C is a constant, which adds to the mass term in p_u^f. This fact is a simple consequence of the structure of the translation group.

In order to solve for $\Phi(x|p)$ we write

$$\gamma_u \Phi = \Phi_u , \quad \gamma_v \Phi = \Phi_v . \tag{7}$$

Inserting (3) into (2), we see that we obtain a solution, if we require

$$p_v \Phi_u + (i \partial_u + p_u^f) \Phi_v - (\gamma_i \pi_i + \kappa) \Phi = \Phi (\not{p} - \kappa) \tag{8}$$

where

$$\pi_i = p_i - \varepsilon a \, a_i . \tag{9}$$

The solution is easily established with the aid of the projection operators

$$P_u = \frac{1}{2} \gamma_v \gamma_u , \quad P_v = \frac{1}{2} \gamma_u \gamma_v . \tag{10}$$

We have

$$P_u \Phi_v = \Phi_v , \quad P_v \Phi_u = \Phi_u , \quad P_u \Phi_u = P_v \Phi_v = 0 . \tag{11}$$

Application of P_v onto Eqn. (8) yields

$$\Phi_u = \frac{1}{2p_v} (\gamma_i \pi_i + \kappa) \gamma_u \Phi_v + \frac{1}{2p_v} \gamma_u \Phi_v (\not{p} - \kappa) . \tag{12}$$

As a consequence, the entire solution can be expressed in terms of Φ_v:

$$\Phi = (P_u + P_v) \Phi = \frac{1}{2} (\gamma_v \Phi_u + \gamma_u \Phi_v) = \frac{1}{2p_v} (\gamma_u p_v - \gamma_i \pi_i + \kappa) \Phi_v + \frac{1}{2p_v} \Phi_v (\not{p} - \kappa) . \tag{13}$$

An equation for Φ_V is obtained by application of P_u onto Eqn. (8) and insertion from Eqn. (12). We have

$$[(i\,\partial_u + p_u^f) - \frac{1}{2p_V}(\kappa^2 + \pi_i\pi_i)]\Phi_V = \frac{1}{2p_V}\Phi_V(p^2-\kappa^2) = 0. \quad (14)$$

Now we observe that the exponential

$$\exp -i(u-u_o)(p_u(u,u_o) - p_u^f) \quad (15)$$

is a solution of this equation if we require

$$p_u(u,u_o) = \frac{1}{u-u_o}\,\frac{1}{2p_V}\int_{u_o}^{u} d\bar{u}\,(\kappa^2 + \pi_i(\bar{u})\pi_i(\bar{u})). \quad (16)$$

Therefore we have

$$\Phi_V(u|p) = \gamma_V\, B \exp -i(u-u_o)(p_u(u,u_o) - p_u^f) \quad (17)$$

where B is a constant matrix. The initial condition (6) implies $\Phi_V(u_o|p) = \gamma_V$ which is fulfilled if we put

$$B = 1 + \gamma_V\, C \quad (18)$$

with another constant matrix C. The final form of our solution takes the form

$$\Phi(u|p) = S(u)\exp -i(u-u_o)(p_u(u,u_o) - p_u^f) \quad (19)$$

where S is determined by insertion of (18) with (19) into Eqn. (13). After some algebra C drops out and we have

$$S(u) = 1 - \frac{\varepsilon a}{2p_v} \gamma_v \gamma_i a_i(u) \ . \tag{20}$$

It is useful to draw a space-time picture in order to understand the boundary behavior of the solution:

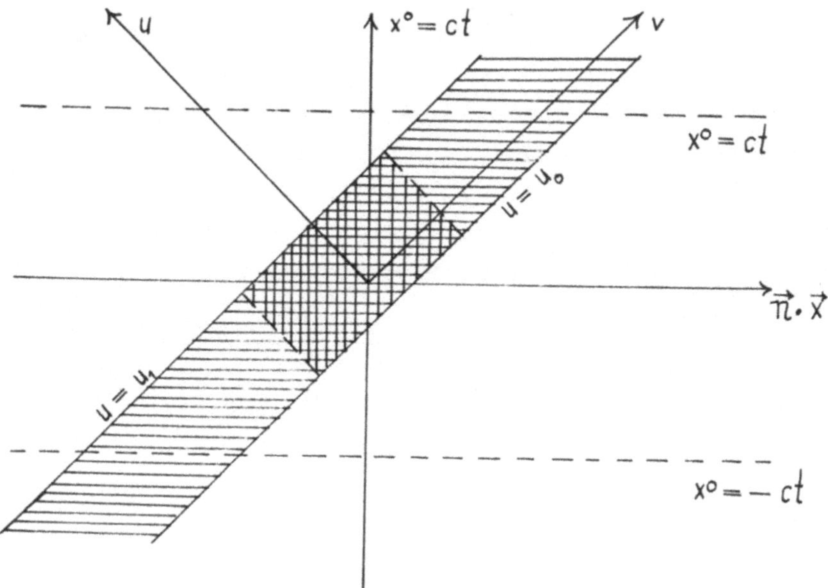

The laser pulse fills the hatched area between the lines $u = u_o$ and $u = u_1$. Outside this region we have a free plane wave. At the boundary $u = u_o$ this matches the Volkov solution, for which p_u depends on u according to Eqn. (16) in the interior domain. Since this domain extends to infinity, it is clear from the picture that the particle is not free (i.e. separated from the laser) in the remote past ($t < -T$) resp. future ($t > +T$). Such a solution can however be easily constructed by superimposing solutions (3) with different values of p_V

$$\phi(x) = \int dp_V \; f(p_V)\phi_p(x) \tag{21}$$

using an appropriate weight function $f(p_V)$ (e.g. a gaussian). The solution will then be concentrated in the cross-hatched area of our diagram and it is clear that it satisfies the correct boundary conditions. Literally one should always use packet solutions of this type (21) (which are analogous to the Wannier functions in solid state physics). We shall be content with Volkov plane waves instead.

We shall now note some properties of our plane Volkov wave. An explicit form of (19) reads

$$\Phi(u|p) = [1 - \frac{\varepsilon a}{2p_V} \gamma_V\gamma_i a_i(u)] \exp\{i \frac{\varepsilon a}{p_V}$$

$$[p_i \int_{u_o}^{u} a_i(\bar{u})d\bar{u} - \frac{\varepsilon a}{2} \int_{u_o}^{u} a_i(\bar{u}) \; a_i(\bar{u})d\bar{u}] \} . \tag{22}$$

Now we include the plane wave exponential rewriting (3) in the form

$$\phi_p(x) = E(x|p)\ u(p) \tag{23}$$

with

$$E(x|p) = \Phi(u|p)\ e^{-i(px)} \tag{24}$$

and define an adjoint solution by

$$\bar{E}(x|p) = \gamma_o\ E^+(x|p)\gamma_o\ . \tag{25}$$

We shall consider p to be an <u>arbitrary</u>, fixed vector here, i.e. we continue the Volkov solution off the mass shell. Then we can show the following properties:

$$\not{\Pi}\ E(x|p) = E(x|p)\not{p}\ , \qquad \bar{E}(x|p)\ \overleftarrow{\not{\Pi}} = \not{p}\ \bar{E}(x|p) \tag{26}$$

$$\frac{1}{(2\pi)^4} \int d^4x\ \bar{E}(x|p)\ E(x|q) = \delta(p-q) \tag{27}$$

$$\frac{1}{(2\pi)^4} \int d^4p\ E(x|p)\ \bar{E}(x'|p) = \delta(x-x') \tag{28}$$

$$\partial^\mu\ \bar{E}(x|p)\ \gamma_\mu\ E(x|p) = 0\ . \tag{29}$$

These relations will prove to be useful later. Similar representations can be found for other external field problems.

Finally we consider the form of the solution for an infinitely extended laser wave. We take the CPPW form (2.6). The integrals in the exponent are elementary. We use polar components for the two-vector p_i writing

$$p_1 = p\cos\tilde{\rho},\ p_2 = p\sin\tilde{\rho},\ p = \sqrt{p_1^2 + p_2^2},\ \tilde{\rho} = \arctan\frac{p_2}{p_1} \tag{30}$$

and expand the exponent in terms of Bessel functions according to

$$\exp i\, B \sin(\xi+\tilde{\rho}) = \sum_{n=-\infty}^{+\infty} Jn(B) \exp i\, n(\xi+\tilde{\rho}) \ . \tag{31}$$

Then the solution can be cast into the form

$$E(x|p) = e^{-i\phi_o} \sum_{n=-\infty}^{+\infty} [C_n - \frac{\varepsilon a}{\sqrt{2}p_v}\gamma_v(\gamma_+ C_{n-1} + \gamma_- C_{n+1})]$$

$$\exp -i[u(p_u^{eff} + nk_u) + vp_v - p_i x_i] \tag{32}$$

where the constant phase ϕ_o is

$$\phi_o = - \frac{\varepsilon^2 a^2}{2(pk)} \xi_o - \frac{\varepsilon ap}{(pk)} \sin(\xi_o + \tilde{\rho}) \tag{33}$$

and we have

$$p_u^{eff} = \frac{1}{2p_v} (\kappa^2+\varepsilon^2 a^2+p_i p_i), \; C_n = Jn[\frac{\varepsilon ap}{(pk)}] \; e^{-in\tilde{\rho}} . \tag{34}$$

Thus the solution consists of an infinite sum of plane waves, where for each term the four-momentum fulfills the condition

$$(p + nk)^2 = \kappa^2 + \varepsilon^2 a^2 \quad . \tag{35}$$

The particle in the field displays therefore a quasi-level structure, where the energies of these levels are determined by (35). This aspect has been stressed long time ago by Zeldovich [2] . Many physical features of particles in laser fields can be understood in terms of these Zeldovich quasi-levels.

The Dirac equation can be solved exactly in more general cases as the one considered here. We list some references, which are of interest in this context:

a) If the particle has in addition to the Dirac moment an anomalous magnetic moment, the interaction of this moment with the external field is the usual Pauli coupling $\mu \, \sigma_{\mu\nu} \, F^{\mu\nu}$. The Dirac equation can be reduced to a simple system of differential equations for functions describing the spin motion in the external field. Exact solutions in closed form are obtainable for linear polarization (only one $a_i(\xi)$) and for the CPPW case (II,6), Becker and Mitter [3], Becker [4]).

b) If a constant magnetic field is present in addition to the laser field, an exact solution may be found (Redmond [5], Oleinik [6]). The level structure is discussed by the latter author (Oleinik [7]).

c) The solution of the Dirac and Klein-Gordon equation can be obtained by purely algebraical methods. This aspect has been stressed by Beers and Nickle [8]. A full account of these techniques is, however, already contained in a much earlier paper by Schwinger [9], in which the propagator is obtained by means of the proper time equations of motion, which are solved algebraically.

d) The case of a constant, crossed field $\vec{E} \perp \vec{B}$ is the limiting case of a very low frequency wave. If we take for instance

$$a_1(\xi) = -\xi, \qquad a_2(\xi) = 0$$

we obtain a crossed field with

$$\vec{E} = \sqrt{2}\omega\, \vec{e}_1, \qquad \vec{B} = \sqrt{2}\omega\, \vec{e}_2 \quad .$$

Thus we have to identify $\sqrt{2}\omega$ with the magnitude of the field strength and $\vec{n}$ with the direction of the Poynting vector in this case. Of course this field has some unrealistic properties: it is never switched off, so that the particle is never free, and the vector potential increases without limit with ξ. Thus the asymptotic properties are questionable (as for a constant homogeneous magnetic field). The theory in a constant, crossed field has been studied extensively by Ritus [10].

IV. THE PROPAGATOR

The electron propagator is a Green's function of the Dirac equation in presence of the external field and fulfils the equations

$$(\not{\Pi}(x)-\kappa)\ G(x,x') = G(x,x')(\overleftarrow{\not{\Pi}}(x')-\kappa) = \delta(x-x'). \qquad (1)$$

Several methods have been used to find a convenient representation of this function. An integral representation has been derived long time ago by Schwinger [9] for linear polarization (only one function $a_i(\xi)$) by means of the proper-time method. The generalization to arbitrary polarization is rather straightforward. We shall, however, apply a different method here, which has been introduced by Ritus [10] in a related context. Using our equations (III, 26-28) we can convince ourselves, that the representation

$$G(x,x') = \frac{1}{(2\pi)^4} \int d^4p\ E(x|p)\ \frac{\not{p}+\kappa}{p^2-\kappa^2+i0}\ \bar{E}(x'|p) \qquad (2)$$

is a solution of Eqn. (1). Thus our propagator is just the free propagator in momentum space, sandwiched between our Bloch-type functions $E(x|p)$. It is easily recognized, that the representation (2) reduces to the ordinary Fourier representation in absence of the laser field, since E is the usual plane wave exponential in this case.

For many problems the form (2) suffices. Nevertheless one might want to have a more explicit form sometimes. If the solution (III, 26 - 28) is inserted, one may convert (2) into an integral representation. We shall write down the result, in spite of the fact, that it is rather lengthy and we have to indroduce a number of abbreviations. We shall write the Green's function as a product

$$G(x,x') = \Phi(x,x')\, G(x-x'|\xi,\xi') \tag{3}$$

where the behaviour under gauge transformations is contained entirely in the first factor, which reads

$$\Phi(x,x') = \exp\left(i\varepsilon a \frac{x_i - x_i'}{\xi-\xi'} \int_{\xi'}^{\xi} a_i(\bar{\xi})\, d\bar{\xi}\right) . \tag{4}$$

This factor is known as Schwinger's line integral and is present for any external field. In general we have

$$\Phi(x,x') = \exp\left(-i\varepsilon \int_{x'}^{x} d\bar{x}^{\mu}\, A_{\mu}^{ext}(\bar{x})\right)$$

where the line integral is to be performed along a straight line (Φ can be defined in a way, which makes it clear, that the definition is path-independent, c.f. Schwinger [9]). The second factor in Eqn. (3) is gauge-invariant and we can write a proper-time representation

$$G(z|\xi,\xi') = \frac{1}{16\pi^2} \int_0^{\infty} \frac{ds}{s^2}\, K\left(\frac{z}{2s}\Big|\xi,\xi'\right) \exp\left(-ism^2(\xi,\xi') - \frac{iz^2}{2s}\right) . \tag{5}$$

In order to specify K and m^2 we introduce the functions

$$T(\xi,\xi') = \frac{\varepsilon^2 a^2}{\xi-\xi'} \int_{\xi'}^{\xi} d\eta \; a_i(\eta)\left(a_i(\eta) - \frac{1}{\xi-\xi'} \int_{\xi'}^{\xi} d\eta' a_i(\eta')\right) \quad (6)$$

$$M_i(\xi,\xi') = - \frac{\varepsilon a}{\xi-\xi'} (a_i(\xi) - a_i(\xi')) \quad (7)$$

$$L_i(\xi,\xi') = \frac{\varepsilon a}{\xi-\xi'} \left(a_i(\xi) + a_i(\xi') - \frac{2}{\xi-\xi'} \int_{\xi'}^{\xi} d\eta a_i(\eta)\right) \quad (8)$$

$$R(\xi,\xi') = (\xi-\xi') \; (L_i L_i - M_i M_i) - \frac{T(\xi,\xi')}{\xi-\xi'} \quad . \quad (9)$$

A collection of formulae for these functions can be found in ref. [11]. The positive definite quantity m^2 is then simply given by

$$m^2(\xi,\xi') = \kappa^2 + T(\xi,\xi') \quad (10)$$

and the kernel is expressed in terms of these functions and the light-like Dirac matrices of appendix 3 in the following way

$$K(p|\xi,\xi') = \not{p} + \kappa + s \sqrt{2}\omega \; Q(p|\xi,\xi') \quad (11)$$

$$\begin{aligned} Q(p|\xi,\xi') = {} & \gamma_V(R-2p_i L_i) + 2p_V \gamma_i L_i + 2\kappa\gamma_V\gamma_i M_i - \\ & -2p_V \; \tau \; \gamma_i M_i + 2i \; \gamma_V \sigma((\xi-\xi')L_i - p_i)\varepsilon_{ij} M_j \; . \end{aligned} \quad (12)$$

Since s appears in K in a simple way, the integral on s could even be performed by means of

$$- \frac{1}{16\pi^2} \int_0^\infty \frac{ds}{s^2} (-is)^n \exp[-i(m^2 s + \frac{z^2}{4s})] = \frac{m^{2-2n}}{2^{n+3}\pi} \frac{H^{(2)}_{1-n}(\sqrt{m^2 z^2 - i0})}{(\sqrt{m^2 z^2 - i0})^{1-n}}$$

(needed here for n = 0,1). This Hankel representation is however not very useful. A more convenient form is obtained by writing

$$G(z|\xi,\xi') = \frac{1}{(2\pi)^4} \int d^4p \; e^{-i(pz)} \; \tilde{G}(p|\xi,\xi') \; . \tag{13}$$

Then we have

$$\tilde{G}(p|\xi,\xi') = i \int_0^\infty ds \; K(p|\xi,\xi') \exp[-is(m^2 - p^2)] =$$

$$= - \frac{\not{p} + \kappa}{m^2 - p^2 - i0} + i\sqrt{2}\omega \frac{Q(p|\xi,\xi')}{(m^2 - p^2 - i0)^2} \; . \tag{14}$$

Thus G resembles an ordinary propagator in momentum space. Note, however, that the "mass" m^2 is space dependent (cf. Eqn. 11)) and the representation (13) is not the full momentum representation! Still one might call m^2 an "effective" mass (which depends both on space and the polarization of the laser; a similar concept is used in solid state physics).

From the explicit form of T one may deduce

$$m^2(\xi,\xi') \to \kappa^2 \qquad \text{for} \quad \xi - \xi' \to 0$$

so that we have free propagation, if the particle spends a short "time" in the field. For an infinitely extended laser wave we may show

$$m^2(\xi,\xi') \to \kappa_*^2 \qquad \text{for} \quad \xi - \xi' \to \infty$$

where the constant effective mass κ_* turns out to be (cf. III, 35)

$$\kappa_*^2 = \kappa^2(1 + \nu^2)$$

for CPPW. For linear polarization we have

$$m^2(\xi,\xi') \to \kappa_*^2 = \kappa^2(1 + \frac{\nu^2}{2}) \qquad \text{for} \quad \xi - \xi' \to \infty$$

and for the constant, crossed field considered above we obtain

$$m^2(\xi,\xi') = \kappa^2(1 + \frac{\nu^2}{12}(\xi-\xi')^2)$$

which increases for large $(\xi - \xi')$ because the vector potential does so.

A number of further comments are here appropriate.

The representation (3) - (5) can also be obtained by means of the proper time method and has been derived in this fashion by Schwinger [9] for linear polarization (only one $a_i(\xi)$). The advantage over the form (2) (which is much simpler) is the explicit presence of the gauge factor (4). Thus the form is useful for all closed loop calculations, where gauge invariance is relevant. In the CPPW case one may write down the full momentum transform of G(x,x'), which consists of an infinite sum, whereby each term has a pole at one of the Zeldovich quasi-levels (III,35). This form is not very convenient for calculations, however. It has been obtained by Reiss and Eberly [12] (their result is exact, though this is not clear from the derivation) and can also be derived by means of eikonal methods (Dittrich [13]). The propagator for a particle having an anomalous magnetic moment can also be calculated by Ritus' method (Becker [4]).

V. FEYNMAN DIAGRAMS

As has been stressed in chapter I, any process taking place in presence of our external field can be described by a Feynman diagram, which differs from the corresponding one in absence of the field only by the expressions to be substituted for the electron lines. Here the diagram has to be read in <u>coordinate</u> space. The vertices and photon lines correspond to the usual expressions. For the electron lines we shall draw double lines with the following rules:

For an external electron line entering a vertex labelled by x^μ we substitute

$$= \phi_p(x) \qquad \text{with } \phi \text{ from (III,22-24)}$$

(correspondingly $\bar{\phi}$ for an outgoing electron line). For an internal line from y to x we substitute

$$= G(x,y) \text{ with G from (IV,2) resp. (IV,3-5).}$$

If we use the E-representation everywhere, another (equivalent) form of a "graphology" emerges. Writing the wave function as

$$\phi_p(x) = E(x|p)u(p) \qquad \text{or}$$

and correspondingly for the propagator

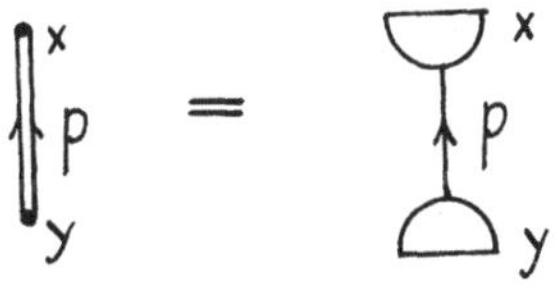

we seen, that we obtain the following rules:

Read the graph in momentum space. Insert the usual expressions from the free theory for electron- and photon lines. Modify, however,

the <u>vertex</u> according to

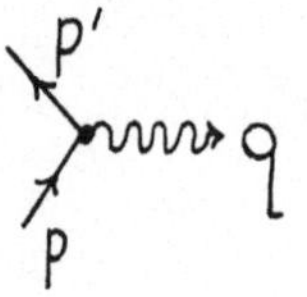

$$\gamma^{\mu}\delta(p'-p+q) \rightarrow v^{\mu}(p';p|q)$$

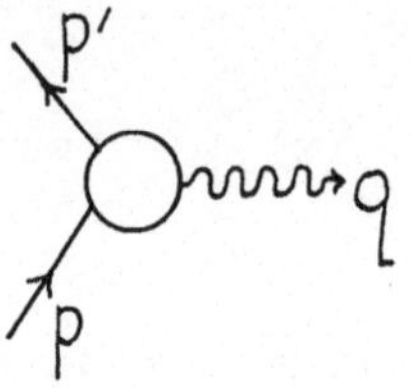

where the modified vertex is

$$v^{\mu}(p';p|q) = \frac{1}{(2\pi)^4}\int d^4x\ \bar{E}(x|p')\gamma^{\mu}\ E(x|p)\, e^{-i(qx)} \quad . \tag{1}$$

This is the only new element in our Feynman rules. If the explicit expression for E is inserted, we can do three of the four integrations in light-like coordinates and obtain

$$v^{\mu}(p';p\ |q)=\delta(p'_v-p_v+q_v)\,\delta(p'_i-p_i+q_i)\cdot\frac{1}{2\pi}\int du M^{\mu}(u)\,e^{-iw(u)} \tag{2}$$

where the phase is

$$w(u) = -\ u(p'_u - p_u + q_u) + W(u) \tag{3}$$

with

$$W(u) = \varepsilon a\left(\frac{p'_i}{p'_v} - \frac{p_i}{p_v}\right)\int_{u_o}^{u} d\bar{u}a_i(\bar{u}) - \frac{\varepsilon^2 a^2}{2}\left(\frac{1}{p'_v} - \frac{1}{p_v}\right)\int_{u_o}^{u} d\bar{u}a_i(\bar{u})a_i(\bar{u}) \tag{4}$$

and the vector M^{μ} has the light-like components

$$M_v = \gamma_v$$

$$M_u = \gamma_u - \frac{\varepsilon a}{2}\left(\frac{1}{p_v'} + \frac{1}{p_v}\right)\gamma_i a_i(u) - \frac{\varepsilon a}{2}\left(\frac{1}{p_v'} - \frac{1}{p_v}\right)\tau\ \gamma_i a_i(u) +$$

$$+ \frac{\varepsilon^2 a^2}{2p_v p_v'} \gamma_v\ a_i(u) a_i(u) \qquad (5)$$

$$M_i = \gamma_i - \frac{\varepsilon a}{2}\left(\frac{1}{p_v'} + \frac{1}{p_v}\right)\gamma_v a_i(u) + \frac{\varepsilon a}{2}\left(\frac{1}{p_v'} - \frac{1}{p_v}\right) i\gamma_v \sigma \varepsilon_{ij} a_j(u).$$

The conservation of the three components p_v, p_i of the total four momentum is manifest from Equ. (2).

It has to be noted, that the modified vertex is invariant under a gauge transformation of the external field

$$A_\mu^{ext} \to A_\mu^{ext} + k_\mu\ \Lambda(\xi)$$

since E picks up an exponential in this case, which cancels out in (1). In order to understand the properties of v under gauge transformations of the quantized (non-external) photon field we note, that

$$q_\mu\ v^\mu(p';p|q) = (\not{p}'-\kappa)\, I(p';p|q) - I(p';p|q)(\not{p}-\kappa) \qquad (6)$$

where

$$I(p';p|q) = \frac{1}{(2\pi)^4} \int d^4x\ \overline{E}(x|p')\ E(x|p) e^{-i(qx)} . \tag{7}$$

Equ. (6) is most easily proved by inserting (1) and observing, that q^μ can be replaced by the derivative of the exponential function under the integral. The surface term obtained after partial integration vanishes because the exponential in $\overline{E}\ E$ oscillates infinitely rapid for $u\to\infty$. From the remainder we obtain Equ. (6) using (III,26).

The kernel I acts like a kind of unit operator upon folding. We have

$$\begin{aligned} v_\mu(p';p''|q'+q'') &= \int d^4p\ v_\mu(p';p|q')\ I(p,p''|q'') = \\ &= \int d^4p\ I(p';p|q')\ v_\mu(p,p''|q'') \end{aligned} \tag{8}$$

$$I(p';p''|q'+q'') = \int d^4p\ I(p';p|q')\ I(p,p''|q'')$$

$$I(p';p''|0) = \delta(p'-p'') \quad . \tag{9}$$

It is easily seen with the aid of Equs. (6) and (8), that a gauge transformation affects an arbitrary graph in the same way as for the conventional theory without external field. If a combination of graphs gives a gauge-invariant contribution to an effect in absence of the external field, this will therefore also be the case for the corresponding effect in presence of the laser field, which is described by the same graphs with modified vertices.

These gauge invariance considerations are useful

in practical applications. If we want to compute for instance a matrix element described by a graph, in which an external photon line attaches at one of the vertices, we have to multiply v_μ with the polarization vector of that photon. If the momentum of the photon is q^μ and its polarization vector is $\tilde{e}^\mu$ (so that $(q,\tilde{e}) = q^2 = 0$) a gauge transformation of the photon field operator amounts to

$$\tilde{e} \to e = \tilde{e} + \lambda q$$

with arbitrary λ. If we take $\lambda = -\tilde{e}_v/q_v$ the new polarization vector has no v-component. Because also $(q,e) = 0$ we have

$$e^\mu = (\frac{1}{q_v} q_i e_i, 0, e_i) \tag{10}$$

and the product with v_μ is considerably simplified, since we may forget about the long expression M_u. If we wish to sum over the two polarization directions in the transition amplitude, this can also be done quite easily. We have then to consider an expression of the type

$$S = \sum_{Pol} (B,e)\ (C,e)$$

where B and C stand for (parts of) the matrix element and its complex conjugate. Because of gauge invariance we may allow e^μ to range over the four coordinate directions. If $B_\perp$ denotes the projection of B in the plane orthogonal to k and q, we have

$$S = -B_{\perp} \cdot C_{\perp} = -(B,C) + \frac{1}{(qk)}(B,k)(C,q) + (B,q)(C,k) \; .$$

In light-like components we obtain for the r.h.s.

$$S = (B_i - \frac{q_i}{q_V} B_V)(C_i - \frac{q_i}{q_V} C_V) \; .$$

The left hand side is however, if we insert (10) for the polarization vector

$$S = \sum_{\text{Pol}} (B_i - \frac{q_i}{q_V} B_V) \varepsilon_i \; (C_j - \frac{q_j}{q_V} C_V) \varepsilon_j \; .$$

Thus we have the following theorem for polarization sums:

If B_i is a 2-vector of the form:

$$\hat{B}_i = B_i - \frac{q_i}{q_V} B_V \tag{11}$$

(correspondingly for C), then

$$\sum_{\text{Pol}} \hat{B}_i \varepsilon_i \; \hat{C}_j \; \varepsilon_j = \hat{B}_i \hat{C}_i \; . \tag{12}$$

In practice the expressions encountered in transition amplitudes contain always vectors of the type (11). This can be seen already from the matrix element. If we observe, that $a_i = \hat{a}_i$ because of $a_V = 0$, most of the terms in $e_{\mu} M^{\mu}$ have already the desired form. The only exception is the contribution of the last term in

M_i, which contains ε_{ij}. Because of reflection invariance there will, however, no single factor $_{ij}$ survive in the transition amplitude, and $\varepsilon_{ij}\,\varepsilon_{kl} = \delta_{ik}\,\delta_{jl} - \delta_{ie}\,\delta_{jk}$.

Finally we want to mention, that all calculations turn out particularly simple, if we specialize to the CPPW case (II.6). There is a group theoretical reason for this fact. In the CPPW case the laser field has an additional symmetry. If we perform a space translation by an arbitrary vector $\vec{b}$, followed by a rotation about the $\vec{k}$-direction by the angle-$(\vec{k}\vec{b})$ (mod 2π), the field remains invariant. This symmetry has many consequences. For instance it can be shown, that the Fourier transform of the <u>exact</u> photon propagator

contains only three terms, which are proportional to $\delta(p-p')$ and $\delta(p-p'\pm 2k)$ respectively (for the proof see ref. [11]). Similar results can be obtained for amplitudes with more external photon lines such as

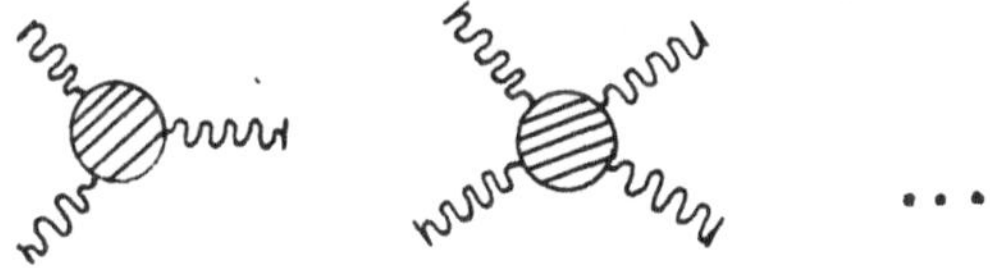

and for electron self-energy diagrams.

VI. SIMPLE PROCESSES

After this rather detailed account of calculational techniques we shall now review the results, which have been obtained so far. We shall consider at first processes, which contain only one vertex Γ_μ. An example for such a process is described by the graph

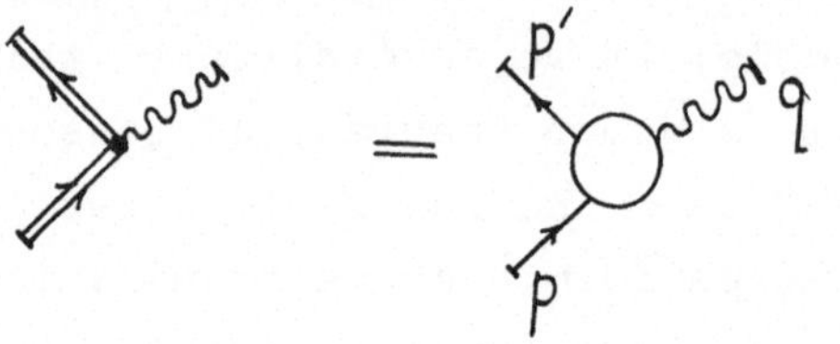

where the outgoing photon is a non-laser quantum. The graph describes a kind of Bremsstrahlung: the electron is deflected in the laser field and emits a secondary quantum. If we consider also the laser as a source of quanta, we have a Compton type process. Therefore the process is usually called high-intensity Compton scattering. The cross section has been calculated by various authors [14], [15]. We shall give the result for the CPPW case. The cross section is a sum of incoherent contributions

$$d\sigma = \sum_{r=1}^{\infty} d\sigma_r \quad . \tag{1}$$

In the laboratory system ($\vec{p} = 0,\ \theta = \sphericalangle\ \vec{k},\vec{q}$) we have

$$\frac{d\sigma_r}{d\Omega} = R_o^2 \left(\frac{\omega_q}{\omega}\right)^2 \Big[\frac{1}{r}(J_{r+1}^2 + J_{r-1}^2) - \frac{2}{r}\left(1 + \frac{1}{\nu^2}\right) J_r^2 + 2\,\frac{\omega\omega_q}{\kappa^2}\,\frac{\sin^4\theta/2}{1+\nu^2\sin^2\theta/2}(J_{r-1}^2 + J_{r+1}^2 - 2J_r^2)\Big] \quad . \tag{2}$$

Here R_o is the classical electron radius and the argument of the Bessel functions J_r is

$$\lambda = 2\nu r \frac{\sin \theta/2 \cos \theta/2}{1 + \nu^2 \sin^2 \theta/2} \quad . \tag{3}$$

The frequency of the emitted quantum ω_q is given by the formula

$$\omega_q \equiv \omega_q(r) = \frac{r\omega}{1 + (\frac{2r\omega}{\kappa} + \nu^2) \sin^2 \theta/2} \quad . \tag{4}$$

For r = 1 and $\nu = 0$ this yields the Compton formula. Thus we see that we have

a) an additional frequency shift due to the last term in Eqn. (4), which depends on the intensity of the laser (and on the polarization: for linear laser polarization we would obtain the same formula with $\nu^2 \to \nu^2/2$). The term survives in the Thomson limit, in which the second term in the denominator is neglected.

b) higher harmonics of the basic frequency for r = 2,3...

In the cross section the term $\sim \frac{\omega\omega_q}{\kappa^2}$ is a relativistic term, which is small, since the laser wave length $1/\omega$ is small compared to the Compton wave length $1/\kappa$. An expansion in powers of the argument λ is valid both for λ small (for all angles) and large (for not too small angles). In the first case we may recover the one-quantum-result for $\nu^2 = 0$ and it turns out, that $(d\sigma_r)_{NR} \sim \nu^{2r}$. The result for an infinitely extended plane laser wave of arbitrary polarization is derived in ref.[14].

In the past the existence of the intensity-dependent frequency shift has been questioned, since it was derived using plane Volkov waves. Neville and Rohrlich [16] have used packet solutions of the type (III,21) and a finite (square) laser pulse. They have shown, that Eqn. (4) holds for the centre of the wave packet, so that the result can be trusted. An experimental verification is still lacking.

Since the typical pattern encountered in Compton scattering (higher harmonics, modified conservation laws) is true for many laser processes, we shall indicate briefly, how this comes about formally. We observe, that the matrix element is proportional to

$$M \sim \delta(p_v' - p_v + q_v)\,\delta(p_i' - p_i + q_i)\,\bar{u}(p')\int du\, e^{-iw(u)}\,\varepsilon_\mu M^\mu(u).u(p) \quad .$$

For circular polarization the exponent can be written in the form

$$w(u) = -K.\xi + \lambda \sin(\xi + \rho) + \text{const.}$$

Here we have used the abbreviations

$$K = \frac{1}{\sqrt{2}\omega}\left[p_u' - p_u + q_u + \frac{\varepsilon^2 a^2}{2}\left(\frac{1}{p_v'} - \frac{1}{p_v}\right)\right]$$

$$\varepsilon a\left(\frac{p_i'}{p_v'} - \frac{p_i}{p_v}\right) = \sqrt{2}\omega\lambda(\cos\rho, \sin\rho) \ , \qquad \xi = \sqrt{2}\omega u$$

(finally λ reduces to (3) in the laboratory system). If we now expand according to

$$e^{i\lambda\sin(\xi+\rho)} = \sum_{n=-\infty}^{+\infty} J_n(-\lambda)e^{in(\xi+\rho)}$$

the integral on u can be performed, keeping in mind, that $\varepsilon_\mu M^\mu(u)$ is linear in $a_\pm(u) = e^{\mp i\xi}/\sqrt{2}$. We obtain δ-functions of the type $\delta(r-K)$ where r is either n or n±1. In the latter case we shift the summation index appropriately, so that the sum runs on r in all terms. The δ-function can be combined with the corresponding ones on the v- and i-components to a four-dimensional one

$$\delta(p'-p+q-k\,[r - \frac{\kappa_*^2-\kappa^2}{2}(\frac{1}{(p'k)} - \frac{1}{(pk)})]) \quad . \tag{5}$$

For the transition amplitude we have to form $M\bar{M}$. If the summation indices in M and $\bar{M}$ differ, the product of the corresponding δ-functions vanishes, so that we retain only one summation (that the sum has to be performed only on positive values of r can be inferred from K > 0). Each summand contains a product of two Bessel functions, whose indices differ from r at most by ±1 (cf. Eqn.(2)). For the square of the δ-function we write as usual $\delta^2 \to \frac{V}{(2\pi)^4}\delta$, where V is the four-dimensional interaction volume, which drops out in the cross section. Thus the result (2) becomes plausible.

In analogy to the usual Compton effect the frequency formula (4) can be derived from the modified energy-momentum balance expressed by the δ-function (5), which is closely related to the Zeldovich quasi-levels. We introduce new momentum vectors

$$\tilde{p}^{\mu} = p^{\mu} - (\ell - \frac{\kappa_*^2 - \kappa^2}{2(pk)})k^{\mu}, \quad \tilde{p}'^{\mu} = p'^{\mu} - (\ell' - \frac{\kappa_*^2 - \kappa^2}{2(p'k)})k^{\mu} \quad . (6)$$

Here ℓ, ℓ' are integers characterizing quasi-levels, since we have

$$(\tilde{p} + \ell k)^2 = (\tilde{p}' + \ell' k)^2 = \kappa_*^2 \quad . \quad (7)$$

The δ-function (5) amounts to

$$\tilde{p}' + q = \tilde{p} + rk \quad (8)$$

with $r = \ell' - \ell$, showing clearly, that a transition between quasi-levels takes place in the process. The frequency formula is then obtained using the mass-shell conditions (7) and $q^2 = 0$.

The same argumentation can be used also for graphs, in which q is not on shell. It has to be observed, however, that then K may not always be positive, so that the range of the summation has to be investigated in detail.

Some processes of similar nature have been investigated in this fashion. We give only a brief account. If the particle has in addition an anomalous magnetic moment μ (which is described phenomenologically by a Pauli term in the Dirac equation), the frequency formula can be obtained from (4) by replacing r by $r_{(i)}$, where i = 0,1,2 and

$$\left.\begin{array}{l} r_o = r \\ r_1 \simeq r - 2g^2 \end{array}\right\} \; r = 1,2,3,\ldots.$$

$$r_2 \simeq r + 2g^2, \quad r = 0,1,2,3\ldots$$

with $g^2 = (\nu\frac{\mu}{2})^2$. Thus the scattered frequencies appear in triplets of very small separation, since g^2 is a small parameter: for electrons $\mu \sim \frac{\alpha}{2\pi}$ is small; for nucleons ν^2 is small because of the large mass. The magnetic contributions are relativistic terms, which add to the last term in (2). A very particular feature is the term with r = 0. Here the scattered quantum has a very low (radio) frequency, since g^2 is small. Unfortunately the corresponding contribution to the cross section is extremely small, so that magnetic Compton scattering is not a very practical method to measure anomalous moments. For the details we refer to [17].

Pair creation by γ-rays interacting with a laser field is described by the graph

and has been discussed by various authors [15], [18]. The corresponding one-quantum process has not been observed because the available photon beams are not intense enough. We shall discuss the results for laser fields later.

If the external photon line is replaced by an external (non-laser-) field

we obtain high intensity corrections to particle scattering by that field resp. to "internal" pair creation in first Born ápproximation.

For Coulomb scattering the cross section has been investigated in refs. [19] and [20]. The specific pattern common to all these processes, i.e. the appearance of an infinite sum in the cross section, is also observed here. In [19] the Volkov solution (for elliptic polarization) is used and the various harmonic terms are discussed. For $r \neq 0$ one could speak of "stimulated Bremsstrahlung", since the electron emits ($r > 0$) or absorbs ($r < 0$) laser photons. The details are quite complicated and we have to refer to the original paper [19].

An interesting fact has been observed by Ehlotzky [20] for the lowest harmonic $r = 0$ (the author has used the Schrödinger equation and circular laser polarization). The Rutherford formula becomes modified by an intensity-dependent term, which is asymmetric with respect to the polarization of the laser, i.e. the cross section for right hand polarization $d\sigma^{(+)}$ differs from the corresponding one for left hand polarization $d\sigma^{(-)}$. The asymmetry parameter is

$$\frac{d\sigma^{(+)} - d\sigma^{(-)}}{d\sigma^{(+)} + d\sigma^{(-)}} = \frac{\nu^2}{\beta} \sin\theta$$

(θ = scattering angle, $\beta = v/c$), if the propagation direction $\vec{n}$ is parallel to the plane formed by the electron momenta $\vec{p}$, $\vec{p}'$ and is zero if $\vec{n}$ is orthogonal to that plane. The asymmetry should be measurable for high intensity lasers.

Pair creation in a Coulomb field has been considered in ref. [21]. From the conservation laws (expressed in terms of the quasi-momenta (6)) one may see, that pair creation is only possible, if r exceeds a certain value r_o, for which

$$\omega\, r_o = 2\kappa_*$$

so that the threshold is given by the effective mass (and increases therefore for growing ν^2). Since we have an energy conservation law only for the quasi-energy, the following pattern is observed: if we characterize the particles by their energies ε_+, ε_- at infinity (after the laser is switched off), the relation between ε_+ and ε_- depends on the angles between the particles and the beam and there are only specific regions of these angles, which are allowed. For the details we refer to [21].

If the external field is a magnetic field, the exact wave functions can be calculated and it is possible to obtain results valid to arbitrary order with respect to the coupling to the laser <u>and</u> the magnetic field. Pair creation by a non-laser photon q^μ has been investigated in this fashion in ref. [7]. The magnetic field B is assumed parallel to the propagation direction $\vec{n}$ of the laser wave. The quasi-level structure is characterized by two integers ℓ resp. r referring to the Landau-resp. Zeldovich structure, and by a spin variable σ, which distinguishes between spin parallel or antiparallel to the magnetic field. As a consequence the transition probability into a pair state is characterized by the quantum numbers ℓ,σ (electron) and ℓ',σ' (positron) and of course by the energies and the momenta of the particles parallel to the field direction. The transition probability is an infinite sum on $r(-\infty < r < +\infty)$. Every summand corresponds to a modified conservation law

$$p'_u + p_u - q_u = k_u(r + \Delta)$$

$$p'_v + p_v - q_v = 0 \ .$$

Here q is the momentum of the initial photon. The quantity Δ is proportional to ν^2. In absence of the magnetic field we have (cf. (5))

$$\Delta(B = 0) = \frac{\kappa_*^2 - \kappa^2}{2} \left(\frac{1}{(p'k)} + \frac{1}{(pk)}\right) \quad .$$

In contrast to this result, Δ displays a resonance structure for $B \neq 0$, i.e. Δ may become arbitrarily large, if

$$(\varepsilon B)^2 = (p'k)^2 \qquad \text{or} \qquad (\varepsilon B)^2 = (pk)^2 \quad .$$

This corresponds to a "cyclotron resonance" for a particle moving in the combined magnetic and laser field. As a consequence there are two possible ways, in which pair production can be achieved by photons of <u>low</u> energy $(\omega_q \leq 2\kappa^2)$

a) by absorption of a sufficient number s of laser quanta, so that the threshold is overcome;

b) by making the field sufficiently strong, so that Δ is large enough.

The latter process can be considered as a multiquantum process, in which laser quanta are emitted and reabsorbed in such a way, that the energy spectrum of the pair is appropriately realigned. For a detailed discussion we refer to [7].

An investigation of the magnitude of the transition probability is rather complicated. After various approximations Oleinik arrives at the result, that process (b) becomes important only for very high ν and that also the probability for (a) is rather small. His calculations refer however to very special values of the parameters involved and to small photon energies ω_q. They should be implemented by numerical analysis.

The results above indicate that one should have a closer look on the corresponding problem for a strong Coulomb field. Unfortunately nobody has given an exact solution for the motion of a particle in a Coulomb <u>and</u> laser field.

VII. HIGHER ORDER PROCESSES

Now we shall briefly review some results obtained so far for processes, in which more than one vertex v_μ is involved. We begin with <u>stimulated Compton scattering</u> described by the graphs

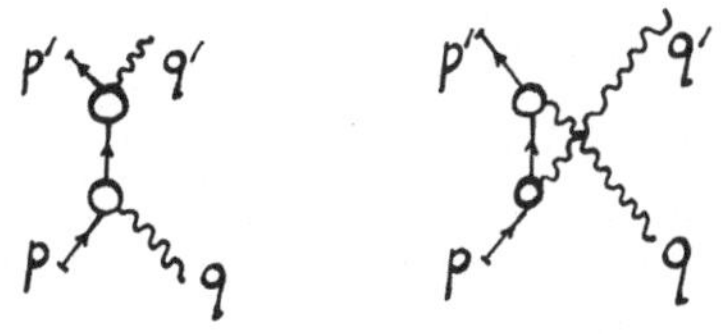

p′ q′ p q   p′ q′ p q

The calculation is due to Oleinik [2], who has (unfortunately) considered the LPPW case, which yields long formulae in general. The frequency shift formula reads in the lab system

$$\omega_q' = \frac{r\omega + \omega_q[1 + (\frac{2r\omega}{\kappa} + \frac{\nu^2}{2}) \sin^2 \theta/2]}{1 + \frac{2\omega_q}{\kappa} \sin^2 \beta/2 + (\frac{2r\omega}{\kappa} + \frac{\nu^2}{2}) \sin^2 \frac{\theta'}{2}} \qquad (1)$$

Here θ resp. θ' refer to the angles of the incoming resp. outgoing photon with the laser beam and β denotes the angle between the two photons. For $\omega_q = 0$ we obtain the formula, which corresponds to (V,4) for linear polarization. For small photon energies (all ω's $\ll \kappa$) we have

$$\omega_q' = \omega_q + r\omega + O(\nu^2) \quad . \qquad (2)$$

We conclude that r has to be positive or zero for $\omega_q < \omega$ (the latter case is only possible for $\omega_q \neq 0$), but we may have also subharmonics ($r < 0$), if $\omega_q > \omega$. If we make ω_q just a little larger than ω and take $r = -1$, the resulting frequency ω_q' is very low and we obtain optical "beat modes".

The cross section contains a product of two propagators and becomes infinite at the poles of each of them. In contrast to ordinary Compton scattering the poles can be reached for physical values of the energy variables involved. A resonance is found, whenever for a definite integer ℓ, ℓ'

$$\omega_q = \frac{\ell\omega}{1 - (\frac{2\ell\omega}{\kappa} - \frac{\nu^2}{2}) \sin^2 \theta/2} \quad \text{or} \quad \omega_q' = \frac{\ell'\omega}{1 + (\frac{2\ell'\omega}{\kappa} + \frac{\nu^2}{2}) \sin^2 \theta'/2}$$

is fulfilled. For low frequencies this gives (3)

$$\omega_q' - \omega_q = (\ell'-\ell)\omega + O(\nu^2) \ . \tag{4}$$

The physical interpretation is clear in terms of quasi-levels: a resonance occurs, if the frequency difference agrees approximately with the energy difference of two quasi-levels. This is in close analogy to resonance scattering of light by atoms.

The resonance infinities can be avoided, if the electron lines are corrected by self-energy insertions. Oleinik [22] has calculated the lowest order self-energy part

$\Sigma =$

approximately for low frequencies and small intensity and has shown, that the quasi-levels obtain a finite width via the imaginary part of Σ. The differential cross section is very large at resonance, but the resonances are very narrow. The total cross section may exceed the ordinary Compton cross section by several orders of magnitude.

Similar results come out for <u>Møller scattering</u>, which has been investigated by the same author [23]. The graph is

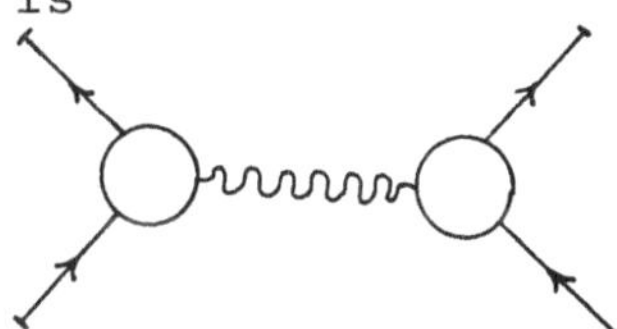

The resonances are here due to the poles of the photon propagator and are avoided, if the photon self-energy part (vacuum polarization) is taken into account. This has been done by Oleinik in an approximative way. An interesting result is, that the effective interaction potential between two electrons in the laser field, as calculated from the Møller matrix element, contains corrections to the Coulomb repulsion, which can be attractive and may under favorite conditions for the momenta of the particles even exceed the Coulomb repulsion. These conditions are, however, hard to meet. Because of the approximations, which have to be done in order to yield a manageable result it is not quite clear, whether two electrons will really be forced by the laser to form a bound pair.

VIII. RADIATIVE CORRECTIONS

Apart from the approximate calculations mentioned above no results are available for the lowest self-energy part (an accurate calculation is in progress for the CPPW case) and the vertex has not been attached at all. For the photon propagator rather detailed results have been derived for the CPPW case. We shall at first define these correction functions and shall then discuss some general results. We begin with the full Green's function G' for the electron (including all radiative corrections) resp. its inverse, defined by

$$\int d^4x'' \, G'^{-1}(x,x'')G'(x'',x') = \delta(x-x') \quad . \tag{1}$$

Then the Schwinger-Dyson equation for G' can be written as

$$G'^{-1}(x,x') = \not{\Pi}(x)\,\delta(x-x')-M(x,x') . \tag{2}$$

Here M is the mass operator, which is related to the self energy Σ by

$$M(x,x') = \kappa\delta(x-x') + \Sigma(x,x') \ . \tag{3}$$

In general Σ contains the dressed electron propagator G' as well as the vertex function $\Gamma_\mu(xy|z)$ and the dressed photon propagator D to be discussed below. To second order in ε we have

$$\Sigma(x,x') = i\ \varepsilon^2\ \gamma^\mu\ G(x,x')\gamma^\nu D^{(o)}_{\mu\nu}(x-x') + \ldots \tag{4}$$

$$= \qquad + \ldots$$

$$\Gamma_\mu(xy|z) = \gamma_\mu\delta(x-y)\delta(z-x) + i\varepsilon^2\gamma^\lambda G(x,z)\gamma_\mu G(z,y)\gamma^\nu D^{(o)}_{\lambda\nu}(x-y) + ..$$

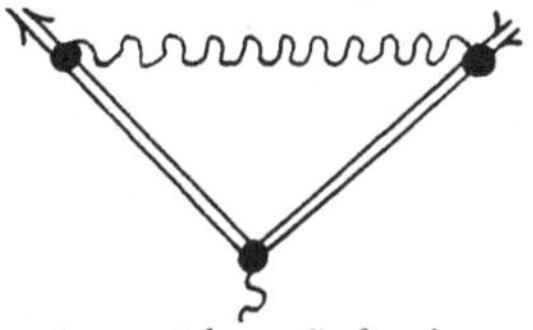

$$= \qquad + \qquad + \ldots . \tag{5}$$

For the full photon propagator the Schwinger-Dyson equation reads

$$D_{\mu\nu}(x,x') = D^{(o)}_{\mu\nu}(x-x') + \varepsilon^2\int d^4z d^4z' D^{(o)}_{\mu\rho}(x-z)\Pi^{\rho\sigma}(z,z')D_{\sigma\nu}(z',x') . \tag{6}$$

To lowest order in ε^2 the vacuum polarization tensor is given by

$$\Pi_{\mu\nu}(x,x') = -i\delta p\gamma_\mu\ G(x,x')\gamma_\nu G(x',x) + \ldots . \tag{7}$$

$$= \qquad + \ldots .$$

As far as Σ and Γ are concerned, some general results can be deduced with Ritus' method. If we represent G' in a similar fashion as G in equ. (IV,2)

$$G'(x,x') = \frac{1}{(2\pi)^4} \int d^4p\, d^4p'\, E(x|p')\, \tilde{G}'(p',p)\bar{E}(x'|p) \tag{8}$$

and use an analogous ansatz for G'^{-1} (kernel $\tilde{G}'^{-1}(p',p)$) and Σ (kernel $\tilde{\Sigma}(p',p)$) we obtain from (1)

$$\int d^4p''\, \tilde{G}^{-1}(p',p'')\, \tilde{G}(p'',p) = \delta(p'-p) \tag{9}$$

and from (2)

$$\tilde{G}'^{-1}(p',p) = (\not{p}'-\kappa)\,\delta(p'-p) - \tilde{\Sigma}(p',p)\ . \tag{10}$$

These equations are in general easier to handle, since $\tilde{\Sigma}$ is diagonal with respect to three of the four light-like momentum components, so that we have to deal with one-dimensional integral equations. To second order we have

$$\tilde{\Sigma}(p',p) = \frac{1}{(2\pi)^4} \int d^4x\, d^4y\, \bar{E}(x|p')\, \Sigma(x,y)\, E(y|p) =$$

$$= \frac{i\varepsilon^2}{(2\pi)^4} \int d^4q\, d^4q'\, v^\mu(p'q|q')\frac{1}{\not{q}-\kappa}\, v^\nu(qp|-q')\tilde{D}^{(o)}_{\mu\nu}(q')+\dots \tag{11}$$

= 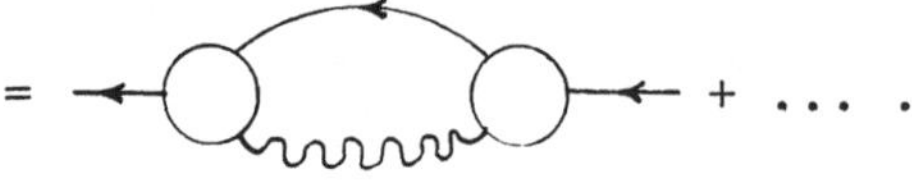 + ...

In light-like coordinates only two integrations are nontrivial (viz (V,2)) and we may read off the appearance of the factor $\delta(p_V'-p_V)\ \delta(p_i'-p_i)$. Regularization can be achieved by taking a regularized photon propagator. The divergencies are in fact only present in the free self-energy part, which is obtained by replacing v^μ (......) by the free vertex $\gamma^\mu\ \delta(........)$, and can be removed by renormalization of the first term in (10). In order to calculate the change of the Zeldovich levels one should insert this expression into (10), then solve for $\tilde{G}'$ from (9), insert the result into (8) and look for the poles of its Fourier transform. A much simpler method starts from the observation, that the unperturbed levels are in fact calculated from the Dirac equation, which may be written as

$$\int G^{-1}(x,x')\ d^4x'\ E(x'|p)u(p) = 0\ .$$

For the corrected functions we may postulate an analogous equation

$$\int G'^{-1}(x,x')\ d^4x'\ E'(x'|p)u'(p) = 0\ . \tag{12}$$

Here u' is a free Dirac spinor with corrected mass

$$(\not{p}-\kappa-\delta\kappa)\ u'(p) = 0\ . \tag{13}$$

($\delta\kappa$ is due to the free self-energy). For the corrected Bloch function we can use an ansatz, which states, that only the u-component of the momentum is changed:

$$E'(x|p) = e^{-iu\delta p_u}\ E(x|p)\ . \tag{14}$$

Then we can obtain from (12) an one-dimensional integral equation for δp_u, which can be solved by iteration. This is in fact the method, which has been used by Oleinik [22].

Now we consider the vertex part Γ_μ. A Ritus representation reads

$$\Gamma_\mu(xy|z) = \frac{1}{(2\pi)^8} \int d^4p\, d^4p'\, d^4q\, E(x|p')\, V_\mu(p'p|q)\, \bar{E}(y|p)\, e^{-i(qz)} . \tag{15}$$

The kernel function is

$$V_\mu(p'p|q) = \frac{1}{(2\pi)^4} \int d^4x\, d^4y\, d^4z\; \bar{E}(x|p')\, \Gamma_\mu(xy|z)\, E(y|p)\, e^{i(qz)} . \tag{16}$$

To second order in ε^2 we have

$$V_\mu(p'p|q) = v_\mu(p'p|q) + \frac{i\varepsilon^2}{(2\pi)^4} \int d^4r\, d^4r'\, d^4q'\, v^\lambda(p'r|q') \frac{1}{\not{r}-\kappa} v_\mu(r,r'|q) \frac{1}{\not{r}'-\kappa} v^\nu(r'p|-q') \cdot \tilde{D}^{(o)}_{\lambda\nu}(q') + \ldots \tag{17}$$

= + + ...

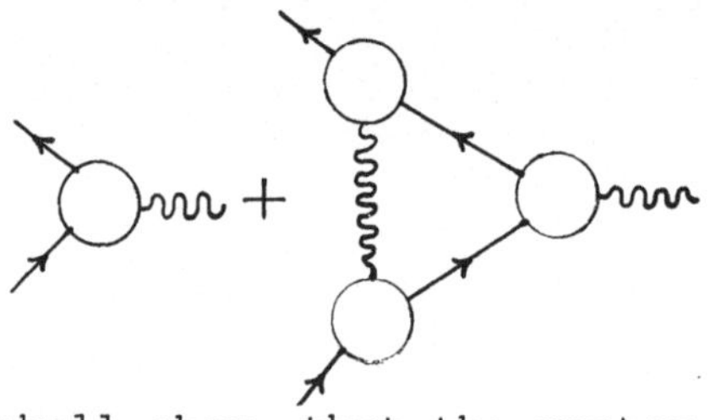

We shall show, that the vertex (15) fulfills the Ward identity to this order. Using (V,6) and (V,8) we obtain after a short calculation

$$q^{\mu}V_{\mu}(p'p|q) = \int d^4r\ [\tilde{G}'^{-1}(p',r)\ I(r,p|q) -$$

$$- I(p',r|q)\ \tilde{G}'^{-1}(r,p)] \tag{18}$$

where V and G'^{-1} are taken up to second order in ε. If we take the derivation of (15) and insert this result, we obtain (again up to order ε^2) Ward's identity

$$i\ \partial^{\mu}(z)\Gamma_{\mu}(xy|z) = (\delta(z-y)-\delta(z-x))\ G'^{-1}(x,y). \tag{19}$$

This shows, that our formalism maintains gauge invariance at least up to second order in ε and to <u>all</u> orders in A_{μ}^{ext} and extends the validity of the corresponding statements of chapter V to graphs containing the lowest radiative corrections.

Finally we shall discuss vacuum polarization. If we consider in addition to the external field a weak radiation field A'_{μ}, Maxwell's equations read

$$\partial^{\mu}(\partial_{\mu}A'_{\nu} - \partial_{\nu}A'_{\mu}) = j_{\nu}^{p}\ (A', A^{ext}) \quad . \tag{20}$$

Here j_{μ}^{p} is the polarization current caused by the external field, which polarizes the virtual electron-positron pairs in the vacuum. In terms of graphs we have

$$j_{\mu}^{p} = \varepsilon\Pi_{\mu} + \varepsilon^2\ \Pi_{\mu\nu}.A'^{\nu} + \varepsilon^3\ \Pi_{\mu\nu\rho}.A'^{\nu}.A'^{\rho} + \ldots$$

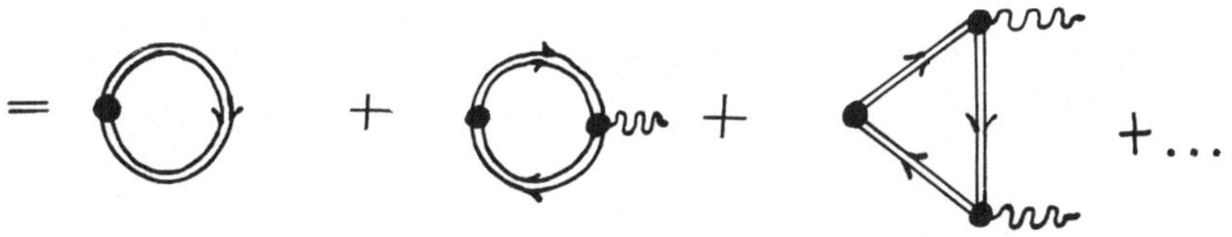

$$= \quad + \quad + \quad + \ldots \tag{21}$$

(the dot in (21) denotes folding). The diagrams with an odd number of vertices are in general different from zero, since Furry's theorem is not valid for our graphology. This is easy to understand: if we expand our vertex in powers of the coupling to the external field, we obtain e.g. for the first term

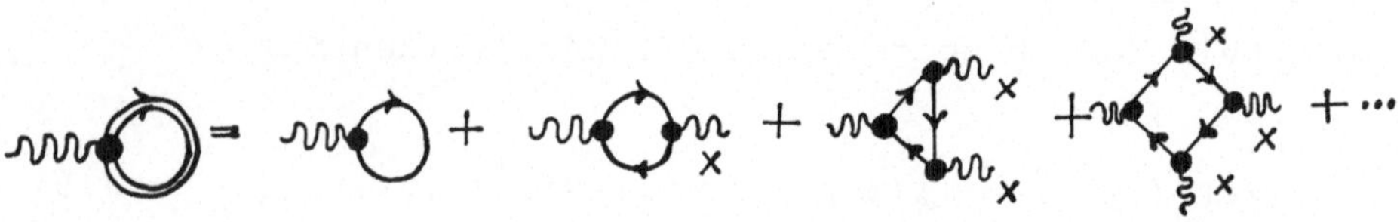

Here the first and third term is zero by Furry's theorem, whereas the other terms survive. For a Coulomb field Π contributes e.g. to the Lamb shift.

If the external field is a laser field, it can be shown exactly, that $\Pi_\mu = 0$, cf. [9], [11], [14]. The third term in (21) would describe the fission of a photon into two or the fusion of two photons. It has not yet been investigated for laser fields. We shall therefore neglect this term, which is one order higher in ε than the second one.

Instead of looking for the solution of Maxwell's equations we can also investigate the photon propagator, which is obtained from equ. (6). We shall at first outline a general method to solve this equation for given $\Pi_{\mu\nu}$ in an arbitrary external field and shall then state the results obtained for a CPPW laser field using the second order expression (7) for $\Pi_{\mu\nu}$. We start from the full momentum transform of $\Pi_{\mu\nu}$

$$\tilde{\Pi}_{\mu\nu}(p,p') = \frac{1}{(2\pi)^4}\int d^4x d^4x' \; e^{i(px)} e^{-i(p'x')} \Pi_{\mu\nu}(x,x') \qquad (22)$$

and look for three 4-vectors $\varepsilon_\mu^{(i)}(p,q)$, $\bar{\varepsilon}_\mu^{(i)}(p,q)$ and three functions $\kappa^{(i)}(q)$ $(i = 1,2,3)$ with the following properties:

(a) ε and $\bar{\varepsilon}$ are transverse

$$(p,\varepsilon^{(i)}(p,q)) = (p,\bar{\varepsilon}^{(i)}(p,q)) = 0 \tag{23}$$

(b) there is an orthogonality relation

$$\int d^4p(\bar{\varepsilon}^{(i)}(p,q),\ \varepsilon^{(j)}(p,q')) = \delta(q-q')\delta^{(i)(j)} \tag{24}$$

(c) $\varepsilon^{(i)}$ resp. $\bar{\varepsilon}^{(i)}$ are right resp. left eigenvectors with eigenvalue $\kappa^{(i)}$

$$\int d^4p'\ \frac{1}{p'^2}\tilde{\Pi}_{\mu\nu}(p,p')\varepsilon^{(i)\nu}(p',q) = \kappa^{(i)}(q)\varepsilon_\mu^{(i)}(p,q) \tag{25a}$$

$$\int d^4p\ \bar{\varepsilon}^{(i)\mu}(p,q)\tilde{\Pi}_{\mu\nu}(p,p')\frac{1}{p'^2} = \kappa^{(i)}(q)\bar{\varepsilon}_\nu^{(i)}(p',q)\ . \tag{25b}$$

In fact the Equs. (25) are integral equations for $\varepsilon,\bar{\varepsilon}$, which have to be solved together with the conditions (23), (24). Of course one will discover, that ε and $\bar{\varepsilon}$ are closely related. If the solution of this problem is obtained, we can write a bilinear expansion for

$$\tilde{\Pi}_{\mu\nu}(p,p') = p'^2\sum_{i=1}^{3}\int d^4q\kappa^{(i)}(q)\varepsilon_\mu^{(i)}(p,q)\bar{\varepsilon}_\nu^{(i)}(p',q)\ . \tag{26}$$

There is no need for a fourth pair ε, $\bar{\varepsilon}$, since $\tilde{\Pi}$ is transverse because of gauge invariance

$$p^{\mu} \tilde{\Pi}_{\mu\nu}(p,p') = \tilde{\Pi}_{\mu\nu}(p,p')p'^{\nu} = 0 . \tag{27}$$

It is then completely trivial to show, that the solution of Dyson's equation (6) in momentum space has the form

$$\tilde{D}_{\mu\nu}(p,p') = \frac{-1}{p^2+i0} \int \frac{d^4q}{1+\varepsilon^2\kappa^{(i)}(q)} \varepsilon_{\mu}^{(i)}(p,q)\bar{\varepsilon}_{\nu}^{(i)}(p',q)$$

$$+ \text{ gauge terms} . \tag{28}$$

As usual the physical properties of the "medium" vacuum + external field are contained in the analytical structure (poles, cuts) of the propagator. This means, that $\kappa^{(i)}(q)$ is the relevant quantity to be looked for. The connection of the poles of the propagator with the optical properties of the medium is exhibited by the solution of Maxwell's equations (20), which reads (if the last term in (21) is dropped, of course)

$$\tilde{A}'_{\mu}(p) = \frac{1}{p^2} \sum_{i} \int d^4q\, \varepsilon_{\mu}^{(i)}(p,q)\, \delta(1+\varepsilon^2\kappa^{(i)}(q))h^{(i)}(q) . \tag{29}$$

Here $\tilde{A}'_{\mu}(p)$ is the Fourier transform of $A'_{\mu}(x)$ and $h^{(i)}$ is an arbitrary function, which determines the shape of the wave packet. Clearly A' receives contributions from the poles of D only. The appearance of the integral on q shows a characteristic property: the modes (whose polarization vectors are the $\varepsilon^{(i)}$) are in general not monochromatic. In concrete cases $\varepsilon^{(i)}$ contains in fact only a few δ-functions, so that there are only few modes for each $\varepsilon^{(i)}$.

For a laser field the integral equations (25) can be solved easily for the CPPW case, since the equations turn out to be algebraic. A transparent form of the propagator is obtained using the transverse vectors

$$g^\mu = p^\mu - \frac{p^2}{(pk)} k^\mu = \left(\frac{2p_+p_-}{p_V} - p_u, p_V, p_+, p_-\right)$$
$$f^\mu = e_+^\mu - \frac{(p,e_+)}{(p,k)} k^\mu = \left(\frac{p_+}{p_V}, 0, e_+, 0\right) \tag{30}$$

and reads

$$D_{\mu\nu}(p,p') = \frac{\delta(p-p')}{p^2 + i0} \left(\text{gauge terms} + \frac{g_\mu g_\nu}{p^2} \frac{1}{1+\varepsilon^2 \kappa^{(1)}(p^2)}\right) +$$
$$+ \delta(p-p')\left(f_\mu f_\nu^* \frac{F(p^{(+)})}{\Delta(p)} + f_\mu^* f_\nu \frac{F(p^{(-)})}{\Delta(p^{(-)})}\right) + \tag{31}$$
$$+ \delta(p^{(+)}-p') f_\mu f_\nu \frac{\varepsilon^2 G(p^{(+)})}{\Delta(p)} + \delta(p^{(-)}-p') f_\mu^* f_\nu^* \frac{\varepsilon^2 G(p^{(-)})}{\Delta(p^{(-)})} .$$

Here we have used the abbreviations

$$p^{(\pm)} = p \pm 2k$$

$$\Delta(p) = p^{(+)2} p^2 (1+\varepsilon^2 \kappa^{(2)}(p))(1+\varepsilon^2 \kappa^{(3)}(p)) \quad . \tag{32}$$

F and G are complicated expressions, which will not be specified here (for details cf. 11). All invariants in (31) are functions of two variables $\sigma = p^2/\kappa^2$ and $\rho = {}^{2(pk)}/\kappa^2$, which are given in terms of one-dimensional integrals of Hankel functions. All results to be given below have been obtained by approximate evaluation of these integrals, which has to be done by combined analytical and numerical methods. The main physical results can be summarized in the following way:

(1) For on-shell photons the first expression in (31) becomes purely longitudinal. Clearly f_μ is related to the helicity of a photon with momentum p. Thus there are only two types of contributions

(a) no frequency change (p = p'), no helicity flip ($f\ f^*$)

(b) frequency change by ± two units k, helicity flip (ff resp. f^*f^*). This structure is due to the additional symmetry realized in the CPPW case, as has been remarked before.

(2) the first expression has a pole at $p^2 = 0$, which does however not correspond to transverse excitations (but which may influence the potential between two electrons).

(3) The other diagonal terms have <u>no</u> pole at $p^2 = 0$, but each of them has a pole in the complex plane with $p^2 \approx O(\varepsilon^2)$. The consequences for dispersion properties will be discussed below.

(4) The non-diagonal terms contain both of these poles.

Starting from the conservation theorems for p_v, p_i and imposing $p^2 = p'^2 = 0$ one would rule out the possibility of a helicity flip as described by these terms. One must not forget, however, that inside the medium $p^2 \neq 0$!

(5) There may be further poles far from the real axis, which are physically not very interesting.

(6) The threshold behaviour differs totally from the one in absence of the laser field (where $\tilde{\Pi}_{\mu\nu}$ has the pair production threshold at $p^2 = 4\kappa^2$ and is real below, complex above that value). Here we may expand $\tilde{\Pi}$ in an infinite series $\tilde{\Pi} = \sum_{n=-\infty}^{+\infty} \Pi^{(n)}$ where each term has a threshold behaviour at $(p + nk)^2 = 4\kappa_*^2$ which is due to the quasi-levels and corresponds to the absorption or emission of n laser quanta. As a whole $\tilde{\Pi}$ is complex everywhere. Of course we should not trust our results for <u>very</u> high n, since then the laser beam may be depleted, cf. sect. II.

Finally we shall discuss the dispersion properties. The solution of Maxwell's equations reads

$$\tilde{A}'_\mu(p) = \frac{1}{p^2} \sum_{i=2,3} [f_\mu \Phi^{(i)}(p)\,\delta(1 + \varepsilon^2 \kappa^{(i)}(p)) +$$

$$+ f^*_\mu \,\hat{\Phi}^{(i)}(p)\,\delta(1 + \varepsilon^2 \kappa^{(i)}(p^{(-)}))] \; . \qquad (33)$$

In general both excitations (i = 2,3) contribute to each helicity and the modes are <u>not</u> monochromatic, as usual in nonlinear optics. In practice it turns out, however, that $\Phi^{(2)} \sim \varepsilon^2 \Phi^{(1)}$ and $\hat{\Phi}^{(1)} \sim \varepsilon^2 \Phi^{(2)}$ so that the mixing $\sim O(\varepsilon^2)$ is small.

Neglecting the mixing terms we have two monochromatic modes with opposite helicity and different dispersion laws, i.e. circular double refraction and dichroism (since $\kappa^{(i)}$ is complex). Thus we may define in this approximation two complex indices of refraction by

$$p^2 = \frac{\omega_p^2}{c^2}\,[1 - (n_{(j)} + i\phi_{(j)})^2] \qquad j = 1,2 \quad .$$

The quantities $\delta_{(j)} = n^2_{(j)} - 1$ give the dispersion behaviour and $\gamma_{(j)} = 2n_{(j)}\phi_{(j)}\,\frac{\omega_p}{c}$ are the linear absorption coefficients. All these quantities depend on the parameters ν^2 and ρ. The real part δ becomes independent of ω_p, if ρ is not too large and we obtain in this case simple power laws

$$\delta_1 \sim \frac{\nu^2 \rho^2}{\omega_p^2} \sim \delta_2 \qquad \frac{\delta_1 - \delta_2}{\delta_1} \sim \rho \qquad (\rho \lesssim 1) \; .$$

Practically δ is very small. For a ND-glass laser with $\nu^2 = 1$ we have $\delta_1 \sim 7.5\text{x}10^{-15}$ so that our medium is very transparent. These results can be obtained also with the lowest approximation in a perturbation expansion with respect to the coupling to the external field (square box diagram). Thus, as far as δ is concerned, perturbation theory is good even at $\nu^2 = 1$!

For the absorption coefficients things are different. These depend on the frequency ω_p and are not so small for higher photon energies. There are also no simple power laws. For the data used above we have for the larger absorption coefficient

$\gamma \sim 0.025 \ cm^{-1}$ for $\rho = 0.5$ (25 GeV photons)

$\gamma \sim 0.25 \ cm^{-1}$ for $\rho = 1$ (50 Gev photons) .

Thus pair production should be observable for these energies and intense lasers. Perturbation theory is not reliable in this case: the square box diagram does not contribute below the lowest threshold!

The behaviour of γ and δ is much more interesting near the thresholds. Unfortunately these are at very high energies:

for $\nu^2=0.25$ at $\rho = 5$(250 GeV) resp.$\rho = 2.5$(125 GeV). The real part changes drastically at these values and we obtain dispersion curves, which look qualitatively like those of ordinary optical media, showing regions of normal and anomalous dispersion. The relation of the real and imaginary part is of course given by the Kramers-Kronig relations.

From the experimental point of view the interesting regions are utopic. If we could build intense x-ray lasers, the behaviour would, however, show up at much lower energies.

APPENDIX 1: NOTATION

For the scalar product of two 4-vectors a^μ, b^μ we shall use the notation

$(a,b) = a_\mu b^\mu = a^\mu b^\nu g_{\mu\nu}$, $(a,a) = a^2$

with the metric

$$g^{\mu\nu} = 0 \text{ f.} \mu \neq \nu,\; g^{oo} = -g^{11} = -g^{22} = -g^{33} = 1,\; \varepsilon^{o123} = +1.$$

The Dirac matrices are defined by

$$\{\gamma^\mu, \gamma^\nu\} = 2g^{\mu\nu},\; \gamma_5 := i\gamma^o\gamma^1\gamma^2\gamma^3,\; \sigma^{\mu\nu} := \frac{i}{2}[\gamma^\mu, \gamma^\nu] .$$

We shall define an adjoint of any Dirac matrix Γ by

$$\bar{\Gamma} := \gamma^o \Gamma^+ \gamma^o .$$

The charge e and the mass m of any particle will occur in the combinations

$$\varepsilon := \frac{e}{\hbar c},\quad \kappa = \frac{mc}{\hbar} \qquad (\varepsilon = -|\varepsilon| \text{ for electrons}) .$$

We use the following abbreviations

$$\Pi_\mu := i\partial_\mu - \varepsilon A_\mu^{ext},\; F_{\mu\nu} = \partial_\mu A_\nu^{ext} - \partial_\nu A_\mu^{ext}$$

$$\not{\Pi} := \gamma_\mu \Pi^\mu$$

and

$$F(x)\overleftarrow{\not{\Pi}} := (-i\partial_\mu - \varepsilon A_\mu^{ext}) F(x) \gamma^\mu .$$

For 3-vectors we use the <u>$\rightarrow$-notation</u>, thus

$p^\mu := (p^0,\vec{p})$

Latin indices are reserved for 2-vectors:

$b_i := (b_1,b_2)$

including a summation convention for the scalar product

$a_i b_i = a_1 b_1 + a_2 b_2$.

The ε-symbol in 2-space is

$$\varepsilon_{12} = -\varepsilon_{21} = 1, \; \varepsilon_{11} = \varepsilon_{22} = 0$$

$$\varepsilon_{ik}\,\varepsilon_{j\ell} = \varepsilon_{ij}\,\varepsilon_{k\ell} - \varepsilon_{i\ell}\,\varepsilon_{kj} = \delta_{ij}\,\delta_{k\ell} - \delta_{i\ell}\,\delta_{kj}$$

$$\varepsilon_{ik}\,\varepsilon_{i\ell} = \delta_{k\ell}$$

$$\varepsilon_{ik}\,\varepsilon_{ik} = 2 \; .$$

For the differential operator Π^μ the following relations are useful

$$[\pi^\mu,x^\nu] = ig^{\mu\nu}, \quad [\pi^\mu,\pi^\nu] = -i\varepsilon F^{\mu\nu}, \quad [\pi^\mu,F^{\alpha\beta}] = i\partial^\mu F^{\alpha\beta}$$

$$[\pi^\mu,\pi^2] = -2i\varepsilon F^{\mu\nu}\pi_\nu + \varepsilon\partial_\nu F^{\mu\nu}$$

$$\not\pi\cdot\not\pi = \pi^2 - \frac{\varepsilon}{2}\sigma_{\mu\nu}F^{\mu\nu}, \pi^2 = -(\Box + 2i\varepsilon A^{ext}_\mu\partial^\mu + i\varepsilon\partial_\mu A^{ext\mu} - \varepsilon^2 A^{ext}_\mu A^{ext\mu}) .$$

APPENDIX 2: LIGHT-LIKE FORMALISM

Let the vectors $\vec{n}$, $\vec{e}_i$ $(i = 1,2)$ form an orthogonal basis in 3-space

$$\vec{n}^2=1,\ \vec{e}_i\cdot\vec{e}_j=\delta_{ij},\ \vec{n}\cdot\vec{e}_i=0,\ \vec{n}\times\vec{e}_1=\vec{e}_2,\ \vec{n}\times\vec{e}_2=-\vec{e}_1,\vec{e}_1\times\vec{e}_2=-\vec{n}\ .$$

Then the 4-vectors

$$n^\mu=\frac{1}{\sqrt{2}}(1,\vec{n}),\ \hat{n}^\mu=\frac{1}{\sqrt{2}}(1,-\vec{n}),\ e_i^\mu=(0,\vec{e}_i)$$

form an orthogonal basis in 4-space, for which we have the relations

$$(n,\hat{n})=1,\ (e_i\cdot e_j)=-\delta_{ij}\ n^2=\hat{n}^2=(n,e_i)=(\hat{n},e_i)=0\ .$$

In contrast to the usually employed basis we have here two light-like basis vectors. Eventually we can also use the combinations

$$e_\pm=\frac{1}{\sqrt{2}}(e_1\pm ie_2),\ e_1=\frac{1}{\sqrt{2}}(e_++e_-),\ e_2=\frac{1}{i\sqrt{2}}(e_+-e_-)$$

(familiar from the description of circular polarization) which fulfil

$$e_+^2=e_-^2=0,\ (e_+\cdot e_-)=-1\ .$$

The metric tensor can be written as

$$g^{\mu\nu}=n^{\mu}\hat{n}^{\nu}+\hat{n}^{\mu}n^{\nu}-e_i^{\mu}e_i^{\nu}=n^{\mu}\hat{n}^{\nu}+\hat{n}^{\mu}n^{\nu}-e_+^{\mu}e_-^{\nu}-e_-^{\mu}e_+^{\nu}$$

and we refer to the coefficients of the decomposition of an arbitrary vector

$$p^{\mu}=n^{\mu}p_u+\hat{n}^{\mu}p_v+e_i^{\mu}p_i=n^{\mu}p_u+\hat{n}^{\mu}p_v+e_+^{\mu}p_-+e_-^{\mu}p_+$$

as the light-like "components" of p^{μ}

$$p^{\mu} = (p_u,p_v,p_i) = (p_u,p_v,p_+,p_-) \ .$$

We have

$$p_u = (\hat{n},p)\ ,\ p_v = (n,p)\ ,\ p_i = -(e_i,p)$$

$$p_{\pm} = -(p,e_{\pm}) = \frac{1}{\sqrt{2}}(p_1 \pm ip_2)\ ,\ p_1 = \frac{1}{\sqrt{2}}(p_+ + p_-)\ ,p_2 = \frac{1}{i\sqrt{2}}(p_+ - p_-)\ .$$

For the coordinate vector it is convenient to define

$$x_u =: v, \qquad x_v =: u$$

so that v is canonically conjugate to p_v, u to p_u. Then the gradient is

$$\partial^{\mu} = (\frac{\partial}{\partial u}, \frac{\partial}{\partial v}, -\frac{\partial}{\partial x_i}) = (\frac{\partial}{\partial u}, \frac{\partial}{\partial v}, -\frac{\partial}{\partial x_-}, -\frac{\partial}{\partial x_+})$$

and we have

$$\Pi^{\mu} = (i\frac{\partial}{\partial u} - \varepsilon A_u^{ext}, i\frac{\partial}{\partial v} - \varepsilon A_v^{ext}, \frac{1}{i}\frac{\partial}{\partial x_i} - \varepsilon A_i^{ext}) =$$

$$=(i\frac{\partial}{\partial u}-\varepsilon A_u^{ext},i\frac{\partial}{\partial v}-\varepsilon A_v^{ext},\frac{1}{i}\frac{\partial}{\partial x_-}-\varepsilon A_+^{ext},\frac{1}{i}\frac{\partial}{\partial x_+}-\varepsilon A_-^{ext}).$$

Any scalar product has the form

$$(p,q) = p_u q_v + p_v q_u - p_i q_i = p_u q_v + p_v q_u - p_+ q_- - p_- q_+$$

$$p^2 = 2p_u p_v - p_i p_i = 2(p_u p_v - p_+ p_-) \ .$$

Note however

$$(p,x) = up_u + vp_v - x_i p_i \ .$$

For 2-vectors the following formulae are useful:

$$p_i q_i = p_+ q_- + p_- q_+, \quad p_i p_i = 2p_+ p_-$$

$$p_i \varepsilon_{ik} q_k = i(p_+ q_- - p_- q_+) \ .$$

The 4-volume element is

$$d^4x = du\, dv\, d^2x$$

where we have for the 2-volume element

$$d^2x = d\mathbf{x}_1\, d\mathbf{x}_2 = i dx_+\, dx_-$$

The hypersurface element is

$$d\sigma^{\mu}(x) = (d^2x\ dv,\ d^2x\ du,\ \varepsilon_{ik}dx_k du dv) .$$

The d'Alembert operator has the form

$$\Box = \partial_{\mu}\partial^{\mu} = 2\frac{\partial^2}{\partial u \partial v} - \frac{\partial^2}{\partial x_i \partial x_i} = 2\left(\frac{\partial^2}{\partial u \partial v} - \frac{\partial^2}{\partial x_+ \partial x_-}\right) .$$

Finally we note for the <u>laser field</u> of sect. II

$$A_u^{ext} = A_v^{ext} = 0,\ A_i^{ext} = a.a_i(\xi),\ \xi = \sqrt{2}\omega.u$$

$$\Pi_{\mu}\Pi^{\mu} = \Pi^2 = -2\frac{\partial^2}{\partial u \partial v} + \frac{\partial^2}{\partial x_i \partial x_i} - 2\varepsilon a\ a_i \frac{\partial}{\partial x_i} - \varepsilon^2 a^2 a_i a_i .$$

The field tensor of the laser field reads

$$F^{\mu\nu} = af_i^{\mu\nu} .\sqrt{2}\omega\ f_i(\xi)$$

where we have

$$f_i(\xi) = \frac{da_i(\xi)}{d\xi},\ f_i^{\mu\nu} = n^{\mu}e_i^{\nu} - n^{\nu}e_i^{\mu} .$$

The dual field tensor

$$F^{*\mu\nu} = \frac{1}{2}\varepsilon^{\mu\nu\rho\sigma} F_{\rho\sigma}$$

fulfills

$$f_i^{*\mu\nu} = \varepsilon_{ij}\, f_j^{\mu\nu}$$

and we have the relations

$$f_i{}^{\mu\alpha} f_{j\alpha}{}^{\nu} = f_i^{*\mu\alpha}\, f_{j\alpha}^{*}{}^{\nu} = n^\mu n^\nu \delta_{ij}$$

$$f_i{}^{\mu\alpha}\, f_{j\alpha}^{*}{}^{\nu} = -n^\mu n^\nu \varepsilon_{ij},\; f_i{}^{\mu\nu} f_{j,\mu\nu} = f_i^{*\mu\nu}\, f_{j\mu\nu}^{*} = 0$$

$$n_\mu\, f_i{}^{\mu\nu} = n_\mu\, f_i^{*\mu\nu} = 0$$

$$\hat{n}_\mu\, f_i{}^{\mu\nu} = e_i^\nu,\; \hat{n}_\mu\, f_i^{*\mu\nu} = \varepsilon_{ij}\, e_j{}^\nu$$

$$e_{i,\mu}\, f_j{}^{\mu\nu} = \delta_{ij} n^\nu,\; e_{i,\mu}\, f_j^{*\mu\nu} = \varepsilon_{ij}\, n^\nu \quad .$$

APPENDIX 3: LIGHT-LIKE DIRAC ALGEBRA

The algebra for the light-like components of γ_μ is particularly simple. It may be used even in cases, where light-like components are not very natural. The reason for the simplicity lies in the fact, that the (u,v)-part and the (i)- or (+-)-part are separate Pauli algebras (like Diracs ρ- and σ-separation). We begin with the (u,v)-part. Here we have

$$\gamma_u^2 = \gamma_v^2 = 0, \qquad \{\gamma_u, \gamma_v\} = 2 \; .$$

Defining a matrix τ by

$$\tau = \frac{1}{2} [\gamma_v, \gamma_u], \quad \tau^2 = 1$$

we have

$$\{\gamma_u,\tau\} = \{\gamma_v,\tau\} = 0$$

and a basis for the (u,v)-part is given by $(1,\gamma_u,\gamma_v,\tau)$. Higher products are reduced to lower ones by means of

$$\gamma_v\gamma_u = 1+\tau, \quad \gamma_u\gamma_v = 1-\tau$$

$$\tau\gamma_u = -\gamma_u\tau = -\gamma_u ,$$

$$\tau\gamma_v = -\gamma_v\tau = \gamma_v$$

$$\gamma_v\gamma_u\gamma_v = 2\gamma_u, \quad \gamma_u\gamma_v\gamma_u = 2\gamma_v \quad .$$

We note in particular

$\gamma_v\Gamma\gamma_v = 0$ if Γ does not contain γ_u

$\gamma_u\Gamma\gamma_u = 0$ if Γ does not contain γ_v .

Projection operators in the (u,v)-subalgebra are

$$P_u = \frac{1}{2}\gamma_v\gamma_u = \frac{1}{2} (1 + \tau), \qquad P_v = \frac{1}{2}\gamma_u\gamma_v = \frac{1}{2}(1-\tau)$$

and we have

$$P_u + P_v = 1, \; P_u - P_v = \tau$$

$$P_u^2 = P_u, \; P_v^2 = P_v,$$

$$P_u P_v = P_v P_u = 0,$$

$$\gamma_u P_v = \gamma_v P_u = P_u \gamma_u = P_v \gamma_v = 0$$

$$\gamma_u P_u = P_v \gamma_u = \gamma_u$$

$$\gamma_v P_v = P_u \gamma_v = \gamma_v$$

$$P_u \tau = P_u, \; P_v \tau = - P_v \; .$$

Now we write down the analogous formulae for the (1,2)-resp. (+-)-part. We have

$$\gamma_1^2 = \gamma_2^2 = -1, \quad \{\gamma_i, \gamma_j\} = -2\delta_{ij}$$

$$\gamma_+^2 = \gamma_-^2 = 0 \; , \quad \{\gamma_+, \gamma_-\} = -2 \quad .$$

Defining a matrix σ by

$$\sigma = \tfrac{1}{2}[\gamma_-, \gamma_+] = i\gamma_1\gamma_2 = \tfrac{i}{2}\varepsilon_{k\ell}\gamma_k\gamma_\ell \; ,$$

we have

$$\sigma^2 = 1, \quad \{\sigma, \gamma_\pm\} = \{\sigma, \gamma_i\} = 0$$

and a basis for the (1,2)-resp. (+-)-part is formed by $(1, \gamma_i, \sigma)$ resp. $(1, \gamma_+, \gamma_-, \sigma)$. Higher products can again be reduced by means of

$$\gamma_i\gamma_j = -\delta_{ij} - i\varepsilon_{ij}\sigma, \quad \sigma\gamma_i = i\varepsilon_{ij}\gamma_j$$

$$\gamma_i\gamma_j\gamma_k = \delta_{ik}\gamma_j - \delta_{jk}\gamma_i - \delta_{ij}\gamma_k$$

$$\gamma_+\gamma_- = -1-\sigma, \ \gamma_-\gamma_+ = -1+\sigma, \ \sigma\gamma_+=-\gamma_+\sigma=\gamma_+, \sigma\gamma_-=-\gamma_-\sigma= -\gamma_-$$

$$\gamma_+\gamma_-\gamma_+ = -2\gamma_+, \ \gamma_-\gamma_+\gamma_- = -2\gamma_-$$

$$\gamma_i\gamma_k\gamma_i = 0, \ \gamma_i\sigma\gamma_i = 2\sigma$$

$\gamma_+\Gamma\gamma_+ = 0$ if Γ does <u>not</u> contain γ_-

$\gamma_-\Gamma\gamma_- = 0$ if Γ does <u>not</u> contain γ_+.

Projection operators are given by

$$P_+ = -\frac{1}{2}\gamma_-\gamma_+ = \frac{1}{2}(1-\sigma), \ P_- =-\frac{1}{2}\gamma_+\gamma_- = \frac{1}{2}(1+\sigma)$$

$$P_++P_-=1, P_+-P_-=-\sigma, P_+^2=P_+, P_-^2=P_-, P_+P_- =P_-P_+ = 0$$

$$P_-\gamma_- = \gamma_+P_- = P_+\gamma_+ = \gamma_-P_+ = 0$$

$$P_-\gamma_+ = \gamma_+P_+ = \gamma_+, \ P_+\gamma_- = \gamma_-P_- = \gamma_-, \ P_\pm\gamma_i = \gamma_i \ P_\pm$$

$$P_+\sigma = \sigma P_+ = -P_+, \ P_-\sigma =\sigma P_- = P_-$$

The commutation relations between elements of different parts are

$$\{\gamma_u,\gamma_i\} = \{\gamma_v,\gamma_i\} = \{\gamma_u,\gamma_\pm\} = \{\gamma_v,\gamma_\pm\} = 0$$

$$[\sigma,\gamma_u] = [\sigma,\gamma_v] = [\tau,\gamma_i] = [\tau,\gamma_\pm] = [\sigma,\tau] = 0$$

$$[P_u,P_\pm] = [P_v,P_\pm] = 0 \ .$$

A complete basis of the Dirac algebra can be written in the form $(1, \gamma_u, \gamma_v, \tau) \otimes (1, \gamma_+, \gamma_-, \sigma)$. In particular a convenient basis consists of the 16 elements

$$\Gamma_A \in (1,\gamma_u,\gamma_v,\gamma_i,i\tau,i\gamma_u\gamma_i,i\gamma_v\gamma_i,\sigma,\gamma_u\sigma,\gamma_v\sigma,i\tau\gamma_i,i\tau\sigma)$$

and fulfills

$$\bar{\Gamma}_A = \Gamma_A, \ \mathrm{Sp}\Gamma_A = 0 \quad \text{for} \quad \Gamma_A \neq 1 \ .$$

The usual basis elements are

$$\gamma_5 = \tau\sigma, \ \gamma_5\gamma^\mu = -n^\mu\gamma_u\sigma + \hat{n}^\mu\gamma_v\sigma + ie^\mu_i \ \varepsilon_{ij}\tau\gamma_j$$

$$\sigma^{\mu\nu} = i\gamma_u\gamma_i \ f^{\mu\nu}_i + i\gamma_v\gamma_i \ (\hat{n}^\mu e^\nu_i - \hat{n}^\nu e^\mu_i)$$

$$- i\tau \ (n^\mu\hat{n}^\nu - \hat{n}^\mu \ n^\nu) + \sigma \ e^\mu_i \ \varepsilon_{ij}e^\nu_j$$

and we have

$$\sigma = \frac{1}{2} \ \sigma_{\mu\nu}e^\mu_i \ \varepsilon_{ij}e^\nu_j, \qquad i \ \gamma_v\gamma_i = - \frac{1}{2} \ \sigma_{\mu\nu}f^{\mu\nu}_i$$

$$\sigma_{\mu\nu}f^{\mu\nu}_i \cdot \sigma_{\alpha\beta} \ f^{\alpha\beta}_j = 0 \ .$$

Traces can be computed most easily by reducing higher products to lower ones first. The trace of any product vanishes, unless it can be reduced to the type $(1,\gamma_u\gamma_v)$. $(1,\gamma_+\gamma_-)$ or $(1,\gamma_u\gamma_v)$ times a product of an even number of γ_i's. For these products we have

$$\mathrm{Sp}\ 1 = \mathrm{Sp}\ \gamma_u\gamma_v = -\ \mathrm{Sp}\ \gamma_+\gamma_- = -\ \mathrm{Sp}\ \gamma_u\gamma_v\gamma_+\gamma_- = 4$$

$$\mathrm{Sp}\ \gamma_i\gamma_j = \mathrm{Sp}\ \gamma_u\gamma_v\gamma_i\gamma_j = -\ 4\ \delta_{ij}$$

REFERENCES

1. D. V. Volkov, Zs. f. Phys. 94 (1935), 250.
2. Ya. B. Zeldovich, JETP 51 (1966), 1492.
3. W. Becker, H. Mitter, J. Phys. A7 (1974), 1266.
4. W. Becker, J. Phys. A8 (1975), 160.
5. P. J. Redmond, Journ. Math. Phys. 6 (1965), 1163.
6. V. P. Oleinik, Ukr. Fiz. Zh. 13 (1968), 1205, ibid. 14 (1969), 2076.
7. V. P. Oleinik, JETP 61 (1971), 27.
8. B. Beers, H. H. Nickle, J. Math. Phys. 13 (1972), 1592; J. Phys. A5 (1972).
9. J. Schwinger, Phys. Rev. 82 (1951), 664.
10. V. J. Ritus, Ann. Phys. (N.Y.) 69 (1972), 555.
11. W. Becker, H. Mitter, to be published.
12. H. Reiss, J. H. Eberly, Phys. Rev. 151 (1966), 1058.

13. W. Dittrich, Phys. Rev. D6 (1972), 2094, 2104.

14. L. S. Brown, T. W. B. Kibble, Phys. Rev. 133 (1964), A705.

15. N. B. Narozhnyi, A. I. Nikishow, V. J. Ritus, JETP 47 (1964), 930.

16. R. Neville, F. Rohrlich, Phys. Rev. D3 (1970), 1692.

17. W. Becker, V. Koch, H. Mitter, to be published.

18. H. Reiss, Phys. Rev. Lett. 26 (1971), 1072; Phys. Rev. D6 (1972), 385.

19. M. M. Denisow, M. V. Fedorow, JETP 53 (1967), 1340.

20. F. Ehlotzky, Acta phys. Austr. 31 (1970), 18, 31; Nuovo Cim. 69B (1970), 72.

21. V. P. Yakovlew, JETP 49 (1966), 318.

22. V. P. Oleinik, JETP 53 (1967), 1997.

23. V. P. Oleinik, JETP 52 (1967), 1049.

Acta Physica Austriaca, Suppl. XIV, 469—470 (1975)

ANALYTIC EXTRAPOLATION TECHNIQUES AND STABILITY PROBLEMS IN DISPERSION RELATION THEORY[x)]

by

S.CIULLI, C.POMPONIU[+)] and I.SABBA STEFANESCU[++)]
Institute for Atomic Physics, P.O.Box 35
Bucharest, Romania

The point we try to make is that in an indirect science like elementary particle physics, it is not sufficient to have a specific description of the world brought by some happy inspiration, but rather it is necessary to optimize among large classes of (preferably among all) possible logically equivalent "revelations". Indeed, although the leading concepts of which every description of nature makes use should bear a very close relation to the experimentally accessible data, in those situations when the basic laws are inherited from other fields, their concepts may prove to be very remote from experiment, and to "measure" them one might have to go through wildly unstable inverse problems (ill posed

---

x) Abstract of Lecture given at XIV. Internationale Universitätswochen für Kernphysik, Schladming, Austria, February 24-March 7, 1975.

+) Now at Institut de Physique Theorique, Dorigny, 1015 Lausanne, Switzerland.

++) Now at Institut für Experimentelle Kernphysik des Kernforschungszentrums, Karlsruhe, West Germany.

problems in the Hadamard sense). Moreover, the instabilities of the inverse laws become especially dangerous when the corresponding "direct laws" are too smooth, as it happens in particle physics whenever we try to cling to classical concepts (Lagrangians, interaction terms, etc.) which were purposedly chosen to produce "good" classical physics laws.

To cope with this situation, one has first to introduce some new (experimentally or theoretically measurable) quantities, extraneous to the original (inherited) theory, with the purpose of delimiting some compact sets inside which stable solutions of the inverse problems can be sought. Then, once the whole problem has been stabilized, there appears a second reason for a rational strategy of concepts and approaches, since mathematically equivalent descriptions are often rendered inequivalent by the ever existing regions where experimental knowledge lacks or is incomplete. As we try to argue in chapter 3, this breaking of tautologies is due to the fact that the randomness of ignorance destroys just those delicate mathematical properties (like, for instance, absolute analyticity of the input) which had rendered the methods equivalent in the ideal case of total knowledge. Therefore it is of practical relevance to find among all the previously tautological methods, that one which is least affected by our limited amount of knowledge.

This paper treats the incidence of these questions in some theoretical and phenomenological problems of particle physics, in which analytic continuation is used at least as an intermediate step.

Acta Physica Austriaca, Suppl. XIV, 471 — 488 (1975)

SOURCE THEORY VIEWPOINTS IN DEEP INELASTIC SCATTERING

by

J. SCHWINGER*)
University of California
Los Angeles, Calif. 90024

The phenomena of deep inelastic electron scattering on single nucleons, viewed as virtual photon absorption, are reexamined with the non-speculative, phenomenological attitude of source theory. The use of a double spectral representation for forward scattering, and of experimental inputs from the low energy resonance region and the high energy real photon diffraction region, lead naturally to the general observed characteristics of deep inelastic scattering. A reasonably successful description of deviations from simple scaling behavior is also presented.

No experiment in the recent history of high energy physics has had more impact on the theoretical community

---

*) Lecture given at XIV. Internationale Universitätswochen für Kernphysik, Schladming, Austria, February 24 - March 7, 1975.

at large than the deep inelastic scattering experiment of the MIT-SLAC collaboration (1). Very high energy electrons are inelastically scattered off individual nucleons, resulting in the production of various nucleonic excited states, or resonances. It is found that, with increasing inelasticity, this resonance structure very quickly blends into a smooth pattern that shows a remarkably simple, scaling, dependence upon the two independent variables of the experiment, which measure the energy and invariant momentum transfer to the nucleonic system. It was the emergence of this scaling behavior that set off an orgy of speculative model building and abstract theorizing,[+)] which has raged unchecked until quite recent experiments on hadronic production in electron-positron collisions dealt a body blow to the confident (but discordant) predictions that accompanied the various speculative viewpoints. Perhaps the time is now propitious for a reassessment of the situation, one that focuses more on correlating experimental facts and less on the urge to speculate about the ultimate constituents of matter. The systematic evolution of particle physics that is based on the epistomological attitude of the last sentence is known as source theory (3). Although it arose in response to the continuing crisis in high energy physics, the major attention for some time has been given to honing its blade on the whetstone of electrodynamics. Here we begin to wield this weapon in the arena for which it was forged.

An inelastic transition of the electron, with the space-like momentum change q^{μ}, creates an electromagnetic

---

[+)] For an instructive exposition of the predominant viewpoints in these matters, together with some citations of the literature, see reference (2).

field $F^{\mu\nu}(q)$ that interacts with the nucleon, of initial time-like momentum p^{μ}. The experiments with which we are concerned work with unpolarized nucleons, and do not detect individual hadronic components of the final state. A useful approach to the evaluation of the desired total transition rate is through the consideration of forward scattering, by the nucleon, of the "photon" of momentum q. The probability for the persistence of this two-particle state, of individual momenta q and p, is diminished below unity by just the required total transition probability. A convenient gauge invariant expression for the forward scattering probability amplitude is given by

$$1 + i\, V\, d\omega_p\, 4e^2m^2 \,\{-\tfrac{1}{2}\, F^{\mu\nu}(-q)\, F_{\mu\nu}(q)\, H_1(q^2,qp) -$$

$$-\tfrac{1}{2}F^{\mu\nu}(-q)(g_{\nu\lambda} + m^{-2}p_\nu p_\lambda)F^{\lambda\kappa}(q)(g_{\kappa\mu}+m^{-2}p_\kappa p_\mu)H_2(q^2,qp)\}, \tag{1}$$

where V represents the space-time interaction volume, e is the electric charge quantum in rationalized units, m is the nucleon mass, and $H_{1,2}(q^2, qp)$ are two functions of the scalar variables that can be formed from the vectors q and p. The latter are constructed with the aid of the metric tensor $g_{\mu\nu}$, which is such that $p^2 = -m^2$, $q^2 > 0$. Since the nucleon is unpolarized, it is completely characterized by its momentum p, which appears in the invariant momentum space measure

$$d\omega_p = \frac{(d\underset{\sim}{p})}{(2\pi)^3}\,\frac{1}{2p^0}\,, \qquad p^0 = (\underset{\sim}{p}^2 + m^2)^{1/2}\,, \tag{2}$$

and in the second of the two possible gauge invariant combinations through the projection tensor $g_{\mu\nu} + m^{-2}p_\mu p_\nu$, which selects vectors orthogonal to p. The functions $H_{1,2}$ also reflect, in their even dependence upon the variable qp, the symmetry between q and -q of the field structure in (1).

The expression of the fields in terms of the vector potential A converts (1) into

$$1 + i\, V\, d\omega_p\; 4e^2\, A^\mu(-q)A^\nu(q)\, [T_{1\mu\nu}H_1 + T_{2\mu\nu}H_2] \;, \tag{3}$$

where the two symmetrical tensors are

$$T_{1\mu\nu} = m^2(q_\mu q_\nu - q^2 g_{\mu\nu}) \;, \quad q^\mu T_{1\mu\nu} = 0 \;, \tag{4}$$

and

$$T_{2\mu\nu} = q^2 p_\mu p_\nu - qp(q_\mu p_\nu + p_\mu q_\nu) + (qp)^2 g_{\mu\nu} +$$

$$+ m^2(q^2 g_{\mu\nu} - q_\mu q_\nu),\; q^\mu T_{2\mu\nu} = p^\mu T_{2\mu\nu} = 0 \;. \tag{5}$$

It is also useful to consider the two distinct polarizations, in the Lorentz gauge, $qA(q) = 0$. For longitudinal polarization (L), A^μ is proportional to a unit time-like vector lying in the q-p plane,

$$A^\mu(q) = \frac{q^2 p^\mu - qp\, q^\mu}{[q^2(m^2 q^2 + (qp)^2)]^{1/2}}\, A(q) \tag{6}$$

and (3) becomes

$$L\colon\; 1 + i\, V\, d\omega_p\; 4e^2\, A(-q)A(q)m^2q^2H_1 \;, \tag{7}$$

while transverse polarization (T) is similarly represented by a unit space-like vector, orthogonal to both q and p, which leads to

$$\text{T: } 1 + i\, V\, d\omega_p\, 4e^2 A(-q) A(q) m^2 [-q^2 H_1 + (q^2+\nu^2) H_2] \, . \qquad (8)$$

Here, we have introduced the symbol

$$\nu = - qp/m \quad , \qquad (9)$$

which gives the electron energy loss in the rest frame of the initial nucleon (laboratory system).

The inferred persistence probabilities, illustrated by

$$\text{L: } 1 - V\, d\omega_p\, 8e^2 A(-q)\, A(q)\, m^2 q^2\, \text{Im}\, H_1 \, , \qquad (10)$$

display the total transition probabilities in the decrement below unity. These probabilities can be expressed as cross sections, through division by the interaction volume V, and by the relative flux for the collision, namely ,+)

$$d\omega_p A(-q)\, A(q)\, 4[m^2 q^2 + (qp)^2]^{1/2} \quad . \qquad (11)$$

This yields ($\alpha = e^2/4\pi$)

$$\sigma_L = (8\pi\alpha/m\nu)[1 + (q/\nu)^2]^{-1/2}\, m^2 q^2\, \text{Im}\, H_1 \qquad (12)$$

and

$$\sigma_T = (8\pi\alpha/m\nu)[1+(q/\nu)^2]^{-1/2} m^2 [(q^2+\nu^2)\, \text{Im}\, H_2 - q^2 \text{Im}\, H_1] , \qquad (13)$$

+) Some authors use a flux definition in which the square root of Eq.(11) is replaced by $-qp - 1/2\, q^2$.

both of which must be positive quantities. The cross section forms make apparent that the H functions have the dimensions of inverse momentum to the fourth power. We also note that

$$\frac{\sigma_L}{\sigma_L + \sigma_T} = \frac{q^2}{q^2 + \nu^2} \frac{\text{Im } H_1}{\text{Im } H_2} , \tag{14}$$

and that on setting $q^2 = 0$, only σ_T survives to become the total photon cross section

$$\sigma_\gamma = (8\pi\alpha/m^2) m^3 \nu \text{ Im } H_2(q^2 = 0, -qp = m\nu) . \tag{15}$$

An important experimental fact is that, with increasing photon energy, the latter approaches an essentially constant limit, the same for proton and neutron, which is represented by

$$\nu/m >> 1: m^3 \nu \text{ Im } H_2(0, m\nu) \cong 1.2 . \tag{16}$$

Also of interest is the manner of approach of the two cross sections, $\sigma_{\gamma n}$ and $\sigma_{\gamma p}$, which can be represented approximately by

$$\nu/m >> 1: \sigma_{\gamma n}/\sigma_{\gamma p} \cong 1 - \frac{1}{4}(m/\nu)^{1/2} . \tag{17}$$

We shall now exhibit double spectral forms for the H functions. It is characteristic of source theory (3) that such spectral forms are inferred by first considering the causal propagation through the system of various excitations. Here, there are two independent inputs associated with the momentum combinations $p \pm q$, both of which are initially timelike. The resulting double spectral forms are

$$H_{1,2}(q^2,qp)=\int \frac{dM_+^2}{M_+^2}\frac{dM_-^2}{M_-^2}\frac{2\,h_{1,2}(M_+^2,\,M_-^2)}{[(p+q)^2+M_+^2][(p-q)^2+M_-^2]}\,, \tag{18}$$

where -i0 is understood in each denominator, and the crossing symmetry between q and -q is expressed by the symmetry of the real, dimensionless weight functions $h_{1,2}$ in the two quadratic mass variables. The process of space-time extrapolation that leads to (18) could be accompanied by an extrapolation of the mass spectrum that supports the weight functions. Rather than relying on specific dynamical models, we shall accommodate our views in this matter to the requirements of experiment. Comparison with experiment is also the basis for omitting possible single spectral forms, which, in this situation, involve the combination

$$\frac{1}{(p+q)^2+M^2}+\frac{1}{(p-q)^2+M^2}=2\,\frac{M^2-m^2+q^2}{[(p+q)^2+M^2][(p-q)^2+M^2]}\,. \tag{19}$$

Apart from the q^2 term in the numerator, this could be incorporated into the double spectral form. The evidence for the absence of the additional q^2 factor will come from experimental results on form factors.*)

---

*) Since it is somewhat aside from our main concern, we only remark on the fact that, in application to the forward Compton scattering amplitude for real photons (Eq.(15)):

$$f=\frac{\alpha}{2m}(2m\nu)^2 H_2(q^2=0,\ -qp=m\nu)\,,\quad \sigma_\gamma=\frac{4\pi}{\nu}\,\mathrm{Im}\,f,$$

the double spectral form can be reduced to a single spectral form, with an additive constant. This combination is such that use of the simplest information supplied by the elastic form factors (Eq. (31)) leads automatically to the correct Thompson amplitude, $-\alpha/m$, for low energy photon scattering by the proton.

Having in mind that the momentum q is absorbed, not emitted, by the nucleon, we infer from (18) that

$$\frac{1}{\pi}\mathrm{Im}\ H_{1,2} = \int \frac{dM_+^2}{M_+^2}\frac{dM_-^2}{M_-^2}\frac{\delta[(p+q)^2+M_+^2]}{(p-q)^2+M_-^2}\,2h_{1,2}(M_+^2,\ M_-^2) =$$

$$= \int \frac{dM_+^2}{M_+^2}\frac{dM_-^2}{M_-^2}\frac{\delta(q^2-2m\nu-m^2+M_+^2)}{q^2+\frac{1}{2}(M_+^2+M_-^2)-m^2}h_{1,2}(M_+^2,M_-^2). \quad (20)$$

An alternative presentation of the latter form is obtained by writing

$$\frac{1}{q^2+\frac{1}{2}(M_+^2+M_-^2)-m^2} = \int_0^\infty \frac{d\zeta}{M_+^2}\, e^{-\frac{q^2}{M_+^2}\zeta\ -\ \frac{M_+^2+M_-^2-2m^2}{2M_+^2}\zeta} \quad . (21)$$

The following definition of new dimensionless functions

$$h_{1,2}(\zeta,m^2/M_+^2) = \pi\int \frac{dM_-^2}{M_-^2}\, e^{-\frac{M_+^2+M_-^2-2m^2}{2M_+^2}\zeta}\ h_{1,2}(M_+^2,M_-^2), \quad (22)$$

then gives

$$\operatorname{Im} H_{1,2}=\frac{1}{(M_+^{\,2})^2}\int_0^\infty d\zeta\, e^{-\frac{q^2}{M_+^{\,2}}\zeta}\, h_{1,2}(\zeta,m^2/M_+^{\,2})\bigg|_{M_+^{\,2}=m^2+2m\nu-q^2} . \tag{23}$$

It should be remarked that the mass m of the argument $m^2/M_+^{\,2}$ not only refers to the nucleon mass, but also characterizes, in order of magnitude, the resonance region. What is being underscored here is the physical hypothesis that, at the level of excitation realized in these experiments, there is, as yet, no significant dependence upon any larger, hypothetical, mass parameter, which would signal the onset of entirely new physical phenomena.

Elastic scattering receives a direct discussion in terms of the two form factors $F_{1,2}(q^2)$ that enter the Dirac matrix current combination

$$\gamma^\mu F_1 - (1/2m)\sigma^{\mu\nu}\, i\, q_\nu F_2 \quad . \tag{24}$$

As defined here, $F_1(q^2 = 0)$ equals the nucleon electric charge (1 or 0, for proton and neutron, respectively), while $F_2(q^2 = 0)$ gives the appropriate anomalous magnetic moment, in nucleon magnetons. The results of this standard calculation can be stated in terms of the longitudinal and transverse cross sections

$$\sigma_{L,T}=(8\pi\alpha/m\nu)\,[1+(q/\nu)^2]^{-1/2}\,\delta\left[\frac{q^2-2m\nu}{m^2}\right]\pi\{L:G_E^{\,2},T:\frac{q^2}{4m^2}G_M^{\,2}\}, \tag{25}$$

where

$$G_E = F_1 - (q^2/4m^2)F_2, \qquad G_M = F_1 + F_2, \tag{26}$$

which is to say that

$$M_+^2 \sim m^2:\ m^2q^2\,\frac{1}{\pi}\mathrm{Im}\ H_{1,2}=\delta\left[\frac{M_+^2}{m^2}-1\right]\{G_E^2,\frac{G_E^2+(q^2/4m^2)G_M^2}{1+(q^2/4m^2)}\}\ . \tag{27}$$

The experimental data on the various elastic form factors, $G_{E,M}$, are all roughly represented by the so-called dipole function,

$$(1 + q^2/m_o^2)^{-2}, \qquad m_o = 0.9\ m\ . \tag{28}$$

In particular, the coefficients of this function in G_M, for proton and neutron, are in the ratio of the respective magnetic moments, which, in magnitude, is approximately 3:2.

For comparison with the constructions of Eqs.(2o-23), we introduce the known excitation spectrum, where the nucleon mass is isolated below the threshold of the continuum. Accordingly, for $M_+^2 \sim m^2$, we have

$$h_{1,2}(M_+^2,M_-^2)=\delta[(M_+^2/m^2)-1]\{h_{1,2}\delta[M_-^2/m^2)-1]+h_{1,2}(M_-^2/m^2)\}$$

$$h_{1,2}(\zeta,m^2/M_+^2) = \pi\ \delta[(M_+^2/m^2) - 1]\ h_{1,2}(\zeta)\ , \tag{29}$$

where

$$h_{1,2}(\zeta)=h_{1,2}+\int_{>m^2}\frac{dM_-^{\,2}}{M_-^{\,2}}e^{-\frac{M_-^{\,2}-m^2}{2m^2}\zeta}\,h_{1,2}(M_-^{\,2}/m^2). \tag{3o}$$

This yields

$$\{G_E^{\,2},\frac{G_E^{\,2}+(q^2/4m^2)G_M^{\,2}}{1+(q^2/4m^2)}\} = h_{1,2}+q^2\int\frac{dM_-^{\,2}}{M_-^{\,2}}\,\frac{h_{1,2}(M_-^{\,2}/m^2)}{q^2+\frac{1}{2}(M_-^{\,2}-m^2)} =$$

$$=\int_0^\infty d\zeta\, e^{-\frac{q^2}{m^2}\zeta}\,h'_{1,2}(\zeta)\ , \tag{31}$$

and we note that the identification of an average $\frac{1}{2}(M_-^{\,2}-m^2)$ with $m_o^{\,2}$ yields $M_- = 1.6\ m = 1.5$ GeV, which is well within the resonance region. Among other inferences from these equations, the constants $h_{1,2}$ are identified, by setting $q^2 = 0$, as the squared nucleon charge, 1 or 0, for proton and neutron (see footnote *)). The introduction of the derivative function of ζ in the last form of (31) involves the following property:

$$h_{1,2}(\zeta = 0) = h_{1,2} + \int\frac{dM_-^{\,2}}{M_-^{\,2}}\,h_{1,2}(M_-^{\,2}/m^2) = 0\ , \tag{32}$$

which is the first of a set of "superconvergence relations" that express the vanishing of the left side in Eq. (31) as $q^2/m^2 \to \infty$. Indeed, to reproduce the $(m^2/q^2)^4$ behavior of the left side for high momenta it is necessary that

$$\zeta << 1: \quad h'_{1,2}(\zeta) \sim \zeta^3 , \tag{33}$$

with the attendant consequences inferred from (3o). It is also this strong decrease of the left side of (31) for large q^2/m^2 that argues against the presence of a single spectral form, with its implied constant multiple of q^2 on the right hand side of that equation, since such a term would be difficult to compensate.

The latter remark refers to elastic form factors, of course, but the weight of evidence about inelastic form factors is that they behave in much the same manner for sufficiently large q^2, which we regard as reasonable justification for the complete omission of the single spectral form in favor of the double spectral form. That the shape of the form factors is essentially universal for sufficiently large q^2 indicates that $h_{1,2}(\zeta,m^2/M_+{}^2)$ generally contains a factor $\bar{h}_{1,2}(\zeta)$, with the characteristics indicated in Eq. (33). And, with decreasing $m^2/M_+{}^2$, the delta function spike of (29) must give way to an increasingly smooth dependence on $m^2/M_+{}^2$. Indeed one could reasonably anticipate that

$$m^2/M_+{}^2 \to 0: \quad h_{1,2}(\zeta,m^2/M_+{}^2) \to \bar{h}_{1,2}(\zeta) . \tag{34}$$

There is, however, a qualification of this statement which originates in the experimental data for $q^2 = 0$.

Let us compare the observed high energy behavior of the photo cross section, as stated in (16), but numerically rounded off to unity, for simplicity, with the construction of (23):

$$M_+{}^2/m^2 >> 1: \int_0^\infty d\zeta \; h_2(\zeta,m^2/M_+{}^2) \cong 2M_+{}^2/m^2 . \tag{35}$$

Obviously, we cannot omit the m^2/M_+^2 dependence of the left side, or, if we were to do so, setting $M_+^2/m^2 = \infty$ for consistency, we would conclude that $\int d\zeta\, \bar{h}_2(\zeta)$ does not converge at the upper limit. The inference is that (34) holds with the additional proviso that $(m^2/M_+^2)\zeta \ll 1$, whereas, in the opposite situation, $h_2(\zeta, m^2/M_+^2)$ behaves something like

$$(m^2/M_+^2)\zeta \gtrsim 1: \quad h_2(\zeta, m^2/M_+^2) \cong e^{-\frac{m^2}{2M_+^2}\zeta} . \tag{36}$$

If, for definiteness, we accept the particular exponential function (36), we can add to Eqs. (33) and (34) the further information that $[(m^2/2M_+^2)\zeta \ll 1]$

$$\zeta \gg 1: \quad \bar{h}_2(\zeta) \cong 1 . \tag{37}$$

Some evidence in favor of (36) and (37) will be adduced later.

The term deep inelastic scattering refers to the region in which both $2\nu/m$ and q^2/m^2 are large, in such a way that the ratio

$$\omega = 2m\nu/q^2 \tag{38}$$

has any value in excess of unity. The essential experimental observations about this region are the following. For both proton and neutron, the cross section ratio σ_L/σ, $\sigma = \sigma_L + \sigma_T$, is quite small, within relatively large experimental errors; the scaling behavior that is characteristic of the region is expressed by

$$\sigma = \frac{4\pi\alpha}{q^2} f(\omega) , \tag{39}$$

where $f(\omega)$ approaches a constant ~ 1 for sufficiently large ω, and vanishes as $\omega \to 1$ in a manner not inconsistent with that of $(\omega-1)^3$; the ratio $f_n(\omega)/f_p(\omega)$ decreases from unity at large ω to a somewhat uncertain limit as $\omega \to 1$.

We first remark that

$$q^2/\nu^2 = (2/\omega)^2 \ (m^2/q^2) \tag{40}$$

is small for the circumstances of interest. Accordingly if $\mathrm{Im}\, H_1$ and $\mathrm{Im}\, H_2$ have similar behaviors in the deep inelastic regions, thereby extending the similarities in the resonance region that are noted in Eq. (27), the cross section ratio (Eq.(14))

$$\frac{\sigma_L}{\sigma} \cong \frac{q^2}{\nu^2} \frac{\mathrm{Im}\, H_1}{\mathrm{Im}\, H_2} \tag{41}$$

will indeed be small, and should continue to diminish with increasing values of q^2/m^2. Then we note, using Eqs.(12) and (13), together with the smallness of q^2/ν^2, that

$$\frac{q^2}{4\pi\alpha}\ \sigma \cong 2m\nu\ q^2\ \mathrm{Im}\, H_2 = \frac{\omega}{(\omega-1)^2} \int_0^\infty d\zeta\ e^{-\frac{\zeta}{\omega-1}}\ h_2(\zeta, m^2/M_+{}^2). \tag{42}$$

The last step applies Eq. (23), with the observation that

$$M_+{}^2 = m^2 + 2m\nu - q^2 \cong (\omega-1)q^2\ . \tag{43}$$

It is at once apparent that the essential independence of the function h_2 on $m^2/M_+{}^2 << 1$, for the finite ζ that is realized by the exponential factor in (42), leads immediately to the scaling behavior recorded in (39) with the identification

$$f(\omega) = \frac{\omega}{(\omega-1)^2} \int_0^\infty d\zeta\, e^{-\frac{\zeta}{\omega-1}} \bar{h}_2(\zeta) = \frac{\omega}{\omega-1} \int_0^\infty d\zeta\, e^{-\frac{\zeta}{\omega-1}} \bar{h}_2'(\zeta) . \tag{44}$$

The second of these forms is conveniently used, first for large ω, where

$$\omega >> 1: \quad f(\omega) = \int_0^\infty d\zeta\, \bar{h}_2'(\zeta) \cong 1 , \tag{45}$$

which employs and vindicates (37), and then as $\omega \to 1$, where small values of ζ dominate and (33) is applied:

$$\omega \to 1: \quad f(\omega) \sim \frac{1}{\omega-1} \int_0^\infty d\zeta \quad e^{-\frac{\zeta}{\omega-1}} \quad \zeta^3 \sim (\omega-1)^3 . \tag{46}$$

Some additional consequences depend explicitly upon the exponential function of (36), which can be approximately joined with the exponential function exhibited in (23) to produce

$$\exp \left[- \frac{q^2 + \frac{1}{2} m^2}{M_+^2} \zeta \right] , \tag{47}$$

replacing $\exp[-\zeta/(\omega-1)]$. Thus, in application to deep inelastic scattering events for which $2m\nu >> q^2 >> \frac{1}{2}m^2$, one could reasonably extrapolate the experimental results of (17) by the substitution

$$\frac{1}{4}\left[\frac{m}{\nu}\right]^{1/2} \cong \frac{1}{2}\left[\frac{\frac{1}{2}m^2}{M_+^{\ 2}}\right]^{1/2} \to \frac{1}{2}\left[\frac{q^2}{M_+^{\ 2}}\right]^{1/2} \cong \frac{1}{2}\,\frac{1}{(\omega-1)^{1/2}} \quad , \tag{48}$$

thereby leading to

$$\omega >> 1: \quad \sigma_n/\sigma_p \cong 1 - \frac{1}{2}\,\frac{1}{(\omega-1)^{1/2}} \quad . \tag{49}$$

Indeed, a not unsatisfactory fit to the data for $\omega > 2$ is obtained in this way. The introduction of the function (47) also changes the large ω evaluation of (45) into (the first of the two forms in (44) is used here)

$$\omega >> 1: \ f(\omega) \to \frac{1}{\omega}\int_0^\infty d\zeta \ \exp\left[-\ \frac{q^2+\frac{1}{2}m^2}{\omega q^2}\ \zeta\right] = \frac{q^2}{q^2+\frac{1}{2}m^2} \ . \tag{50}$$

This gives an account of deviations from scaling behavior for small values of $q^2 \sim \frac{1}{2}m^2$, one that is in qualitative accord with the trend of the data.+) More generally, if (47) is used, and the approximations of (42) and (43) avoided, one replaces (42) by

---

+) We recall here the hypothesis that the experiments under discussion do not reach such high excitation levels, as represented by a characteristic mass $M_o >> m$, that new phenomena come into play. The significant presence of an additional variable in (23), namely $M_+^2/M_o^2 \cong (\omega-1)(q^2/M_o^2)$, would show itself through a deviation from scaling behavior for large values of q^2.

$$\frac{q^2}{4\pi\alpha}\,\sigma \cong \left[1 + \frac{q^2}{\nu^2}\right]^{1/2} \frac{2m\nu}{2m\nu + \frac{3}{2}m^2}\,\frac{q^2}{q^2 + \frac{1}{2}m^2} f(\omega_S) \;, \tag{51}$$

where

$$\omega_S = \frac{2m\nu + \frac{3}{2}m^2}{q^2 + \frac{1}{2}m^2} \;, \tag{52}$$

which combines a description of scaling deviations with the suggestion that ω_S is a better scaling variable than ω. Combinations similar to ω_S have already been used, to some advantage, in widening the scaling region (1). Here we also find that the introduction of ω_S in (49) extends the range of an acceptable fit to the data down to $\omega_S > 1.5$. We also remark, as a purely empirical observation, that the replacement of $\omega_S - 1$ by $\omega_S - \frac{3}{4}$, which does not conflict with the large ω status of (49), gives a quite respectable account of the whole range of present data: $20 > \omega > 1.14$. If the empirical formula is taken seriously, it predicts a null limit for σ_n/σ_p as the scaling variable approaches unity.

We have now seen that the general characteristics of deep inelastic scattering emerge as a reasonable interpolation between the known properties of the low energy resonance region and of the high energy diffractive region. And in doing this we have shunned the widespread practice of hanging such phenomenological correlations on the scaffolding of some speculative dynamical model. It is thereby emphasized that whatever understanding has been achieved cannot be adduced as evidence in favor of a particular model.

There is one significant point that remains to be

mentioned. The slow exponential decrease of $h_2(\zeta, m^2/M_+^2)$ that is detailed in Eq. (36) is not what one might have expected from (22) on the basis of the physical spectrum. It appears that, under the high energy circumstances of diffractive scattering, the dominant support of the double spectral weight function occurs for $M_+^2 + M_-^2 - 2m^2 \sim m^2$, implying large <u>negative</u> values of M_-^2. While we have alluded briefly to the possibility of an extrapolation of the spectrum, the danger of serious conceptual problems posed by negative M^2 requires further investigation.

I am indebted to Wu-Yang Tsai, Kimball Milton, and Lester DeRaad for helpful comments on the manuscript. This work was supported in part by the National Science Foundation.

REFERENCES

1. Friedman, J.I., and Kendall, H.W. (1972), Deep Inelastic Electron Scattering, in Annual Reviews of Nuclear Science (Annual Reviews, Inc., Palo Alto, Cal.), Vol. 22, pp. 2o3-254.

2. Feynman, R.P. (1972), Photon-Hadron Interactions (W.A. Benjamin, Reading, Mass.).

3. Schwinger, J. (197o and 1973), Particles, Sources, and Fields (Addison-Wesley, Reading, Mass.), Vols. I and II.

Acta Physica Austriaca, Suppl. XIV, 489 — 520 (1975)

VACUUM POLARIZATION, SOURCE- AND OTHERWISE+

by

W. Dittrich
Institut für Theoretische Physik
der Universität Tübingen
W. Germany

1. INTRODUCTION

Let me first review some of the history of the phenomenon called vacuum polarization (VP).

Anyone who has been lecturing on quantum electrodynamics (QED) knows that Coulomb's law is only correct under macroscopic or atomic conditions. But it ceases to be valid if the charges penetrate far inside the Compton wavelength $\frac{\hbar}{mc}$. The effective potential seen by the electron as a result of VP effects differs considerably from the pure $\frac{1}{r}$-Coulomb potential. In scattering processes as well as in bound state problems one has to take into account the existence of VP effects, most notably in the energy displacement formula for the hydrogen atom. In the nonrelativistic approximation, the hydrogenic S-states are lowered; in particular, the 2 $S_{1/2}$ level by 27 Mc/sec relative to

+ Seminar given at XIV. Internationale Universitätswochen für Kernphysik, Schladming, Austria, February 24-March 7, 1975.

the 2 $P_{1/2}$ level. Since the agreement between theory and experiment is within 0.2 Mc/sec, this constitutes direct proof of the vacuum polarization.

There are other observations which point directly to VP effects, e.g., the difference between the g-factor measurements of electron and muon.

Lately it has become a mayor industry to compute the VP in presence of external fields to all orders in the strength of the inducing field, e.g., in the case of an intense magnetic field or laser field. Only briefly, however, will I make contact with those problems. I will instead concentrate on some of the methods which are now available to attack the above mentioned problems. Here then are some of the possibilities for computing VP effects in spinor QED, namely (a) Schwinger's proper-time method[1], (b) Pauli-Villars' regularization technique[2], (c) Källén's procedure via dispersion relations[3] and finally, last not least, (d) Schwinger's source method[4].

In particular I want to elaborate on the first and last methods, underlining the technical advantages of source-versus operator field theory. It must be emphasized that the techniques employed here are really not new; however, their use (functional - as well as sourcewise) is still limited to a rather small group of people. It is therefore the purpose of these remarks to present the underlying formalisms to a wider class of physicists.

2. VACUUM POLARIZATION AND GAUGE INVARIANCE 1951[1]

Perhaps the most compact approach to our problem begins by writing down the generating functional

$$Z\{\eta,\bar{\eta}\} = \frac{1}{N}\exp\{i\bar{\eta}G_+[A]\eta\}\exp L[A], \qquad (2.1)$$

where $\eta,\bar{\eta}$ represent anti-commuting c-number sources. $A_\mu(x)$ denotes an arbitrary external electromagnetic field, and

$$G_+[A] = G_+(1-e\gamma.A\ G_+)^{-1} \qquad (2.2)$$

$$L[A] = -\ \mathrm{Tr}\ \ln\ (1-e\gamma.A\ G_+)^{-1} \qquad (2.3)$$

are the Green's function and the closed loop factor, respectively. Tr operates both on the coordinate and spinor space. The expression as given by (2.1) may be considered the mutilated form of the full QED generating functional which reads[5]

$$Z\{\eta,\bar{\eta},j\} = \exp\{\frac{i}{2}jD_+j\}\exp\{-\frac{i}{2}\ \frac{\delta}{\delta A}D_+\frac{\delta}{\delta A}\}\exp\{i\bar{\eta}G_+[A]\eta\}\frac{1}{N}\exp L[A],$$

$$A_\mu(x) = \int(dx')\ D_+(x-x')j_\mu(x')\ .$$

Clearly our reduced form for $Z\{\eta,\bar{\eta}\}$ neglects dressing effects as well as linkages between the electron's Green's function and the closed loop factor. N is a normalization constant denoting the vacuum persistence amplitude; G_+ is the free electron propagator. In short,

$$\langle 0_+|0_-\rangle^{A\neq 0}_{\eta=\bar{\eta}=0} = N, \qquad Z|_{\eta=\bar{\eta}=0} = 1,$$

therefore

$$N = \exp L[A] \equiv \exp \{i W[A]\}.$$

Thus, our generating functional reduces to

$$Z = \exp \{i\bar{\eta}\, G_+ [A]\,\eta\}$$

which can be used to derive all electrons Green's functions in presence of $A_\mu(x)$. The lowest one is the propagator

$$-\frac{1}{i}\frac{\delta}{\delta\bar{\eta}(x)}\frac{\delta}{\delta\eta(y)} Z\Big|_o = G_+(x,y|A).$$

$G_+(x,y|A)$ satisfies an inhomogeneous differential equation, namely

$$[m + \gamma\cdot(\frac{1}{i}\partial - eA)]G_+(x,y|A) = \delta(x-y) . \tag{2.4}$$

The related integral equation reads

$$G_+(x,y|A) = G_+(x-y) + \int(du)G_+(x-u)e\gamma.A(u)G_+(u,y|A) .$$

The quantity of actual interest here is the current induced in the vacuum by the external field $A_\mu(x)$:

$$<j^A_\mu(x)>=\lim_{y\to x} ie\ tr[\gamma_\mu G_+(x,y|eA)]\exp\{-ie\int_y^x d\bar{x}^\mu A_\mu(\bar{x})\},$$

where the limit is taken space-like and symmetrically. With the aid of the integral representation

$$L[A]= -\int_0^e de' \; tr \; [\int (dx)\gamma.A(x)G_+(x,x|e'A)]$$

one finds

$$\frac{\delta L \; [A]}{\delta A^{\mu}(\xi)} = i \; \langle j^A_\mu(\xi)\rangle = -e \; tr [\; \gamma_\mu \; G_+(x,x|A) \;].$$

Introducing the symbolic operator

$$\pi_\mu = \frac{1}{i}\partial_\mu - eA_\mu,$$

the Green's function equation (2.4) can be cast into an algebraic operator equation

$$(m + \gamma\pi) \; G_+ \; [A] = 1 \tag{2.5}$$

Inverting equation (2.5) yields

$$G_+[A] = \frac{1}{m+\gamma\pi} = (m-\gamma\pi) \; \frac{1}{m^2-(\gamma\pi)^2}$$

or

$$G_+ \; [A] = (m-\gamma\pi) i \int_0^\infty ds \; e^{-is(m^2-i\varepsilon)} \; e^{is(\gamma\pi)^2} \; ,$$

a proper-time parametrization.

In defining the induced current through the effect of the test field $\delta A_\mu(x)$

$$[\delta W] = \int (dx)\ \delta A_\mu(x)\ \langle j^\mu(x)\rangle$$

$$= \int (dx)\ \delta A_\mu(x)\, ie\ \mathrm{tr}[\gamma^\mu\ G_+(x,x|A)]$$

we want to exhibit $[\delta W]$ as a total differential, i.e., $[\delta W] = \delta[W]$ to within a constant. Since Tr./. = $\int(dx)$ tr./., we have

$$[\delta W] = ie\ \mathrm{Tr}\ [\gamma^\mu \delta A_\mu\ G_+[A]] =$$

$$= (ie)i \int_0^\infty ds\ e^{-ism^2}\ \mathrm{Tr}[\{\delta(\gamma A)\}(m\!\!\!/ - \gamma\pi)e^{is(\gamma\pi)^2}]$$

$$\mathrm{tr}\ [\mathrm{odd}\ \gamma\text{'s}] = 0$$

$$= -\int_0^\infty ds\ e^{-ism^2}\ \mathrm{Tr}\ [\{\delta(\gamma\pi)\}(\gamma\pi)e^{-is(\gamma\pi)^2}]$$

having used $-e\delta(\gamma A) = \delta(\gamma\pi)$.

If we then employ the relation Tr(AB) = Tr(BA) as well as

$$\delta[e^{is(\gamma\pi)^2}] = is\int_0^1 d\lambda \exp\{is(\gamma\pi)^2(1-\lambda)\}\delta[(\gamma\pi)^2]\exp\{is(\gamma\pi)^2\lambda\}$$

we obtain indeed, to within an additive constant,

$$iW[A]=L[A]=-\frac{1}{2}\int_0^\infty \frac{ds}{s}\, e^{-ism^2}\, Tr[U(s)],\quad U(s) = e^{-iHs} \tag{2.6}$$

with

$$H = -(\gamma\pi)^2 = \pi^2 - \frac{e}{2}\sigma_{\mu\nu}F^{\mu\nu}, \qquad \sigma_{\mu\nu} = \frac{i}{2}[\gamma_\mu,\gamma_\nu] \quad .$$

If we retain only the lowest order dependence of U(s) on A^2 we find

$$\begin{aligned} &Tr[U(s) - U_o(s)] = \\ &=-isTr[U_o(s)H']+i^2\int_0^s dt(s-t)Tr[U_o(s-t)H'U_o(t)H'] +\ldots \end{aligned} \tag{2.7}$$

where $$U_o(s) = e^{is\partial^2}$$

and $$H' = -e(\frac{1}{i}\partial.A + A\frac{1}{i}\partial) - \frac{e}{2}\sigma F + e^2 A^2 .$$

Hence, the first term on the right-hand-side of eq. (2.7) is given by

$$-isTr[U_o(s)H']=-isTr[e^{is\partial^2}(e^2A^2-e(\frac{1}{i}\partial.A+A\frac{1}{i}\partial)-\frac{e}{2}\sigma F)] \quad .$$

The traces are most naturally calculated in momentum representation. Using

$$\mathrm{Tr}[e^{is\partial^2}A^2] = 4\int(dp)e^{-isp^2}\int(dk)\ A_\mu(k)\ A^\mu(-k)\ ,$$

$$\mathrm{Tr}[e^{is\partial^2}(\tfrac{1}{i}\partial.A+A.\tfrac{1}{i}\partial)] = 2\ A_\mu(0)\ \int(dp)p^\mu\ e^{-isp^2} = 0\ ,$$

and

$$\int(dp)\ e^{-isp^2} = -\frac{i\pi^2}{s^2}\ ,\ s > 0\ ,$$

we find $$-is\ \mathrm{Tr}[U_o(s)H'] = -4\frac{e^2\pi^2}{s}\int(dk)A_\mu(k)A^\mu(-k)\ , \quad (2.8)$$

a result which, by itself, is not gauge invariant and leads to a divergent integral $\int_o \frac{ds}{s^2}$ in $L^{(2)}$.

The second term in $\mathrm{Tr}[U(s) - U_o(s)]$ yields the contribution

$$i^2e^2\int_o^s dt(s-t)\mathrm{Tr}[e^{i(s-t)\partial^2}(\tfrac{1}{i}\partial.A+A\tfrac{1}{i}\partial)e^{it\partial^2}(\tfrac{1}{i}\partial A+A\tfrac{1}{i}\partial)] + \quad (2.9)$$

$$+\ i^2(\tfrac{e}{2})^2\int_o^s dt(s-t)\mathrm{Tr}[e^{i(s-t)\partial^2}\sigma.F\ e^{it\partial^2}\sigma.F]\ . \quad (2.10)$$

The term (2.10) depends only on $F_{\mu\nu}$, i.e., it is manifestly gauge invariant. The evaluation of (2.9) yields after some changes of variables

$$(2.9)=-4ie^2\pi^2 \int_{-1}^{+1} dv(1-v)\int (dk)A^\mu(k)A^\nu(-k)[\tfrac{i}{2s}\delta_{\mu\nu}-\tfrac{1}{4}v^2k_\mu k_\nu]\cdot$$
$$\cdot e^{-is(1-v^2)\frac{k^2}{4}}$$

It is convenient to perform an integration by parts to rewrite the $\delta_{\mu\nu}$-term

$$\int_{-1}^{+1} dv(1-v)e^{-is(1-v^2)\frac{k^4}{4}} = 1 -is\frac{k^2}{4}\int_{-1}^{+1} dv\ v^2 e^{-is(1-v^2)\frac{k^2}{4}} \tag{2.11}$$

The contribution of the "1"-term of (2.11) is effectively

$$"1"=-4ie^2\pi^2(\tfrac{i}{s})\delta_{\mu\nu}\int(dk)A_\mu(k)A^\mu(-k)=4e^2\pi^2\tfrac{1}{s}\int(dk)A_\mu(k)A^\mu(-k),$$

which cancels the term (2.8), leaving us with a gauge invariant result.

The remainder of (2.9) is just

$$\text{rem.}(2.9)= -ie^2\pi^2\int_{-1}^{+1} dv\ v^2 e^{-is\frac{k^2}{4}(1-v^2)} (k^2\delta_{\mu\nu}-k_\mu k_\nu)A^\mu(k)A^\nu(-k). \tag{2.12}$$

If we add to this result the $\sigma.F$ contribution (2.10)

$$(2.10) = \frac{ie^2\pi^2}{2}\int_{-1}^{+1} dv\ \int(dk)F_{\mu\nu}(k)F^{\mu\nu}(-k)e^{-is\frac{k^2}{4}(1-v^2)},$$

and use the relation $2(k_\mu k_\nu - k^2\delta_{\mu\nu})A^\mu(k)A^\nu(-k) = -F_{\mu\nu}(k)F^{\mu\nu}(-k)$, the entire contribution to eq. (2.7) becomes

$$\mathrm{Tr}[U(s)]-\mathrm{Tr}[U_o(s)] = ie^2\pi^2 \int_{-1}^{+1} dv \int (dk)\,(k^2\delta_{\mu\nu}-k_\mu k_\nu)\cdot A^\mu(k)A^\nu(-k)\,e^{-is\frac{k^2}{4}(1-v^2)}\,(1-v^2) \quad ,$$

and hence to second order

$$L^{(2)} = -\frac{1}{2}\int_o^\infty \frac{ds}{s}\, e^{-ism^2}\, 2ie^2\pi^2 \int_o^1 dv\,(1-v^2) \int (dk)\,e^{-is\frac{k^2}{4}(1-v^2)}\cdot (k^2\delta_{\mu\nu}-k_\mu k_\nu)A^\mu(k)A^\nu(-k) \quad .$$

The s-integral diverges logarithmically near $s \sim 0$. We can isolate this singularity by writing

$$\int_o^\infty \frac{ds}{s}\, e^{-ism^2} \int_o^1 dv\,(1-v^2)\,e^{-is\frac{k^2}{4}(1-v^2)} =$$

$$= \frac{2}{3}\int_o^\infty \frac{ds}{s}\, e^{-ism^2} - \frac{k^2}{2}\int_o^1 dv\; v^2\left(1-\frac{v^2}{3}\right)\left[m^2 - \frac{k^2}{4}(1-v^2)\right]^{-1} \quad .$$

In terms of $M^{(2)}_{\mu\nu}(k)$, where

$$L^{(2)} = \frac{i}{2}(2\pi)^4 \int (dk)\,A^\mu(k)\; M^{(2)}_{\mu\nu}(k)\,A^\nu(-k)$$

we have, changing $s \to -is$,

$$M^{(2)}_{\mu\nu}(k) = \{-\frac{e^2}{12\pi^2}\int_0^\infty \frac{ds}{s} e^{-ism^2} + \frac{k^2e^2}{16\pi^2}\int_0^1 dv \frac{v^2(1-\frac{v^2}{3})}{[m^2+\frac{k^2}{4}(1-v^2)]}\}\cdot$$

$$[k^2\delta_{\mu\nu}-k_\mu k_\nu] \; . \qquad (2.13)$$

From the generating functional in the absence of fermion sources[5],

$$Z^{(2)}\{j\}\Big|_{\eta=\bar{\eta}=0} = \frac{1}{N} e^{L^{(2)}[\frac{1}{i}\frac{\delta}{\delta j}]} \; e^{\frac{i}{2}jD_+j}$$

$$= \frac{1}{N} e^{-\frac{i}{2}\int\frac{\delta}{\delta j}M^{(2)}\frac{\delta}{\delta j}} \; e^{\frac{i}{2}jD_+j}$$

or

$$Z^{(2)}\{j\}\Big|_{\eta=\bar{\eta}=0} = \frac{1}{N} e^{\frac{i}{2}\int j^\mu (D_+\frac{1}{1-M^{(2)}D_+})_{\mu\nu} j^\nu} \; e^{\frac{1}{2}\mathrm{Tr}\,\ln(1-M^{(2)}D_+)^{-1}}$$

one obtains the dressed propagator

$$D'_{+\mu\nu} = \frac{1}{i}\frac{\delta}{\delta j_\mu}\frac{\delta}{\delta j_\nu} Z\Big|_0 \; , \text{ i.e.}$$

$$D'_{+\mu\nu} = [D_+(1 - M^{(2)}D_+)^{-1}]_{\mu\nu}$$

corresponding to the summation over all (improper) bubbles

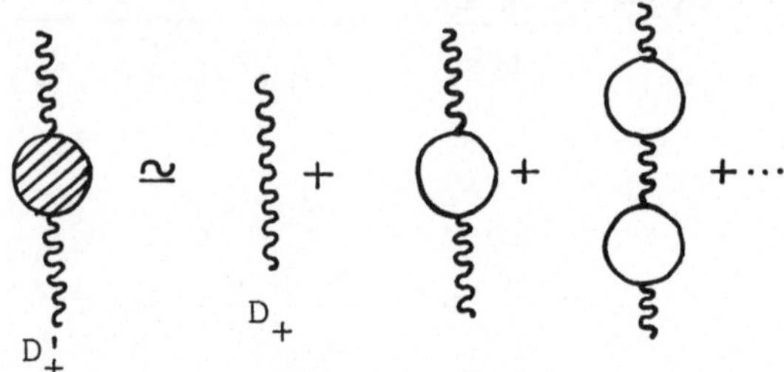

After charge renormalization the dressed photon propagator becomes (in the Landau gauge

$$D'_{+\mu\nu} = (\delta_{\mu\nu} - \frac{k_\mu k_\nu}{k^2})\ D'_+(k))$$

$$D_+^{',-1}(k) = k^2(1 - \frac{\alpha}{4\pi}\ k^2 \int_0^1 dv\ \frac{v^2(1-\frac{v^2}{3})}{[m^2+\frac{k^2}{4}(1-v^2)]})$$

or after introducing the variable M^2 via $v = (1 - \frac{4m^2}{M^2})^{1/2}$

$$D'_+(k) = \frac{1}{k^2-i\varepsilon}\ \frac{1}{1-k^2 \int_{4m^2}^{\infty} dM^2\ \frac{a(M^2)}{k^2+M^2-i\varepsilon}} \tag{2.14}$$

where

$$a(M^2) = \frac{\alpha}{3\pi}\frac{1}{M^2}\left(1 + \frac{2m^2}{M^2}\right)\left(1 - \frac{4m^2}{M^2}\right)^{1/2} .$$

By retaining only the first terms in an expansion of the inverse expression that appears in (2.14), we get

$$D'_+(k) = \frac{1}{k^2 - i\varepsilon} + \int dM^2 \frac{a(M^2)}{k^2+M^2-i\varepsilon} ,$$

a single spectral form which we want to re-derive in the next chapter using pure source techniques.

3. MODIFIED PHOTON PROPAGATION FUNCTION $\bar{D}_+$[4]

We start with the QED action

$$W = \int (dx)\,[J^\mu A_\mu + \eta\gamma^0\psi - \tfrac{1}{4}F^{\mu\nu}F_{\mu\nu} - \tfrac{1}{2}\psi\gamma^0(\gamma^\mu(\tfrac{1}{i}\partial_\mu - eqA_\mu) + m)\psi] . \quad (3.1)$$

The vacuum amplitude (V.A.) that refers to the primitive interaction is

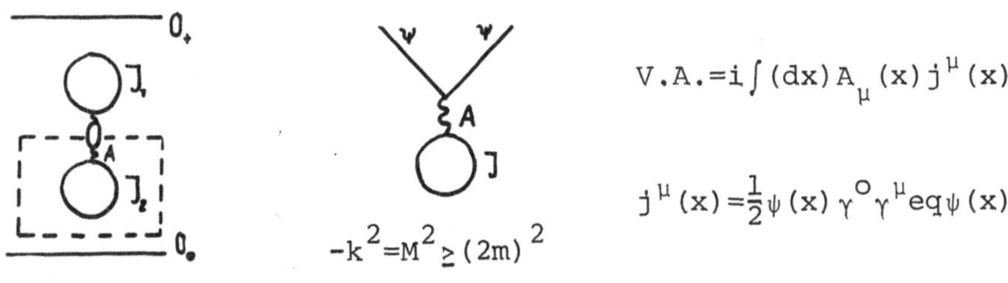

0+
J1
A
J2
0-
ψ
ψ
A
J
$-k^2 = M^2 \geq (2m)^2$
$V.A. = i\int (dx) A_\mu(x) j^\mu(x)$
$j^\mu(x) = \frac{1}{2}\psi(x)\gamma^0\gamma^\mu eq\psi(x)$

Consider the bottom part of the diagram where the external source emits a virtual photon which decays into two real oppositely charged spin 1/2 particles, e.g., e^+, e^-. $A_\mu(x)$ describes the virtual photon while $\psi(x)$ $\psi(x)$ represents the two real fermions.

Emission (and absorption) of two non-interacting particles is described by the quadratic term of

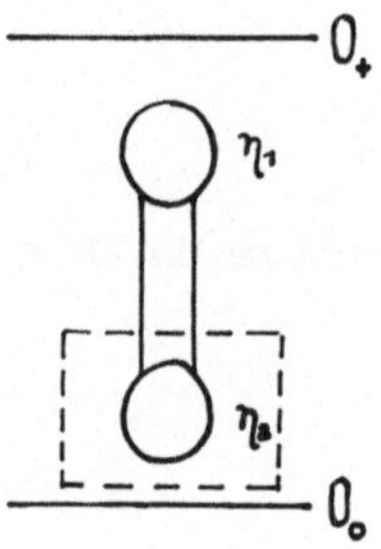

$$<0_+|0_->=e^{i\int \eta_1(x)\gamma^0 G_+(x-x')\eta_2(x')} =$$

$$= e^{i\int \psi_1(x')\gamma^0 \eta_2(x')}$$

$$V.A.=-\frac{1}{2}\int \psi_1(x)\gamma^0\eta_2(x)\eta_2(x')\gamma^0\psi_1(x)$$

wich has to be identified with

$$V.A.=\frac{i}{2}\int \psi_1(x)\gamma^0\ eq\gamma.A(x)\psi_1(x)\quad .$$

Comparison yields the effective source in emission

$$i\eta(x)\eta(x')\Big|_{\text{eff.em.}} = e\ q\ \gamma^\mu\gamma^0 A_\mu(x)\delta(x-x')$$

or

$$i\eta(x)\eta(x')\gamma^0\Big|_{\text{eff.em.}} = \delta(x-x')\ e\ q\gamma^\mu\ A_\mu(x)\quad .$$

η carries spin and charge indices, i.e. 4x2 = 8 component object.

In momentum space:

$$i\eta(p)\eta(p')\gamma^{o}\Big|_{\text{eff.em.}} = e\, q\, \gamma^{\mu}A_{\mu}(k), \tag{3.2}$$

where $k = p + p'$.

Re-writing $\exp\{i\int\eta_1(x)\gamma^{o}G_{+}(x-x')\eta_2(x')\}$ in momentum space using the projection matrix $(m-\gamma p)$ for spin 1/2 particles, i.e.,

$$\exp\{i\int\eta_1(x)\gamma^{o} i\int d\omega_p e^{ip(x-x')}(m-\gamma p)\eta_2(x')\},\ d\omega_p = \frac{1}{(2\pi)^3}\frac{d^3\vec{p}}{2p^{o}},$$

then yields

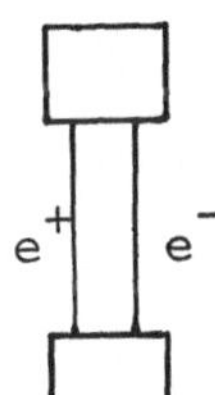

$$\exp\ \{-\int d\omega_p \eta_1(-p)\gamma^{o}(m-\gamma p)\eta_2(p)\}$$

Exchange of two particles between sources is then given by picking the quadratic term only:

$$\text{V.A.} = \tfrac{1}{2}\int d\omega_p d\omega_{p'}\eta_1(-p)\gamma^{o}(m-\gamma p)\eta_2(p)\eta_2(p')\gamma^{o}(m+\gamma p')\eta_1(-p').$$

Using

$$\eta_1(-p)_a M_{ab}\eta_1(-p')_b = -M_{ab}\eta_1(-p')_b\eta_1(-p)_a = -\text{tr}[M\eta_1(-p')\eta_1(-p)],$$

where a,b are indices for the 8-dimensional components η, the trace therefore being an 8-dimensional one.

$$V.A.=\frac{1}{2}\int d\omega_p d\omega_{p'} \operatorname{tr}[(m-\gamma p)\eta_2(p)\eta_2(p')\gamma^o(-m-\gamma p')\eta_1(-p')\eta_1(-p)\gamma^o].$$

After inserting the effective sources (3.2), we obtain

$$V.A.=-\frac{1}{2}\int d\omega_p d\omega_{p'} \operatorname{tr}[(m-\gamma p)eq\gamma A_2(k)(-m-\gamma p')eq\gamma A_1(-k)]$$

$q^2 = 1$, $\operatorname{tr}_2(q^2) = 2$; here tr_2 only in charge space:

$$q = \begin{pmatrix} o & -i \\ i & o \end{pmatrix} .$$

Inserting the unit factor $1 = (2\pi)^3 \int dM^2 d\omega_k \delta(k-p-p')$ we then obtain

$$V.A. = -e^2 \int dM^2 \; d\omega_k \; A_1^\mu(-k) \; I_{\mu\nu}(k) \; A_2^\nu(k) \quad ,$$

where

$$I_{\mu\nu}(k)=(2\pi)^3 \int d\omega_p d\omega_{p'} \delta(k-p-p') \operatorname{tr}_4[\gamma_\mu(m-\gamma p)\gamma_\nu(-m-\gamma p')] \; .$$

From Bose-Einstein statistics we infer the symmetry property $I_{\mu\nu}(k) = I_{\nu\mu}(k)$, which also can be proved by some trace properties in

$$\operatorname{tr} \; [\gamma_\mu(m-\gamma p)\gamma_\nu(-m-\gamma p')].$$

Using $\gamma_5^2 = -1$, we get

$$-\mathrm{tr}[\gamma_5\gamma_\mu\ (m-\gamma p)\gamma_\nu(-m-\gamma p')\gamma_5] = \mathrm{tr}[\gamma_\mu(-m-\gamma p)\gamma_\nu(m-\gamma p')] =$$

$$\{\gamma_5, \gamma_\mu\} = 0$$

$$= \mathrm{tr}[\gamma_\nu(m-\gamma p')\gamma_\mu(-m-\gamma p)] = \mathrm{tr}[\gamma_\nu(m-\gamma p)\gamma_\mu(-m-\gamma p')]$$

since the integral is symmetrical under exchange of $p \leftrightarrow p'$.

Therefore we have indeed $I_{\mu\nu}(k) = I_{\nu\mu}(k)$. $I_{\mu\nu}(k)$ carries a gauge invariant structure as can be seen by making a gauge transformation

$$A_\mu(k) \rightarrow A_\mu(k) + i\ k_\mu\lambda(k)$$

and demonstrating that $k^\mu\ I_{\mu\nu}(k) = 0$.

$\gamma k = \gamma(p + p')$, therefore

$$\mathrm{tr}\ [\gamma . k(m-\gamma p)\gamma_\nu(-m-\gamma p')] =$$

$$= \mathrm{tr}\{[(\gamma p+m) + (\gamma p'-m)]\ (m-\gamma p)\gamma_\nu(-m-\gamma p')\} = 0$$

because $p^2 + m^2 = 0 = p'^2 + m^2$.

The symmetrical tensor constructed from the vector k^μ is

$$I_{\mu\nu}(k) = (g_{\mu\nu} - \frac{k_\mu k_\nu}{k^2})\ I(M^2)\ .$$

To find $I(M^2)$, look at the trace of $I_{\mu\nu}(k)$:

$$I^\mu_\mu(k) = 3I(M^2) = \int d\omega_p d\omega_{p'} (2\pi)^3 \delta(p+p'-k)\, tr_4[\gamma^\mu(m-\gamma p)\gamma_\mu(-m-\gamma p')]\, .$$

Using

$$\gamma^\mu\gamma_\mu = -4, \qquad \gamma_\mu\ \gamma\cdot p\gamma^\mu = +2\gamma\cdot p$$

$$\frac{1}{4} tr\gamma_\mu\gamma_\nu = -g_{\mu\nu}, \quad tr\ \gamma_\mu = 0$$

$$k = p+p', \quad k^2 = -M^2 = -2m^2+2pp',$$

$$(2\pi)^3 \int d\omega_p d\omega_{p'} \delta(p+p'-k) = \frac{1}{(4\pi)^2} (1 - \frac{4m^2}{M^2})^{1/2}$$

we obtain

$$I(M^2) = \frac{4}{3}(M^2 + 2m^2)\ \frac{1}{(4\pi)^2} \sqrt{1 - \frac{4m^2}{M^2}}\ .$$

$$V.A. = -e^2 \int dM^2 d\omega_k A_1^\mu(-k)\, (g_{\mu\nu} + \frac{k_\mu k_\nu}{M^2})\, \frac{4}{3} (\frac{1}{(4\pi)^2} (M^2+2m^2)$$

$$\sqrt{1 - \frac{4m^2}{M^2}}\ A_2^\nu(k)\, . \qquad (3.3)$$

Since this expression is gauge invariant, we have $k_\mu \, J^\mu = 0$ after having inserted

$$A_2^\nu(k) = -\frac{1}{M^2} J_2^\nu(k), \quad A_1^\mu(-k) = -\frac{1}{M^2} J_1^\mu(-k) \; .$$

Hence

$$V.A. = i \frac{\alpha}{3\pi} \int \frac{dM^2}{M^2} \left(1 + \frac{2m^2}{M^2}\right) \sqrt{1 - \frac{4m^2}{M^2}} \; id\omega_k \, J_1^\mu(-k) J_{2\mu}(k)$$

$$J_1^\mu(-k) = \int J_1^\mu(x) e^{ikx}(dx) \qquad J_{2\mu}(k) = \int e^{-ikx'} J_{2\mu}(x')(dx')$$

$$i \int d\omega_k \; e^{ik(x-x')} = \Delta_+(x-x';M^2)$$

$$V.A. = i \frac{\alpha}{3\pi} \int \frac{dM^2}{M^2} \left(1 + \frac{2m^2}{M^2}\right) \sqrt{1 - \frac{4m^2}{M^2}} \int (dx)(dx') J_1^\mu(x) \Delta_+(x-x';M^2) \cdot J_{2\mu}(x') \tag{3.4}$$

which modifies the photon propagation function according to

$$\bar{D}_+(k) = \frac{1}{k^2 - i\varepsilon} + \frac{\alpha}{3\pi} \int_{(2m)^2}^{\infty} \frac{dM^2}{M^2} \left(1 + \frac{2m^2}{M^2}\right) \sqrt{1 - \frac{4m^2}{M^2}} \; \frac{1}{k^2 + M^2 - i\varepsilon} \tag{3.5}$$

$$= \frac{1}{k^2 - i\varepsilon} + \int dM^2 \frac{a(M^2)}{k^2+M^2-i\varepsilon}, \quad a(M^2) = \frac{\alpha}{3\pi}\frac{1}{M^2}\left(1+\frac{2m^2}{M^2}\right)\sqrt{1-\frac{4m^2}{M^2}} \ .$$

The function $a(M^2)$ is real and non-negative. For $M \gg 2m$, the integral behaves like

$$\int \frac{dM^2}{M^2}\frac{1}{k^2+M^2}$$

and there is no question about its existence from threshold up to infinity.

There are several problems where the modified propagation function can be used. Here is one example.

The dynamics has changed the original vacuum amplitude

$$\langle 0_+|0_-\rangle^J = \exp\{\tfrac{i}{2}\int (dx)(dx')J^\mu(x)D_+(x-x')J_\mu(x')\}$$

of freely propagating photons into

$$\langle 0_+|0_-\rangle^J_j = \exp\{\tfrac{i}{2}\int (dx)(dx')J^\mu(x)\bar{D}_+(x-x')J_\mu(x')\}$$

with $\bar{D}_+(x-x') = D_+(x-x') + \int dM^2 a(M^2)\Delta_+(x-x';M^2)$.

For a weak extended source this yields immediately the production for a single pair by the source

$$|\langle 0_+|0_-\rangle|^2 = 1 - \int \frac{(dk)}{(2\pi)^4} J^\mu(-k) \ \mathrm{Im}\ \bar{D}_+(k)J_\mu(k)$$

where

$$\text{Im}\ \bar{D}_+(k) = \pi a(M^2) = \frac{\alpha}{3}\frac{1}{M^2}\left(1 + \frac{2m^2}{M^2}\right)\sqrt{1 - \frac{4m^2}{M^2}},$$

$$-k^2 = M^2 > (2m)^2 .$$

The probability of producing a single pair is then

$$\frac{\alpha}{3}\int \frac{(dk)}{(2\pi)^4}\frac{1}{M^2}\left(1 + \frac{2m^2}{M^2}\right)\sqrt{1 - \frac{4m^2}{M^2}}\ J^\mu(-k)\ J_\mu(k)\ , \tag{3.6}$$

which has its counterpart in the probability P of the vacuum remaining unchanged

$$P = |N^{(2)}[A]|^2 = |e^{iW^{(2)}}|^2 = e^{-2\ \text{Im}W^{(2)}} = 1-2\ \text{Im}\ W^{(2)}+\ldots$$

$$A^\mu(k) = \frac{1}{k^2}J^\mu(k) = D_+(k)J^\mu(k)\ , \tag{3.7}$$

if the terminology of chapter 2 is used. The production probability for an electron-positron pair by the external field thus obtained is

$$2\ \text{Im}\ W^{(2)} = (3.6)$$

which is Schwinger's old result[1].

There are other examples where the modified photon propagation function is necessary, e.g., in the change of the Coulomb potential, in the calculation of the lepton g-factor anomalies, in the π^o-decay, etc.

Notice that we arrived at our various results by using the field-source relation in its first approximation (3.7). However, the action-principle can tell us what the field-source relation is to the next-order correction. First let us re-write our expression (3.3), i.e.,

$$V.A. = -e^2\int dM^2 I(M^2)\, d\omega_k A_1^\mu(-k)\left(g_{\mu\nu} + \frac{k_\mu k_\nu}{M^2}\right)A_2^\nu(k) \tag{3.8}$$

by introducing fields

$$-\frac{1}{2M^2}F_1^{\mu\nu}(-k)F_{2\mu\nu}(k) = A_1^\mu(-k)\left(g_{\mu\nu} + \frac{k_\mu k_\nu}{M^2}\right)A_2^\nu(k)$$

from

$$F_1^{\mu\nu}(-k) = (-ik^\mu A^\nu + i\, k^\nu A^\mu)_1$$

$$F_{2\mu\nu}(k) = (i\, k_\mu A_\nu - i\, k_\nu A_\mu)_2 \;.$$

Then eq. (3.8) is converted into

$$\begin{aligned} V.A. &= i\, e^2\int\frac{dM^2}{M^2}\, I(M^2)\left(-\tfrac{1}{2}\right)F_1^{\mu\nu}(-k)\, i d\omega_k\, F_{2\mu\nu}(k) \\ &= i\int dM^2 M^2\; a(M^2)\left(-\tfrac{1}{2}\right)\, F_1^{\mu\nu}(-k)\, i d\omega_k\, F_{2\mu\nu}(k)\;, \end{aligned} \tag{3.9}$$

which, when space-time extrapolated, i.e.,

$$i\int d\omega_k e^{ik(x-x')} \to \int\frac{(dk)}{(2\pi)^4}\,\frac{e^{ik(x-x')}}{k^2+M^2-i\varepsilon} + \text{contact terms}$$

removal of causal indices 1 and 2, and picking up a factor 1/2, leads to

$$\text{V.A.}= i\int\frac{(dk)}{(2\pi)^4}dM^2 a(M^2)\,(-\tfrac{1}{4})F^{\mu\nu}(-k)F_{\mu\nu}(k)\,[\frac{1}{k^2+M^2-i\varepsilon} + \text{c.t.}] \quad .$$

The complete action is given by adding the initial electromagnetic action

$$\int \frac{(dk)}{(2\pi)^4}\,[A^{\mu}(-k)J_{\mu}(k) - \tfrac{1}{4}F^{\mu\nu}(-k)F_{\mu\nu}(k)] \ .$$

To obtain the original pole structure of the freely propagating photon we have to construct the contact terms in such a way that

$$\text{for } k^2 = 0: \qquad [\frac{1}{k^2 + M^2 - i\varepsilon} + \text{c.t.}] = 0 \ .$$

The minimum choice of the c.t. terms is, in configuration space,

$$\Delta_+(x-x';M^2) - \frac{1}{M^2}\delta(x-x') = \frac{1}{M^2}\,\partial^2\Delta_+(x-x';M^2)$$

or, written in momentum space

$$\frac{1}{k^2 + M^2 - i\varepsilon} - \frac{1}{M^2} = - \frac{k^2}{M^2} \frac{1}{k^2 + M^2 - i\varepsilon} \quad .$$

Hence the modified action is given by

$$W = \int (dx) \; [J^\mu(x) \; A_\mu(x) - \tfrac{1}{4} F^{\mu\nu}(x) \; F_{\mu\nu}(x)] \; - \tag{3.10}$$

$$- \int dM^2 a(M^2) (-\tfrac{1}{4}) \int (dx)(dx') \partial^\lambda F^{\mu\nu}(x) \Delta_+(x-x';M^2) \partial'_\lambda F_{\mu\nu}(x')$$

which is no longer a local interaction.

The new field-source relation is now given by

$$\bar{A}_\mu(k) = \bar{D}_+(k) J_\mu(k)$$

where

$$\bar{D}_+(k) = \frac{1}{k^2 - i\varepsilon} \; \frac{1}{1 - k^2 \int dM^2 \dfrac{a(M^2)}{k^2 + M^2 - i\varepsilon}} \quad . \tag{3.11}$$

If $a(M^2)$ is sufficiently small, we can expand to get

$$\bar{D}_+(k) \cong \frac{1}{k^2 - i\varepsilon} + \int dM^2 \frac{a(M^2)}{k^2 + M^2 - i\varepsilon} \tag{3.12}$$

which, for single pair exchange makes use of $a(M^2) = (3.5)$ where $M > 2m$.

The exact single spectral form of the modified photon propagation function should read

$$\bar{D}_+(k) = \frac{1}{k^2-i\varepsilon} + \int dM^2 \frac{A(M^2)}{k^2+M^2-i\varepsilon} \tag{3.13}$$

with a different positive weight factor $A(M^2)$.

$A(M^2)$ is easily determined by a comparison of imaginary parts of (3.11) and (3.13) for $k^2 = M^2$, using the relation

$$\frac{1}{M'^2-M^2-i\varepsilon} = P\frac{1}{M'^2-M^2} + i\pi\delta(M'^2 - M^2) \ .$$

This yields

$$A(M^2) = \frac{a(M^2)}{[1-M^2 P\int dM'^2 \frac{a(M'^2)}{M^2-M'^2}]^2 + [\pi M^2 a(M^2)]^2}$$

which is positive.

The next (huge) step in the direction of application consists in studying the modified photon propagation function in presence of an external prescribed field to all orders in the coupling of the latter. This is the subject of Prof. Mitter's talk[6] where the external field is taken a laser field. There is also a closed-form solution for the vacuum polarization in an intense, homogeneous magnetic field.[7,8] However, nobody has ever tried to solve the problem for the familiar Coulomb potential. The problem one encounters in all those processes is perhaps most simply formulated with the aid of the vacuum persistance amplitude, which reads here

$$\langle 0_+ | 0_- \rangle^{A^{ext} \neq 0}_{j=0} = N = e^{L[\frac{1}{i}\frac{\delta}{\delta j} + A^{ext}]} \, e^{\frac{i}{2} j D_+ j}\Big|_{j=0} \tag{3.14}$$

$$A = \int D_+ j: \quad = \exp\{-\frac{i}{2}\frac{\delta}{\delta A} D_+ \frac{\delta}{\delta A}\} \exp\{L[A + A^{ext}]\}\Big|_{A=0} \quad .$$

If we expand in A but keep all orders in A^{ext}, we obtain

$$G_+[A+A^{ext}] = G_+[A^{ext}] + (\frac{\delta}{\delta A} G_+[A+A^{ext}])_{A=0} \, A + \ldots$$

$$= G_+[A^{ext}] + G_+[A^{ext}] e\gamma^\mu A_\mu G_+[A^{ext}] + \frac{1}{2} G_+[A^{ext}] e\gamma^\mu A_\mu G_+[A^{ext}] \cdot \\ \cdot e\gamma^\nu A_\nu G_+[A^{ext}] + \ldots \quad .$$

Likewise

$$L[A+A^{ext}] = L[A^{ext}] + (\frac{\delta}{\delta A} L[A+A^{ext}])_{A=0} \; A + \ldots$$

$$= -\mathrm{Tr} \ln(1 - e\gamma A^{ext} G_+)^{-1} - e \; \mathrm{tr}[\gamma_\mu G_+[A^{ext}] \cdot$$

$$\cdot A^{\mu} - \frac{e^2}{2} \mathrm{tr}[\gamma_{\mu} G_+[A^{ext}]\gamma_{\nu} G_+[A^{ext}]]A^{\mu}A^{\nu} + \ldots.$$

$$= L[A^{ext}] + i\langle j_{\mu}\rangle^{A^{ext}} A^{\mu} + M_{\mu\nu}^{A^{ext}} A^{\nu} + \ldots \equiv \sum L^{(n)} A^n.$$

Hence, terms linear, quadratic, etc. in A appear in the expansion

$$\langle O_+|O_-\rangle^{A^{ext}} = \exp\{-\frac{i}{2} \frac{\delta}{\delta A} D_+ \frac{\delta}{\delta A}\} \exp\{L^{(1)}[A^{ext}]A + L^{(2)}[A^{ext}] \cdot A^2 + \ldots\}\Big|_{A=O}$$

With the aid of formulae like

$$\exp\{-\frac{i}{2} \frac{\delta}{\delta A} D_+ \frac{\delta}{\delta A}\} \exp\{ifA\}\Big|_{A=O} = \exp\{\frac{i}{2} fD_+ f\}$$

and

$$\exp\{-\frac{i}{2} \frac{\delta}{\delta A} D_+ \frac{\delta}{\delta A}\} \exp\{\frac{i}{2} AMA\}\Big|_{A=O} = \exp\{\frac{1}{2} \mathrm{Tr}\ \ln(1-D_+M)^{-1}\}$$

we can generate the series of loops

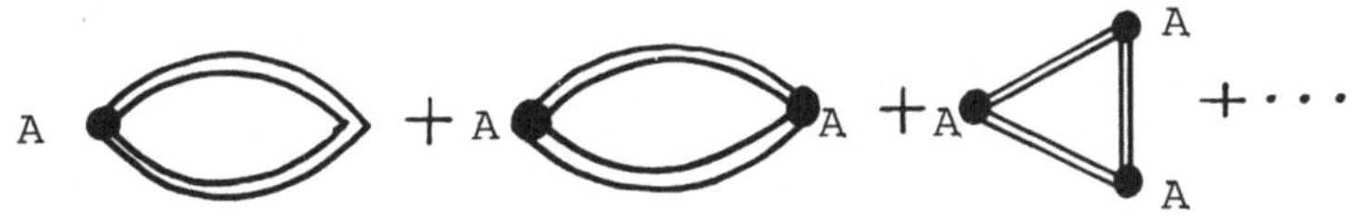

A + A A + A A A + ...

which is also reflected in

$$<j_\mu(x)>_A^{A^{ext}\neq 0} = \lim_{y\to x} i\, e\, \mathrm{tr}[\gamma_\mu\, G_+(x,y|A+A^{ext})] =$$

$$= \lim_{y\to x} i\, e\, \exp\{-ie \int_y^x d\bar{x}^\mu A_\mu^{ext}(\bar{x})\}\mathrm{tr}[\gamma_\mu G_+(x,y|A^{ext})] +$$

$$(\equiv <j_\mu(x)>^{A^{ext}})$$

$$+ \int (du)\, M_{\mu\nu}(x,u|A^{ext})\, A^\nu(u) + \ldots$$

where the vacuum polarization tensor is given by

$$M_{\mu\nu}(x,y|A^{ext}) = ie^2\, \mathrm{tr}[\gamma_\mu\, G_+(x,y|A^{ext})\gamma_\nu G_+(y,x|A^{ext})].$$

One can show that $<j_\mu(x)>^{A^{ext}}$ vanishes if A^{ext} is chosen a laser field; hence the vacuum persistence amplitude $<0_+|0_->^{A^{ext}}$ as well as $<j_\mu(x)>_A^{A^{ext}}$ start out with the ordinary second-order vacuum graph.

We can also use Schwinger's expression (2.6) to write

$$L[A+A^{ext}] = iW[A+A^{ext}] = -\frac{1}{2}\int_0^\infty \frac{ds}{s}\, e^{-ism^2}\, \mathrm{Tr}\, [e^{-iHs}]\ ,$$

which is to be computed up to order A^2, however, to all orders in A^{ext}. In the sequel we want to use the latter representation for $L\,[A + A^{ext}]$.

The Hamiltonian that governs the time development

is now given by

$$H = -(\gamma\tilde{\pi})^2 \equiv H_o + H' ,$$

with

$$H_o = -(\gamma\pi)^2 = \pi^2 - \frac{e}{2}\sigma F^{ext}, \qquad \pi_\mu = p_\mu - eA_\mu^{ext}$$

and

$$H' = -e(\pi A + A\pi) - \frac{e}{2}\sigma F + e^2 A^2 . \tag{3.15}$$

Whereas in chapter 2 the operator representing the free system was given by $U_o(s) = e^{-is(\frac{1}{i}\partial)^2}$, we have, in presence of an external field,

$$U_o(s) = e^{-isH_o} = e^{is(\gamma\pi)^2} .$$

With prescribed field A^{ext}: $U(s) = e^{-isH} \equiv e^{is(\gamma\tilde{\pi})^2}$.

Following the same steps as in chapter 2, we need to compute

$$Tr\,[U(s)] - Tr\,[U_o(s)] = -is\,Tr[U_o(s)H'] + \\ + i^2 \int_o^s dt(s-t)Tr[U_o(s-t)H'U_o(t)H'] + \ldots \quad . \tag{3.16}$$

As mentioned before, the program outlined in eqs. (3.14 - 3.16) has been carried out in ref. 7 for a strong magnetic field. For the laser field the contribution to eq.(3.16) starts out with

$$-is\ Tr[U_o(s)H']=-is\ Tr[e^{is(\gamma\pi)^2}(e^2A^2-e(\pi A+A\pi)-\frac{e}{2}\sigma F)]$$

where it is now convenient to take the Trace in configuration space:

$$Tr\ [e^{is(\gamma\pi)^2}\ A^2] = \int(dx)tr_4\ [\langle x(s)|x\rangle\ A^2(x)]\ .$$

If one uses

$$\langle x(s)|x'\rangle=C(x,x')\frac{1}{s^2}\exp\{i\frac{(x-x')^2}{4s}\}\exp\{-is\frac{1}{\xi-\xi'}\cdot$$

$$\int_{\xi'}^{\xi} d\eta[e^2A^{ext2}-eF_i^{ext}\ \frac{1}{2}\sigma f^i]\}\ \exp\{is(\frac{1}{\xi-\xi'}\int_{\xi'}^{\xi} d\eta eA^{ext})^2\}$$

$$C(x,x') =-i(4\pi)^{-2}\exp\{ie\int_{x'}^{x} d\bar{x}^{\mu}A_{\mu}^{ext}(\bar{x})\},\ \xi = wx,\ w^2=0$$

$$F_{\mu\nu}^{ext}=f_{\mu\nu}^{i}F_{i}^{ext}(\xi)$$

$$f_{\mu\nu}^{i}=(w_{\mu}\varepsilon_{\nu}^{i}-w_{\nu}\varepsilon_{\mu}^{i})$$

which can be calculated by Schwinger's proper-time method or eikonal approximation[9], we obtain

$$-is\ Tr[e^{is(\gamma\pi)^2}A^2] =-is\ 4\ \frac{1}{i(4\pi)^2}\ \frac{1}{s^2}\int(dx)A^2(x)$$

$$=-is\ 4\frac{1}{i(4\pi)^2}\frac{1}{s^2}(2\pi)^4\int(dk)A_\mu(k)A^\mu(-k)\ ,$$

which is precisely the result obtained in eq. (3.8). Since $\langle x(s)|\pi_\mu|x\rangle = 0$, we find

$$-is\ \mathrm{Tr}\ [e^{is(\gamma\pi)^2}(-e)(\pi A + A\pi)] = 0\ .$$

Furthermore

$$-is\ \mathrm{Tr}\ [e^{is(\gamma\pi)^2}(-\frac{e}{2})\sigma F] = i\frac{e^2}{8\pi^2}\int(dx)F^{\mu\nu}(x)F^{ext}_{\mu\nu}\ .$$

which reduces to zero after performing an integration-by-parts:

$$\int(dx)F^{ext}_{\mu\nu}(\partial^\mu A^\nu-\partial^\nu A^\mu) = -2\int(dx)A^\nu(\partial^\mu F^{ext}_{\mu\nu}) \to 0\ ,$$

since the source j^{ext}_μ operates far away from the region where the laser field is effective.

The remaining terms in (3.16) can be evaluated in the same way. Details of the calculation will be published elsewhere. We also refer to ref. 6 where various modified propagation functions together with other processes are treated in presence of an external laser field.

REFERENCES

1. J. Schwinger, Phys. Rev. 82, 664 (1951).

2. W. Pauli and F. Villars, Revs. Modern Phys. 21, 434 (1949).

3. G. Källén, Helvetia Physica Acta 25, 417 (1952).

4. J. Schwinger, e.g., in "Particles, Sources, and Fields", Vol. II, Addison-Wesley Publ. Co., Reading, Mass. 1973.

5. H. M. Fried, "Functional Methods and Models in Quantum Field Theory", MIT Press, Cambridge, Mass., 1972.

6. H. Mitter, cf. these proceedings

7. Wu-yang Tsai, T. Erber, Phys. Rev. D10, 492 (1974).

8. S. Adler, Am. Phys. (N.Y.) 67, 599 (1971).

9. W. Dittrich, Phys. Rev. D6, 2104 (1972).

Acta Physica Austriaca, Suppl. XIV, 521 — 548 (1975)

SPONTANEOUS MASS GENERATION, RENORMALIZATION GROUP AND SOLVABLE U(N)-SYMMETRIC MODELS+

by

H. RÖMER
Physikalisches Institut der
Universität Bonn

INTRODUCTION

Recently, the problem of giving masses to the gauge mesons which appear in the theoretically attractive gauge theories without spoiling renormalizability and asymptotic freedom has aroused some interest in the mechanism of spontaneous mass generation. The main idea of this mechanism is that masses might be generated by an instability of the perturbation theoretical vacuum, a state for which the expectation values of all the fields of the theory vanish: In some situations the real ground state of a system may not be identical with the vacuum of perturbation theory, some field Φ of the theory develops a non-vanishing expectation value in the real ground state, which results in attributing masses to apparently massless theories.

+ Seminar given at XIV. Internationale Universitätswochen für Kernphysik, Schladming, Austria, February 24 - March 7, 1975.

There are three versions of the mechanism of spontaneous mass generation:

a) The Higgs-Kibble mechanism[1].Here, Φ is a scalar (in order not to destroy Poincaré invariance) field with a mass term of unphysical sign. It is already in the corresponding classical theory that the ground state is characterized by a non-vanishing value of Φ.

b) The Coleman E. Weinberg mechanism[2]. This time, Φ is a scalar massless field. The classical ground state has vanishing fields, the instability of the vacuum is due to higher order quantum corrections. In the so-called one loop approximation S. Coleman and E. Weinberg demonstrated, that, starting from massless scalar quantum electrodynamics one ends up with massive scalar and vector fields.

In both of the versions a) and b) one explicitely has to introduce scalar fields into the theory, which implies some arbitrary ness and will probably destroy asymptotic freedom, if there are enough scalar fields to give masses to all but one of the vector fields[3]. In the

c) Implicite mechanism, Φ is a scalar field which does not explicitely appear in the Lagrangean but rather is a functional of the other fields. This mechanism has been invisaged by several authors[4,5]. In particular, D. Gross and A. Neveu[6] demonstrated spontaneous mass generation in U(N)-invariant four fermion interaction theories with space-time dimension D = 2 for the limit $N\to\infty$, in which the theories become exactly solvable.

We[7] further investigated spontaneous mass generation. Especially we consider the case in which the Callan Symanzik function[8,9] $\beta(g)$ has more than one zero and argue, that if the value of the coupling constant lies between two zeros of β spontaneous mass generation does not occur. This may give rise to some doubt, whether the results which were obtained for $N\to\infty$ are reliable for finite N as well, as far as spontaneous mass generation is concerned. A four fermion model of R. Dashen and Y. Frishman[10], which is exactly solvable for every N and one special value of the coupling constant yields a second zero of β which is not obtained in the $N \to \infty$ approximation, and it is difficult to see how to obtain it from higher orders in $1/N$[11]. One possibility is that some coefficient in the $\frac{1}{N}$ expansion has a singularity in the coupling constant[11]. But even for regular coefficients there may be a second zero of $\beta(N,g)$ which tends to infinity for $N \to \infty$ and yet, prevents spontaneous mass generation for every finite N.

In the following two sections we shall first define the notion of spontaneous mass generation and make some remarks about mass generation and effective potential, before we state our rule, that spontaneous mass generation should be impossible, if the value of the coupling constant lies between two zeros of the coefficient function β.

Then this rule will be confirmed in several U(N)-symmetric models, which in the limit $N \to \infty$ are solvable. A U(N)-symmetric Φ^4-theory for space time dimension $D = 4-\varepsilon$, for which β has an U.V.-stable zero at the origin and a positive I.R. stable zero of order ε will be especially illustrative for this purpose.

1. SPONTANEOUS MASS GENERATION AND EFFECTIVE POTENTIAL

The fundamental tool for discussing the stability of the vacuum is the effective potential of Y. Nambu and G. Jona Lasinio[12,13]. Consider a renormalizable field theory characterized by a Lagrangean

$$L = L(\phi, \frac{\partial\phi}{\partial x}) \tag{1.1}$$

and certain boundary conditions for the vertex functions $\Gamma^{(n)}$, e.g.

$$\Gamma^{(2)}(p^2)\Big|_{p^2 = m^2} = 0$$

$$\Gamma^{(2)}(p^2)\Big|_{p^2=-\mu^2} = -i\,(m^2 + \mu^2) \tag{1.2}$$

$$\Gamma^{(4)}\Big|_{p_i \cdot p_k=-\frac{1}{3}\mu^2(4\delta_{ik}-1)} = -i\,g \qquad \text{for a } \Phi^4\text{-theory.}$$

For reasons of simplicity we suppress all indices which distinguish between different fields. The effective potential $V(\phi)$ gives the minimal energy per volume in configurations $|\phi\rangle$, for which the field $\Phi(x)$ has the expectation value

$$\langle\phi|\,\Phi(x)\,|\phi\rangle = \phi = \text{const.} \tag{1.3}$$

as a function of ϕ. It fulfils $V(o) = 0$ and can be obtained from the generating functional $W[J]$ of connected τ-functions by a functional Legendre transformation. In terms of the vertices $\Gamma^{(n)}$ it is given by[2]

$$V(\phi) = i \sum_{n=0}^{\infty} \frac{1}{n!} \phi^n \Gamma^{(n)}(p_1,\dots,p_n)\Big|_{p_j=0} \quad . \tag{1.4}$$

In the classical limit it is simply the classical potential

$$V_{cl}(\phi) = - L(\phi,0) \quad . \tag{1.5}$$

If V has an absolute minimum at $\phi = \phi_M$ with $V(\phi_M)<0$, then the perturbation theoretical vacuum is no longer the ground state of the theory, a state $|\phi_M>$ has to be regarded as a ground state instead. By performing a shift $\Phi = \Phi' + \phi_M$ one arrives at a theory with vanishing expectation value of Φ' in the real ground state. The generating functional for the vertex functions $\tilde{\Gamma}^{(n)}$ and the effective potential of the new theory are obtained by the same shift. The choice of $|\phi_M>$ as a ground state breaks every symmetry which does not leave ϕ_M fixed (spontaneous symmetry breaking).

In ref.14 it is shown how the effective potential can also be defined for composite non-elementary fields.

Now we can precisely state what spontaneous mass generation means. We call (a component of) the field Φ apparently massless if

$$\Gamma^{(2)}(p^2)\Big|_{p^2=0} = 0 \tag{1.6}$$

i.e. if the two-vertex, calculated with respect to the perturbation theoretical vacuum has the mass shell zero at $p^2 = 0$. The apparently massless field Φ is said to have a spontaneously generated mass m^2, if V has a minimum at $\phi_M \neq 0$ and if the two-vertex $\tilde{\Gamma}^{(2)}$, which corresponds to the new ground state $|\phi_M\rangle$, obeys

$$i\,\tilde{\Gamma}^{(2)}(p^2)\Big|_{p^2=0} > 0 \qquad \text{and} \tag{1.7}$$

$$\tilde{\Gamma}^{(2)}(p^2)\Big|_{p^2=m^2} = 0\,. \tag{1.8}$$

Eq. (1.7) can also be written

$$\frac{\partial^2 V}{\partial \phi^2}\Big|_{\phi=\phi_M} > 0\,, \tag{1.7a}$$

as to be seen from Eq. (1.4) and the discussion following it. In many cases spontaneous mass generation will be associated to spontaneous breaking of a symmetry which calls for masslessness.

Consider now a field theory with formally massless fields only. (This means that the mechanism a), mentioned in the Introduction is excluded.) Furthermore, let us confine to theories with just one dimensionless coupling constant. If any mass m is spontaneously generated, it can only depend on the coupling constant g and on the scale parameter μ, which enters in the normalization conditions of the type of Eq. (1.2) in order to fix the normalization of the two-point function and the definition of the coupling constant. The introduction of the scale parameter μ is unavoidable, since in the absence of any scale the vertices $\Gamma^{(n)}$ cannot be normalized because of their infrared divergences. The dependence of the vertices $\Gamma^{(n)}$ on this parameter μ is described by the renormalization group equations

$$\{\mu \frac{\partial}{\partial\mu} + \beta(g) \frac{\partial}{\partial g} - n\gamma(g)\} \Gamma^{(n)}(\{n\},\mu,g) = 0 \quad \text{with} \tag{1.9}$$

$$\beta = \mu \frac{\partial}{\partial\mu} g . \tag{1.10}$$

A change of the arbitrary parameter μ cannot lead to any observable effect, and, indeed, Eq. (1.9) shows that it can always be compensated by an accompanying redefinition of the coupling constant and the wave function normalization. A spontaneously generated mass $m(\mu,g(\mu))$ is directly observable. So, it cannot depend on μ at all, and one derives the fundamental equation[6]

$$\begin{aligned}\mu\frac{d}{d\mu} m(\mu,g(\mu)) &= (\mu \frac{\partial}{\partial\mu} + \frac{\partial g}{\partial\mu} \frac{\partial}{\partial g}) m (\mu,g(\mu)) \\ &= (\mu \frac{\partial}{\partial\mu} + \beta(g) \frac{\partial}{\partial g})m(\mu,g) = 0 .\end{aligned} \tag{1.11}$$

This equation can also be obtained in a more formal way be acting with the operator $(\mu\frac{\partial}{\partial\mu} + \beta\frac{\partial}{\partial g})$ upon the identity (1.8):

$$\tilde{\Gamma}^{(2)}(p^2,\mu,g)\Big|_{p^2=m^2(\mu,g)} = 0$$

and employing the R. G. equation.

The generalization of Eq. (1.11) to a theory with several coupling constants is obvious:

$$\{\mu \frac{\partial}{\partial\mu} + \sum_{i=1}^{r} \beta_i(\underset{\sim}{g}) \frac{\partial}{\partial g_i}\} m^2(\mu,\underset{\sim}{g}) = 0 . \tag{1.11a}$$

For every coupling constant g_i one has a coefficient function β_i which can depend on all the coupling constants.

We remark here, that it turns out to be difficult to calculate the spontaneously generated mass from the effective potential[7].

The effective potential can be defined in terms of the vertices $\Gamma^{(n)}$ and, accordingly, also fulfils a R.G. equation which envolves the coefficient functions β and γ. Normally, and especially for massless theories it is more convenient to define the normalization of V not in terms of the vertices but more directly by conditions like[2]

$$\begin{aligned} \frac{\partial^2 V}{\partial\phi^2}\Big|_{\phi=0} &= 0 \\ \frac{\partial^4 V}{\partial\phi^4}\Big|_{\phi=M} &= g \quad . \end{aligned} \tag{1.12}$$

(For concreteness we gave in (1.12) the normalization conditions for a massless Φ^4 theory.)

Again, in a massless theory the introduction of a dimensional parameter M cannot be avoided because of infrared divergences.

The dependence on M is governed by a R. G. equation[2]

$$\{M \frac{\partial}{\partial M} + \tilde{\beta}(g)\frac{\partial}{\partial g} - \tilde{\gamma}\phi \frac{\partial}{\partial \phi}\} \quad V(\phi, M, g) = 0 \; . \tag{1.13}$$

The coefficient functions $\tilde{\beta}$ and $\tilde{\gamma}$ agree only in lowest nontrivial order with the coefficients β and γ respectively[2,6]. Of course, a spontaneously generated mass m, taken as a function of M and g this time, has to satisfy the equation

$$(M \frac{\partial}{\partial M} + \tilde{\beta}\frac{\partial}{\partial g}) \; m \; (M,g) = 0 \; . \tag{1.14}$$

If one tries to calculate m from the effective potential, two tentative definitions in terms of V seem to be especially natural:

$$m_M^2(M,g) = [\frac{\partial^2 V}{\partial \phi^2} (\phi, M, g)]_{\phi=\phi_M(M,g)} \tag{1.15}$$

$$m_s^2(M,g) = \left.\frac{\partial^2 V_{cl}}{\partial \phi^2}\right|_{\phi=\phi_M(M,g)} \tag{1.16}$$

where V_{cl} is taken from Eq. (1.5) and ϕ_M denotes the position of the minimum of V:

$$\left.\frac{\partial V(\phi,M,g)}{\partial\phi}\right|_{\phi\,=\,\phi_M(M,g)} = 0 \,. \tag{1.17}$$

For sure, in view of Eqs. (1.7), (1.7a) a positive m_M^2 is sufficient for spontaneous mass generation in a formally massless theory.

m_M^2 and m_s^2 may be good approximations to the generated mass in certain situations, but in general both m_M^2 and m_s^2 fail to satisfy the R.G. equation (1.14), as can be seen in the following way:

By differentiating Eq. (1.13) with respect to ϕ we get

$$\{M\frac{\partial}{\partial M} + \tilde{\beta}\frac{\partial}{\partial g} - \tilde{\gamma} - \tilde{\gamma}\phi\frac{\partial}{\partial\phi}\}\frac{\partial V}{\partial\phi} = 0 \,. \tag{1.18a}$$

$$\{M\frac{\partial}{\partial M} + \tilde{\beta}\frac{\partial}{\partial g} - 2\tilde{\gamma} - \tilde{\gamma}\phi\frac{\partial}{\partial\phi}\}\frac{\partial^2 V}{\partial\phi^2} = 0 \,. \tag{1.18b}$$

Acting with $(M\frac{\partial}{\partial M} + \tilde{\beta}\frac{\partial}{\partial g})$ upon Eq. (1.17) yields, together with (1.18) the identity

$$m_M^2(M\frac{\partial}{\partial M} + \tilde{\beta}\frac{\partial}{\partial g} + \tilde{\gamma})\,\phi_M(M,g) = 0 \,, \tag{1.19}$$

which is a R.G. equation for ϕ_M if $m_M^2 \neq 0$ i.e. if spontaneous mass generation occurs. By applying $(M\frac{\partial}{\partial M} + \tilde{\beta}\frac{\partial}{\partial g})$ on (1.15) and using (1.19) and again (1.18) we obtain a R.G. equation

$$(M \frac{\partial}{\partial M} + \tilde{\beta}\frac{\partial}{\partial g} - 2\tilde{\gamma})\ m_M^2(M,g) = 0\ . \tag{1.20}$$

For $\tilde{\gamma} \neq 0$ (1.20) does not agree with the desired equation (1.14).

The result (1.20) is understandable because $m_M^2 = \Gamma^{(2)}(p^2,M,g)\big|_{p^2=0}$ and $\tilde{\Gamma}^{(2)}$ is multiplicatively renormalized by a change of M.

By employing eq. (1.19) it is easy to show that m_s^2, too, does not obey eq. (1.14), unless very special relations between $\tilde{\beta}$ and $\tilde{\gamma}$ are true.

It is, however, possible to define a mass

$$m_{inv}^2(M,g) = m_M^2(M,g)\, e^{-2\int_{g_o}^{g} \frac{\tilde{\gamma}(g')}{\tilde{\beta}(g')}dg'}\ , \tag{1.21}$$

which, indeed, fulfils eq. (1.14). Since via (1.13) $\tilde{\beta}$ and $\tilde{\gamma}$ can be calculated from V, $m_{inv}^2(M,g)$ is determined in terms of V up to an unknown multiplicative constant independent from M and g.

We now come to the solution of the fundamental equation (1.11) (or (1.14)). For dimensional reasons $m(\mu,g)$ has to be of the form

$$m(\mu,g) = \mu \cdot f(g) \tag{1.22}$$

and the function f(g) can be determined up to a multiplicative constant:

$$m(\mu,g) = c\,\mu\, e^{-\int_{g_o}^{g} \frac{dg'}{\beta(g')}} \tag{1.23}$$

c and g_o have to be fixed by other considerations. As m^2_{inv} from (1.21) satisfies (1.14), the true spontaneously generated mass m^2 and m^2_{inv} have to be proportional.

The asymptotic behaviour of the Green's functions is determined by the zeros of β[9].

A negative β'(g) in the vicinity of a zero corresponds to an ultraviolet-(U.V.-) stable fixed point, and a positive β'(g) near a zero belongs to an infrared-(I.R.-) stable fixed point. (If β has a zero of even order, it will be U.V.-stable when being approached from one side and I.R.-stable from the other side, because β' changes sign at the zero)

The dependence of m^2 on the coupling constant g is completely fixed by Eq. (1.23). If β has a zero at $g = g_1$:

$$\beta(g) = -\,a\,(g-g_1)^r + O((g-g_1)^r)\ , \tag{1.24}$$

then, in the vicinity of g_1, a spontaneously generated mass m(μ,g) will behave like

$$m(\mu,g) \sim \mu\,|g-g_1|^{\frac{1}{a}} \tag{1.25a}$$

for r = 1 and like

$$m(\mu,g) \sim \mu\, e^{\frac{-1}{a(r-1)(g-g_1)^{r-1}}} \tag{1.25b}$$

for $r > 1$.

In any case, a spontaneously generated mass vanishes, if a zero g_1 is approached from an U.V.-stable side. $g_1 = 0$ is always a zero of β; if it is U.V.-stable, the generated mass tends to zero when the coupling is switched off, as to be expected.

On the other hand, $m(\mu,g)$ tends to infinity, if an I.R.-stable zero of β is approached (and if $c \neq 0$ in eq. (1.23)).

It is argued in ref. 6, that spontaneous mass generation is impossible, if the origin $g = 0$ is I.R.-stable, because it is not acceptable, that $m(\mu,g)$ tends to infinity if g approaches zero; $m(\mu,g)$ has to vanish identically in this case.

We argue, that a mass cannot be generated whenever the coupling constant one starts with lies on the I.R.-stable side of any zero of β. This has to be so, since m tends to infinity, when the coupling constant is moved towards that zero, whereas at the zero all Green's functions are power behaved with normally non-integer powers. (This can be seen directly from the R.G. equation (1.9)). So, at the zero the Green's functions have cuts in the kinematical variables, which start at the origin, which means that one has a zero mass theory.

Especially, if β has two zeros, there should be no spontaneous mass generation for any value of the coupling constant which lies between the two zeros.

The reason for this is the fact that the zeros of β alternate in their character: If $g_1 < g_2$ are two zeros of β with no other zero of β between g_1 and g_2, then g_2 is U.V.-(I.R.-) stable from the left if and only if g_1 is I.R.-(U.V.-) stable from the right. So, g with $g_1<g<g_2$ always lies on the I.R. stable side of one zero.

The only way to circumvent this conclusion is to assume that g_1 is U.V.-stable and that there is a singularity of β in the interval $[g_1, g_2]$, at which β changes sign[11]. In this case, both g_1 and g_2 could be U.V.-stable.

The scheme, that spontaneous mass generation is impossible if g lies between two zeros of β subsequently will be illustrated by several examples.

2. U(N)-SYMMETRIC MODELS AND THE LIMIT $N\to\infty$

The first model we consider has the Lagrangean

$$L = i\,\bar{\psi}\,\gamma\partial\psi + \frac{\lambda}{2N}\,(\bar{\psi}\psi)^2 \qquad (2.1)$$

and space-time dimension D = 2. ψ is a Dirac spinor which transforms like the Quark representation under U(N). L is U(N)-symmetric and invariant under the discrete γ_5-transformation $\psi\to\gamma_5\psi$, which guarantees that the theory is massless unless the γ_5 symmetry is spontaneously broken. In the limit $N\to\infty$ the theory is exactly solvable[6].

An implicite breaking of the γ_5-symmetry and spontaneous mass generation could be due to a non-vanishing ground state expectation value of the composite field $\bar{\psi}\psi$, which is odd under the γ_5-transformation. In ref.6 the local coupling $(\bar{\psi}\psi)^2$ in (2.1) is replaced by the exchange of a scalar meson σ of infinite mass, so that (2.1) becomes the equivalent Lagrangean

$$L' = i\,\bar{\psi}\partial\gamma\psi - \frac{1}{2}\,\sigma^2 - \sqrt{\frac{\lambda}{N}}\,\sigma\,\bar{\psi}\psi \quad . \tag{2.2}$$

The fields σ and $\bar{\psi}\psi$ are closely related, the same holds true for their respective effective potentials[6]. The bare σ-propagator in momentum space is simply (-i). The R.G. equation

$$\{\mu\,\frac{\partial}{\partial\mu} + \beta(\lambda)\,\frac{\partial}{\partial\lambda} - n_\sigma\gamma_\sigma - n_\psi\gamma_\psi\}\;\Gamma^{(n_\sigma,n_\psi)} = 0 \tag{2.3}$$

for the vertex $\Gamma^{(n_\sigma,n_\psi)}$ of n_σ σ-fields and n_ψ ψ-fields allows to determine β, γ_σ and γ_ψ e.g. in terms of $\Gamma^{(1,2)}$, $\Gamma^{(0,2)}$, and $\Gamma^{(2,0)}$. The behaviour of a Feynman graph F for large values of N is a power $N^{E(F)}$ in N, where $E(F) = L_F - O_F$. Here O_F denotes the order of F in the coupling constant λ and L_F is the number of closed fermion loops in F. One easily convinces oneself that for $N\to\infty$ to leading order in N $\Gamma^{(0,2)}$ agrees with the bare ψ-propagator, which implies $\gamma_\psi = 0$. Also $\Gamma^{(1,2)}$ is given by its bare contribution to leading order in N, which has as a consequence that $\beta(\lambda) = 2\lambda\gamma_\sigma(\lambda)$.

Finally, $\Gamma^{(2,0)}$ is in the same limit determined by the graph

σ - - ◯ - - σ with one ψ-bubble. So for $N\to\infty$ the full σ-propagator D_σ is given by

$$D_\sigma = \frac{-i}{1 + \frac{\lambda}{2\pi} \ln \frac{(-p^2)}{\mu^2}} , \qquad (2.4)$$

where we normalized to $D_\sigma \Big|_{p^2 = -\mu^2} = -i$. (2.4) leads to

$$\beta(\lambda) = 2\lambda\gamma_\sigma(\lambda) = -\frac{\lambda^2}{\pi} \qquad (2.5)$$

for $N \to \infty$. $\beta(\lambda)$ has a zero at the origin $\lambda = 0$, which is U.V.-stable from the right. (Negative values of λ are unphysical). The theory is asymptotically free. This is only possible since γ_σ, the anomalous dimension of the σ-field, is negative, contrary to the usual statement that positivity of the norm restricts anomalous dimensions to be positive. But σ is only an auxiliary field; the normalization of D_σ involves only one normalization condition: $D_\sigma \Big|_{p^2=-\mu^2} = -i$, while a finite mass propagator has to satisfy a mass shell condition and a second condition like $\frac{\partial D^{-1}}{\partial p^2}$ to be some fixed number, independent of λ. This last condition cannot be imposed on D_σ, and this leaves open the possibility of a negative γ_σ, since multiplying a propagator by some function $f(\lambda)$ changes γ to $\gamma - \frac{1}{2}\beta f'$.

The normalization of D_σ is more like the normalization of a four point function, and, indeed, the fermion four point function is essentially given by λD_σ up to terms which carry the necessary spinor indices.

One notices, that D_σ in eq. (2.4) has an unphysical tachyon pole at $(-p_t^2) = \mu^2 e^{-\frac{2\pi}{\lambda}}$, which also shows up in the fermion four point function. For $\lambda \to 0$ its position tends to zero as it has to do, because the theory is asymptotically free and $(-p_t^2)$ as an observable quantity has to fulfil the equation $(\mu \frac{\partial}{\partial \mu} + \beta(\lambda) \frac{\partial}{\partial \lambda})$ $(-p_t^2(\mu,\lambda)) = 0$, which can in fact readily be verified.

It is shown in ref. 6, that the appearance of the tachyon pole is indicative for spontaneous symmetry breaking; it emerges as an artefact, because one expanded with respect to the perturbation theoretical vacuum instead of the real ground state. The computation of the effective potential $V(\sigma)$ also confirms[6] the assertion, that there is spontaneous mass generation and breaking of the γ_5-symmetry by the implicite mechanism mentioned under c) in the Introduction.

The tachyon pole in D_σ can also be detected as a singularity in the effective coupling constant $\bar{\lambda}(\lambda,t)$, which is defined by

$$\frac{\partial \bar{\lambda}(\lambda,t)}{\partial t} = \beta(\bar{\lambda}); \; \bar{\lambda}(\lambda,0) = \lambda, \text{ where } t = \tfrac{1}{2}\ln \frac{(-p^2)}{\mu^2} \tag{2.6}$$

Eq. (2.6) can be solved to give

$$t = \int_{\lambda}^{\bar{\lambda}(\lambda,t)} \frac{d\lambda'}{\beta(\lambda')} , \tag{2.7}$$

and with $\beta(\lambda)$ from (2.5) we get

$$\bar{\lambda}(\lambda,t) = \frac{\lambda}{1 + \frac{\lambda}{\pi} t} . \tag{2.8}$$

The tachyon pole in (2.8) comes from the convergence of the integral

$$\int_{\lambda}^{\infty} \frac{d\lambda'}{\beta(\lambda')}$$

(compare Eq. (2.7)), which means that there is an infinite $\bar{\lambda}$ for some finite value of t, and is produced by the quadratic behaviour of $\beta(\lambda)$ in λ for $N\to\infty$.

It is not surprising, that the same tachyon pole appears in the effective coupling λ and in the fermion four-vertex $\Gamma^{(o,4)}$. The R.G. equation (2.3) together with the normalization condition

$$\Gamma^{(o,4)}\Big|_{p_i \cdot p_j = -\frac{1}{2}(\delta_{ij} + \frac{1}{2})\mu^2} = i\,\frac{\lambda}{N} \tag{2.9}$$

is sufficient to determine $\Gamma^{(o,4)}$ for all symmetry points with s = t = u:

$$\Gamma^{(o,4)}\Big|_{p_i \cdot p_j = -\frac{1}{2}(\delta_{ij} + \frac{1}{2})s} = \frac{i}{N}\,\bar{\lambda}(\lambda,t)\,e^{-4\int_o^t \gamma_\psi(\bar{\lambda}(\lambda,t'))dt'} \; ;$$

$$\left(t = \frac{1}{2}\ln\frac{(-s)}{\mu^2}\right) . \tag{2.10}$$

(We suppressed all spinor indices). γ_ψ vanishes in the limit $N\to\infty$, so, the singularities of $\bar{\lambda}$ and $\Gamma^{(o,4)}$ should coincide.

In the limit $N\to\infty$ β is a simple monomial in the coupling constant with one single U.V.-stable zero at the origin. Spontaneous mass generation occurs and, indeed, it is allowed according to the rule we gave in the preceeding chapter, because there is no I.R.-stable zero of β.

It is, however, not necessarily true, that this result also holds for finite values of N, since contributions of higher order in 1/N might produce a second zero λ_1 of β, which for $0 < \lambda < \lambda_1$ would forbid spontaneous mass generation.

Also by inspection of the behaviour of the effective coupling constant $\bar{\lambda}$ it can be observed, how a second zero λ_1 of β prevents spontaneous mass generation. From Eq. (2.7) it is evident, that for $0 < \lambda < \lambda_1, \bar{\lambda}(\lambda,t)$ varies only between 0 and λ_1, i.e. $\bar{\lambda}(\lambda,t)$ is bounded as a function of t, there is no tachyon singularity in $\bar{\lambda}$ and $\Gamma^{(o,4)}$ and therefore no sign of spontaneous mass generation.

Even if the second singularity λ_1, generated by terms of higher order in 1/N tends to infinity for $N\to\infty$, we have a bounded $\bar{\lambda}(\lambda,t)$ for every finite value of N. Take as an example

$$\beta = -\frac{\lambda^2}{2\pi}\left(1 - \frac{f(\lambda)}{N}\right) + O\left(\frac{1}{N^2}\right) \quad . \tag{2.11}$$

There will be no tachyon pole in $\bar{\lambda}$, if the equation $f(\lambda) = N$ has a positive solution. The function β, as taken from Eq. (2.5) will be a good approximation only as long as $f(\bar{\lambda}) << N$. In no case it will be reliable for discussing the tachyon mechanism for finite N, since the appearance of the tachyon is an effect of the behaviour of β for very large λ.

The situation is even more stringent for the two dimensional four fermion coupling model of R. Dashen and Y. Frishman[10] which (in a slightly simplified version) is given by the Lagrangean

$$L_{D.F.} = i\bar{\psi}\gamma\partial\psi - \frac{G}{2N}\sum_{a=0}^{N^2-1} j_\mu^a \, j^{a\mu}, \qquad \text{where} \tag{2.12}$$

$$j^{a\mu} = \frac{1}{2}\,\bar{\psi}\gamma^\mu \lambda^{(a)}\psi \qquad (a = 0,\ldots,N^2-1). \tag{2.13}$$

ψ is a spinor which again transforms like the Quark representation of U(N), and $\lambda^{(a)}$ are the U(N)-generators, normalized according to

$$\mathrm{Tr}\,(\lambda^{(a)}\lambda^{(b)}) = 2\,\delta^{ab} \quad . \tag{2.14}$$

The Lagrangean is invariant under the continuous γ_5-transformation $\psi \to e^{i\alpha\gamma_5}\psi$, which forbids a mass term.

The authors of ref.10 showed that there exist conformal invariant solutions only for G = 0 and for $G = G_o(N) = \frac{4\pi N}{N+1}$ (for every N) and explicitely constructed the solutions.

For the coefficient function $\beta(G,N)$ this means, that β has to have a zero at $G = G_o(N) \underset{N\to\infty}{\to} 4\pi$, and that one should expect to find this zero also in the limit $N\to\infty$.

By a Fierz transformation (2.12) can be reformulated to

$$L'_{D.F.} = i\bar{\psi}\gamma\partial\phi + \frac{G}{4N} \{(\bar{\psi}\psi)^2 - (\bar{\psi}\gamma_5\psi)^2\} \tag{2.15}$$

which is equivalent to

$$L_{\sigma,\pi} = i\bar{\psi}\gamma\partial\psi - \frac{1}{2}(\sigma^2 + \pi^2) + \sqrt{\frac{G}{2N}}(\sigma\bar{\psi}\psi + i\pi\bar{\psi}\gamma_5\psi) \tag{2.16}$$

with infinite mass scalar and pseudoscalar fields σ and π.

This Lagrangean can be subject to the same analysis as (2.2) with the result

$$\beta(G) = -\frac{G^2}{2\pi} \qquad \text{for} \qquad N\to\infty \ . \tag{2.17}$$

So, in the $N\to\infty$ limit $\beta(G)$ is again a monomial in the coupling constant, and there is no trace of the second zero, which should be present after all. Terms of higher order in 1/N have to contribute in a singular way to produce it[11]. From the analysis in leading order in N alone, one expects spontaneous mass generation, but this expectation is seriously questioned here because the limit has some evident peculiarities.

To provide for models which show a second zero of β also to leading order in N, we go over to the non-integer space-time dimension $D = 2+\varepsilon$ ($\varepsilon > 0$). Both for the model (2.2)[6] and for the Dashen-Frishman model[7] the coefficient has a second zero at some positive value of the coupling constant, which is of order ε. The origin is I.R.-stable, and the second zero turns out to be U.V.-stable. According to our rule, and already because of the I.R.-stability of the origin there should be no spontaneous mass generation for values of the coupling constant between the two zeros. A detailed analysis confirms this for (2.2)[6] and analogously for (2.12). For values of the coupling constant in that interval the σ-propagator has no tachyon pole and the effective potential has no minimum different from $\sigma = 0$.

Before we come to our last example we should like to add a short remark about the possible influence of new induced couplings, which could change the interactions (2.2) and (2.12), which contain only one coupling constant. There are no induced couplings to leading and next-to-leading order in N. For the leading order in N this is directly evident, and in next-to-leading order there are only two diagrams which possibly contribute to induced couplings:

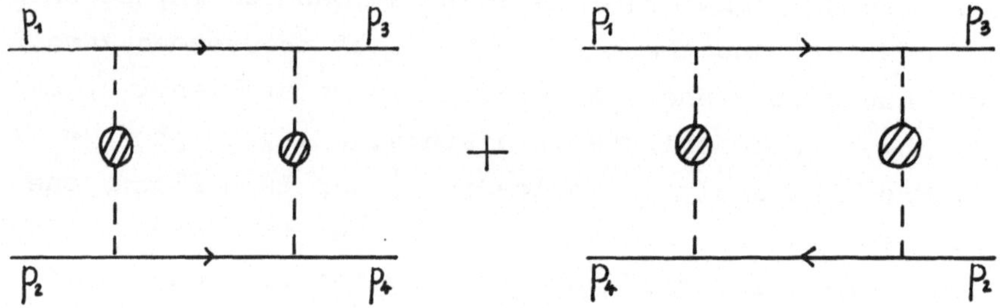

But the sum of these two diagrams is finite, and thus there are no new interactions induced also in next-to-leading order. This suffices to justify our discussion of (2.2) and (2.12) as theories with one coupling constant.

Finally, we consider a U(N)-invariant Φ^4 model with space-time dimension $D = 4-\varepsilon$ $(\varepsilon > 0)$ in the limit

$N\to\infty$. Φ is a Lorentz-scalar field with N components. The Lagrangean

$$L = \frac{1}{2}\,\partial\Phi^{*}\,\partial\Phi - \frac{\lambda}{2N}\,\mu^{\varepsilon}\,(\Phi^{*}\Phi)^{2} \tag{2.18}$$

can be transformed by introduction of an infinite mass scalar auxiliary field σ with one component to the equivalent Lagrangean

$$L_{\sigma} = \frac{1}{2}\,\partial\Phi^{*}\partial\Phi + \frac{1}{2}\,\sigma^{2} - \sqrt{\frac{\lambda}{N}}\,\mu^{\varepsilon/2}\,\Phi^{*}\Phi\ . \tag{2.19}$$

The factor μ^{ε} in eq. (2.18) serves to keep the coupling constant λ dimensionless. Notice that the term quadratic in σ in Eq. (2.19) does not have the proper sign of a physical mass term. This is unavoidable, if one wants to reproduce the physical sign of the Φ^4 interaction term with a hermitian coupling $\sigma\Phi^{*}\Phi$. The sign of the interaction in (2.18) has to be negative to ensure that the classical energy is bounded from below. Because of the sign of the σ^2-term the bare σ-propagator in momentum space is given by (+i). To leading order in N the Φ-propagator and the $\sigma\Phi^{*}\Phi$ vertex are again given by their corresponding bare expressions, and the coefficients γ_{σ} and β are solely determined by the σ-propagator D_{σ}. The only higher order contribution to D_{σ}^{-1} in the limit $N\to\infty$ is provided by the one Φ-loop graph

σ — — —◯— — — σ

The corresponding Feynman amplitude for unphysical dimension

$$\Gamma^{(2)}(p^2) = \text{const.}\lambda.\text{Reg} \int \frac{d^D \ell}{(2\pi)^D} \frac{1}{\ell^2 (p-\ell)^2} \qquad (2.20)$$

can be evaluated according to the method of G. 't Hooft and M. Veltman[15]. With the "natural" normalization condition

$$D_\sigma \Big|_{p^2=-\mu^2} = +\ i \qquad (2.21)$$

one finds

$$D_\sigma = \frac{1}{1 + \frac{\lambda}{8\pi^2 \varepsilon} c(\varepsilon)\left[\left(\frac{\mu^2}{-p^2}\right)^{\varepsilon/2} - 1\right]} , \qquad (2.22)$$

where

$$c(\varepsilon) = \frac{\frac{1}{2}\pi\varepsilon}{\sin\frac{\pi\varepsilon}{2}} \frac{\Gamma^2(1-\varepsilon)}{\Gamma(2-2\varepsilon)\Gamma(1-\varepsilon/2)} 2^\varepsilon \pi^{\varepsilon/2} \underset{\varepsilon\to 0}{\to} 1 . \qquad (2.23)$$

This can be used to find the coefficients β and γ_σ in the limit $N\to\infty$:

$$\gamma_\sigma(\lambda) = \frac{\lambda}{16\pi^2} c(\varepsilon) \qquad (2.24)$$

$$\beta(\lambda) = -\varepsilon\lambda + \frac{\lambda^2}{8\pi^2} c(\varepsilon) . \tag{2.25}$$

β has a positive second zero at the point $\lambda_0 = \frac{8\pi^2 \varepsilon}{c(\varepsilon)}$. The zero at the origin $\lambda = 0$ is U.V.-stable, λ_0 is I.R.-stable. So, our example is an especially suitable test of our rule, which claims that spontaneous mass generation should be impossible, if the coupling constant lies between two zeros of β. If the behaviour of β at the origin were alone decisive in this respect, one would expect to find spontaneous mass generation, because the origin is U.V.-stable.

Eq. (2.22) teaches us that, indeed, for $0 < \lambda < \lambda_0$ there is no tachyon pole in D_σ and thus no sign of spontaneous mass generation, whereas for $\lambda > \lambda_0$, there is a tachyon pole at

$$(-p_t^2) = \mu^2 \left[1 - \frac{8\pi^2 \varepsilon}{c(\varepsilon)\lambda}\right]^{-2/\varepsilon} \tag{2.26}$$

whose position tends to infinity for $\lambda \to \lambda_0 + 0$, as to be expected for an I.R.-stable fixed point.

As a further test we compute the effective potential $V(\sigma)$ of the auxiliary field σ. In the $N\to\infty$ limit it is merely given by the tree graph and the one loop graphs:

$V(\sigma)$: - - - + - - -O- - - + ...

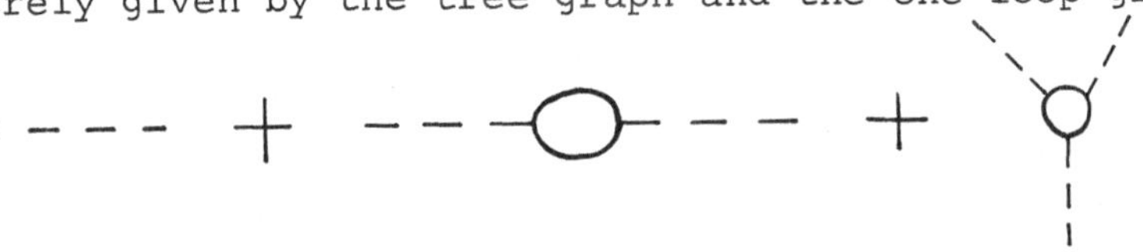

Eq. (1.4) shows how to compute $V(\sigma)$ in terms of these graphs. At first sight, the one loop graphs with more than two external legs seem to be non-leading in N, they have to be kept, however, because of their I.R.-divergences of increasing degree, which sum up[2] in the effective potential to a mild logarithmic behaviour:

$$V(\sigma) = -\frac{1}{2}\sigma^2 + \frac{N}{2}\,\text{Reg} \sum_{r=1}^{\infty} \int \frac{d^D k}{(2\pi)^D} \frac{(-1)^r}{r} \left(\frac{\sqrt{\lambda}\mu^{\varepsilon/2}\sigma}{\sqrt{N}\, k^2}\right)^r$$

$$= -\frac{1}{2}\sigma^2 + \frac{N}{2}\,\text{Reg} \int \frac{d^D k}{(2\pi)^D} \ln\left(1 + \frac{\sqrt{\lambda}\mu^{\varepsilon/2}\sigma}{\sqrt{n}\, k^2}\right) . \tag{2.27}$$

The integration is understood over Euclidean momenta. The renormalization is performed in the usual way[2] by first integrating over a finite volume $k^2 \leq \Lambda^2$ and adding counter terms $A + B\sigma + C\sigma^2$ in such a way, that the normalization conditions

$$V(0) = \left.\frac{\partial V}{\partial \sigma}\right|_{\sigma=0} = 0 \tag{2.28a}$$

$$\left.\frac{\partial^2 V}{\partial \sigma^2}\right|_{\sqrt{\frac{\lambda}{N}}\sigma = \mu^{2-\varepsilon/2}} = -1 \tag{2.28b}$$

are fulfilled, and then letting $\Lambda \to \infty$. The final result is

$$V(\sigma) = -\frac{1}{2}\sigma^2 + \frac{N\pi}{\frac{D}{2}\Gamma\left(\frac{D}{2}\right)(4\pi)^{D/2}} \frac{1}{\sin\frac{\pi D}{2}} \cdot$$

$$\cdot \{ (\sqrt{\tfrac{\lambda}{N}}\sigma\mu^{\varepsilon/2})^{D/2} - \tfrac{1}{2}(\sqrt{\tfrac{\lambda}{N}}\,\sigma)^2\, \tfrac{D}{2}(\tfrac{D}{2} - 1)\} \quad . \tag{2.29}$$

To look for a minimum of $V(\sigma)$, we compute

$$V'(\sigma) = - \sigma \frac{\sqrt{N\lambda}\ \pi}{\Gamma(\frac{D}{2})\,(4\pi)^{D/2}\sin\frac{\pi D}{2}} \{\mu^{\varepsilon/2}(\sqrt{\tfrac{\lambda}{N}}\sigma\mu^{\varepsilon/2})^{D/2-1} -$$

$$-\sqrt{\tfrac{\lambda}{N}}\sigma\ (\tfrac{D}{2} - 1)\} \tag{2.30}$$

and find, that there is no zero of V' away from the origin $\sigma = 0$ for $0 \leq \lambda < \lambda_1$ with

$$\lambda_1 = 8\pi^2\varepsilon \frac{\sin\frac{\pi\varepsilon}{2}}{\frac{\pi\varepsilon}{2}} \Gamma(1 - \tfrac{\varepsilon}{2})\,(4\pi)^{-\frac{\varepsilon}{2}} \tag{2.31}$$

in accordance with our assertion about the impossibility of spontaneous mass generation. For $\lambda > \lambda_1$ there is a minimum of $V(\sigma)$, which goes to infinity for $\lambda \to \lambda_1 + 0$ as it should. A thorough analysis analogous to the one given in Ref. 6 for $D = 4$ shows, however, that this minimum does not correspond to mass generation. The essential reason is, that one should actually consider the effective potential of $\Phi^*\Phi$ rather than $V(\sigma)$.

REFERENCES

1. P. W. Higgs, Phys. Lett. 12, 132 (1964); T. Kibble, Phys. Rev. 155, 1554 (1967).
2. S. Coleman, E. Weinberg, Phys. Rev. D7, 1888 (1973).
3. D.J. Gross, F. Wilczek, Phys. Rev. Lett. 30, 1343 (1973); Phys. Rev. D8, 3633 (1973).
4. J. M. Cornwall, R.E. Norton, Phys. Rev. D8, 3338 (1973).
5. R. Jackiw, K. Johnson, Phys. Rev. D8, 2386 (1973).
6. D.J. Gross, A. Neveu, Phys. Rev. D10, 3235 (1974).
7. Y. Frishman, H.Römer, S.Yankielowicz, Weizman Institute WIS-74/45-Ph.
8. M. Gell-Mann, F. Low, Phys. Rev. 95, 1300 (1954).
9. C. G. Callan, Jr., Phys. Rev. D2, 1541 (1970).
   K. Symanzik, Comm. Math. Phys. 18, 227 (1970);
   Springer Tracts in Modern Phys., Vol. 57 (1971).
10. R. Dashen, Y. Frishman, Phys. Lett. 46B, 439 (1973).
11. H. Römer, H.R. Rubinstein, S.Yankielowicz, Nucl. Phys. B79, 285 (1974).
12. Y. Nambu, G. Jona Lasinio, Phys. Rev. 122, 345 (1961).
    G. Jona Lasinio, Nuovo Cim. 34, 1790 (1964).
13. R. Jackiw, Phys. Rev. D9, 1686 (1974).
14. J.M. Cornwall, J. Jackiw, E. Tomboulis, Phys. Rev. D10, 2428 (1974).
15. G. 't Hooft, M. Veltman, Nucl. Phys. B44, 189 (1972).

Acta Physica Austriaca, Suppl. XIV, 549—565 (1975)

A SECOND LOOK AT RELATIVISTIC WAVE EQUATIONS+)

by

H. BIRITZ
School of Physics, Georgia Institute of Technology
Atlanta, Georgia 3o332, USA

1. INTRODUCTION

Although it has been known for a long time that there exist more general theories besides the Klein-Gordon and Dirac equations, they were mainly considered a mathematical curiosity and no serious attempts were made to determine their physical content like mass spectra, magnetic moments or form factors. Such an attitude was permissible perhaps thirty years ago when only a few elementary particles were known and it still could be believed that only the simplest (=irreducible) mathematical structures should be used for their description. In the meantime it has been learned from experiment that most particles have a rather complicated internal structure and can exist in various excited states. Hence we see no cogent reason why, for

+) Seminar given at XIV. Internationale Universitätswochen für Kernphysik, Schladming, Austria, February 24 - March 7, 1975.

example, only the Dirac equation should be used for the description of spin $\frac{1}{2}$ - particles.

Obviously we do not want to start with the most general relativistic wave equation. Today we want to report on the first explicit calculations with finite-component, manifestly covariant wave equations of the form

$$(- i \beta^{\mu}\partial_{\mu} + 1)\psi(x) = 0. \tag{1}$$

To get a real mass spectrum with a complete set of solutions we require β^{o} to be hermitean; with the unit matrix as mass term we are excluding massless particles from the spectrum. The wave function is assumed to transform according to an arbitrarily given finite-dimensional representation $D(g)$ of the quantum mechanical Lorentz group $SL(2,C)$, $\psi'(x') = D(g)\psi(x)$. Such wave equations describe a whole spectrum of particles with different spin. We do not want to impose any constraints to project out a single spin value - on the contrary, we are deliberately looking for theories containing a whole spectrum of states as suggested by the observed spectrum of elementary particles and their excited states.

We want to discuss the mass spectra and parities of such general wave equations, the magnetic moments for minimal coupling, and the renormalization of the axial-vector coupling constant. Here we can only present the results - for more details we refer to a series of papers now in press in the Physical Review; there also further references may be found.

Apart from the philosophical prejudice mentioned

above, it was technical difficulties which prevented a detailed study of the general wave equation (1): the multidimensional matrices β^{μ} make explicit calculations more and more difficult. Whereas in the well-known case of the Dirac equation all the calculations can be done without an explicit representation of the Dirac matrices, employing only their simple commutation and anticommutation relations, it does not seem possible to develop similar algebraic techniques for the general wave equation: the algebra generated by the β^{μ} does not close as the β-matrices contain arbitrary parameters; besides, from the few special examples considered up to now, it seems fair to say that the algebraic relations of the β^{μ} are much too intricate to be of any help in actual calculations. A convenient explicit realization of the β^{μ} is therefore essential; this realization is closely connected to the standardization of the transformation law D(g) of the wave function under the Lorentz group.

It is well-known that all the inequivalent, finite-dimensional, irreducible representations of the Lorentz group can be written as the direct product of two rotation matrices, $D^{AB}(g) = D^{A}(g) \otimes D^{B}(g)^{\dagger -1}$, A and B being any integers or halfintegers. This standardization of the irreducible representations of the Lorentz group has the advantage that the decoupling coefficients for the Lorentz group are now simply products of two ordinary Clebsch-Gordan coefficients (CGC's) of the rotation group. Considering the highly developed state of the art of the rotation group, it is only natural to make use of a parametrization which fully exploits this close connection between the representations of the Lorentz group and those of the rotation group. In this way all the algebraic problems encountered in the study of the general wave equation (1) can be reduced to problems of angular momentum analysis. As every finite-dimensional representation of the Lorentz group is completely reducible, we may assume the transformation law D(g) to be already given as a direct sum of irreducible representations. Thus we write $D(g) = \sum \oplus D^i(g)$, where the index i = 1,2,.., N denotes the various irreducible representations contained in D(g); each irreducible component $D^i(g)$ is labelled by a pair of angular momenta, $i = (A_i, B_i)$. The matrices β^μ have to form a vector in a manifestly covariant theory, $D^{-1}(g)\ \beta^\mu\ D(g) = L(g)^\mu{}_\nu\ \beta^\nu$. Labelling the rows and columns of β^μ according to the decomposition of D(g) into its irreducible components $D^i(g)$, i = 1,2,..,N, it follows immediately from Schur's lemma that, up to a numerical factor, the block of matrix elements (β^μ_{ik}) connecting the i-th with the k-th irreducible component of D(g) is given by a product of two CGC's of the rotation group:

$(\beta^\mu_{ik}) = b_{ik}$ 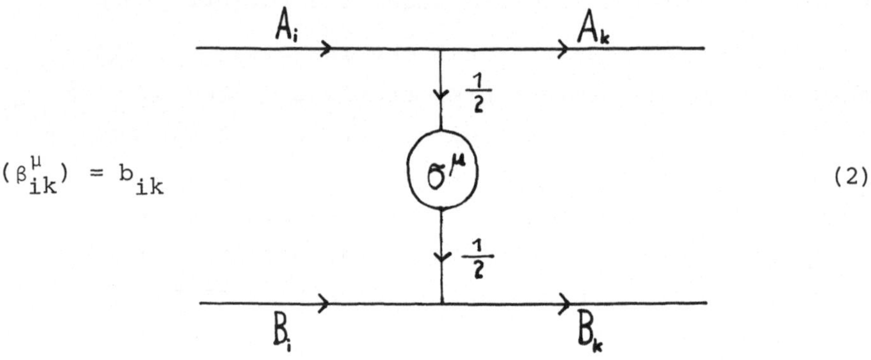 (2)

Here each vertex denotes a CGC and the σ^μ are the usual Pauli matrices, transforming between cartesian and spherical tensor components.

We see that the representation D(g) necessarily has to be reducible. It has to contain at least one pair of irreducible representations i,k with $A_i = A_k \pm 1/2$, and, independently, $B_i = B_k \pm 1/2$. We shall call such a pair of representations interlocking or linked. Thus the fundamental building blocks for the β-matrices are products of two CGC's, the only non-vanishing matrix elements being those connecting two interlocking representations. For each interlocking pair i,k in the transformation law D(g) there is a free parameter b_{ik} in the β-matrices (the hermiticity of β^o requires b_{ik} to be proportional to b^*_{ki}, * denoting the complex conjugate). Clumsy as this expression for the β^μ may seem, we found it the most convenient one for all practical purposes. It has the advantage that all the algebraic problems can be reduced to problems of the recoupling of angular momenta, for which elegant and powerful graphical techniques have already been developed.

2. MASSES AND PARITIES

With the only exception of the Dirac equation, the general wave equation contains more than one particle (and the corresponding antiparticle), some possibly also having infinite mass. The mass spectrum is determined by the inverse eigenvalues of the matrix β^o,

$$\beta^o\, u(\alpha) = \lambda_\alpha\, u(\alpha),\ \lambda_\alpha = 1/m_\alpha \quad . \tag{3}$$

Here α stands for a triplet of indices to label the different solutions, $\alpha = (s\ s_z\ \rho)$: s and s_z denote the spin and its z-component, and ρ is a degeneracy parameter to distinguish solutions with the same spin. It is easy to see that with every λ_α also $-\lambda_\alpha$ is a solution of the eigenvalue equation (3). We define the diagonal matrix β_5 by $(\beta_5)_{ik} = \delta(i,k)\ (-1)^{2A_i}\ 1_i$, where 1_i denotes the unit matrix in the $(2A_i+1)(2B_i+1)$-dimensional space of the representation $i = (A_i, B_i)$. As the only non-vanishing matrix elements of β^μ are those connecting two interlocking representations i and k, β_5 anticommutes with all the β^μ.

The spinors $u(\alpha)$ describing a particle of spin s at rest transform under rotations r according to

$$D(r)\ u(s\ s_z\ \rho) = u(s\ s_z'\ \rho)\ D^{(s)}(r)_{s_z'\ s_z}\ , \tag{4}$$

$D^{(s)}(r)$ being the usual rotation matrix. In accordance with the decomposition of $D(g)$ into a direct sum of irreducible representations $D^i(g)$, and with arranging the matrix elements of β^μ into the blocks (β^μ_{ik}), we break up the spinor $u(\alpha)$ into $(2A_i+1)(2B_i+1)$-component column vectors $u_i(\alpha)$. Their components are proportional to CGC's because of their transformation property under rotations:

$$[u_i(\alpha)]_{a_i b_i} = z_i(s\rho)\ \langle A_i\ a_i\ B_i\ b_i\ |\ s\ \ s_z\rangle\ . \tag{5}$$

The N constants $z_i(\alpha) \equiv z_i(s\rho)$ are determined by the eigenvalue equation (3). Using our explicit expression (2) for the matrix β^o, we find that for a fixed spin s the eigenvalue problem of β^o can be reduced to an eigenvalue problem of a N-dimensional matrix $\Lambda(s)$, N being the number of irreducible representations contained in the transformation law

D(g) of the wave function:

$$\Lambda\ (s)\ \vec{z}(\alpha) = \lambda_\alpha \vec{z}(\alpha)\ . \tag{6}$$

The matrix elements of the "reduced mass matrix" $\Lambda(s)$ are given by Racah 6j-coefficients:

$$\Lambda_{ik}(s) = b_{ik}(-1)^{\phi}\sqrt{2} \begin{Bmatrix} A_i & B_i & s \\ B_k & A_k & 1/2 \end{Bmatrix}\ , \tag{7}$$

with the phase $\phi = A_i + B_k + s + 1/2$. This reduced mass matrix acts on the "reduced wave function" $\vec{z}(\alpha)$ with the N components $z_i(\alpha)$. As already familiar from the atomic and nuclear shell models, the dependence on the magnetic quantum numbers s_z, a_i, b_i,.... can be completely factored out of the dynamical equations; the magnitude of the spin s and the other angular momenta A_i, B_i,... enter in the form of 6j-symbols or higher recoupling coefficients. For hermitean β^o the reduced mass matrix also has to be hermitean.

For every eigenvector $u(\alpha)$ of β^o belonging to the eigenvalue λ_α there is a corresponding reduced wave function $\vec{z}(\alpha)$ which is an eigenvector of the reduced mass matrix $\Lambda(s)$ to the same eigenvalue λ_α. The inverse is not true. The N-dimensional hermitean matrix $\Lambda(s)$ has always N nontrivial and linearly independent eigenvectors which, however, do not all lead to linearly independent eigenvectors of β^o, as some of the CGC's in eq.(5) may be identically zero. β^{o} need not have exactly N states for every spin in its spectrum, the various branches of its mass spectrum starting and ending at different spin values. This does not cause any problems as we always can choose a suitable basis of the eigenvectors $\vec{z}(\alpha)$ such that the supernumerary, i.e. linearly dependent vectors $u(\alpha)$ vanish identically.

Qualitatively, we find that the mass spectrum of a wave equation containing N irreducible components in its transformation law D(g) consists of N/2 positive branches of finite length, plus the corresponding **antiparticles**; for odd N one branch lies at infinity corresponding to vanishing eigenvalues of β^{o}. The various branches of the mass spectrum may overlap, some of them may have discontinuities and infinities. The spin dependence within one branch is determined by a certain function of Racah coefficients. The free parameters b_{ik} in the β-matrices are overdetermined by the mass spectrum, i.e. there are in general more states in the theory than there are free parameters. Thus not every arbitrarily given mass spectrum can be fitted by such a wave equation.

For the simplest examples of wave equations containing only two linked representations 1 and 2, the positive mass spectrum consists of one single branch, the spin dependence of which is rigidly fixed by a Racah coefficient. There is one free parameter b_{12} in the theory, which may be used to set arbitrarily one mass in the spectrum; all the other masses are then uniquely determined. We find two different types of wave equations:

Type I: 1 = (A,B), 2 = (A+1/2, B+1/2) leads to a mass spectrum increasing with spin, whereas

Type II: $1 = (A,B)$, $2 = (A+1/2, B-1/2)$ gives a decreasing mass spectrum.

We shall see that these two types of wave equations differ also in other physical properties like the magnetic moments etc. Wave equations containing more than two irreducible components in the transformation law D(g) show a more flexible mass spectrum as there are more free parameters in the theory.

Compared to, say, the observed nucleon resonances, the mass spectrum of the general wave equation(1) looks physically more reasonable than those of most infinite-component wave equations considered up to now; not to mention the linearly rising, single valued Regge trajectories. Qualitatively, the observed hadron resonances seem to form various (presumably infinitely many) branches of finite length, as we have found them for these general wave equations. It remains to be seen whether this qualitative agreement can be made a quantitative one.

Representing parity in the usual way by a unitary transformation, $\psi'(x') = V\psi(x)$, $x' = (x_o, -\vec{x})$, the parity matrix V has to satisfy the group property $V\, D(g)\, V^{-1} = D(g)^{\dagger -1}$. Thus to accomodate parity, the transformation law D(g) has to be pseudo-unitary. As $D^{AB}(g)^{\dagger -1}$ is equivalent to $D^{BA}(g)$, D(g) has to contain with every irreducible component $i = (A_i, B_i)$ also its conjugate representation $\bar{i} = (B_i, A_i)$. In general this can only be accomplished by adding to D(g) its corresponding conjugate representations, i.e. by a doubling of the degrees of freedom. This parity doubling has far from trivial consequences for the mass spectrum of the general wave equation: the spectrum of the parity-doubled theory does not simply consist of doublets of particles having opposite parity. The splitting of these parity doublets is the rule rather than the exception. With certainty we can only predict the parity-doubled theory to contain twice as many states as the theory without parity; we have much freedom how to split these doublets and even to change their parity eigenvalues.

To determine the intrinsic parities of the various particles described by the wave equation, we do not need the full parity matrix V. We only have to know the effect of the parity transformation on the reduced wave functions $\vec{z}(\alpha)$. For a fixed spin s, it is given by the reduced parity matrix $V(s)$ with the elements

$$V(s)_{ik} = (-1)^{s-F} \, v_i \, \delta(\bar{i},k) \quad , \tag{8}$$

with the number $F = 0$ for bosons, $F = 1/2$ for fermions. The v_i are spin-independent constants with $|v_i| = 1$ and $v_{\bar{i}} = v_i^*$, * denoting the complex conjugate. The reduced mass-matrix $\Lambda(s)$ and the reduced parity matrix $V(s)$ have to commute, giving further conditions on the constants v_i. We remark that in general there exist more than one inequaivalent solutions for the parity matrix V. Physically this means that we have a certain freedom in choosing the relative parities of the various branches in the mass spectrum. The parities within one branch alternate like $(-1)^{s-F}$, except at singularities: there the particles immediately to the right and left of the infinity have the same intrinsic parity.

3. MAGNETIC MOMENTS

As one of the first interesting problems, we have calculated the magnetic moments predicted by the general wave equation (1) with minimal coupling to an external electromagnetic field. Defining the g-factor as usual, $\vec{\mu} = g(e/2m)\,\vec{S}$, $\vec{\mu}$ being the magnetic moment and $\vec{S}$ the angular momentum matrices for spin s, we obtain for the g-factor $g(\alpha)$ of a particle in the state $\alpha = (s\ s_z\ \rho)$:

$$g(\alpha)\ \vec{S} = m_\alpha\ u^\dagger(\alpha)\ \vec{K} \times \vec{\beta}\ u(\alpha)\ , \tag{9}$$

or, multiplying with $\vec{S}$,

$$s(s+1)\ g(\alpha) = m_\alpha\ u^\dagger(\alpha)\ (\vec{J} \times \vec{K}) \cdot \vec{\beta}\ u(\alpha)\ . \tag{10}$$

Here $\vec{J}$ and $\vec{K}$ are the generators for rotations and Lorentz transformations (boosts) of the transformation law D(g); the mass m_α of the state α appears for dimensional reasons because of our standardization (1) of the wave equation. On the left hand side of eq. (10) we have suppressed the unit matrix in the (2s + 1)-dimensional spin space. With the help of our explicit realization (2) of the matrices $\vec{\beta}$, the g-factor can be evaluated as a sum over 12j-symbols. We obtain as final result

$$g(\alpha) = \sum_{i,k} f_{ik}(\alpha)\ g_{ik}(s)\ , \tag{11a}$$

the double sum going over all interlocking pairs of representations contained in D(g). In this formula

$$f_{ik}(\alpha) = m_\alpha \; z_i^*(\alpha) \quad \Lambda_{ik}(s) \; z_k(\alpha) \tag{11b}$$

and

$$g_{ik}(s) = \frac{(\Delta_i - \Delta_k)^2 - s(s+1)}{s(s+1)} \tag{11c}$$

with

$$\Delta_i = A_i(A_i + 1) - B_i(B_i + 1) \; . \tag{11d}$$

The kinematical weight factors $f_{ik}(\alpha)$ are computed from the reduced wave function $\vec{z}(\alpha)$ and the reduced mass matrix $\Lambda(s)$. They allow to describe in a phenomenological way systems with internal structure. These weight factors have the symmetry properties $f_{ik} = f_{ki} = f_{\bar{i}\bar{k}}$, and satisfy the sum rule $\sum f_{ik} = 1$; they are real but not necessarily positive.

The numbers $g_{ik}(s)$ have a simple physical significance: they are the g-factors predicted by the wave equation corresponding to the single interlocking pair i,k of irreducible representations. For wave equations of

<u>Type I:</u> $i = (A,B)$, $k = (A + 1/2, B + 1/2)$, we find the g-factor always to be negative, $-1 \leq g_{ik}(s) \leq -1/(s+1)$, whereas for

Type II: $i = (A,B)$, $k = (A + 1/2,\ B - 1/2)$, the g-factors are always positive, $1/s \leq g_{ik}(s) < \infty$, there being no upper limit on the magnitude of the magnetic moment.

Only for a very special class of wave equations do we get the result $g(s) = 1/s$ conjectured some time ago.

Of special interest are the results for spin $\frac{1}{2}$-particles. To obtain a particle of spin 1/2 with finite mass in a wave equation corresponding to the linked pair of representations i and k, both the triplets of angular momenta $(\frac{1}{2}, A_i, B_i)$ and $(\frac{1}{2}, A_k, B_k)$ have to satisfy the triangle condition. This is possible in two different ways:

Type I: $i = (A, A + \frac{1}{2})$, $k = (A + \frac{1}{2}, A + 1)$ gives a g-factor of $g_{ik}(s = \frac{1}{2}) = -\frac{2}{3}$, independent of A. (12a)

Type II: $i = (A, A + \frac{1}{2})$, $k = (A + \frac{1}{2}, A) = \bar{i}$. Here we obtain

$$g_{ik}(s = \tfrac{1}{2}) = \frac{(4A + 3)^2 - 3}{3} . \qquad (12b)$$

Numerically:

A	0	$\frac{1}{2}$	1	$\frac{3}{2}$	2	...
g_{ik}	2	$7\frac{1}{3}$	$15\frac{1}{3}$	26	$39\frac{1}{3}$	...

The case A = 0 corresponds of course to the Dirac equation. Here the magnetic moment depends rather strongly on A, for example, A = 20 giving a g-factor of about 23oo!

The general wave equation does not predict a unique value for the magnetic moment of a particle with spin s; according to eq. (11a) the g-factor does not only depend on the type of wave function (i.e. on the specific transformation law D(g) of the wave function under the Lorentz group), but also on the details of the mass spectrum (i.e. on the free parameters b_{ik} in the β^{μ}).

4. RENORMALIZATION OF THE AXIAL-VECTOR COUPLING CONSTANT

The second static physical observable which is sensitive to an internal structure is the renormalization of the axial-vector coupling constant in beta-decay. We write for spin 1/2-states at rest:

$$C\, u^{\dagger}(\alpha)\ \beta^{\mu}\ (1 + \beta_5)\ u(\alpha)\Big|_{s\,=\,1/2} = u_D^{\dagger}\ (C_V\gamma^{\mu} + C_A\ \gamma^{\mu}\ \gamma_5)u_D . \tag{13}$$

Here u_D and the γ's are the usual Dirac spinors and matrices. Observing that $\vec{J}_D \cdot \vec{\gamma} = \frac{3}{2}\ \gamma^{o}\gamma_5$, with $\vec{J}_D$ denoting the generators for rotations of the Dirac theory, we obtain for the ratio

$$C_A/C_V = \frac{2}{3}\, m_{\alpha}\quad u^{\dagger}(\alpha)\ \ (\vec{J}\cdot\vec{\beta})\ \ \beta_5\ u(\alpha)\ . \tag{14}$$

A calculation very similar to the one of the magnetic moments leads to the result

$$C_A/C_V = \sum_{i,k} f_{ik}(\alpha)\, C_{ik} \ , \tag{15a}$$

with the same weight factors as in eq. (11b), and

$$C_{ik} = (-1)^{2A_k} \frac{2}{3} (\Delta_i - \Delta_k) \ , \tag{15b}$$

where again $\Delta_i = A_i(A_i + 1) - B_i(B_i + 1)$.

The C_{ik} describe the renormalization of the coupling constant for theories corresponding to the single interlocking pair i,k. As before, we find qualitatively different results for the two types of wave equations:

Type I: $i = (A, A + 1/2)$, $k = (A + 1/2, A + 1)$ gives $C_{ik} = (-1)^{2A+1} \frac{1}{3}$ i.e. $|C_A/C_V| = 1/3 < 1$ independent of A.

Type II: $i = (A, A + 1/2)$, $k = (A + 1/2, A) = \bar{i}$. Here $C_{ik} = (-1)^{2A} \frac{1}{3}(4A + 3)$, giving a ratio $|C_A/C_V|$ always larger than one. Numerically

A	0	$\frac{1}{2}$	1	$\frac{3}{2}$	2	...
$\lvert C_{ik}\rvert$	1	$1\frac{2}{3}$	$2\frac{1}{3}$	3	$3\frac{2}{3}$	...

5. DISCUSSION

It is perhaps worthwhile to mention that all the wave equations (1) considered above (with non-singular β^o) lead to acceptable quantum field theories without the difficulties found for some higher spin equations. For all these wave equations we can define an invariant scalar product leading to a positive definite metric in Fock space; when second quantized in the usual way, these theories have positive definite energy and are local, i.e. satisfy the microcausality condition. Furthermore, these equations can be derived from a Lagrangian and are causal, at least when minimally coupled to an external electromagnetic field. Thus, from a purely theoretical point of view, all these equations are as good as the Klein-Gordon or Dirac equations. They are dynamically inequivalent, the numerical values for the observables of a single particle depending on the masses and parities of the other states in the spectrum.

These wave equations offer the possibility to describe in a phenomenological way particles with an internal structure like the proton, using as input (some of) the masses and parities of the spectrum of its excited states. Physically this is quite plausible as we expect the interaction which gives rise to the nucleon resonances also to be responsible for the structure of the proton as observed in its weak and electromagnetic interactions. Calculations with wave equations containing $N = 4$ irreducible representations are encouraging. We can get the observed values for $g \sim 5.6$ and $|C_A/C_V| \sim 1.2$ of the proton at the expense of a somewhat unrealistic mass spectrum. However, this result is inconclusive: because of the many known nucleon resonances it is not quite clear which ones (or what suitable averages) to use in the two branches of the mass spectrum. To obtain a more realistic model, we want to study wave equations with at least five branches in the mass spectrum, as there are already five spin $\frac{1}{2}$-states known for the nucleon. The phenomenological description of the nucleon which might be achieved in this way is certainly not a simple one; the real question is whether such a description can be made an accurate one.

To conclude, we have learned that the wave function or field operator we can associate with a given particle is not uniquely determined by its spin alone. We need further experimental information like the mass spectrum and parities of the multiplet to which this particle belongs. The problem of constructing relativistic interactions is therefore not only to couple certain fields in a Lorentz invariant way, but first, and perhaps most important of all, to select the correct type of wave function for a given particle. This choice is determined by experiment and not by reasons of mathematical convenience or simplicity. Hence we find it important to study such wave equations in great detail to learn more about the structure of particles they describe.

Acta Physica Austriaca, Suppl. XIV, 567 — 580 (1975)

STRUCTURAL QUESTIONS IN QUANTUM FIELD THEORY+

by

R. HAAG
DESY, Hamburg

1. INTRODUCTION

You will realize that the title of my series of lectures arose from an emergency situation: I had to give a title at a time at which I was very uncertain about the choice of subject matter and style that would be most suitable to this meeting. After coming here and listening to a number of talks I was most fascinated by the aspect that physics, although it is commonly and justly classified among the exact sciences, draws very heavily in its development on undercurrents of a metaphysical nature and that there is a great diversity of those. To encounter several of them side by side in an unhurried atmosphere is perhaps the greatest benefit which the participants of this school may gain.

In particular I whole heartedly support Prof.Schwinger's plea for tolerance in the sense that it is quite unpredictable which attitude at any given time may prove most fruitful.

+ Lecture given at XIV. Internationale Universitätswochen für Kernphysik,Schladming,Austria,February 24-March 7, 1975.

A person engaged in creative research must have a strong faith in his own ideology, he must be convinced about the abilities of the horse on which he places his bets but he should not carry this so far as to think that all other horses are lame. I have heard great physicists express the view that there can be no inspiration in physics besides that provided by experiment. Equally great physicists have maintained that it is the continuous reexamination of the basic concepts which leads to progress. Still other great physicists have taken the beauty of mathematical structures as a guiding principle and operated on the belief that God certainly must have made use of them in building the world.

Of course there is some common ground. A minimal knowledge in all the three pillars: experiment, philosophy (conceptual structure), mathematics is required from anybody who wants to give a useful contribution. For elementary particle physics such a lower limit might be: in mathematics a familiarity with the main ideas and techniques existing at the end of the 19th century. In the conceptual structure, the physical principles, one has to reach at least the status existing around 193o. As for experiments a lower limit of knowledge should include the principal results known in 195o. I have chosen these lower limits not without reflection but with the purpose of shocking you into considering the large scale aspects of our science. Obviously there has been a tremendous development in mathematics during the past 80 years and some of it is or will be very pertinent in the development of particle physics. Similarly there is a tremendous amount of experimental information and theoretical digest of it resulting from the work of the past 25 years. But to make progress on a specific problem only a very small part of the one or the other stores of information is needed.

It is remarkable that no so much progress has been made in the past 40 years in the area of the third pillar, the basic concepts, the physical principles. And this is not because of a lack of attempts and ideas. Many interesting ideas have been proposed but all really bold ideas turned out to be unsuccessful or premature whereas the conceptual structure of 1930 turned out to be much more elastic and adaptable than one had any right to expect. What has been achieved is a somewhat deepened understanding of the principles known in 1930 and even in that respect there are gaps; outstanding is still a convincing fusion between General Relativity and Quantum Physics. What I want to do in today's lecture is to describe the conceptual structure (leaving out General Relativity) in a language which I find most economical. To put this in the right perspective let me emphasize beforehand that I expect, as probably many of you do, a major advance of fundamental theory within the next 10 or 2o years, an advance which may well render the frame I am about to describe obsoletely.

The principles which I want to incorporate are those of Special Relativity and Quantum Physics. Special Relativity means on the one hand the invariance of the laws under Lorentz transformations and space-time translations (together: invariance under the Poincaré group). On the other hand it means that no physical effect can travel faster than light i.e. it gives a sharpened form to the Faraday-Maxwell principle of locality in contrast to theories which allow action at a distance. To emphasize this point we may call the resulting framework that of "local, relativistic quantum physics". The principles of Quantum Physics are less clear cut in the sense that it is difficult to distinguish essential and peripheral features. Certainly one is making probability predictions and therefore needs statistical ensembles, traditionally called "states" of a system. One may then distinguish pure states and mixtures and for most purposes concern oneself only with the former. Secondly, there are experimental apparatus by which one can change a state and extract information. Among these one is usually concerned in particular with such idealized instruments which correspond to the concept of an "observable" in the sense of Dirac[1] and von Neumann[2]:

One should keep in mind that actual instruments are often not of this nature and that it may be more natural to formulate Quantum Physics in terms of a more primitive notion of apparatus e.g. the notion of "operation" instead of "observable" as sketched in[3]. Nevertheless we shall focus on observables here, mainly because the concept of compatible observables gives a very simple and direct translation from the physical principle of locality to the mathematical structure.

Usually one describes pure states by vectors in a Hilbert space and observables by self-adjoint operators acting in this space. A less familiar formulation is the description in terms of C^*-algebras. This is somewhat deeper and allows one to understand some phenomena which in the Hilbert space description have to be fed in somewhat ad hoc. I shall describe this here, giving first a brief mathematical account of what a C^*-algebra is, then show how it is used in Quantum Physics in general, then give the adaption to local, relativistic Quantum Physics and point out the type of question for which this point of view is more appropriate than the Hilbert space version.

2. WHAT IS A C^*-ALGEBRA?

Consider first bounded linear operators acting in a Hilbert space H. The set of these is usually denoted by $B(H)$. Within this set one can perform the following algebraic operations

1) αA (multiplication of the element $A \in B(H)$ by a complex number α)

2) $A+B$ addition

3) AB multiplication

Together: $B(H)$ is an "algebra over the field of complex numbers".

Furthermore: To every $A \in B(H)$ there is an adjoint (Hermitean conjugate) in $B(H)$. We denote it by A^*. The operation

4) $A \to A^*$

is sometimes called an involution because of the relation

$$(AB)^* = B^*A^*.$$

So $B(H)$ may be called an algebra with involution or, shorter, a *-algebra.

Furthermore: Each $A \in B(H)$ has a norm $\|A\|$, defined by

$$\|A\| = \sup_{\Psi} \frac{\|A\Psi\|}{\|\Psi\|}$$

where the norm symbols on the right hand side means the lengths of the vectors. This operator norm defines a topology on B(H) and, in fact, B(H) is complete with respect to it. So B(H) is also a Banach space and may be called a Banach *-algebra or B^*-algebra. One finds that the norm has the property

5) $||A^*A|| = ||A||^2$

Consider next any subset $A \subset B(H)$ which is closed under the algebraic operations 1) - 4) and closed in the norm topology. Such a subset is the prototype of a C^*-algebra. (It is obviously a B^*-algebra and, because the norm has the property 5) and C follows B in the alphabet it is called a C^*-algebra).

The most important step comes now. We may disconnect the entire structure from Hilbert space and consider A as an abstract algebra, a set of objects induced with the algebraic structure 1) - 4) and a norm satisfying 5). This step is analogous to the one in group theory where instead of a concrete transformation group of some space one focuses the attention on the group structure alone and considers the group defined by its algebraic relations (multiplication table in that case). In this way we arrive at an abstract C^*-algebra. It will be obvious from the analogy with groups that this point of view becomes important if and only if there exist inequivalent representations of the same abstract C^*-algebra A by concrete operator algebras in Hilbert spaces. One comment should be added: One might be inclined to consider other topological algebras and indeed they do enter into some physical problems. However a norm topology with the special property 5) is in some sense most closely related to the algebraic structure 1) - 4) (see the theory of the "minimal regular norm"). Loosely speaking, the norm, if it exists at all, is already determined by the algebraic structure 1) - 4).

C^*-algebras in Quantum Physics

Emphasizing the algebraic structure (e.g. commutation relations) as the primary thing and Hilbert space only as a representation vehicle is close to the spirit of the approach of Heisenberg-Born-Jordan and Dirac's concept of q-numbers. A full fledged, precise formulation of this point of view was first given by I.E. Segal[4]. The scheme is as follows:

Consider an abstract C^*-algebra $\mathcal{A}$ as the primary object. The "observables" are supposed to be mathematically represented by the self-adjoint elements of $\mathcal{A}$, i.e. $A \in \mathcal{A}$ with $A^* = A$. One notes that the spectrum of a self adjoint element can be algebraically defined.

What are the states? A C^*-algebra determines a dual object, the set of all bounded linear forms. Among these the normalized, positive ones correspond to states. Specifically a "state" ω is any expectation functional which assignes a number $\omega(A)$ to each $A \in \mathcal{A}$ such that

i) $\omega(\alpha A_1 + \beta A_2) = \alpha\omega(A_1) + \beta\,\omega(A_2)$ (linearity)

ii) $\omega(A^*A) \geq 0$ (positivity)

iii) $\omega(I) = 1$ (normalization) .

This set contains, of course, mixtures as well as pure states. It is a convex set i.e. if ω_1 and ω_2 are states then $\omega = \lambda\,\omega_1 + (1-\lambda)\,\omega_2$ is also a state when $0 \leq \lambda \leq 1$. This represents the mixture of ω_1 and ω_2 with the weights

λ and $1 - \lambda$. The extremal points of this convex set i.e. those which can not be decomposed into a nontrivial mixture are the pure states.

Comment: In contrast to usage in the usual treatment of Quantum Mechanics $\mathcal{A}$ contains only <u>bounded</u> observables, no quantities like the position <u>or</u> momentum whose spectrum runs out to infinity. From the point of view of principles this makes no difference because $\mathcal{A}$ contains all bounded functions of these unbounded observables e.g. all spectral projections. The use of unbounded observables is often very useful computationally just as the use of non-square integrable functions is in the Hilbert space version, but it can be avoided or else dealt with by associating some idealized elements.

<u>Representations and families of states</u>

Given a representation $A \to \pi(A)$ (the abstract element A is represented by the bounded linear operator $\pi(A)$, acting in a Hilbert space $\mathcal{H}$). Pick any unit vector $\Psi \in \mathcal{H}$ then

$$\omega_{\Psi}(A) = (\Psi, \pi(A)\Psi)$$

has the properties i), ii), iii) and hence is a state over $\mathcal{A}$. Thus a representation π furnishes a family of states, the so called vector states of the representation.

Conversely, given any state ω one can construct a representation π_{ω} in which ω appears as a vector state. If ω is pure then π_{ω} will be irreducible. This is done by the Gelfand-Naimark-Segal construction (GNS-constr.)

which occurs like the daily bread in almost any paper using C^*-algebra techniques.

Now the essential feature is that the pure states over A decompose into disjoint families, each family being associated (as vector states) with one equivalence class of irreducible representations of A.

This shows that the C^*-algebra approach becomes important in precisely those situations in which A possesses several inequivalent irreducible representations. This happens typically in all systems with infinitely many degrees of freedom (Quantum Field Theory, Quantum Statistical Mechanics of infinite media) whereas in the Quantum Mechanics of a finite number of canonical variables one has essential uniqueness of the representation and hence there is no point there to use the algebraic approach.

At the same time we see that the algebraic approach gives a natural explanation to the phenomenon of superselection rules: No coherent superposition of state vectors which belong to different families is possible. This phenomenon was first pointed out in[5] in the example of states with integer and half integer spin.

Local, relativistic Quantum Physics

Add now to the structure the principle of locality. Then instead of observables in general we will talk of local observables which can be measured in a specified region O of space-time. This leads to the replacement A by a net of local algebras: To each region O of space-time we have a subalgebra A(O) generated by the observables of that region. The essential observation is here that giving the net i.e. the correspondence

(1) $\quad O \to A(O)$

defines the physical theory, completely including the interpretation. This can be understood by remembering that the correspondence between a particular instrument and its mathematical representative (selfadjoint element of A) can never be a priori known but only tested by the relations which result if various such instruments are placed in different geometric arrays. Thus the starting point is always only the knowledge about the localization of the apparatus.

Of course, if we wanted to construct a theory by arbitrarily making the assignment (1) this "theory" would not have the slightest ressemblance to nature. The principles which should at least be respected are

a) relativistic causality:
if O_1 and O_2 lie space like to each other then the observables in $A(O_1)$ are compatible with those in $A(O_2)$ and hence these two algebras should commute with each other.

b) primitive causality (existence of a dynamical law):
if O_i is a set of regions such that $\bigcup O_i$ has no causal complement (no point space-like to all O_i) then the algebra generated by all the $A(O_i)$ must already be the full algebra A.

c) Poincaré invariance: To each Poincaré transformation L there should be an automorphism α_L of A satisfying the group property $\alpha_{L_1}\,\alpha_{L_2} = \alpha_{L_1 L_2}$ and having the correct geometric action:

$$\alpha_L(A(O)) = A(LO).$$

d) Stability: In the simplest form: existence of a dynamical ground state, the vacuum.

Comments. Comparing this frame with conventional Quantum Field Theory we note:

1) The physical interpretation is based exclusively on the space-time placement of observables. Emphasis on the local algebra not on individual elements of it. Thus a specific net (one theory) may be generated in a variety of ways using different types of fields (e.g. different fields in one Borchers class). The fields play a role analogous to coordinates, they have no intrinsic significance.

2) The net of abstract algebras is considered to be the primary object of the theory. A representation of it in a Hilbert space selects a specific subset of states which may be of interest for a particular set of experiments whereas other representations may be more suitable for other questions about the same physical system. This is of particular importance in statistical mechanics where the superselection labels which distinguish families are quantities like temperature and chemical potential. In elementary particle physics the states of interest are more restricted: We may confine our attention to that class S_o of states which all look like the vacuum state when tested by observables in very far away regions of 3-dimensional space at any finite given time. Still even among this restricted class of states there are superselection rules operating. They are related to charge quantum numbers and govern the distinction between particle types with different statistics (Bose, Fermi, para).

3) The net of observables is local in the sense of commutativity at far distances, not anticommutativity. It is

therefore a not obvious problem how to obtain in this frame the existence of particles which are not Bosons. It is very gratifying that this problem can indeed be solved and that it turns out that local commutativity of the observables is indeed the reason for the permutation symmetry of any kind for multiparticle states and that the types of statistics allowed are limited to para-Bose and para-Fermi of some order p and that the structure of the observable algebra alone determines the statistics of the particles described by the theory.

Statistics and charge quantum numbers

In 1953 H.S. Green[6] proposed commutation relations for quantum fields which described particles with a more complicated permutation symmetry than that of Bose- or Fermi statistics. A para-Fermion of order p has n-particle wave functions which are restricted only by the requirement that they cannot be symmetrized in more than p arguments. Similarly for a para-Boson there is a limit in the number of variables in which the wave function can be antisymmetric.

If one wants to analyze the structure of such a theory within the frame sketched above the first question is: Which expressions in the para-field can be chosen to generate the net of local observables. This was analyzed in[7] with the following result:

A para-Fermi field of order p allows a maximal net of local observables A_o which can also always be described by starting from a system of p ordinary Fermi fields and then selecting the observables as these functions of the set of Fermi fields which are invariant under the internal symmetry group SO(p) for even p and O(p) for odd p. Since the internal symmetry has to be exact and any observable is invariant under it we may speak of a "gauge group" instead of an internal symmetry. If a smaller algebra than A_o is the observable algebra then the gauge group is correspondingly bigger. A case of particular interest is that where U(p) is the gauge group. In that case an irreducible representation of the para-field algebra contains all states of physical interest. For smaller gauge groups (larger observable algebras) an irreducible representation of the para-field algebra is not enough to obtain all physical states (all families of states over A which belong to S_o) and hence the description in terms of a para-field is then not very convenient.

One may approach the question of statistics directly from the structure of the observable algebra[7,8,9]. One finds that the existence of superselection rules within S_o is related to the existence of localized morphisms of the observable algebra. A morphism of A is a mapping $A \to \rho(A)$ from A into A conserving the algebraic structure. It is called localized in O if it acts trivially on observables space-like to O ($\rho(A) = A$ if A is in the space-like complement of O). A trivial example of such a morphism is the "inner automorphism" σ_u induced by a unitary element $U \in A(O)$ by

$$\sigma_u(A) = U A U^*.$$

Such automorphisms are, of course, irrelevant for the superselection rules. We may divide localized morphisms into equivalence classes, calling ρ_1 and ρ_2 equivalent if $\rho_2 = \sigma_u \rho_1$ for some U. The nontrivial equivalence classes (ρ not equivalent to the identity transformation) are in correspondence with the superselection rules. Physically they correspond to charge quantum numbers. They have a composition law (addition of charges), a conjugation and to each charge is attached a "statistics quantum number"

$\lambda = \pm \frac{1}{p}$ (p integer).

The significance of λ is the following: in a family of states with charge n (corresponding to the morphism ρ^n) one has permutation symmetry of para-Bose type with order p if λ is positive, para-Fermi of order p if λ is negative.

REFERENCES

1. P.A.M. Dirac, Principles of Quantum Mechanics.
2. J. von Neumann, Math. Grundlagen der Quantenmechanik.
3. R. Haag and D. Kastler, J.Math. Phys. 5, 848 (1964).
4. I.E. Segal, Ann. Math. 48, 930 (1947).
5. Wick, A.S. Wightman and E.P. Wigner, Phys.Rev.88, 101 (1952).
6. H.S. Green, Phys. Rev. 90, 270 (1953).
7. H.J. Borchers, Commun. Math. Phys. 1, 281 (1965).
8. P. Federbush, J.Math. Phys. 9, 1718 (1968).
9. S. Doplicher, R. Haag and J. E. Roberts, Commun. Math. Phys. 13, 1 (1969); 15, 173 (1969); 23, 199 (1971); 35, 49 (1974).

Acta Physica Austriaca, Suppl. XIV, 581 — 629 (1975)

STOCHASTIC PROCESSES AND QUANTUM THEORY+)

by

J. R. KLAUDER
Bell Laboratories
Murray Hill, New Jersey 07974

INTRODUCTION

The concepts of probability theory and quantum theory have been closely intertwined ever since these subjects were developed, and quantum theory has often been the beneficial recipient of such an interchange. Quantum theory expressed in imaginary time becomes the theory of generalized diffusion in real time and this provides useful insight in either a differential equation formulation or a path-integral formulation. More recently, covariant field theories have been re-examined in the context of imaginary time, and such Euclidean field theories have been defined by means of functional integrals for some super-renormalizable models, and, at least heuristically, have long been studied for special renormalizable models.[1] Whether or not such methods will ultimately prove essential, it certainly cannot be denied that they possess an enormous appeal and provide considerable

+) Lecture given at XIV. Internationale Universitätswochen für Kernphysik, Schladming, Austria, February 24 - March 7, 1975.

intuitive insight.

The subject matter that we deal with in these notes pertains to quantum theory in the same spirit as described above. More specifically, we shall analyze a variety of stochastic processes, namely, real time diffusion phenomena, which are analogues of imaginary time quantum theory for certain examples and also analogues of covariant imaginary time quantum field theory. While direct study of the true problems that arise in a specific covariant quantum field model are paramount, the problems encountered in such studies are sufficiently complicated that study of analogous problems in alternative models can be of considerable value. In an effort to reach the *core* of some of the essential problems that face contemporary quantum field theory, there is every reason to strip away certain features in order to make the difficulties as clear and self-evident as possible. It is this guiding principle, which we freely admit, that motivates our discussion and the problems we consider.

In the next section we elaborate some standard properties involving probability measures and stochastic variables. The section after that is devoted to a simple class of examples that includes one of the clearest cases of a discontinuous perturbation.

In the concluding section, we develop the fact that certain stochastic theories actually exhibit divergences that simulate those of covariant quantum field theory and present examples of both renormalizable and nonrenormalizable behavior. An example of the latter type is proposed and completely solved. We conclude with some challenging questions ripe for further study.

DIFFUSION, MEASURES, AND STOCHASTIC VARIABLES

Schrödinger's equation for a single particle, i.e.

$$i \frac{\partial\psi}{\partial t} = H\psi = \{ -\frac{1}{2} \frac{\partial^2}{\partial x^2} + V(x)\}\psi \quad , \tag{1}$$

is one of the cornerstones of quantum theory, while the generalized diffusion equation,

$$\frac{\partial f}{\partial t} = -Hf = \{ \frac{1}{2} \frac{\partial^2}{\partial x^2} - V(x)\}f \tag{2}$$

is one of the cornerstones of stochastic theory. The close relation between these equations is self-evident and accounts for the rich cross-fertilization that has taken place between these distinct but intimately related fields of physics. Evidently, most of the essential properties of the Schrödinger equation may be examined through a study of the generalized diffusion equation; and starting from this simple point of view we shall develop a rather complete picture of imaginary time quantum mechanics as well as quantum field theory.

A fundamental solution to (2) was given by Kac (following ideas of Feynman) in the form of an integration over paths.[2] We first introduce standard Wiener measure appropriate to conventional Brownian motion. Let C denote the space of continuous paths $x(t)$, $0 \leq t \leq T$, constrained so that $x(0) = 0$. Define a normalized Gaussian measure on such paths with mean zero and covariance $\min(t,t')$; this is standard Wiener measure, μ_W.[3] In symbols

$$\langle x(t)\rangle_W = \int x(t)d\mu_W(x) = 0 ,$$

$$\langle x(t)x(t')\rangle_W = \int x(t)x(t')d\mu_W(x)$$

$$= \min(t,t') ,$$

which for normalized Gaussian measures gives all the needed information to define μ_W. The expression

$$f(y,T) = \int \delta(x(T)-y)e^{-\int_0^T V(x(t))dt} d\mu_W(x)$$

is a solution of (2) subject to the initial condition $f(y,0) = \delta(y)$; this is the Feynman-Kac Formula.

Figuring significantly in this expression is the path space measure

$$d\nu(x) = e^{-\int_0^T V(x(t))dt} d\mu_W(x) ,$$

which is generally not a probability measure since ν is generally not normalized. The normalized form of this expression,

$$d\mu_Y(x) = N_T e^{-\int_0^T V(x(t))dt} d\mu_W(x) , \tag{3}$$

where N_T is chosen so that $\int d\mu_Y(x) \equiv 1$, is a new, generally non-Gaussian probability measure on the space C of continuous paths. The measure μ_Y gives different weight to paths as compared to that of μ_W, which is just the content of (3).

We can view this property in another way. Let W(t) denote a stochastic process (time-dependent random variable) distributed according to the probability measure μ_W. Likewise, let Y(t) denote a stochastic process distributed according to the measure μ_Y. What then is the relation between Y(t) and W(t)? This question can be answered in several interesting ways.

In a recent paper it has been shown that Y(t) and W(t) may be related according to the equation

$$Y(t) = \int_0^t a(Y(s),s)ds + W(t) \tag{4}$$

where a(x,t) is connected to the potential V(x,t) (here assumed to depend on t as well as x).[4] This connection is based on the relation

$$a(x,t) = \frac{\partial}{\partial x} \ln B(x,t)$$

where B(x,t) is the unique, positive definite solution of the equation

$$\frac{\partial B}{\partial t} = -\frac{1}{2}\frac{\partial^2 B}{\partial x^2} + V(x,t)B$$

subject to the final condition $B(x,T) \equiv 1$. These relations establish a connection between the path space probability measures and the associated stochastic processes. Indeed it is convenient to regard this connection as a map from W paths to Y paths through Eq. (4). It is conventional and especially convenient to state (4) as a differentiable relation (Itô equation),[5] i.e.

$$dY(t) = a(Y(t),t)dt + dW(t) \; , \tag{5}$$

which is the mathematical formulation of the Langevin equation of physics often written in the form $\dot{Y}(t) = a(Y(t),t) + \dot{W}$. While this form is heuristically equivalent, such an equation is less preferred since W(t), although continuous, is nowhere differentiable.

The differential equation (5) provides a convenient pictorial relation for the incremental development of Y(t) for a given W(t). If we regard W(t) as the process corresponding to the "free" particle

$$H_o = - \frac{1}{2} \frac{\partial^2}{\partial x^2} \; ,$$

then Y(t) corresponds to the "interacting" particle $H = H_o + V$, and a(Y,t) mediates between the two pictures and especially a = 0 whenever V = 0. In the language of quantum theory, the drift term a(Y,t), represents the effects of the potential in the interaction picture (intermediate between the Schrödinger and Heisenberg pictures). This view can be quite convincingly developed.[4]

Before illustrating these various relations, it is worth indicating still another property of stochastic variables. Let U(t) denote a stochastic process that fulfills the differential equation

$$dU(t) = U(t)a(W(t),t)dW(t)$$

where W(t) is the Wiener process defined before and we choose U(0) = 1 as initial condition. Since W(t) is a stochastic process, the solution to the Itô equation is not as given by ordinary calculus but is given rather generally by[6]

$$U(t) = e^{\int_{t'}^{t} a dW - \frac{1}{2}\int_{t'}^{t} a^2 ds} U(t') .$$

As a stochastic integral (of a nonanticipating functional), U(t) is a <u>martingale</u>,[7] a term applied to any stochastic variable that fulfills the general property

$$E(U(t)|F_s) = U(s), \; t \geq s .$$

Here $E(.|F_s)$ denotes conditional expectation, and F_s represents the fact that all knowledge of the process is prescribed for times less than or equal to s. (Technically F_s denotes a σ-algebra.) Applied to our case we learn that

$$E(U(t)|F_0) = U(0) \equiv 1$$

for all t, $0 \leq t \leq T$. Observe that F_0 is "empty", no conditioning since the processes are fixed at t = 0. In particular, then

$$E(U(T)) = E(U(T)|F_o) = 1.$$

If we use the relation between a and V it may be shown[4] that

$$U(T) = e^{\int_o^T a dW - \frac{1}{2}\int_o^T a^2 dt}$$

$$= N_T e^{-\int_o^T V(W(t),t)dt} ,$$

which is just the expression of interest (cf., Eq.(3)). Thus we observe that we have derived a stochastic variable U(t), which is at the same time a martingale, that is also an object of considerable physical interest. Clearly, U(t) also corresponds to the interaction picture since $V \equiv 0$ trivially implies that $U(t) \equiv 1$.

Some Additional Explanation

Stochastically Equivalent Processes

A stochastic process, say W(t), is called a (standard) Wiener process if it has a Gaussian distribution with mean zero and covariance min(t,t'). However, in quantum theory specifying all the correlation functions uniquely determines the representation only up to unitary equivalence, and there is a corresponding notion in stochastic theory.[6,7] For example, the process

$$\tilde{W}(t) = kW(t/k^2)$$

for any $k \neq 0$ is also a standard Wiener process. Such variables as $W(t)$ and $\tilde{W}(t)$ are termed stochastically equivalent. Another less obvious example stochastically equivalent to $W(t)$ is given by

$$\hat{W}(t) = tW(1/t)$$

since $E(\hat{W}(t)) = 0$ and

$$E(\hat{W}(t)\hat{W}(t')) = tt'\min(t^{-1},t'^{-1}) = \min(t,t') \ .$$

Clearly, then, the realization of a stochastic variable is not necessarily unique, and the general prescription for stochastic equivalence of any two variables $Y_1(t)$ and $Y_2(t)$ is simply

$$\text{Prob.}\{Y_1(t_k)<a_k, k=1,\ldots,K\} = \text{Prob.}\{Y_2(t_k)<a_k, k=1,\ldots,K\},$$

$0 \leq t_1 < t_2 \ldots < t_K \leq T$; i.e., stochastically equivalent variables always have identical finite distributions.[8]

Rudiments of Itô Calculus

Suppose $F(x,t)$ is a C^2 function and that $x(t)$ is a C^1 function. Then the usual chain rule gives

$$dF(x(t),t) = F_x(x(t),t)dx(t) + F_t(x(t),t)dt$$

where $F_x = \partial F/\partial x$, etc. Evidently one has the formula $[x_k \equiv x(t_k)]$

$$\int dF(x,t) = \lim \Sigma \{F_x(x_k,t_k)(x_{k+1}-x_k)+F_t(x_k,t_k)(t_{k+1}-t_k)\}$$

since dx and dt are infinitesimals of the same order. For arguments that are stochastic variables this rule is not generally correct as is easily seen for a Wiener process W(t). First, note that $E((W_{k+1}-W_k)^2) = t_{k+1}-t_k$, where $W_k \equiv W(t_k)$, so that $E((dW)^2) = dt$. Second, one may easily show that

$$E((W_{k+1}-W_k)^2A) \rightarrow E((dW)^2A) = dtE(A)$$

for any stochastic variable A, and thus we have the identity $(dW)^2 = dt$. Guided by this fact, it is plausible to adopt the chain rule

$$dF(W(t),t) = F_x(W(t),t)dW(t) + \frac{1}{2}F_{xx}(W(t),t)(dW(t))^2$$
$$+ F_t(W(t),t)dt \ .$$

As a simple example, consider

$$dW^2(t) = 2W(t)dW(t) + (dW(t))^2$$
$$= 2W(t)dW(t) + dt \ ,$$

which therefore implies

$$2\int_0^T W(t)dW(t) = \int_0^T dW^2(t) - \int_0^T dt$$

$$= W^2(T) - T$$

contrary to the classical (nonstochastic) behavior. As another example consider the equation

$$dU(t) = U(t)dW(t)$$

which has the solution

$$U(t) = e^{W(t)-\frac{1}{2}t}U(0)$$

since

$$dU(t) = U(t)dW(t) + \tfrac{1}{2}U(t)(dW(t))^2 - \tfrac{1}{2}U(t)dt = U(t)dW(t).$$

The rules quoted are due to Itô and we note that they are based on <u>nonanticipating</u> integrands;[6] e.g., one interpretes

$$\int \Phi(W(t))dW(t) = \lim \Sigma\ \Phi(W(t_k))[W(t_{k+1})-W(t_k)]$$

where only dW involves the "future" in each term. The Itô calculus extends to other and multiple stochastic processes, say $X_1, X_2, \ldots, X_p$, with the general rule that

$$dX_i dX_j = c_{ij}dt$$

for some stochastic variables c_{ij}, and

$$dX_i dX_j dX_k = dX_i dX_j dt = (dt)^2 \equiv 0 .$$

Martingales

Consider the Wiener process W(t) and observe that

$$E(W(t)|F_s) = E(W(t)-W(s)|F_s) + E(W(s)|F_s)$$

$$= W(s)$$

since the first term has mean zero. Thus W(t) is a martingale. However, $W^2(t)$ is not a martingale, while $W^2(t)-t$ is one; we leave the argument as an exercise.

Let $X(t) \equiv \int_0^t \Phi(Y)dY(t)$ denote a stochastic integral. Then

$$E(X(t)|F_s) = E(X(t)-X(s)|F_s) + X(s) \; .$$

Due to the definition of a stochastic integral involving non-anticipating functions (or functionals), the first mean vanishes leaving

$$E(X(t)|F_s) = X(s) \; ;$$

in other words, every stochastic integral is a martingale.

Although the implications for quantum theory are not clear we note that martingales enjoy several additional properties.[6] If U(t) denotes a general martingale, then

$$\text{Prob.}\ (\sup_{0\leq t\leq T} |U(t)| > e^c) \leq e^{-c} \quad ,$$

and for any $\alpha > 1$,

$$E\Big(\sup_{0\le t\le T} |U^{\alpha}(t)|\Big) \le \Big(\frac{\alpha}{\alpha-1}\Big)^{\alpha} E(|U^{\alpha}(T)|) \ .$$

ILLUSTRATIVE EXAMPLE

We choose an extremely simple example so as to illustrate certain features as clearly as possible.+) In particular, we adopt

$$V(x,t) = \tfrac{1}{2}\lambda f(t) x^2 \ , \tag{6}$$

simply a quadratic perturbation. This also leads to a Gaussian measure and the relevant formulas may be calculated fairly explicitly.[8] We present the basic relations here as a means to illustrate discontinuous perturbations.

Consider, first, the stochastic equation

$$dU(t) = h(t)U(t)W(t)dW(t)$$

where $W(t)$ is a Wiener process, and $h(t)$ is a nonrandom function. The solution to this stochastic equation subject to $U(0) = 1$ reads

$$U(t) = e^{\int_0^t h(t)W(t)dW(t) - \frac{1}{2}\int_0^t h^2(t)W^2(t)dt} \ . \tag{7}$$

+) The main results in this example are based on unpublished work of H.Ezawa, L.A. Shepp and the author.

We next wish to recast the first term using the Itô calculus, namely the formula

$$d(\tfrac{1}{2}hW^2) = \tfrac{1}{2}\dot{h}W^2dt + hWdW + \tfrac{1}{2}hdt$$

since $(dW)^2 = dt$. Consequently,

$$U(t) = \exp\{\tfrac{1}{2}h(t)W^2(t) - \tfrac{1}{2}\int_0^t [(\dot{h}+h^2)W^2 - h]dt\} .$$

Let us compare this with the desired expression based on (6), i.e., with the expression

$$N_T e^{-\frac{1}{2}\lambda\int_0^T f(t)W^2(t)dt} .$$

To achieve equality with $U(T)$, we set

$$\dot{h}(t) + h^2(t) = \lambda f(t)$$

and insist that

$$h(T) = 0 ;$$

then as a consequence we learn that

$$N_T = e^{\frac{1}{2}\int_0^T h(t)dt} .$$

If we introduce $\dot{g}(t)/g(t) = h(t)$, then these relations read

$$\ddot{g}(t) = \lambda f(t) g(t) \ ,$$

$$\dot{g}(T) = 0$$

and $N_T = [g(T)/g(0)]^{\frac{1}{2}}$. Clearly any scale for g can be chosen insofar as these equations are concerned.

Next, let us ask for the form of the Y paths, which fulfill the equation

$$dY(t) = h(t) Y(t) dt + dW(t) \ .$$

The general solution may be readily given as

$$Y(t) = \hat{g}(t) Y(0) + \hat{g}(t) W(\int_0^t \hat{g}^{-2}(s) ds)$$

where

$$\hat{g}(t) \equiv e^{\int_0^t h(s) ds} = g(t)/g(0) \ .$$

If we insist that $Y(0) = 0$ and choose $g(0) = 1$, then it follows that

$$Y(t) = g(t) W(\int_0^t g^{-2}(s) ds) \ ,$$

where $g(t)$ is the solution given earlier. (Indeed according to remarks made earlier any other choice for $g(0)$ leads to an equivalent process.) Observe that this relation for $Y(t)$ just involves a scaling plus a time transformation of the Wiener process, a very simple map indeed between the W paths with the measure μ_W and the Y paths with the measure μ_Y.

<u>Remark</u>

The preceding relation between Y(t) and W(t) holds whether or not the corresponding measures are equivalent. According to (7) for t = T, equivalence of μ_Y and μ_W is ensured if and only if

$$\text{Prob.}\ \{\int_0^T h^2(t)W^2(t)dt < \infty\} = 1 ,$$

which is the existence criterion for both terms in the exponent.[7] In turn, this criterion is equivalent to the condition $\int_0^T h^2(t)tdt < \infty$.[8] Since $h = \dot{g}/g$ and $g > 0$ for all t, equivalence of the measures holds if and only if $t\dot{g}^2 \in L^1(0,T)$.

Special Cases

We now specialize the foregoing even further to a certain class of examples where roughly speaking $f(t) = (T-t)^{-\alpha}$, $\alpha > 0$. Consider first $\alpha = 2$, and introduce $f_\varepsilon(t) = (T-t+\varepsilon)^{-2}$, $\varepsilon > 0$, where ε is a parameter at our disposal (soon to go to zero). The equation for g_ε reads

$$\ddot{g}_\varepsilon(t) = \lambda(T-t+\varepsilon)^{-2}g_\varepsilon(t) ,$$

and has the solution

$$g_\varepsilon(t) = (T-t+\varepsilon)^{\gamma_-} - \frac{\gamma_-}{\gamma_+}\varepsilon^{\gamma_- - \gamma_+}(T-t+\varepsilon)^{\gamma_+} ,$$

where

$$\gamma_{\pm} \equiv \frac{1}{2} \pm \frac{1}{2}\sqrt{1 + 4\lambda} \ .$$

This solution fulfills the boundary condition $\dot{g}_{\varepsilon}(T) = 0$ and is normalized so that

$$\lim_{\lambda \downarrow 0} g_{\varepsilon}(t) \equiv 1$$

since $\gamma_- \to 0$ when $\lambda \downarrow 0$. On the other hand, if we wish to study $\varepsilon \downarrow 0$ when $\lambda > 0$, then we must adopt

$$g_{\varepsilon}(t) = (T-t+\varepsilon)^{\gamma_+} - \frac{\gamma_+}{\gamma_-}\varepsilon^{\gamma_+-\gamma_-}(T-t+\varepsilon)^{\gamma_-} \ .$$

When $\varepsilon \downarrow 0$, $\lambda > 0$, we have

$$g_{\varepsilon}(t) \to (T-t)^{\gamma_+} \ .$$

In this case the Y paths are explicitly given by

$$Y(t) = (T-t)^{\frac{1}{2}(1+\sqrt{1+4\lambda})} W((T-t)^{-\sqrt{1+4\lambda}} - T^{-\sqrt{1+4\lambda}}) \ .$$

Observe as $t \to T$, the factor in front vanishes while the effective time goes to infinity. From the general relation

$$E(Y^2(t)) = g^2(t)E(W^2(\int_0^t g^{-2}(s)ds)) = g^2(t)\int_0^t g^{-2}(s)ds$$

it follows in our case that

$$E(Y^2(t)) = (T-t)[1-(1-t/T)^{\sqrt{1+4\lambda}}]$$

which vanishes at $t = 0$, as it must, and at $t = T$, which is clearly new (and in fact it vanishes linearly).

Now consider the limit $\lambda \downarrow 0$. In that case

$$Y(t) = (T-t)W(t/T(T-t)) \quad .$$

Alternatively, we could have incorporated T into the normalization of g in such a way that

$$Y(t) = (1-t/T)W(t/(1-t/T))$$

which is stochastically equivalent to our previous relation. Next observe for $t < s$ that

$$E(Y(t)Y(s)) = t(1-s/T)$$

using either expression for $Y(t)$.

Finally, one further realization provides the greatest insight of all. Let $\tilde{W}(t)$ be a standard Wiener process and define

$$\tilde{Y}(t) = \tilde{W}(t) - (t/T)\tilde{W}(T)$$

which has zero mean and covariance, for $t < s$, given by

$$E(\tilde{Y}(t)\tilde{Y}(s)) = t(1-s/T) \quad ,$$

as is readily computed. Thus $Y(t)$ and $\tilde{Y}(t)$ are equivalent stochastic variables. Evidently $\tilde{Y}(0) = 0$, $\tilde{Y}(T) = 0$, and $\tilde{Y}(t)$ acts rather like a Wiener process in between. The process $\tilde{Y}(t)$ is called a pinned Wiener process (or the Brownian bridge).[6]

Now let us summarize. Starting with a Wiener process $W(t)$ and measure μ_W we introduced the related $Y(t)$ process and measure μ_Y according to the relation

$$d\mu_Y(x) = N_T e^{-\frac{1}{2}\lambda\int_0^T f_\varepsilon(t)x^2(t)dt} d\mu_W(x)$$

and, correspondingly,

$$Y(t) = g_\varepsilon(t)W(\int_0^t g_\varepsilon^{-2}(s)ds) \ .$$

In the particular case that

$$f_\varepsilon(t) = (T-t+\varepsilon)^{-2}$$

an explicit solution was found such that

$$\operatorname*{w-lim}_{\varepsilon\downarrow 0} \mu_Y = \mu_{Y,0}$$

which is a measure inequivalent (in fact mutually singular) to Wiener measure. (This follows since $tf_\varepsilon(t)$ fails to converge in $L^1(0,T)$.) Finally, we learned that

$$\operatorname*{w-lim}_{\lambda\downarrow 0} \lim_{\varepsilon\downarrow 0} \mu_Y = \operatorname*{w-lim}_{\lambda\downarrow 0} \mu_{Y,0} = \mu_{W,0} \ ,$$

the measure for a pinned Wiener process, not the Wiener process at all. A perturbation introduced and then removed has changed the system; it has left an indelible imprint on the system.

The foregoing example is especially nice since it can be explicitly calculated in terms of very simple functions. We now quote some results of a more extensive study with $f(t) = (T-t)^{-\alpha}$, $\alpha > 0$, and with various choices for $f_\varepsilon(t)$.

Positive Regularizations

Suppose we study $f(t) = (T-t)^{-\alpha}$ and choose

$$f_\varepsilon(t) = (T-t+\varepsilon)^{-\alpha}$$

in direct analogy with the analysis for $\alpha = 2$ given above. Although the analytic behavior is not so simple when $\alpha \neq 2$, the basic results may be simply stated:

$$\operatorname*{w-lim}_{\lambda\downarrow 0} \lim_{\varepsilon\downarrow 0} \mu_Y = \mu_W \quad , \ \alpha < 1 \ ,$$

$$\operatorname*{w-lim}_{\lambda\downarrow 0} \lim_{\varepsilon\downarrow 0} \mu_Y = \mu_{W,0} \quad , \ \alpha \geq 1 \ .$$

In words: For $\alpha < 1$ the interaction can be entirely removed; for $\alpha \geq 1$ the interaction leaves an idelible imprint changing a Wiener process into a pinned Wiener process as we explicitly found for $\alpha = 2$.

Alternative Regularizations

Again consider the formally singular case

$$f(t) = (T-t)^{-\alpha} ,$$

which we now "regularize" in the following manner,

$$\begin{aligned} f_\varepsilon(t) &= (T-t)^{-\alpha} , && 0 \leq t < T-\varepsilon , \\ &= -K\delta(T-t-\varepsilon) , && t \approx T-\varepsilon , \\ &= 0 , && T-\varepsilon < t \leq T , \end{aligned}$$

where the intent of the notation at least is clear. Given this choice of $f_\varepsilon(t)$, set $g_\varepsilon(0) = 1$ (say), choose $\dot{g}_\varepsilon(0)$ any value $O(\lambda)$, and integrate forward adjusting K as need be so that $\dot{g}_\varepsilon(t) \equiv 0$ for all $t > T-\varepsilon$. This is accomplished by choosing

$$\lambda K = \dot{g}_\varepsilon(T-\varepsilon-0)/g_\varepsilon(T-\varepsilon) .$$

Clearly $f_\varepsilon(t) \to f(t) = (T-t)^{-\alpha}$ for $0 \leq t < T$, but since we now consider a different form of regularization potentially different results may arise. Let us now study that question.

The solution $g_\varepsilon(t)$ is well defined and generally involves Bessel functions. Imagine forming $g(t) = \lim g_\varepsilon(t)$ and studying the Y process,

$$Y(t) = g(t)\ W\ (\textstyle\int_0^t g(s)^{-2} ds) , \tag{8}$$

or a stochastically equivalent process. In all cases, for small t, $Y(t) \approx W(t)$, while for intermediate t values, Y(t) may not be identical to a Wiener process but only differs in a quantitative sense, such that as $\lambda \downarrow 0$ it approaches a Wiener process. Only for $t \lesssim T$ is their some potential <u>qualitative</u> variation. We quote next the <u>qualitative</u> behavior of the Y process for $t \lesssim T$ based on (8). For $0 < \alpha < 2$ and $(2-\alpha)^{-1}$ nonintegral, it may be shown that

$$Y(t) \approx [1+O(\lambda)]W(t) + \sum_{j=1}^{[(2-\alpha)^{-1}]-1} k_j \lambda^j (1-t/T)^{(2-\alpha)j} W(T) ,$$

where k_j are α-dependent numerical coefficients that may be determined. For $\alpha < 3/2$, the sum is absent, and so present interest centers on $3/2 < \alpha < 2$. Then the qualitative features are clear: Y(t) does not vanish at $t = T$, but it behaves differently than a Wiener process near $t = T$ because of the fractional powers in the coefficient of W(T). Specifically, we know that

$$E((W(T)-W(t))^2) = T - t ,$$

while

$$E((Y(T)-Y(t))^2) \approx [1+O(\lambda)](T-t) - 2T[1+O(\lambda)]\Sigma k_j \lambda^j (1-t/T)^{(2-\alpha)j+1}$$

$$+ T\{\Sigma k_j \lambda^j (1-t/T)^{(2-\alpha)j}\}^2$$

where the <u>leading</u> temporal behavior as $t \to T$ is (T-t) whenever $2(2-\alpha) > 1$, while for $2(2-\alpha) < 1$ it is given (in part) by

$$E((Y(T)-Y(t))^2) \approx T\, k_1 \Sigma k_j \lambda^{j+1} [(T-t)/T]^{(2-\alpha)[1+j]} + \ldots,$$

where the sum runs from 1 to $[(2-\alpha)^{-1}]-1$. All of the illustrated terms (generally, plus others) dominate temporally over the Wiener process since for each j one has

$$(2-\alpha)[1+j] \leq (2-\alpha)[(2-\alpha)^{-1}] < 1$$

under our assumptions.

In all such cases $(3/2 < \alpha < 2)$ we deal with mutually singular measures, even for different λ values, but it is clear that

$$\lim_{\lambda \downarrow 0} Y(t) = W(t) ,$$

namely that the interaction is removed as $\lambda \downarrow 0$ and a Wiener process emerges.

We omit a parallel discussion for cases where $(2-\alpha)^{-1}$ is an integer, i.e., where $\alpha = (2m-1)/m$, $m = 1,2\ldots$, for which the essentials are the same, but logarithmic terms arise.

Let us restudy the case $\alpha = 2$ in some detail with the new form of regularization. The solution for $g_\varepsilon(t)$

becomes

$$g_\varepsilon(t) = a(T-t)^{\gamma_-} + b(T-t)^{\gamma_+}$$

where a and b are constants such that $a \to 1$, $b \to 0$ as $\lambda \downarrow 0$, and

$$aT^{\gamma_-} + bT^{\gamma_+} = 1 \text{ (say)} .$$

For $T-\varepsilon \leq t \leq T$,

$$g_\varepsilon(t) = g_\varepsilon(T-\varepsilon) = a\varepsilon^{\gamma_-} + b\varepsilon^{\gamma_+} ,$$

and clearly $\dot{g}_\varepsilon(T) = 0$ as needed. The solution for $Y(t)$ can be given and the ε limit taken. Near $t = 0$, $Y(t) \approx W(t)$, while near $t = T$ the essential behavior of $Y(t)$ is represented by

$$Y(t) \approx [1+O(\lambda)](1-t/T)^{\gamma_-} W(t) .$$

Here we encounter a quite different behavior than previously found. Almost every path <u>diverges</u> at $t = T$ because $\gamma_- = \frac{1}{2}[1 - \sqrt{1+4\lambda}] < 0$. Since $Y(T) = \pm\infty$, almost surely, the paths are <u>not</u> almost surely continuous; on the other hand, for any continuous function $s(t)$, $0 \leq t \leq T$, the variable

$$\int_0^T s(t)Y(t)dt$$

is well defined, provided $\gamma_- > -1$ (i.e., $\lambda < 2$), and converges with probability one as $\lambda\downarrow 0$ to

$$\int_0^T s(t)W(t)dt \ .$$

Thus in this elementary distribution sense we may say that $Y(t) \to W(t)$, a.s., as $\lambda\downarrow 0$, and that

$$\underset{\lambda\downarrow 0}{\text{w-lim}}\ \mu_Y = \mu_W$$

even though, in this particular case, it is necessary to extend the space of support for μ_Y outside the space C of continuous functions.

We leave it as an exercise for the reader to consider the case $\alpha > 2$ and to show in each such case that <u>no</u> regularization can be found so that $\mu_Y \to \mu_W$; rather, the only convergent choice satisfies

$$\underset{\lambda\downarrow 0}{\text{w-lim}}\ \lim_{\varepsilon\downarrow 0} \mu_Y = \mu_{W,0} \ ,$$

the pinned Wiener process. The central reason for this behavior arises because: For any $\lambda > 0$, $g(t) = \lim g_\varepsilon(t)$ generally diverges at $t = T$ faster than any inverse power; to accommodate this, each test function $s(t)$ must vanish at $t = T$ faster than any power so that

$$\lim_{\lambda\downarrow 0} \int s(t)Y(t)dt$$

is defined for only a restricted class of test functions; and extension of the limiting stochastic variable to a sufficiently large test function space (e.g., continuous s(t)) involves arbitrariness and ambiguity as usual. The only unambiguous alternative possibility arises when g(t) is proportional to that unique solution which is nonsingular at t = T. The consequence of this choice is that $Y(t) \to \tilde{W}(t)-(t/T)\tilde{W}(T)$ as $\lambda \downarrow 0$; i.e.,

$$\underset{\lambda \downarrow 0}{\text{w-lim}}\ \mu_Y \to \mu_{W,0} .$$

Remarks

We make three additional remarks. A generalization of the foregoing example has been studied by Streit in the form of space-dependent mass perturbations of a relativistic free field.[9] This problem may also be put into the framework of Gaussian random processes, and some results analogous to those presented here have already been obtained.

In addition, a related, but more difficult (non-Gaussian) one-dimensional problem, where the potential is given by

$$V(x) = \lambda|x-c|^{-\alpha}$$

has been rather carefully studied elsewhere with general results that parallel those discussed here in many respects.[10] The interested reader will find the present discussion to be a useful introduction to the more involved case of the anharmonic oscillator treated there.

The elementary example treated here is, in the author's view, a miniature version of what may be expected in nonlinear quantum field theory.[11] In particular, for a scalar field in self interaction: The cases with $\alpha < 2$ correspond to super-renormalizable theories; the case with $\alpha = 2$ to renormalizable theories; and the cases with $\alpha > 2$ to non-renormalizable theories. The discontinuous behavior of the perturbation for $\alpha > 2$, no matter what the regularization, is proposed as representative of the behavior of nonrenormalizable interactions in quantum field theory, and certain soluble model field theories support this view.[12]

STATIONARY STOCHASTIC PROCESSES WITH APPLICATION TO QUANTUM FIELDS

Stationary Distributions for Quantum Mechanics

The Wiener process $W(t)$ is evidently nonstationary in view of the requirement that $W(0) = 0$, and cannot be made stationary (for reasons of divergence like those that exclude covariant massless scalar fields in two-dimensional field theory). The Ornstein-Uhlenbeck process[13] is the stationary process that generalizes the Wiener process (in the field theory analogy it includes a nonzero mass). To define this process, let C now denote the space of continuous functions $x(t)$, $-\infty < t < \infty$, and introduce the Gaussian probability measure $\mu_F(x)$ characterized by

$$<x(t)>_F = \int x(t)\, d\mu_F(x) = 0$$

$$<x(t)x(t')>_F = \int x(t)x(t')d\mu_F(x)$$

$$= \frac{1}{2} e^{-|t-t'|} \quad .$$

The condition for stationarity is that the characteristic functional is invariant under the test-function transformation $s(t) \to s(t+\tau)$, for all τ and all $s(t)$ (say continuous functions of compact support). This condition is evidently fulfilled by μ_F.

To introduce interactions one considers the probability measures

$$d\mu(x) = N_T e^{-\lambda\int_{-T}^{T} x^p(t)dt} d\mu_F(x)$$

where $\lambda > 0$, p even, and we have chosen only a single monomial for illustrative purposes. As $T\to\infty$ these measures weakly converge to a new, stationary, generally non-Gaussian probability measure closely related to the generalized diffusion equation or the Schrödinger equation, Eqs. (1) and (2), for which

$$H = - \frac{1}{2} \frac{\partial^2}{\partial x^2} + \frac{1}{2} x^2 + \lambda x^p \; .$$

If p = 2 - corresponding to a mass shift - then the new process is also normal with a covariance

$$\frac{1}{2}(1+2\lambda)^{-\frac{1}{4}} \exp\left(-\sqrt{1+2\lambda}\ |t-t'|\right) .$$

It would be nice to be able to characterize these stationary processes directly without the need for temporal cut-offs. In principle, this characterization should be possible in the framework of stochastic equations, but as yet the necessary formalism does not exist (or at least is not known to this author). In the case that p = 2, the map of stochastic variables is straightforward: If $F_1(t)$ denotes the Ornstein-Uhlenbeck process, then the stationary stochastic process $F_{1+2\lambda}(t)$ is given by the distributional integral equation[+)]

$$F_{1+2\lambda}(t) = \int K_\lambda(t-t')F_1(t')dt'$$

where the kernel $K_\lambda(t) \equiv 0$, $t < 0$, and otherwise

$$K_\lambda(t) = \delta(t) - (\sqrt{1+2\lambda} - 1)e^{-\sqrt{1+2\lambda}\ t} .$$

In differential form one has the relation

+) The map from F_1 to $F_{1+2\lambda}$ illustrated here is just that of a linear filter in electric circuit theory, where $K_\lambda(t)$ denotes the impulse response of the filter.

$$dF_{1+2\lambda}(t) + F_{1+2\lambda}(t)dt = dF_1(t) + F_1(t)dt$$

which establishes the Markov character of the map from $F_1(t)$ to $F_{1+2\lambda}(t)$.

The foregoing analysis of stochastic processes pertains to quantum mechanics. We now make one "minor" change that makes everything pertain to covariant imaginary time quantum field theory in n space-time dimensions, n = 2,3,4, ...

Generalized Stationary Distributions for Quantum Field Theory

Let us embed the Ornstein-Uhlenbeck process in a one parameter family of zero-mean, Gaussian stationary distributions μ_ξ with the formal covariance

$$\langle x(t)x(0)\rangle_\xi \equiv (2\pi)^{-1} \int (\omega^{2\xi} + 1)^{-1} e^{-i\omega t} d\omega \quad ,$$

where $0 \leq \xi \leq 1$ and $\omega^{2\xi} \equiv |\omega^{2\xi}|$.[+)] If $\xi = 1$ we deal with the Ornstein-Uhlenbeck process, but not otherwise. For $\xi > \frac{1}{2}$, the measure μ_ξ has support on C, while for $\xi \leq \frac{1}{2}$ this is not the case. In particular Kolmogorov's condition for a continuous stochastic process,[7] i.e. the existence of parameters $\alpha > 0$, $\beta > 0$, and $C < \infty$ such that

---

+) Equally good for our purposes would be to choose $(\omega^2+1)^{-\xi}$ in place of $(\omega^{2\xi}+1)^{-1}$.

$$<|x(t)-x(t')|^{\alpha}>_{\xi} \leq C|t-t'|^{1+\beta} \quad ,$$

is easily verified for any $\xi > \frac{1}{2}$. In fact it may be shown that with probability one

$$|x(t)-x(t')| \leq C|t-t'|^{\xi-\frac{1}{2}} |\ell n|t-t'||^{1+\varepsilon}$$

for any $\varepsilon > 0$ with C some constant.[14] When $\xi \leq \frac{1}{2}$ we deal with a generalized stochastic process (not point wise defined) and x(t) is properly regarded as a distribution (and should be smeared with a test function $s(t) \in L^2(R)$, which is adequate for all $\xi \geq 0$).

As distributions, local products of x(t) are not well defined; and we appeal to the standard remedy of Wick ordering (related to a multiple Itô integral[15]). First let

$$x(t_{\varepsilon}) = \varepsilon^{-1} \int_t^{t+\varepsilon} x(t')dt'$$

which is well defined for all $\varepsilon > 0$ and for any ξ process. The expression

$$:e^{ax(t_{\varepsilon})}: \equiv e^{ax(t_{\varepsilon})-\frac{1}{2}a^2<x^2(t_{\varepsilon})>_{\xi}}$$

can be viewed as the generating function for the various ordered products

$$:x^p(t_{\varepsilon}): \quad , \qquad p = 1,2,3, \ldots \quad .$$

In view of the form of the generating functional these products may be expressed with help of the Hermite polynomials. In any case, one may construct the variable

$$D_\varepsilon(h) = \int h(t) \; :x^p(t_\varepsilon): \, dt$$

for h(t) a continuous test function of compact support. It follows that

$$\langle D_\varepsilon^2 \rangle_\xi = p! \iint h(t)h(t') \langle x(t_\varepsilon)x(t'_\varepsilon) \rangle_\xi^p \, dt dt' \;,$$

and convergence as $\varepsilon \downarrow 0$ arises if and only if the covariance is locally L^p.

It may be estimated for asymptotically small $|t|$ that

$$\langle x(t)x(0) \rangle_{\frac{1}{2}} \simeq -(2\pi)^{-1} \ln|t| \quad , \qquad \xi = \tfrac{1}{2} \;,$$

$$\langle x(t)x(0) \rangle_\xi \simeq c_\xi |t|^{2\xi-1} \quad , \qquad 0 < \xi < \tfrac{1}{2} \;,$$

where c_ξ is a nonvanishing positive constant; for $\xi = 0$

$$\langle x(t)x(0) \rangle_0 = \tfrac{1}{2}\delta(t) \;.$$

Ignoring $\xi = 0$ temporarily, the integrability in question requires that

$$p < (1-2\xi)^{-1} .$$

In fact this is just the condition that D_ε converges in mean as $\varepsilon \downarrow 0$ so that whenever $p < (1-2\xi)^{-1}$ there exists a local distributional product $:x^p(t):$ and

$$D_\varepsilon(h) \to \int h(t) :x^p(t): dt .$$

Conversely, if $p \geq (1-2\xi)^{-1}$ it may be shown that this prescription does not define a local distributional product.

With this much apparatus at our disposal let us indicate the formal expression of interest and relate it to covariant quantum field theory in n space-time dimensions.[16] Choose $\lambda > 0$, p even and consider the formal expression

$$d\mu(x) = N_T e^{-\lambda\int_{-T}^{T}:x^p(t):dt} d\mu_\xi(x) .$$

In the study of correlations of this expression, as expanded into a power series in the coupling constant λ, there arise terms such as

$$\int_{-T}^{T}\cdots\int_{-T}^{T} dt_1\ldots dt_m \langle x(s_1)\ldots x(s_r):x^p(t_1):\ldots:x^p(t_m):\rangle_\xi .$$

The analysis of each such term is greatly facilitated by (Feynman) graphs completely analogous to those used for scalar field theory.[17] The covariance (two-point function) plays the role of propagator, and in momentum space it reads $(\omega^{2\xi} + 1)^{-1}$. Divergence or convergence of a graph may be found by cursory examination of the integrand. Generally, any graph with L internal loops and P internal propagators with $L-2P\xi > 0$ diverges as $\Omega^{L-2P\xi}$, with $L-2P\xi = 0$ diverges as $\ell n\Omega$, or with $L-2P\xi < 0$ converges, where Ω is a frequency cutoff. In the corresponding n-dimensional covariant quantum field theory (in Euclidean space, or after Wick rotation), the same graph with $nL-2P > 0$ diverges as Λ^{nL-2P}, with $nL-2P = 0$ diverges as $\ell n\Lambda$, or with $nL-2P < 0$ converges, where Λ is a momentum cutoff. Identical divergence behavior in these two theories is established by choosing $\Lambda = \Omega^{\xi}$ and $\xi = 1/n$. In other words, there is a one-to-one relation between convergence and divergence, even the same ordering of divergences, in the field theory and stochastic theory, or simply noise theory as we shall call it.

Divergent noise theory graphs need renormalization counterterms just as in field theory,[17] and in view of the divergence equivalence, it is not difficult to imagine the type and number of counterterms needed, and that there are models that are super-renormalizable, renormalizable (requires field strength renormalization) and nonrenormalizable. Although $p < (1-2\xi)^{-1}$ is the condition for existence of a local distributional product, it should not be too surprising that renormalization techniques allow one to go further; and that super-renormalizable behavior arises for $p < 2/(1-2\xi)$, renormalizable for $p = 2/(1-2\xi)$, and nonrenormalizable for $p > 2/(1-2\xi)$. In the familiar case of $n = 4$ and thus $\xi = \frac{1}{4}$, one observes that $p = 4$ is the maximum renormalizable power, as should be familiar from field theory.[17] Values of $\xi \neq 1/n$ may be taken as analogues of nonintegral space-time dimension, and it is amusing to observe that for $\xi = 0.49$, say, models with exponents $p \leq 100$ are renormalizable while those with $p \geq 102$ are nonrenormalizable. On the other hand, the simplest noise theories to study would be $p = 4$, and $\xi = \frac{1}{3}, \frac{1}{4}$ and 1/5, to probe the three basic types of interactions.

We emphasize once more: Simply in the context of noise theory (stationary stochastic processes) one can find problems that accurately simulate the divergence difficulties of n-dimensional covariant quantum field theory. To understand these divergences in noise theory is to understand them in quantum field theory!

Hard Core Picture for Nonrenormalizability

It is heuristically helpful to give yet another argument that models with $p > 2/(1-2\xi)$ are nonrenormalizable, and indeed to see what might be the nature of the difficulty in such cases. In the language of path space integration, the characteristic functional for the λx^p interaction based on the "free" ξ process may be formally written as

$$\mathcal{N}\int e^{i\int s(t)x(t)dt-\lambda\int x^p(t)dt-\frac{1}{2}\int(\omega^{2\xi}+1)|\tilde{x}(\omega)|^2 d\omega}\,Dx\ , \qquad (9)$$

where the notation is fairly standard.[+)] With the inter-

+) In an equation such as (9), one clearly sees how minor is the change from "quantum mechanics" ($\xi = 1$) to "quantum field theory" ($\xi = \frac{1}{2},\frac{1}{3},\frac{1}{4},\ldots$), at least in principle.

action term absent this expression is taken to define the characteristic functional for the normal ξ process based on the measure μ_ξ introduced earlier. In some way (fairly standard,[18] but not spelled out here), the quadratic terms in the exponent determine the support properties of the Gaussian measure μ_ξ. One knows that finiteness of the quadratic expression holds only on a set of zero μ_ξ measure. Nevertheless, let us ignore this fact and examine the consequences of the condition

$$W_\xi \equiv \frac{1}{2}\int (\omega^{2\xi} + 1)|\tilde{x}(\omega)|^2 d\omega < \infty .$$

First, if $\xi > \frac{1}{2}$, we easily deduce that each path is continuous (since $\tilde{x}(\omega) \in L^1(R)$) and fulfills

$$|x(t)-x(t')| \le K_\xi |t-t'|^{\xi-\frac{1}{2}} \left(\int (\omega^{2\xi} + 1)|\tilde{x}(\omega)|^2 d\omega\right)^{\frac{1}{2}}$$

where K_ξ is a strict numerical constant. Compared with the correct result, this estimate is very nearly right lacking only the logarithmic factor. For $\xi \le \frac{1}{2}$, the paths are not continuous, for consider the counterexample with

$$\tilde{x}(\omega) = (1 + |\omega| \quad |\ln|\omega||^\sigma)^{-1}$$

where $\frac{1}{2} < \sigma < 1$. For all $\xi \le \frac{1}{2}$,

$$\int (\omega^{2\xi} + 1)|\tilde{x}(\omega)|^2 d\omega < \infty$$

while

$$x(0) = \int \tilde{x}(\omega)\, d\omega = \infty \quad .$$

Again this qualitative feature of the support is correctly given by the estimate.

Next let us compare the criteria for finiteness of the two expressions

$$w_\xi \equiv \frac{1}{2} \int (\omega^{2\xi} + 1) |\tilde{x}(\omega)|^2 d\omega$$

$$v \equiv \int |x(t)|^p dt \quad .$$

Here we have generalized p to an arbitrary positive parameter. As we have observed, $w_\xi < \infty$ gives a reasonable qualitative characterization of the support of μ_ξ. Now it may be shown,[19] provided

$$p \leq 2/(1-2\xi) \quad ,$$

that

$$v \leq K\, w_\xi^{p/2} \quad ,$$

where K is a path-independent finite constant. If $p > 2/(1-2\xi)$ no such bound holds, and there are uncountably many paths x(t) with $w_\xi < \infty$ and $v = \infty$.

Consider the following plausible implication of these facts. If $p \leq 2/(1-2\xi)$, the support properties based on $w_\xi + \lambda v$, $\lambda > 0$, and w_ξ are qualitatively similar and as $\lambda \downarrow 0$ (outside the integral in (9)) we expect the characteristic functional to pass to the free one, i.e., the measures involved to weakly converge to μ_ξ. If $p > 2/(1-2\xi)$, however, the support properties based on $w_\xi + \lambda v$, $\lambda > 0$, and w_ξ are qualitatively different. Some of the paths allowed by w_ξ can be effectively forbidden, or projected out in the presence of v. In this sense, we may characterize the behavior of the interaction term, partly, as that of a "hard core". And as such, turning off the coupling $\lambda \downarrow 0$ does <u>not</u> restore the free theory (free measure μ_ξ), but instead the measures weakly converge to an alternative, pseudofree measure $\mu_{PF\xi}$, say, which incorporates the effects of the hard core.

Although the present problem is more complicated, the principles involved here are just the same as those for the elementary example of the preceding section. A similar condition may be postulated in covariant quantum field theory[11] based on more conventional multiplicative inequalities and Sobolev embedding theorems.[20] Here we illustrate these ideas within stochastic theory - within noise theory, no less - in hopes that the reader can see these issues clearly.

Another point deserves comment. In perturbation theory, there is a genuine distinction between super-renormalizable and renormalizable models having to do with field strength renormalization. Even at the level of classical inequalities such a distinction seems also to exist. The bound cited before may be refined[19] to yield

$$v^{2/p} \leq \delta \int \omega^{2\xi} |\tilde{x}(\omega)|^2 d\omega + M(\delta) \int |\tilde{x}(\omega)|^2 d\omega, \; p < \frac{2}{(1-2\xi)},$$

$$v^{2/p} \le K \int \omega^{2\xi} |\tilde{x}(\omega)|^2 d\omega + \delta \int |\tilde{x}(\omega)|^2 d\omega, \; p = \frac{2}{(1-2\xi)} ,$$

where δ is an arbitrary positive number, M a constant depending on δ, and K an irreducible fixed number (estimated to be 16). For super-renormalizable cases, a bound exists that involves arbitrarily little of the $\omega^{2\xi}$ term, while for renormalizable cases there is no such possibility.

This subsection has been devoted to heuristic arguments suggesting that nonrenormalizable interactions with $p > 2/(1-2\xi)$ may act as discontinuous perturbations, and once introduced, unalterably change the free theory. The difficulties seem to escalate as ξ gets smaller, and computationally the most divergent situation arises when $\xi = 0$. Qualitatively, every argument presented in this section still applies for $\xi = 0$ where

$$w_o = \int |\tilde{x}(\omega)|^2 d\omega = \int x^2(t)dt .$$

Here the comparison of interest is between $w_o = \int x^2(t)dt$ and $v = \int |x(t)|^p dt$, and one does not need any sophisticated estimates in this case. In particular, we suspect that every interaction with $p > 2$ is nonrenormalizable and should be a discontinuous perturbation. Remarkably enough, in this case we can prove these qualitative remarks are true!

Shot Noise as a Nonrenormalizable Theory

For $\xi = 0$ the free process is δ-correlated, and

actually corresponds to the familiar white noise.[21] This process has independent values for each time, and reference to (9) specialized to $\xi = 0$ suggests that the inclusion of a local interaction maintains the feature of independent values at each time. Thus, in particular, every characteristic functional of this type necessarily has the general form[21]

$$e^{-\int dt L[s(t)]} \quad ,$$

where (for even L as we need) the most general function L is given by

$$L[s] = as^2 + \int_{|u|>0} [1-\cos(us)]\, d\sigma(u) \quad ,$$

where $a \geq 0$ and σ is a positive measure satisfying

$$\int_{|u|>0} u^2/(1+u^2)\, d\sigma(u) < \infty \ .$$

White noise is characterized by $\sigma = 0, a > 0$ and describes the free theory (with various masses). Our interest focusses on the non-Gaussian term, and we choose $a \equiv 0$ and σ absolutely continuous, specifically

$$d\sigma(u) = C^2(u)du \ .$$

In other words, let us consider

$$e^{-\int dt \int \{1-\cos[us(t)]\} C^2(u)du} = \int e^{i\int s(t)x(t)dt}\, d\mu_C(x) \quad ,$$

which defines a probability measure $\mu_C(x)$ and thereby (up to stochastic equivalence) a stochastic variable $S(t)$ closely related to shot noise.[22]

The statistics of $S(t)$ are fundamentally different than those of a Gaussian variable. It may be shown that Wick products are meaningless for $S(t)$, but there is an acceptable alternative. In particular it may be shown that[23]

$$S_R^p(t) = \lim_{\varepsilon \downarrow 0} \varepsilon^{-1} \left[\int_t^{t+\varepsilon} S(t')dt'\right]^p \tag{10}$$

exists as a distribution for all $p \geq 1$. Indeed, if $p \geq 2$ one learns that

$$E(e^{i\int g(t) S_R^p(t)dt}) = e^{-\int dt \int [1-e^{iu^p g(t)}] C^2(u)du} ,$$

and the same techniques of calculation[23] show one, for $\lambda > 0$, $g \geq 0$, and p even, that

$$E(e^{i\int s(t)S(t)dt-\lambda\int g(t)S_R^p(t)dt})$$

$$= \int e^{i\int s(t)x(t)dt-\lambda\int g(t)x_R^p(t)dt} d\mu_C(x)$$

$$= e^{-\int dt \int [1-e^{ius(t)-\lambda u^p g(t)}] C^2(u)du} .$$

Next, let us normalize this relation by dividing by the

same expression with $s(t) \equiv 0$. In the resulting quotient we may take the limit $g(t) \to 1$ without difficulty to find the characteristic functional

$$e^{-\int dt \int \{1-\cos[us(t)]\} \; e^{-\lambda u^p} \; C^2(u)\,du} \tag{11}$$

Here, then, is an explicit computation of the introduction of a p-th order interaction into this model, and it only remains to select a suitable function $C(u)$.

Let us once again focus on the formal problem solved by (11), namely

$$\eta \int e^{i\int s(t)x(t)dt - \lambda\int x_R^p(t)dt - \int x_R^2(t)dt} \; \mathcal{D}x \; , \tag{12}$$

where it is here recognized that even the square of the path in the quadratic term necessarily also involves the shot noise prescription in (10). With this simple fact understood it becomes easy to find $C^2(u)$.

Set $p = 2$ and denote (12) by $\hat{C}_{(1+\lambda)}(s)$. By a simple scaling it is clear that $\hat{C}_{(1+\lambda)}(s) = \hat{C}_{(1)}(s/\sqrt{1+\lambda})$. Equation (11) gives another relation for $\hat{C}_{(1+\lambda)}(s)$. Equating these, we find that the relation

$$\int [1-\cos(us)]\, e^{-\lambda u^2} \; C^2(u)\,du = \int [1-\cos(us/\sqrt{1+\lambda})]\, C^2(u)\,du$$

must hold for all s. In turn, this implies for all $u \neq 0$ and $\lambda \geq 0$ that

$$e^{-\lambda u^2} C^2(u) = \sqrt{1+\lambda}\, C^2(\sqrt{1+\lambda}\,u) ,$$

and, since $C^2(u)$ is symmetric, it follows that

$$C^2(u) = k\,\frac{e^{-u^2}}{|u|} ,$$

where k is an undetermined positive constant. It is easy to see that k just represents an arbitrariness in the units of t, and without loss of generality we choose $k = 1$ in what follows.

We now summarize. The formal path integral corresponding to the case $\xi = 0$ has the complete solution given by

$$e^{-\int dt \int \{1-\cos[us(t)]\}|u|^{-1} e^{-\lambda u^p - u^2} du}$$

which implicitly defines the interacting measure. Passage to the limit $\lambda \downarrow 0$ does <u>not</u> result in the free (white noise) measure, but rather in the pseudofree measure as characterized by

$$e^{-\int dt \int \{1-\cos[us(t)]\}|u|^{-1} e^{-u^2} du} .$$

The interacting cases are continuous perturbations about the pseudofree theory; and in fact they clearly admit an asymptotic expansion in the parameter λ.

What could be a neater demonstration of the relevance of nois theory than to find ready-made in the shot noise of electric circuit theory the stochastic process that solves a nonrenormalizable model! The shot noise solution has immediate implications for an n-dimensional quantum field model[24] as we next observe.

Remark

In the context of n-dimensional Euclidean quantum field theory a model has been proposed in which all the space-time gradients are dropped leaving only the mass term in the free action.[25] If a momentum space cutoff is introduced into the interaction term, a solution continuously connected to the free theory may be constructed,[25] but the cutoff cannot be removed.[26] On the other hand, an analysis completely parallel to that given in the shot noise case can be used to derive a completely acceptable solution for such model field theories without any cutoffs whatsoever.[24]

CONCLUSION

We have attempted to illustrate several interesting problems in stochastic theory that have relevance for problems in quantum theory. The most familiar connection between such problems lies in the path space integration formulation, but we feel that further development and exploitation of the techniques of stochastic equations should be valuable,[4,27] and we would encourage interested readers to undertake such studies. Discontinuous perturbations arise in a great variety of problems, as we have attempted to make clear, and it should be possible to develop interacting theories starting from pseudofree theories in the language of stochastic equations "driven" by the stochastic process of the pseudofree theory rather than that of the free theory. Moreover, perhaps there are other noise processes, like shot noise, that can be proposed as suitable processes to characterize other nonrenormalizable theories (for which $p > 2/(1-2\xi)$, $\xi > 0$); if so, such processes could conceivably tell us a great deal about nonrenormalizable quantum field theories. Clearly, many interesting questions remain in the fascinating relationship of stochastic processes and quantum theory.

ACKNOWLEDGEMENTS

We thank H. Ezawa, L. A. Shepp and L. Streit for fruitful discussions and continued interest.

REFERENCES

1. G. Velo and A. Wightman, Eds., Springer Lecture Notes in Physics, Vol. 25 (Springer-Verlag, 1973). Some basic references to applications of probability methods to quantum theory may be traced from contributions and references in Constructive Quantum Field Theory.
2. M. Kac, Proc. Second Berkeley Symposium on Mathematical Statistics and Probability (Univ. of California, 1951) p. 189.
3. E. Nelson, Dynamical Theories of Brownian Motion (Princeton Univ. Press, 1967); T. Hida, Stationary

Stochastic Processes (Princeton Univ. Press, 1970); P. Lévy, Le Mouvement Brownien (Gauthier-Villars, 1954); K. Itô and H. P. McKean, Jr., Diffusion Processes and their Sample Paths (Academic Press, 1965).

4. H. Ezawa, J.R. Klauder and L.A. Shepp, Ann. of Phys. (NY) 88, 588 (1975).

5. K. Itô, Mem. Am. Math. Soc., No.4, 1 (1951); J.L.Doob, Stochastic Processes (Wiley & Sons, 1953).

6. H.P. McKean, Jr., Stochastic Integrals (Academic Press, 1969); J.L. Doob, Stochastic Processes (Wiley & Sons, 1953).

7. A.V. Skorohod, Studies in the Theory of Random Processes (Addison-Wesley, 1965).

8. L.A. Shepp, Ann. Math. Statist. 37, 321 (1966).

9. L. Streit, On Interactions Which Can't be Turned Off, Bielefeld preprint, Dec. 1974.

10. H. Ezawa, J.R. Klauder and L.A. Shepp, J. Math. Phys. 16, 783 (1975).

11. J.R. Klauder, Acta Physica Austriaca Suppl. XI, 341 (1973);[+)] Phys. Lett. 47B, 523 (1973).

12. For a brief survey see J.R. Klauder, Soluble Models and the Meaning of Nonrenormalizability, Proceedings of the International Symposium on Mathematical Problems in Theoretical Physics, Kyoto, Japan, January 23-29, 1975 (to be published).

[+)] See the note at the end.

13. G.E. Uhlenbeck and L.S. Ornstein, Phys. Rev. 36, 823 (1930); M.C. Wang and G. E. Uhlenbeck, Rev. Mod. Phys. 17, 323 (1945); J.L. Doob, Ann. Math. 43, 351 (1942). All three articles are reprinted in Selected Papers in Noise and Stochastic Processes, N. Wax, Ed. (Dover, 1954).

14. I.I. Gihman and A.V. Skorohod, The Theory of Stochastic Processes I (Springer-Verlag, 1974), p. 194.

15. K. Itô, J. Math. Soc. Japan 3, 157 (1951).

16. The material of this section is based on J.R.Klauder, Phys. Lett. 56B, 93 (1975).

17. See, e.g., J.D. Bjorken and S.D. Drell, Relativistic Quantum Fields (McGraw-Hill, 1965).

18. T.Hida, Stationary Stochastic Processes (Princeton Univ. Press, 1970); A.V. Skorohod, Integration in Hilbert Space (Springer-Verlag, 1974).

19. V.P. Il'in and V.A. Solonnikov, Trudy MIAN SSSR 66, 2o5 (1962). We thank Prof. Solonnikov for helpful correspondence regarding this work.

20. O.A. Ladyzenskaja, V.A. Solonnikov and N.N. Ural'ceva, Linear and Quasi-Linear Equations of Parabolic Type, Trans. of Math. Mono., Vol. 23 (American Math. Society, 1968).

21. I.M. Gel'fand and N.Ya. Vilenkin, Generalized Functions, Vol. 4: Applications of Harmonic Analysis, translated by A. Feinstein (Academic Press, 1964).

22. See, e.g., S.O. Rice, Bell System Tech. Journal 24, 46 (1945), reprinted in Selected Papers on Noise and Stochastic Processes, N.Wax, Ed. (Dover, 1954).

23. See, e.g., J.R. Klauder, Acta Physica Austriaca, Suppl. VIII, 227 (1971); Lectures in Theoretical Physics, Vol. XIVB, W. Britten, Ed. (Colorado Associated Univ. Press, 1974), p. 329; G.C. Hegerfeldt and J.R. Klauder, Nuovo Cimento 19A, 153 (1974).

24. J.R. Klauder, On Model Fields with Independent Values at Every Space-Time Point, Acta Physica Austriaca (in press).

25. E.R. Caianiello and G. Scarpetta, Nuovo Cimento 22A, 448 (1974); Nuovo Cimento Letters 11, 283 (1974).

26. W. Kainz, Nuovo Cimento Letters 12, 217 (1975).

27. See also, E. Nelson, Dynamical Theories of Brownian Motion (Princeton Univ. Press, 1967); L. Guerra and P. Ruggiero, Phys. Rev. Lett. 31, 1022 (1973).

<u>Added Note:</u> Please observe the following:

Important Errata for

Field Structure Through Model Studies:
Aspects of Nonrenormalizable Theories
Acta Physica Austriaca Suppl. XI, 341 (1973)

<u>Footnote, page 349:</u>

Instead of	Read
...is one without...	...is not without....

<u>Last 7 lines of page 35o and first 4 lines of page 351:</u>

where $\gamma \geq 0$ and $\lambda \equiv \gamma(\gamma+1)$. The first solution is bounded and locally L^2, while the second solution is unbounded and locally L^2 only for $\gamma < \frac{1}{2}$, i.e., for $\lambda < \frac{3}{4}$. For $\lambda \geq \frac{3}{4}$ only the first solution is admissible. (This behavior contrasts with that for $\alpha < 2$ where both solutions are admissible for all $\lambda > 0$.) Physically it seems reasonable to choose only those solutions that are admissible for <u>all</u> $\lambda > 0$. As a consequence we reject the second solution and are left with the only possible form $\psi \sim x^{1+\gamma}$.*) This means that for even and odd solution we have, respectively,

---

*) <u>Note Added Subsequently:</u> For $\alpha = 2$ it now appears preferable to exploit both solutions when $\lambda < \frac{3}{4}$ so that as $\lambda \downarrow 0$ one arranges for $\alpha = 2$ to fall into the "free" rather than "pseudo free" category. (This viewpoint would alter Fig.1 correspondingly.) No such ambiguity exists for $\alpha > 2$ and those cases remain unchanged.

Acta Physica Austriaca, Suppl. XIV, 631 — 642 (1975)

CONVEXITY PROPERTIES OF COULOMB SYSTEMS[+]

by

W. THIRRING
Institut für Theoretische Physik
der Universität Wien

Usually when one calculates the eigenvalues $E_j(\alpha)$ of a Hamiltonian $H = H_o + \alpha V$, $E_j(O)$ being known, one attempts a Taylor expansion: $E_j(\alpha) = E_j(O) + \alpha E_j'(O) + \ldots$. Unfortunately, even when this series converges, there is no garanty that the first few terms will be close to $E_j(\alpha)$. For instance, if $E_j(\alpha)$ is a rapidly oscillating function, a linear or parabolic approximation will evidently not be very good. However, for the ground state this cannot happen because of the

<u>Theorem I:</u>

If H has m-1 eigenvalues below the beginning $E_m(\alpha)$ of the essential spectrum, then

[+] Abstract given at XIV. Internationale Universitätswochen für Kernphysik, Schladming, Austria, February 24-March 7, 1975.

$$\varepsilon_n(\alpha) = \sum_{k=1}^{n} E_k(\alpha), \qquad n = 1,2\ldots, m$$

are concave functions of α.

Proof:

According to the theorem of Ky-Fan

$$\varepsilon_n(\alpha) = \inf_{H_n} \mathrm{Tr}_{H_n} H$$

where H_n is a n-dimensional subspace of the domain of H and Tr_{H_n} is the partial trace in H_n. Now $\mathrm{Tr}_{H_n} H$ is linear in α and the infimum of linear functions is concave.

Remarks:

1.) If H is a jointly concave function of several variables α_i, so is $\varepsilon_n(\alpha_i)$. This happens in particular if $H = \sum_i \alpha_i H_i$

2.) For concave functions the right and left derivatives exist and we have the Hellmann-Feynman theorem

$$\frac{d\varepsilon_n}{d\alpha}\bigg|_{\ell,r} = \sum_{j=1}^{n} \langle\alpha \mp \varepsilon, j|V|\alpha \mp \varepsilon, j\rangle$$

where $H|\alpha,j\rangle = E_j(\alpha)|\alpha,j\rangle$

3.) The concavity also carries over for the free energy

$$F(\alpha) = - \ln \mathrm{Tr}\, e^{-H(\alpha)}.$$

We shall illustrate the usefullness of these properties by a few examples from atomic and molecular physics. They are taken from Ref. 5 where further information can be obtained. It is obvious that similar arguments can be used in any quantummechanical problem.

Applications

1.) Atoms (1,2)

The Hamiltonian is

$$H = \sum_{i=1}^{N} \frac{p_i^2}{2m} - \sum_{i=1}^{N} \frac{\alpha_o}{|x_i|} + \sum_{i>j} \frac{\alpha}{|x_i - x_j|}, \quad \begin{matrix} \alpha_o = Z e^2 \\ \alpha = e^2 \end{matrix},$$

and because of dimensional reasons the ε_n will be of the form

$$\varepsilon_n = m\, \alpha_o^2\, f\left(\frac{\alpha}{\alpha_o}\right).$$

ε_n must be jointly concave in $\frac{1}{m}$, α_o, α, which implies that the matrix of the second derivatives of ε_n must be negative definite. (To ensure its existence one may have to approximate the ε_n by smooth concave functions.) The corresponding condition is reduced to the statement

$f'' \leq \frac{f'^2}{2f}$ or that even

$-\sqrt{-E_n}$ is concave in α. This observation can be used to improve bounds for E_n linear in α to parabolic bounds. As illustration we plot in fig. 1,2,3 for N = 2 the lowest energies with the quantum numbers L = 0, S = 0, L = 0, S = 1, and L = 1, S = 0 and L = 1, S = 1. The lower bounds are obtained by an application of two-dimensional projections (3,4) and the upper ones using trial functions of the form $e^{-\alpha r_1 - \beta r_2} \pm e^{-\alpha r_2 - \beta r_1}$. The experimental points show clearly the concavity of $-\sqrt{-f}$.

2.) Molecules (5)

The Hamiltonian in the Born-Oppenheimer Approximation is

$$H = \sum_{i=1}^{N} \frac{p_i^2}{2m} - \sum_{i=1}^{N} \frac{\alpha Z_1}{|x_i|} - \sum_{i=1}^{N} \frac{\alpha Z_2}{|x_i - R|} + \alpha \sum_{i>j} \frac{1}{|x_i - x_j|}$$

$\vec{R}$ is the distance between the nuclei and provides an external length. Thus the ground state energy will be of the form

$$E = \frac{f(m\alpha R^2)}{mR^2} \quad .$$

Here concavity in $(\frac{1}{m}, \alpha)$ gives the same information as concavity in α. It is illustrated in fig. 4 where we have plotted $R^2E(R)$ for H_2. It can be used to obtain a simple upper limit for the vibrational energy of the nuclei: A concave function is below any of its tangents. Taking the one at

$$R_o, \left.\frac{\partial E}{\partial R}\right|_{R=R_o} = 0, \qquad \text{we have}$$

$$E(R) \leq E(R_o) \frac{R_o^2}{R^2} (1 + 2 \frac{R-R_o}{R_o}) \quad .$$

This upper bound is a combination of $\frac{1}{R}$ and $\frac{1}{R^2}$ - potentials (Kratzer-potential) and hence the Schrödinger equation for the nuclei admits an analytic solution.

3.) Mass-dependence

The Born-Oppenheimer approximation corresponds to neglecting the kinetic energy of the nuclei. Adding $\frac{p_1^2}{2M_1} + \frac{p_2^2}{2M_2}$ to the molecular Hamiltonian the energy must now be of the form $m\alpha^2 f(\frac{m}{M})$. Since it is concave in

$(\frac{1}{m}, \frac{1}{M})$, we find $f'' \leq \frac{2f'^2}{f}$, or $\frac{1}{f}$ convex, or $\frac{1}{|f|}$ concave. Now we have

$$\frac{\partial}{\partial \frac{m}{M}} \frac{1}{|f|} \Bigg|_{\frac{m}{M} = 0} = \frac{f'(0)}{f^2(0)}$$ and concavity tells us

$$\frac{1}{|f|} \leq \frac{1}{|f(0)|} + \frac{m}{M} \frac{f'(0)}{f(0)^2}$$ or

$$f(\tfrac{m}{M}) \leq \frac{f(0)}{1 - \frac{f'(0)}{f(0)} \frac{m}{M}} .$$ $f(0)$ is our previous

result and $f'(0)$ = <kinetic energy of nuclei>.

Special cases:

a) H-atom. In the center of mass system one calculates $f'(0)/f(0) = -1$. Thus first order perturbation theory gives the upper limit $f \leq f(0)(1-\frac{m}{M})$ which by concavity is improved to $f \leq \frac{f(0)}{1+m/M}$. This cannot be further improved since we know that it is exactly the effective mass correction.

b) He-atom. Here the exact result is unknown and

the above method gives

$$f(\frac{m}{M}) \leq \frac{f(0)}{1+2\frac{m}{M}} .$$

This is good enough to prove the binding of $e^- \mu^+ e^-$ but not of $e^- e^+ e^-$.

c) H_2. In our application 2) we deduced an upper bound potential for which the Schrödinger equation for the nuclei can be solved analytically. The result is

$$E \leq \frac{E(R_o)}{(\sqrt{1+x} + \sqrt{x})^2}, \quad x_i = \frac{1}{4R_o^2 M |E(R_o)|} .$$

x is $\sim \frac{m}{M}$ and we see that $|1/E|$ is already concave in this variable. So here no further improvement results.

In summary we can say that concavity is easier to verify than analyticity and has numerically more useful consequences.

REFERENCES

1. W. Thirring, Acta Physica Austr. Suppl. XI, 493 (1973).

2. T.K. Rebane, Opt. Spec. (USSR) 34, 488 (1973).

3. A. Weinstein, W. Stenger, Intermediate Problems for Eigenvalues, Academic Press, New York (1972).

4. W. Thirring, Vorlesungen über Mathematische Physik, T7.

5. H. Narnhofer, W. Thirring, Acta Physica Austr. 1975, Festschrift für P. Urban.

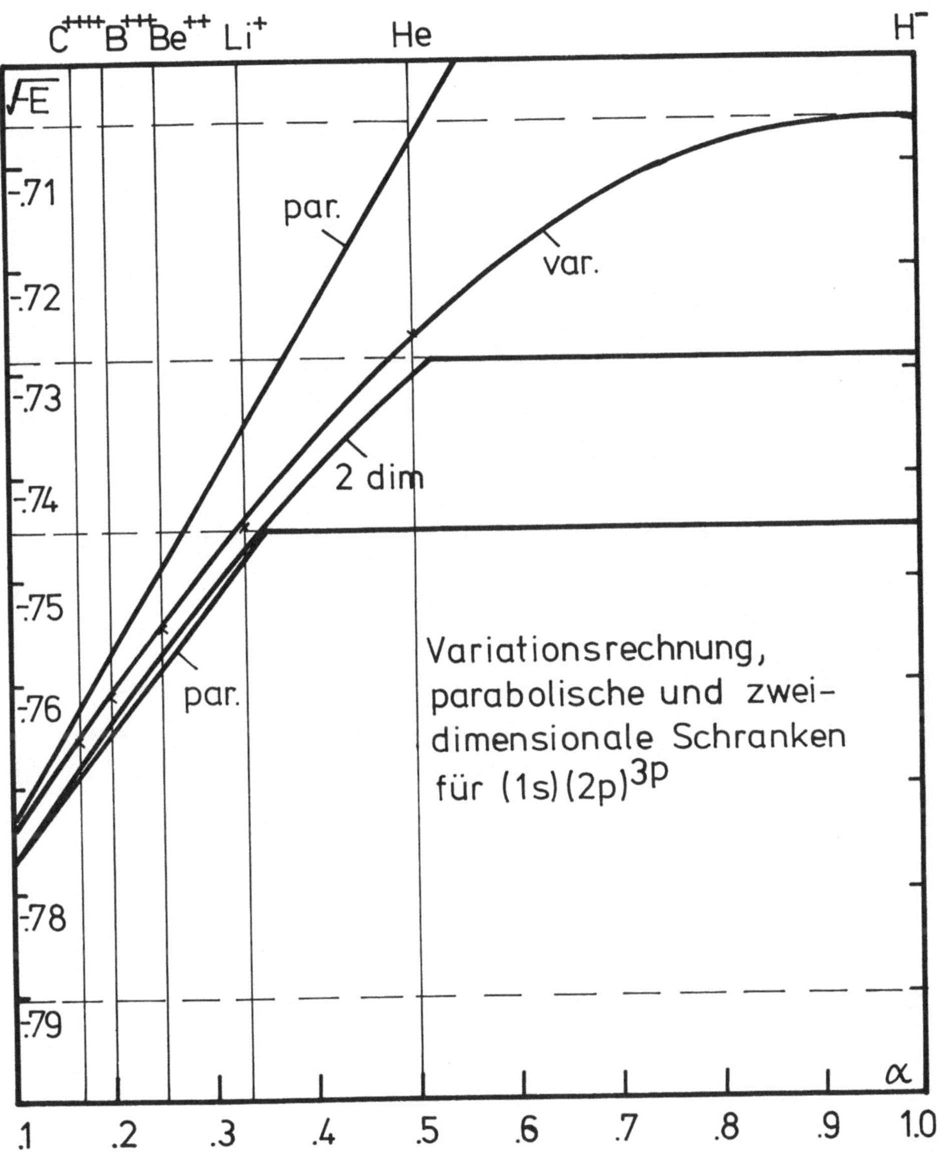
C++++ B+++ Be++ Li+ He H−
√-E
-.71
-.72
-.73
-.74
-.75
-.76
-.78
-.79
par.
var.
2 dim
par.
Variationsrechnung,
parabolische und zwei-
dimensionale Schranken
für (1s)(2p)3P
α
.1 .2 .3 .4 .5 .6 .7 .8 .9 1.0

Fig. 1

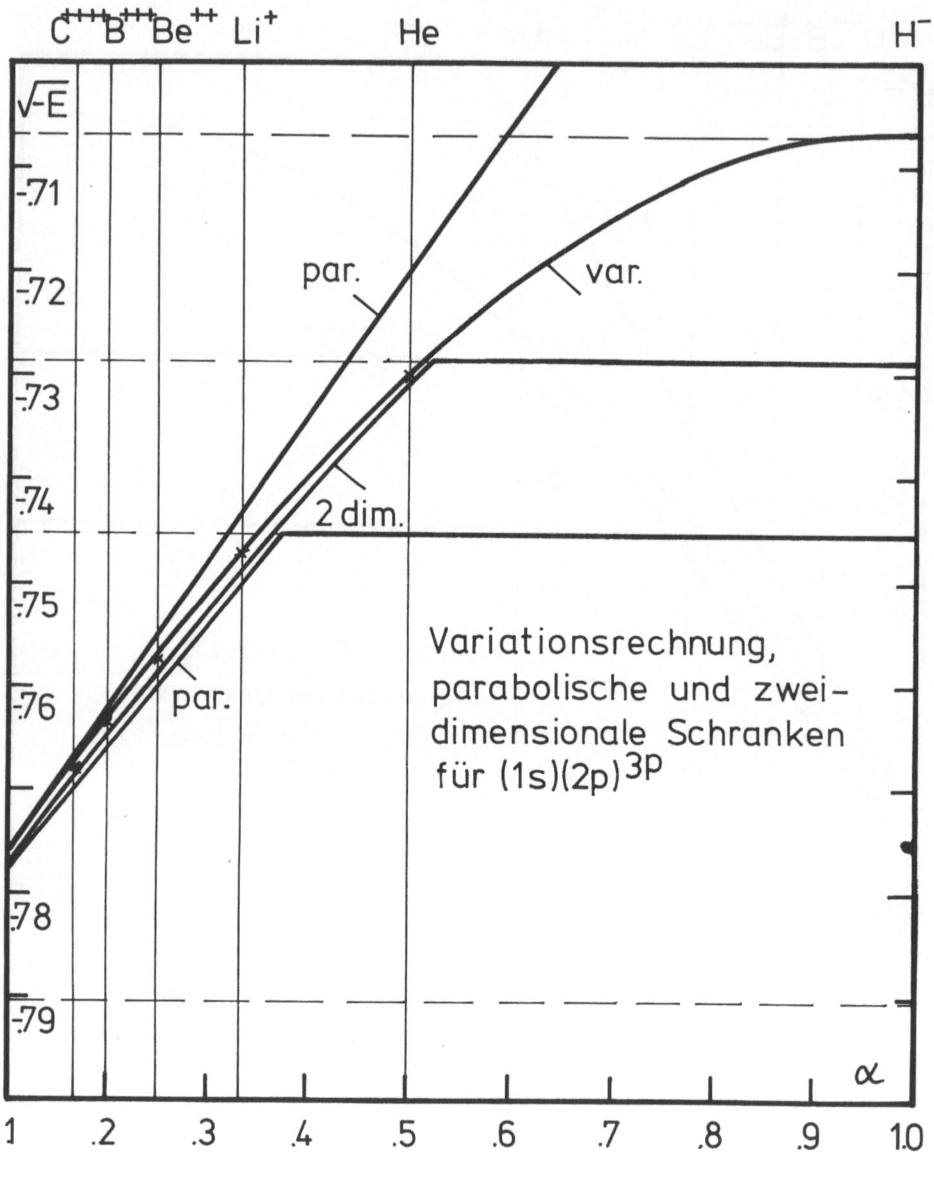

C⁺⁺⁺ B⁺⁺⁺ Be⁺⁺ Li⁺ He H⁻
√-E
-.71
-.72
-.73
-.74
-.75
-.76
-.78
-.79
par.
var.
2 dim.
par.
Variationsrechnung, parabolische und zweidimensionale Schranken für (1s)(2p)³P
α
.1 .2 .3 .4 .5 .6 .7 .8 .9 1.0

Fig. 2

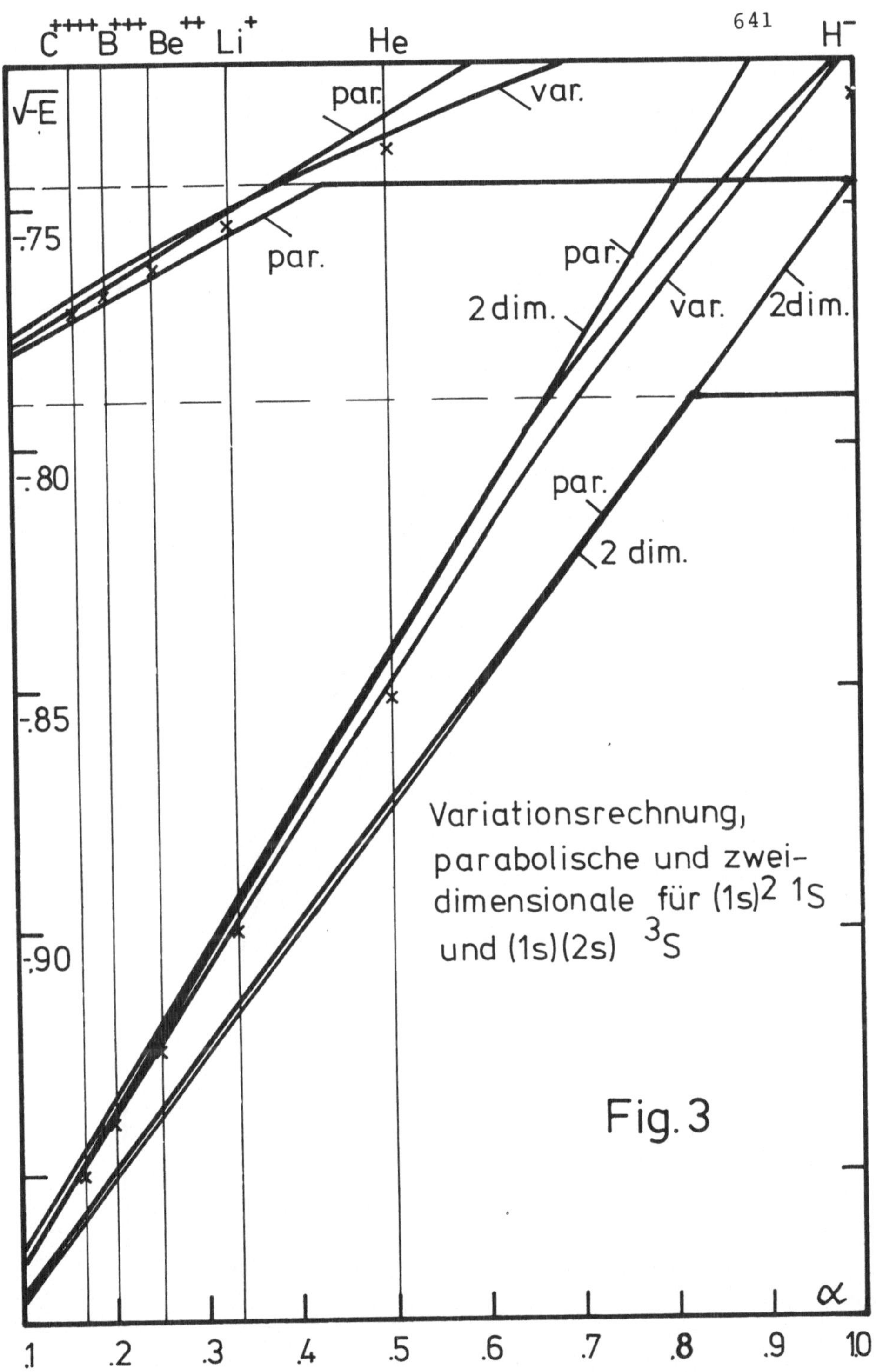

C++++ B+++ Be++ Li+
He
H−
√-E
par.
var.
par.
-.75
par.
2dim.
var.
2dim.
-.80
par.
2 dim.
-.85
-.90
Variationsrechnung, parabolische und zweidimensionale für (1s)2 1S und (1s)(2s) 3S
Fig. 3
α
.1
.2
.3
.4
.5
.6
.7
.8
.9
1.0

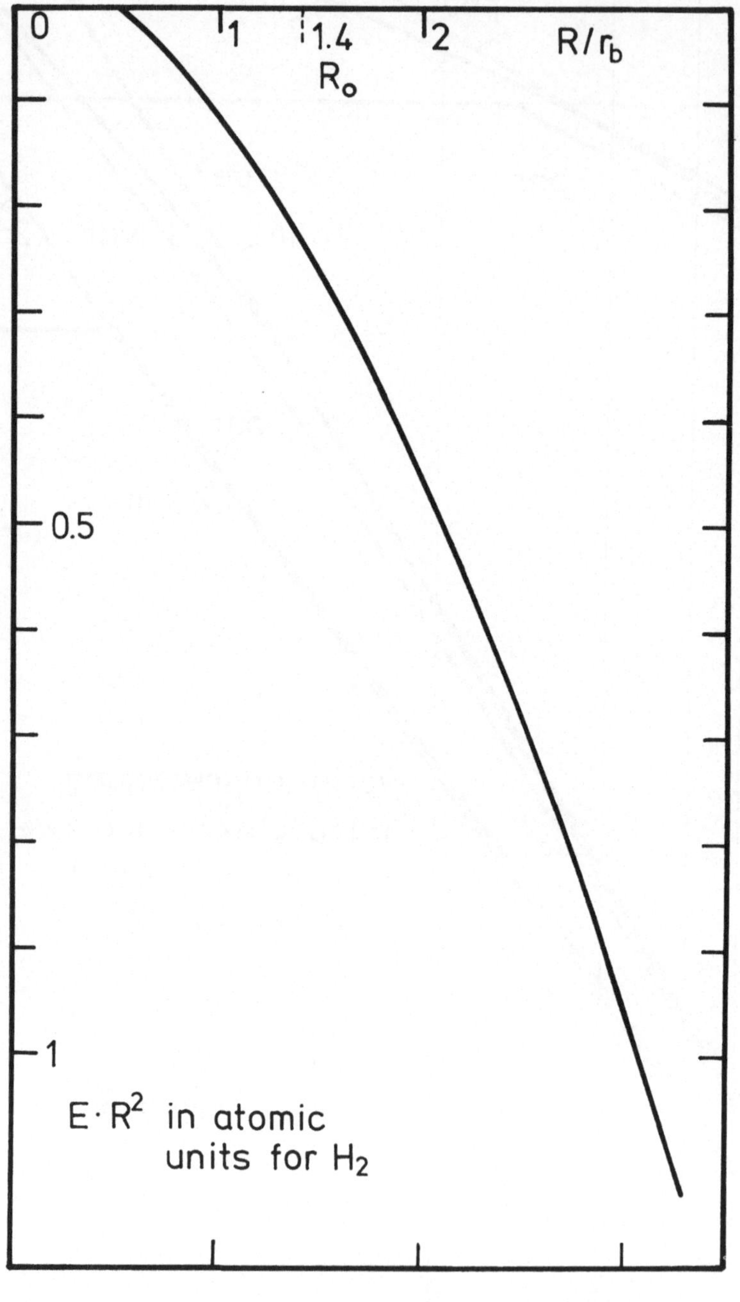

Fig. 4

Acta Physica Austriaca, Suppl. XIV, 643—666 (1975)

FINITE CONFORMAL TRANSFORMATIONS IN LOCAL QUANTUM FIELD THEORY+

by

W. RÜHL
Universität Kaiserslautern
Fachbereich Physik

1. INTRODUCTION

The problem of finite conformal transformations in local quantum field theory.

The group of conformal transformations consists of inhomogeneous Lorentz transformations, dilations

$$x_g^\mu = \lambda x^\mu , \; \lambda > 0 \tag{1}$$

and special conformal transformations

$$x_g^\mu = \frac{x^\mu + b^\mu x^2}{1 + 2bx + b^2 x^2} \, . \tag{2}$$

+ Seminar given at XIV. Internationale Universitätswochen für Kernphysik, Schladming, Austria, February 24-March 7, 1975.

The transformations (2) may map finite points of Minkowski space M_4 into infinity, and spacelike into timelike vectors and vice versa. Thus for a conformal transformation g, an ansatz for a scalar operator field $\Phi(x)$:

$$U_g \Phi(x) U_g^{-1} = \mu(x,g) \Phi(x_g) \qquad (3)$$

with a number-multiplier μ may lead to two difficulties:

a) If $x_g = \infty$, we obtain the operator field $\Phi(\infty)$, that is not defined. But this can probably be cured by the use of test functions that drop off sufficiently if $x \to \infty$.

b) a relative spacelike pair of points x_1, x_2, $(x_1 - x_2)^2 < 0$, may be transformed into a relative timelike pair. Therefore applying (3) to the commutator

$$[\Phi(x_1), \Phi(x_2)] = 0 \quad \text{for} \quad (x_1 - x_2)^2 < 0 \qquad (4)$$

implies a trivial field $\Phi(x)$.

This problem of reconciling locality with conformal invariance is known since a long time. The main steps towards a solution seem to have been made during the last year. This solution can be formulated in two ways. Either we replace (3) by a "nonlocal" transformation formula (see (57)) or we introduce quantum fields on an extended Minkowski space ("Geometric approach"). Both approaches are equivalent. The aim of this seminar is to show this for the case of the Thirring model.

The starting points of any discussion of conformal invariant fields is the assumption that we have a quantum field theory in the sense of the Wightman axioms that is "weakly conformal invariant", namely

$$(\alpha) \qquad U_g |0\rangle = |0\rangle , \qquad (5)$$

(β) (3) is valid in infinitesimal form.

The basic idea of the geometric approach is to implement Minkowski space M_4 by compactification and constructing the universal covering space:

$$M_4 \quad \overrightarrow{\text{compact.}} \quad \bar{M}_4 \quad \overrightarrow{\text{covering}} \quad \bar{M}_4^{uc} .$$

The space $\bar{M}_4^{uc}$ consists of a countably infinite number of sheets over M_4.

The notion of "relative spacelike pairs of points" can be extended onto $\bar{M}_4^{uc}$. The conformal group has to be enlarged to the universal covering group SU(2,2) that acts transitively on $\bar{M}_4^{uc}$. The first task is then to show that relative spacelike pairs of points on $\bar{M}_4^{uc}$ form a transitive set under $SU(2,2)^{uc}$.

Next we may try to define operator fields $\Phi(\tilde{x})$ on $\bar{M}_4^{uc}$ ($\tilde{x}$ over x on M_4), i.e. on an infinitely sheeted Minkowski space, such that locality is defined by

$$[\Phi(\tilde{x}_1),\Phi(\tilde{x}_2)] = 0 \qquad \text{if } \tilde{x}_1, \tilde{x}_2 \qquad (6)$$

are relative spacelike on $\bar{M}_4^{uc}$.

This is assured by the theorem of Mack and Lüscher[1]: Any weakly conformal invariant field theory allows the definition of fields $\Phi(\tilde{x})$ on $\bar{M}_4^{uc}$ with the locality property (6) that is invariant under finite transformations (3) of $SU(2,2)^{uc}$ (where $\tilde{x}_g$ is now not necessarily on the same sheet of $\bar{M}_4^{uc}$ as $\tilde{x}$). Whereas Lüscher and Mack have proven only an existence theorem, we want finally to show how the operator field on $\bar{M}_4^{uc}$ really looks in a special solvable model.

We consider a point $\tilde{x} \in \bar{M}_4^{uc}$ over $x \in M_4$ lying on the <u>n' th sheet.</u> Operator fields on two sheets n,n' over the same point x may be denoted then

$$\Phi(\tilde{x}) = \Phi_n(x), \ \Phi(\tilde{x}') = \Phi_{n'}(x) \ , \ x \in M_4 \ .$$

They are connected by a unitary transformation, belonging to an element g of $SU(2,2)^{uc}$

$$U_g \Phi_n(x) U_g^{-1} = \Phi_{n'}(x) \tag{7}$$

This element g is obtained by n'-n times circling around the generating curve of the homotopy group on $SU(2,2)/Z_4$ which generates the center of $SU(2,2)^{uc}$, too:

$$U_g = Z^{n'-n} \ , \ Z \text{ representing the generator.} \tag{8}$$

The unitary operator $Z = e^{if(Q)}$ depends on a "charge" operator Q. The special nature of this "charge" Q and the function f (Q) depends on the specific model, is a <u>dynamic</u>

rather than a <u>kinematic</u> feature of that model. We shall discuss it for the case of the Thirring model (following the work of Kupsch, Rühl and Yunn[2]).

We mention that the problem <u>a</u> has not yet been understood completely.

2. COMPACTIFICATION AND COVERING

Let us define for $x \in M_4$

$$\mathbf{X} = x^o \sigma_o + \vec{x}\,\vec{\sigma} \qquad (1 \equiv \sigma_o) \tag{9}$$

$$U = \frac{1 + iX}{1 - iX}, \quad X = i\,\frac{1 - U}{1 + U} \tag{10}$$

$$U = e^{i\tau} u, \qquad u \in SU(2) \tag{11}$$

$$u = u^o \sigma_o + i\,\vec{u}\,\vec{\sigma} \tag{12}$$

$$u^{o2} + \vec{u}^2 = 1 . \tag{13}$$

Thus we have a bijective mapping of M_4 into the matrix set U(2), i.e. on those matrices $U \in U(2)$ for which

$$\det (1 + U) \neq 0 . \tag{14}$$

A compactification of M_4 is obtained by closing this set in U(2), it results the whole U(2). This compactification can be endowed with the standard matrix topology of U(2). By this closure those matrices U are adjoined to M_4 that obey

$$\det (1 + U) = 0 \tag{15}$$

so that one eigenvalue of U is -1. A subset of U(2) with a fixed eigenvalue $e^{i\alpha} \neq -1$ corresponds to the cone

$$(x - (2\mathrm{tg}\,\frac{\alpha}{2}, 0,0,0))^2 = 0 \tag{16}$$

in M_4. So we may call our compactification "adjoining the infinite light cone" ($\alpha \to \pi$).

The eigenvalues of U

$$U = u_1 \begin{bmatrix} e^{i\phi_1} & 0 \\ 0 & e^{i\phi_2} \end{bmatrix} u_1^{-1}, \quad u_1 \in SU(2) \tag{17}$$

are directly related with the causal property of x:

$$(x)^2 = \mathrm{tg}\,\frac{\phi_1}{2} \cdot \mathrm{tg}\,\frac{\phi_2}{2} \tag{18}$$

$$x^0 = \frac{1}{2} (\mathrm{tg}\,\frac{\phi_1}{2} + \mathrm{tg}\,\frac{\phi_2}{2}) \tag{19}$$

$$|\vec{x}| = \frac{1}{2}|\mathrm{tg}\,\frac{\phi_1}{2} - \mathrm{tg}\,\frac{\phi_2}{2}| \quad . \tag{20}$$

We see that $\phi_1 \geq \phi_2$, $-\pi < \phi_{1,2} < +\pi$ is sufficient to parametrize the image of M_4 (Figure 1).

Note that in the ϕ_1, ϕ_2-plane

1. dilations and homogeneous Lorentztransformations leave quadrangles invariant that are bounded by lines ($k_{1,2}$ are integers)

$$\phi_1 = k_1\pi$$
$$\phi_2 = k_2\pi$$

2. translations leave quadrangles invariant bounded by

$$\phi_1 = (2k_1 + 1)\pi$$

$$\phi_2 = (2k_2 + 1)\pi$$

3. special conformal transformations leave the quadrangles

$$\phi_1 = 2k_1\pi$$

$$\phi_2 = 2k_2\pi$$

invariant.

The corners of the invariant quadrangles are fixed points in particular.

The parameter τ in (11) is related to $\phi_{1,2}$ by

$$\tau = \frac{1}{2}(\phi_1 + \phi_2) \quad . \tag{21}$$

If we maintain the restriction

$$2\pi \geq \phi_1 - \phi_2 \geq 0$$

and extend τ by

$$-\infty < \tau < +\infty$$

(uniformizing the function arg $(\det U)^{\frac{1}{2}}$) we obtain the u.c. space $\bar{M}_4^{uc}$. Moreover we have

$$\begin{aligned} \det(1 + U) &= 1 + \operatorname{Tr} U + \det U \\ &= 2e^{i\tau}\left(\cos\tau + \cos\frac{\phi_1 - \phi_2}{2}\right) \\ &= 4e^{i\tau}\cos\frac{\phi_1}{2}\cos\frac{\phi_2}{2} \end{aligned} \tag{22}$$

or

$$e^{-i\tau}\det(1 + U) > 0 \quad \text{on} \quad M_4 \tag{23}$$

instead of (14).

Two points x_1, x_2 are spacelike relative to each other if

$$(x_1 - x_2)^2 = \det(X_1 - X_2) < 0 \quad . \tag{24}$$

But

$$\det (X_1 - X_2) = - \frac{\det 2 (U_2 - U_1)}{\det (1+U_1) \det (1+U_2)} \tag{25}$$

so that we may characterize spacelike pairs on M_4 by

$$\arg \det (U_2 - U_1) = \tau_1 + \tau_2 \ . \tag{26}$$

In turn, let $U_1 = 1$, $U_2 = U$, $\tau_1 = 0$. Then

$$\arg \det (U - 1) = \tau + \arg (-\sin \frac{\phi_1}{2} \cdot \sin \frac{\phi_2}{2}) \tag{27}$$

for arbitrary ϕ_1, ϕ_2. Therefore we have $\arg \det (U - 1) = \tau$ for all points in the triangle

$$0 \geq \phi_2 \geq \phi_1 - 2\pi$$

$$0 \leq \phi_1 \leq 2\pi \qquad .$$

On the other hand this is the manifold of points reached by any spacelike point on M_4 by a special conformal transformation. Thus we can use (26) as a definition of spacelike pairs of points (provided we prove its invariance under $SU(2,2)^{uc}$) and achieve by this that those points of the first and the (-1)st sheet which lie over timelike points on the zeroth sheet and can be reached from the spacelike points of the zeroth sheet by a special conformal transformation are now "spacelike" as points on $\bar{M}_4^{uc}$.

The invariance of definition (26) is proved directly

from the relation

$$e^{-i\tau_1' - i\tau_2'} \cdot \det(\tilde{U}_2' - \tilde{U}_1')$$
$$= |\det(AU_1+B)|^{-1} \cdot |\det(CU_2+D)|^{-1} e^{-i\tau_1 - i\tau_2} \qquad (28)$$
$$\cdot \det(U_2 - U_1)$$

with $\tilde{U}_{1,2} \rightarrow \tilde{U}_{1,2}'$ on $\bar{M}_4^{uc}$ under $SU(2,2)^{uc}$ and the meaning of the matrices A,B,C,D is explained below.

Our final aim is to show that spacelike pairs of points form a transitive subset of $\bar{M}_4^{uc} \times \bar{M}_4^{uc}$. To show this we can consider the action of $SU(2,2)^{uc}$ on $\bar{M}_4^{uc}$. The group SU(2,2) consists of matrices

$$M = \begin{pmatrix} A, & B \\ C, & D \end{pmatrix} \quad , \text{ (A,B,C,D } 2 \times 2 \text{ matrices)} \qquad (29)$$

$$M\,H = H\,M^{\dagger,-1} \quad , \qquad H = \begin{pmatrix} -E_2, 0 \\ 0, +E_2 \end{pmatrix} \qquad (30)$$

with E_n the n x n unit matrix. SU(2,2) acts on U(2) by

$$U' = (A\,U + B)\,(CU + D)^{-1}$$
$$= (A^{\dagger} + UB^{\dagger})^{-1}\,(C^{\dagger} + UD^{\dagger}) \qquad (31)$$

The maximal compact subgroup consists of the matrices

$$K = \begin{pmatrix} e^{i\alpha}u_1 & 0 \\ 0 & e^{-i\alpha}u_2 \end{pmatrix} \qquad , 0 \leq \alpha < \pi \tag{32}$$

$u_{1,2} \in SU(2)$.

For a special element K we have instead of (31)

$$U' = e^{2i\alpha} u_1 U u_2^{-1} \quad . \tag{33}$$

The center of SU(2,2) consists of the matrices $e^{i\frac{\pi}{2}k} \cdot E_4$, $k = 0,1,2,3$.

The path

$$K_\alpha = \begin{pmatrix} e^{i\alpha}E_2, & 0 \\ 0 & , e^{-i\alpha(1+2\sigma_3)}E_2 \end{pmatrix} \tag{34}$$

goes through the central elements for $\alpha = 0, \frac{\pi}{2}, \pi, \frac{3}{2}\pi$.

The projection of the contour K_α, $0 \leq \alpha \leq \pi/2$ on $SU(2,2)/Z_4$ cannot be contracted to a point and thus is the generator of the homotopy group of $SU(2,2)/Z_4$. Consequently $K_{\frac{\pi}{2}}$ is the generator of the center of $SU(2,2)^{uc}$. Inserting (34) into (33) shows that $K_{\frac{\pi}{2}}$ maps the n-th sheet on the n+1-st sheet of $\bar{M}_4^{uc}$ (we denote $\bar{M}_4^{uc} \to M_4$)

$$\tilde{U}' \to U' = e^{+2i\alpha(1+\sigma_3)} \cdot U \Big|_{\alpha = \frac{\pi}{2}} \tag{35}$$

i.e.

$$\phi_1' = \phi_1 + 2\pi$$

$$\phi_2' = \phi_2 + 0 \, . \tag{36}$$

Now we return to the subset of spacelike pairs of points of $\bar{M}_4^{uc} \times \bar{M}_4^{uc}$. Given $\tilde{U}_2$ we define

$$\tilde{U}_2 \to U_2 = u_2 \begin{bmatrix} e^{i\phi_{2,1}} & 0 \\ 0 & e^{i\phi_{2,2}} \end{bmatrix} u_2^{-1} \tag{37}$$

and

$$K(U_2) = \begin{bmatrix} e^{-i\frac{\tau_2}{2} - i\frac{\phi_{2,1}}{2}\sigma_3} u_2^{-1}, & 0 \\ 0 & , e^{+i\frac{\tau_2}{2} - i\frac{\phi_{2,2}}{2}\sigma_3} u_2^{-1} \end{bmatrix} . \tag{38}$$

$K(U_2)$ maps $\tilde{U}_2$ on E_2 (the origin in M_4) and $\tilde{U}_1$ on $\tilde{U}_1'$. The latter point lies spacelike relative to E_2 because of the invariance of the causal ordering. By a special conformal transformation we can map $\tilde{U}_1'$ (keeping E_2 fixed!) onto U_1''

$$\phi''_{1,1} = \frac{\pi}{4}, \ \phi''_{1,2} = -\frac{\pi}{4}, \ u_1'' = E_2 \ ,$$

(parameters as in (37)) on the zeroth sheet. So we have shown that any spacelike pair can be mapped on the standard pair $x_1 = (0,0,0,0)$, $x_2 = (0,0,0,1)$ and transitivity follows.

Requiring that the commutator (6) vanishes for spacelike pairs on $\bar{M}_4^{uc}$ is therefore an invariant requirement and entails vanishing on spacelike pairs of M_4 in the usual sense.

The relevance of the universal covering group in conformal invariant quantum field theory has been emphasized and the representation have been studied by the author in[3]. The invariant causal ordering on the universal covering space of compactified Minkowski space has first been studied by I. Segal[4]. Its importance has been emphasized in a series of papers[5]. Its proper role has become clear finally in Ref. 1.

3. THE THIRRING MODEL

The Thirring model[6] deals with a conformally covariant quantum field $\phi(x)$ in two-dimensional spacetime satisfying

$$-i\partial_\mu \gamma^\mu \phi(x) = g\gamma^\mu \ N[\phi J_\mu](x) \tag{39}$$

$$J_\mu(x) = N[\phi^\dagger \gamma^0 \gamma_\mu \phi](x) \tag{40}$$

$$\tilde{J}_\mu(x) = N\,[\phi^\dagger \gamma^o \gamma^5 \gamma_\mu \phi](x) \tag{41}$$

The problem to define the normal product N [...] is part of the problem to construct a solution to this model.

Note that in 2-dimensional space-time the spin s can take on all real non-negative values and that for any s > O the field can be chosen as a two-component ("spinor") object. The dimension d of ϕ is defined as $-\frac{1}{2}$ times the homogenity degree of the two-point function of ϕ, considered as a homogeneous distribution in x. It satisfies

$$d \geqq s \tag{42}$$

by positivity, d = s belongs to free fields.

The Minkowski space M_2 can be decomposed

$$M_2 = R_1(x_+) \otimes R_1\ (x_-)$$

such that the conformal group decomposes into two direct factors

$$SL(2,R)/Z_2 \quad \otimes \quad SL(2,R)/Z_2$$

each one transforming one variable

$$x_\pm \to x'_\pm = \frac{\sigma_\pm\ x_\pm + \tau_\pm}{\xi_\pm\ x_\pm + \eta_\pm} \tag{43}$$

$$x_\pm = x_o \pm x_3\ . \tag{44}$$

The universal covering $\bar{M}_2^{uc}$ of M_2 is obtained in the three steps

a. by the mapping

$$\phi_{\pm} = 2 \text{ arctg } x_{\pm}, \quad -\pi < \phi_{\pm} < \pi \tag{45}$$

b. compactification, viz. adjoining the points

$$\phi_{\pm} = \pi$$

c. "rolling off" the unit circles on R_1 such that

$$-\infty < \phi_{\pm} < +\infty$$

$\bar{M}_2^{uc}$ has the same topological structure as M_2! $\phi_{\pm}$ transforms as

$$e^{i\phi} \to e^{i\phi'} = \frac{\alpha e^{i\phi} + \beta}{\bar{\beta} e^{i\phi} + \bar{\alpha}} \tag{46}$$

where α, $\bar{\alpha}$, β, $\bar{\beta}$ are linear combinations of σ, τ, ξ, η (see Ref. 2).

$$z = \begin{pmatrix} -1 & 0 \\ 0 & -1 \end{pmatrix} \quad \varepsilon \quad SL(2,R)$$

generates the homotopy group of $SL(2,R)/Z_2$ and the center of the universal covering group of $SL(2,R)/Z_2$.

Moreover

$$\phi \underset{z}{\rightarrow} \phi' = \phi + 2\pi . \tag{47}$$

Thus $z_+ \in SL(2,R)(x_+)$ maps the (n_+,n_-) sheet of $\bar{M}_2{}^{uc}$ on the $(n_+{+}1,n_-)$ sheet and $z_- \in SL(2,R)\ (x_-)$ maps (n_+,n_-) on $(n_+,n_- + 1)$. We expect that in the Thirring model $z_\pm$ are represented by operators that depend on the conformally invariant charges

$$Q = \int dx_3\ J_o(x) \tag{48}$$

$$\tilde{Q} = \int dx_3\ \tilde{J}_o(x) . \tag{49}$$

The Thirring model is solved by explicitly constructing the field operators ϕ in the Fock space of the free spin $-\frac{1}{2}$ field. There is an infinity of solutions that are all isomorphic in the sense that they lead to the same Wightman functions, namely Klaiber's Wightman functions[7 *)]:

$$\langle O|\phi(x_1)\ldots\phi(x_n)\phi^+(y_1)\ldots\phi^+(y_n)|O\rangle$$

$$= (2\pi)^{-n}\, e^{-\pi i\,[dn + (\sum_i N^i_{1,o})(\sum_j N^j_{2,o})]} \cdot$$

$$\prod_{i<j} (\bar{x}^i_+ - x^j_+)^{c^i_+ c^j_+} \prod_{i'<j'} (\bar{y}^{i'}_+ - y^{j'}_+)^{c^{i'}_+ c^{j'}_+} \prod_{i,j} (\bar{x}^i_+ - y^{j'}_+)^{-c^i_+ c^{j'}_+} .$$

---

*) f(x) (f($\bar{x}$)) denotes the boundary value from above (below).

$$\prod_{i<j} (\bar{x}_-^i - x_-^j)^{c_-^i c_-^j} \prod_{i'<j'} (\bar{y}_-^{i'} - y_-^{j'})^{c_-^{i'} c_-^{j'}} \prod_{i,j'} (\bar{x}_-^i - y_-^{j'})^{-c_-^i c_-^{j'}} \tag{50}$$

where

$$c_+^i = \frac{1}{2} [(\alpha-1) - (\beta -1) \gamma_5^i] \tag{51}$$

$$c_-^i = \frac{1}{2} [(\alpha-1) + (\beta -1) \gamma_5^i] \tag{52}$$

and α, β label the solutions of the Thirring model (precisely: the isomorphy classes). The coupling constant g is

$$g = \pi (\alpha - \beta) . \tag{53}$$

We define moreover

$$N_+ = C_+^2, \; N_- = C_-^2, \; \delta = C_+ C_- \tag{54}$$

and the corresponding values

$$N_{+,o} = \frac{1}{2} (1 - \gamma_5) \tag{55}$$

$$N_{-,o} = \frac{1}{2} (1 + \gamma_5) \tag{56}$$

for the free fields ($\alpha = \beta = 0$ or $\alpha = \beta = 2$). We have then the following transformation formula for the field $\phi(x)$ (for any fixed solution in one isomorphy class α, β)

$$U_g \phi(x) U_g^{-1} = (\xi_+ x_+ + \eta_+)^{-N_+} (\xi_- x_- + \eta_-)^{-N_-}$$

$$\phi(x_g) \exp i\{[\sum_{\sigma=\pm} (N_\sigma - N_{\sigma,O})(Q+\tilde{Q}\gamma^5) + \delta(Q-\tilde{Q}\gamma^5)]\chi_\sigma(x)\} \qquad (57)$$

with

$$\chi_\sigma(x) = \arg(\xi_\sigma x_\sigma + \eta_\sigma) \qquad (58)$$

(understood as boundary value of an analytic function in x with $\mathrm{Im} x_\sigma > 0$ that is real analytic on the u.c.g.)

for any g in the uc.g. of the conformal group U_g is unitary and leaves the vacuum invariant. It is a symmetry transformation of the theory. From this formula we read off

$$d = \frac{1}{2}(N_+ + N_-) \qquad (59)$$

$$\mp s\,\gamma^5 = \frac{1}{2}(N_+ - N_-) \qquad (60)$$

depending on whether sign $(\alpha - 1)\cdot(\beta - 1) = \pm 1$. The constants d,s,g are related correspondingly by

$$d = \pm s + \frac{g^2}{4\pi^2} \qquad (61)$$

(In the case $\alpha = 0$, $\beta = 2$ or $\alpha = 2$, $\beta = 0$ we have

$$\begin{aligned} N_+ &\to N_{-,O} \\ N_- &\to N_{+,O} \end{aligned} \qquad (62)$$

but this is not a free field solution, in the sense that it satisfies the field equation with $g = \mp 2\pi$).

From the above transformation formula it can be seen that the center of the u.c.g. of the conformal group is generated by

$$Z_{\pm} = \exp i \frac{\pi}{4} [(\alpha-1)\, Q_{\mp}\, (\beta-1)\, \tilde{Q}\,]^2 \, . \tag{63}$$

From the expression for the interacting currents (Ref. 2, equ. (3.72)) one finds that interacting chiral charges can be introduced by

$$q_{\pm} = \frac{1}{2}(q \mp \tilde{q}) = \frac{1}{2}[(1-\alpha)Q \mp (1-\beta)\tilde{Q}]$$

where the interacting scalar or pseudoscalar charges $q, \tilde{q}$ are defined analogously to $Q, \tilde{Q}$ in (48), (49). Then $Z_{\pm}$ can be expressed simply by

$$Z_{\pm} = e^{+i\pi q_{\pm}^2} \, . \tag{64}$$

On the other hand we have

$$f(q_{\pm})\phi(x) = \phi(x)\; f(q_{\pm} + C_{\pm}) \tag{65}$$

so that interchange with the field $\phi(x)$ yields the exponents $C_{\pm}$ (51), (52). One finds this way

$$Z_{\pm}\phi(x) Z_{\pm}^{-1} = \phi(x) \; \exp\{i\pi\; (2q_{\pm}C_{\pm} + C_{\pm}^2)\} \tag{66}$$

$$= \phi(x)\exp\{-i\pi[N_{\pm}(Q+\tilde{Q}\gamma_5-1)+\delta(Q-\tilde{Q}\gamma_5)]\} \quad .$$

We then define

$$\Phi_{n_+n_-}(x) = Z_+^{n_+} Z_-^{n_-} \phi(x) Z_-^{-n_-} Z_+^{-n_+} \tag{67}$$

$$\Phi(\tilde{x}) = \Phi_{n_+ n_-}(x) \tag{68}$$

$$\text{if } \tilde{x} \text{ over } x \text{ on the } (n_+,n_-)\text{-sheet of } \bar{M}_2^{uc} \; ,$$

and get finally

$$U_g \Phi(\tilde{x}) U_g^{-1} = |\xi_+ x_+ + \eta_+|^{-N_+} \, |\xi_- x_- + \eta_-|^{-N_-} \, \Phi(\tilde{x}_g) \quad . \tag{69}$$

It is, however, more imaginative to define the operator field (on the sheet (o,o))

$$f(\phi) = (\cos\frac{\phi_+}{2})^{-N_+} \cdot (\cos\frac{\phi_-}{2})^{-N_-} \phi(x) \tag{70}$$

$$(x_+,x_-) = (\mathrm{tg}\,\frac{\phi_+}{2},\ \mathrm{tg}\,\frac{\phi_-}{2}) \, , \quad -\pi < \phi_{\pm} < +\pi \quad .$$

Then the n-point function comes out to be

$$\langle 0|f(\phi_1)\ldots f(\phi_n) f^+(\psi_1)\ldots f^+(\psi_n)|0\rangle = \tag{71}$$

= the same expression as (50) with

$\bar{x}^i_+ - x^j_+$ replaced by $\sin \frac{1}{2} (\bar{\phi}^i_+ - \phi^j_+)$ etc.

This expression can be easily continued analytically in the halfplanes of the variables

$$\Delta\phi^i_\pm = \phi^i_\pm - \phi^{i+1}_\pm \quad , \; i = 1,\ldots,n-1$$

$$\Delta\psi^i_\pm = \psi^i_\pm - \psi^{i+1}_\pm \quad , \; i=1,\ldots,n-1 \tag{72}$$

$$\chi_\pm = \phi^n_\pm - \psi^1_\pm \quad .$$

The product of the lower half planes of all these variables is a domain of analyticity into which the Wightman functions can be continued so that they in turn appear as boundary values from below along the piece

$$-\pi < \phi^i_\pm, \psi^j_\pm < + \pi$$

of the whole boundary. The boundary values at other points of the real axis of variables $\phi^i_\pm$ or $\psi^j_\pm$ are connected with the field operators on the higher sheets (n_+, n_-), $n^2_+ + n^2_- \neq 0$, of $\bar{M}^{uc}_2$.

In particular we can define the field f on the sheet (n_+, n_-) by the unitary transformation

$$Z_+^{n_+} Z_-^{n_-} f(\phi) Z_-^{-n_-} Z_+^{-n_+} = f(\phi_+ + 2\pi n_+, \phi_- + 2\pi n_-) \tag{73}$$

which then implies for all n_+, n_-

$$f(\phi_+ + 2\pi n_+, \phi_- + 2\pi n_-) = \left(\cos \frac{\phi_+}{2}\right)^{-N_+} \left(\cos \frac{\phi_-}{2}\right)^{-N_-} \Phi_{n_+ n_-}(x) \tag{74}$$

$$-\pi < \phi_\pm < \pi$$

and

$$U_g f(\phi) U_g^{-1} = |\alpha_+ e^{i\phi_+} + \beta_+|^{-N_+} \, |\alpha_- e^{i\phi_-} + \beta_-|^{-N_-} f(\phi_g) \tag{75}$$

for arbitrary ϕ and g.

The result that a conformal invariant field theory can be formulated by means of field operators on $\bar{M}^{uc}$, whose Wightman functions represent boundary values of an analytic function over a common domain is due to Lüscher and Mack[1]. The formulation of conformal invariant field theories by means of field operators on $M_{2(4)}$ with a transformation behaviour such as (57) involving an operator-multiplier has first been obtained in a model calculation by Schroer and Swieca[8] and independently for the Thirring model in Ref. 2. It was the aim of this work to emphasize and work out the equivalence of both formulations.

ACKNOWLEDGEMENT

The author thanks M. Lüscher for pointing out to him the simple formula (64) for the operators $Z_\pm$.

REFERENCES

1. M. Lüscher, G. Mack, Global conformal invariance in quantum field theory, Institut für Theoretische Physik der Universität Bern, Preprint, August 1974.
G. Mack, Lectures presented at the 1974 Bonn summer school.

2. J. Kupsch, W. Rühl, B.C. Yunn, Conformal invariance of quantum fields in two-dimensional space time, Annals of Physics, to appear.

3. W. Rühl, Comm. Math. Phys. 30, 287 (1973) and 34, 149 (1973).

4. I. Segal, Bull. Am. Math. Soc. 77, 958 (1971).

5. I.T. Todorov, Conformal invariant quantum field theory with anomalous dimensions, CERN preprint TH 1697 (1973); in particular the Appendix,
W. Rühl, Conformal kinematics, Teheran lectures, Kaiserslautern Preprint, September (1973),
D. H. Mayer, Conformal invariant causal structures on pseudo-Riemannian manifolds, Aachen Preprint, April 1974.
T. H. Go, Some remarks on conformal invariant theories formulated on some four-Lorentz manifolds. Aachen Preprint, June 1974.

6. W. Thirring, Annals of Physics 3, 91 (1958).

7. B. Klaiber, in "Quantum Theory and Statistical Physics, Lectures in Theoretical Physics", Vol X - A, p. 141, (A.O. Barut and W.E. Brittin, Eds.), Gordon and Breach, New York, 1968.

8. B. Schroer, J.A. Swieca, Phys. Rev. D10, 480 (1974).

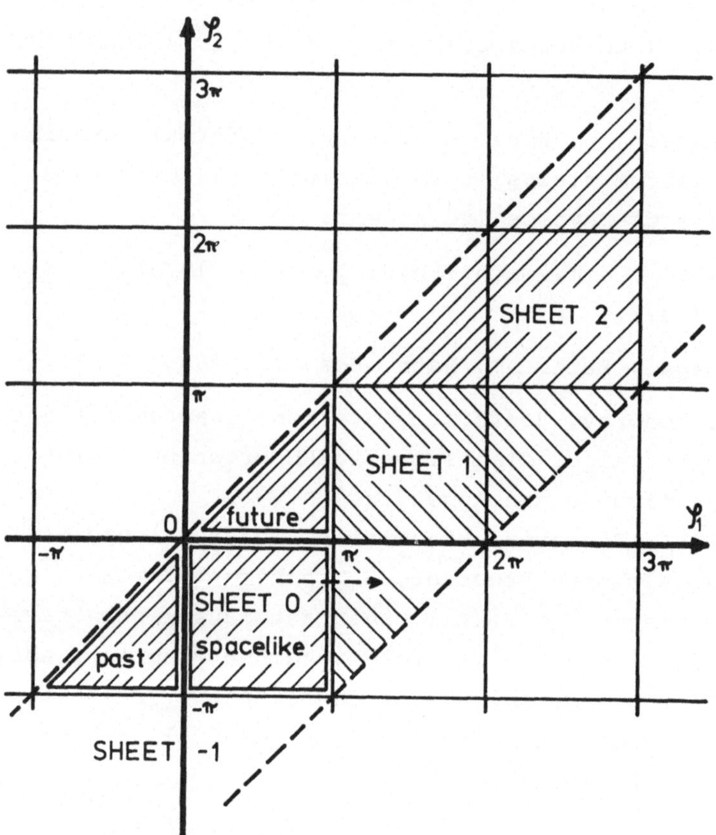

The image of M_4 in the eigenphase plot of U(2).
--➤ denotes a special conformal transformation.

Acta Physica Austriaca, Suppl. XIV, 667 — 681 (1975)

SUMMARY+

by

H. PIETSCHMANN
Institut für Theoretische Physik
Universität Wien

When I was asked by Prof. Urban to give the summary of this years Schladming Winter School, I was at the first moment a little bit discouraged by the tremendous task to summarize so many lectures with so many different topics. But I soon found out, that it cannot be the aim of a summarizer to just repeat in short what has been said in the lectures. Thus I will try to give you my own personal view of what has happened during these very fine two weeks here in Schladming. I will now and then add a little bit of outside information, which I find relevant to our topics.

It must be clear from the outset, that I shall not be able to give you profound views on mathematics, experiments and phenomenology. I shall follow the advice given by R. Haag in his lecture, that you may very well ride your own horse, but you should not think that the other horses are lame. Let me comment a little bit on our situation in phy-

---

+ Summary given at XIV. Internationale Universitätswochen für Kernphysik, Schladming, Austria, February 24-March 7, 1975.

sics: following R. Haag, I will distinguish three separate approaches: the mathematical approach, the conceptual basis, and phenomenology. (It is fairly clear, that any other separation may be just as good). The notorious trouble is, that each representative of any of these fields always thinks, that his own field is the most important one. This is certainly not correct! However, phenomenology is distinguished from the other fields by the fact, that it is really the most important one!

Having disclosed to you which horse I am riding personally, I must confess, that I will not be able to comment too much on the very fine lectures on mathematical physics during this school. I apologize for this, in particular to the lecturers who did such an excellent and outstanding job. But all of those lectures were very nice and self-contained, so those of you who were listening got out their content with no trouble.

We also had a lecture on the conceptual basis, it was the part of J. Schwinger's lectures devoted to the "sources". Needless to say, that Schwinger's presentation was absolutely admirable and as far as the content is concerned, I do not have to add anything. However, it seems to me, that whenever you talk about the conceptual basis, you have mixed feelings: on the one hand you have a feeling, that it is absolutely necessary to have a conceptual basis; J. Schwinger made it quite clear, that the way he got his ideas for the phenomenology in deep inelastic scattering stems from the way he treats the conceptual basis, the sources. On the other hand, once you go to applications, it seems that the conceptual basis plays little role. In other words, if you want to compute energy levels of a certain atom, it makes little or no difference, whether you believe in the Copenhagen interpretation of quantum mechanics or any other competitive interpretation. At this level, you just don't feel the conceptual basis any more; I want to repeat however, that it is very important to have a conceptual basis in order to have new ideas at the frontier.

The bulk of the lectures on which I am going to comment in more details, can be collected under the main heading of

STUDY OF STRUCTURE OF CURRENTS.

Needless to repeat, that right now my own personal interest centers around this field. We have had a large number of very nice lectures on this topic. Let us begin with the lecture by H. Mitter on quantum electrodynamics in laser fields, although this is not done to study the structure of the currents, it is rather the other way round: the structure of the currents is given and you want to apply it in laser fields. This is a general observation; if you want to find the basic underlying structure, you have to go to the most simple physical system of course; but once you have the underlying structure - or have it to an extent which is far reaching enough that you can go to applications - you can go and test all the possible predictions of this underlying structure. Part of the excitement in physics stems from this possibility and we have heard in the lectures of Mitter about many extremely interesting applications of quantum electrodynamics in laser field and we shall hope that one day it should be possible to transport one of the powerful lasers to one of the high energy accelerators in order to perform these experiments.

Next let me mention the two series of lectures given by A. Bartl and H. Rollnik. First of all, I would like to say, that it was very nicely arranged how the lecturers split the subject and avoided duplication of material. To me, it was rather surprising that in the region beyond the resonances the Born-approximation is apparently the best thing we have and it works almost better than anything more elaborate. Like in quantum electrodynamics in laser fields, in photoproduction you do not study the structure of the currents either; you rather study the structure of the produced products. It was therefore very often a matter of discussion, whether photoproduction belongs to the section of electromagnetic interactions or rather to strong interactions in conferences. Indeed, the language is more close to those of hadron spectroscopy. However, we should not forget that photoproduction very often serves also as a tool to answer basic questions about the structure of the current. The question of absence of iso-tensor contributions to the electromagnetic current was studied with photoproduction experiments and during this conference we learned, that the question whether the new ψ-particle is an intermediate boson or not was to a good extent cleared up by photoproduction, which showed that the photoproduction cross-section for the ψ-meson is so big that the ψ cannot be identified with an intermediate boson.

Let me now turn to the actual study of the structure of the currents. The most simple matrix element of the current one can find, is the matrix element with only one particle in the left and in the right state of the matrix element. What you then get in the Lorentz covariant decomposition are, of course, the well-known form factors. Let me just spend a few minutes on the history of form factors:

In the years before 1960 the electromagnetic structure of nucleons was experimentally investigated and in those days there was no underlying theoretical concept. Therefore, the experimentally obtained data points were fitted with all sorts of curves: Exponentials, Lorentzians, Gaussians and so on. There was no theoretical preference. When Frazer and Fulco undertook to study the structure of the nucleons by means of dispersion relations, they were able to stir an appreciable excitement, because they actually predicted the existence of a vector meson. In those days, one had 16 different elementary particles (not counting antiparticles separately) and the prediction of a new one - even with qualitatively different quantum numbers like that of a vector meson - was something discussed about in all laboratories. Soon after that, a particle was produced in a helium bubble chamber provisionally called the ABC-particle (after the experimentalists Abachian, Booth and Crow). Lateron this particle was called ω but it died. It is not the particle we know now under the name ω. After the pioneering work of Frazer and Fulco people started to analyze the form factors by means of the so-called Clementel-Villi form, which is simply a sum of poles. And in those days, one put in all the vector mesons and you have heard in the seminars of G. Höhler that from the simple fact that one has done it so long one still does it in our days: one tries to put all the vector mesons including the ϕ into a pole description of the form factors. This is done inspite of the fact, that in another lecture you can hear strong arguments that the decays of the ψ-mesons are so strongly suppressed for the same reason that the ϕ is not coupled to the nucleons; I was therefore very glad that G. Höhler pointed out this inconsistency very clearly.

When the higher energy regions came into play, the

time of the nice and simple description of form factors was over. Experimentalists told us that the form factors have to be described by the famous dipole fit. If you look through a magnifying glass, you find it is not exactly the dipole fit; there is a little deviation. Many theoreticians (including your speaker) rushed with many ideas to explain the deviation from the dipole fit. After all the inventiveness was exhausted, you had so many different theoretical fits, that at any given value of q^2 (invariant four momentum transfer) you had at least one fit which practically went through the data. The same fit would not do at other values of q^2 but there you would find another theoretical fit which would do the trick. In other words, the theoretical predictions fill a whole diagram almost densly so that the predictive power is practically nil. Therefore it was extremely nice to hear in the lectures of Höhler that the more serious attempts to study the nucleon form factors have been taken up again and seem to lead to rather good understanding, although things have become much more difficult. In a way it is sad, that the time of the simple understanding - which you cou could easily store in your memory - is over, but after all it is very important to have an understanding of this matrix element because it is the most simple one. And how can we even hope to understand the more complicated one if we do not understand the most simple one!

Proceeding step by step, we would now go to the next more complicated matrix element: the one with two particles in the final state. But this is a matrix element which you investigate in photoproduction, for instance, and as I have said before, it does not really belong to the study of the electromagnetic current; it rather belongs to the study of the final state.

As always in physics, counting goes: 1, 2, 3,... ∞. Therefore we jump right away to the inclusive matrix element, where you have one particle in the initial state and an arbitrary number of particles in the final state. This is of course the matrix element contributing to deep inelastic scattering. It leads me back to the lectures of J. Schwinger. Personally, I cannot talk about these lectures without at least mentioning the fascination which J. Schwinger was able to radiate by the mere way of phrasing the sentences and putting the words in place.

J. Schwinger discarded the notion of partons and quarks, at least for the largest part of his lectures. By using a different - kinematical singularity free - combination for the structure functions and by using his basic concepts of sources (which, as I have said before, enters only in the argumentation which leads to the answer rather than in the actual calculations) he arrived at an interesting phenomenological analysis of these structure functions. I feel compelled to add a word at this place for those of you who are not really working in this field. You remember that J. Schwinger showed a figure with the experimental points of F_N/F_P (the scaling function for the neutron divided by the scaling function for the proton). This ratio is plotted against x or ξ or $1/\omega$ (whatever language you prefer), which is the scaling variable. Points in this graph drop from 1 essentially on a straight line down to almost exactly 1/4. Now in J. Schwinger's model, he would fit those points with a straight line going down to 0. Obviously, this is the most natural way to fit those experimental data. On the other hand, the parton model gives you a lower limit of 1/4 for this ratio. So if the parton model is correct, the experimental points have to make a rather sharp turn at 1/4 and go at least vertically off or even go up again. If they go below 1/4, the predictions of the quark-parton-model are not met and the model is ruled out. Therefore I believe it will be extremely exciting to see how this curve goes on.

In the study of the structure of currents, we do not only want to investigate the decompositon of matrix elements; what may be even more important to know, is the structure of the currents themselves. This brings me to the lectures of M. Gourdin who talked on neutral currents. It is interesting to recall that in last years summary of the Schladming Winter School W. Frazer expressed his surprise, that nobody seemed to doubt the existence of neutral currents, which he said was not unequivocally accepted everywhere. He said, that it seemed that the confidence in neutral currents is an inverse function of the distance from CERN. Within the last year, the situation has drastically changed in the sense that there cannot be any doubt any more on the existence of neutral currents. Therefore, the next relevant question is the question as to their structure. M. Gourdin investigated neutral currents in the best way we have at hand at the moment: by means of phenomenology without betting on one of the many models which are offered in the literature.

Let us contemplate a little bit! What do we want to find out about the neutral currents. The first question is of course, whether they are <u>currents</u> at all. By "current" one normally means an object which is of vector nature (including of course vector axial vector mixture) as opposed to, for instance, scalar pseudoscalar or tensor structures. There are speculations, that the neutral currents might be scalar or pseudoscalar or tensor. In these cases, you would have funny things like neutrinos flipping their spin and π^0 decaying into neutrino-antineutrino pairs. I think it is fair, that at the present time we leave these possibilities out and start from the assumption that the neutral current is of vector (axial vector) nature. Naturally, if you want to challenge this assumption I would not have too strong arguments.

The next question of utmost importance about the neutral current is whether it violates parity or not. Just a week before the Schladming meeting was opened, an "International Symposium on Interaction Studies in Nuclei" was closed in Mainz. This symposium was partly dedicated to parity violation in nuclear physics. For these experiments it is of course of top importance to know whether neutral currents exist or not because it turns out that the contribution to parity violating nuclear forces from the neutral current is much stronger than those from charged currents. Therefore, if the neutral current violates parity, parity violation in nuclear forces should be much stronger than otherwise. Now you could say, since we have discarded the first question on the assumption that the neutrino never flips its spin, at least part of the neutral current must be parity violating because a two-component neutrino automatically does it. This is of course true. However, there are very respectable models in the literature, where the neutrino part (or the leptonic part) of the weak neutral current violates parity whereas the hadronic weak neutral current does not. In fact, the hadronic weak neutral current may even be proportional to the electromagnetic current. If this were the case, it would be also the end for experiments looking for parity violation in atomic transitions, for instance, in Caesium atoms. It would be a tremendous thing of course to determine contributions of neutral weak currents in atomic physics.

The third question we can ask about the neutral current is: what are its isospin transformation properties? This is certainly a question to be answered by experiments. M. Gourdin in his lectures concentrated only on deep inelastic scattering. He told you, that at the moment there is no conclusive evidence from those experiments. However, I might add that from exclusive experiments, to be specific, one pion production, we do get information on the isospin properties of the weak neutral current in a much easier fashion. Since the neutrino hits a target of isospin 1/2 (proton or neutron), one can analyze the final state (the pion-nucleon system) in terms of isospin wave functions. This is the experiment made at the Argonne National Laboratory. On the basis of a couple of dozen events one cannot conclude very much yet, but at the moment the situation is funnily so, that at the place where the Δ resonance of isospin 3/2 should show up, there is a dip in the effective mass plot; that is to say there is no event at the mass of the Δ. So it looks as if the neutral current could be an isoscalar. However, I want to repeat that this is on the basis of very few events and I would like to remind you what I have said on the ABC particle. So I would not be surprised if it turns out to be just a statistical fluctuation.

At this point I would like to take the liberty to add something which has not been discussed during this fine Winter School: The properties of the charged weak currents. Even there we are not absolutely sure about its isospin (or rather G parity) properties. It is the notorious question of second class currents. At the International Symposium on Interaction Studies in Nuclei in Mainz - which I have mentioned before - D. Wilkinson reported good news on second class currents: there seems to be no evidence. On the other hand, F. Calaprice reported results of an experiment on ^{19}Ne in which he found evidence of second class currents as strong as the first class currents. So again the question is not cleared up and we have to wait for more and better experiments.

The question which is usually asked first about the weak neutral current comes only as number 4 on my list: The question whether it is one of the currents suggested by one of the gauge models of unified weak and electromagnetic interactions. Of course, if it is a neutral current of a gauge model, many physicists including myself would prefer it to be the Salam-Weinberg current amended by the Glashow-Iliopoulos-Maiani-scheme. Why is that so? This is for the simple reason that this model is the most economic model in the sense, that it introduces the least number of new particles. And we should not forget that it was, is, and should always be the aim of progress in physics to look for the most simple way to describe nature. Only if we are forced by experiments should we give up this most economic model.

Only when all of those questions have been positively answered can we ask the final and fifth question: What is the value of the Salam-Weinberg mixing angle? Why is it then, that there is a tremendous number of papers extracting the value of the Salam-Weinberg angle from experiments? Well, it is of course important to know this value for predicting other experiments. It is the only free parameter in the model and after it has been fixed, predictions can be made without further freedom of parameters. Therefore I think it is legitimate to try to obtain a value for the Salam-Weinberg mixing angle, if one only keeps in mind that it is done on the working hypothesis that all four other questions have been answered. From a systematic and logical point of view, one has to answer the questions one after the other. In fact, it is not yet completely clear whether the various data from deep inelastic scattering are consistent at all with the Salam-Weinberg model. Even if they are inconsistent, one can of course extract a value for this angle which will then represent the "point of minimal inconsistency".

Before we leave the subject of neutral weak currents proper, let me throw in a historical detail which I think is characteristic: When Wolfgang Pauli invented the neutrino, he said to the astronomer Walter Baade: "I have done a terrible thing today, something which no theoretical physicist should ever do. I have suggested something that can never be verified experimentally." Doesn't this quotation show the proper attitude of a conscious theoretical physicist? I think in our days we are much too liberal in introducing new things without ever really thinking twice how and whether at all they can be checked experimentally. On the other hand it shows the tremendous progress we have made. Walter Baade made immediately a bet with Wolfgang Pauli that the neutrino will one day be discovered. Not only was the neutrino discovered, we are even able to study its detailed interaction and in our days we even investigate interactions where the lepton in the final state is again a neutrino.

You all know that the discovery of neutral weak currents gave a tremendous boost to elementary particle physics because it predicted new rather stable particles in order to explain the selection rule that a neutral current never changes strangeness. In this connection, the lectures of J. Schwinger come up again and I remind you on the way he produced the ψ's by avoiding the Cabibbo

gymnastics. I think this in itself is a very interesting approach because it offers an alternative to beaten tracks. Also I would clearly point out, that J. Schwinger actually predicted the ψ's before they were discovered and he got their properties right. Moreover, I think it is also healthy to know that the standard approach is not always necessarily the only one which is able to explain data. I point this out in particular because one can stretch the Cabibbo theory so far as to try to compute the mass of the neutral intermediate vector boson from slight deviations in the determination of the Cabibbo angle from nuclear decay, hyperon decay and muon decay. I don't want to critizise those attempts, I only want to point out that there exist alternatives in the literature. Personally, I think that the Cabibbo theory is just beautiful and that it is a great surprise that it works as well as it actually does. (Inspite of the approximations which go into this theory).
I think it would be worthwhile, to try to find clear-cut experimental distinctions - even in the ordinary field of weak decay processes - between the various alternatives.

J. Schwinger used his approach to insist that you do not have to use quarks at all and you do not have to introduce an new quantum number in order to explain the metastability of the new particles. Personally, I would think that if you go to the purely phenomenological level you can argue in the following way: Neutral currents exist! They obey the selection rule $\Delta S = 0$. A new selection rule leads to a new quantum number. The lightest particle carrying this quantum number must be metastable. The only question which is still open is, of course, what kind of quantum number you will have. This depends on the particular kind of model you are using. But I am quite convinced, that if you look a little bit more closely into the scheme of J.Schwinger, you may be able to extract (possibly in a hidden way) some kind of a new quantum number also.

This leaves us with the question what kind of quantum number is it? It was extensively discussed in the lectures of J. Ellis and H. Schopper. Is it color? Is it charm? Presently, it seems that color is not the favored model. Charm is certainly a little bit ahead in the race. Very fortunately, this question can be definitely answered experimentally. You just have to find out whether there are particles which carry the quantum number charm. Everybody who believes that the charm scheme is the right one, also believes that the new particles found so far are zero-charm particles made out of charm-anticharm. Well, there are some indications and various experimental pieces fall into the right place so that we can hope we will have an answer, let me say before the next Schladming meeting. In the lectures of H. Schopper we have all seen that experimentalists are really doing a wonderful job and they are preparing fascinating equipment to answer those rather difficult questions.

It is of course intriguing to ask, have charmed particles been found already? You remember the story about the Ω^- particle. You know that the Ω^- was one of the crucial predictions of the SU(3)-scheme. In 1963, it was announced that the Ω^- had been found after one was searching for it in order to check this scheme. Everybody was extremely happy. And after several of them had been found, people went back and looked into the literature. Sure enough, three Ω^- particles altogether had been reported. But before one had any understanding, he couldn't just make sense out of the discovery (there is a nice paper by L. Alvarez in the Physical Review D8, 702 (1973) about this history). I am sure some of you have heard rumors - even printed in daily newspapers - that charmed particles have actually been found. But, at the end of my summary, I do not want to come too close to science fiction but I must say, that we are living again in a tremendously exciting time of physics. I also think I have closed a circle, because (stepping down from my own horse - phenomenology - which every now and then one has to do) I would say progress in physics is a spiral process, which if viewed from the axle just seems like a circle. You have the conceptual basis, you have the mathematics and the experiments and they link into each other and there would be no progress if one of them would be left out.

One thing remains to be done which I am doing with great personal pleasure: I think I am speaking in the name of all the participants when I say that this Schladming Winter School again was a very successful and fruitful one and it was certainly due to the immens effort of Prof. Urban and his organizing committee. So we all thank them very cordially. Prof. Urban has organized this Winter School in such a wonderful way and with such a personal touch that we all hope he will go on to organize this fine school for many more years to come.